Modern RF Product Development

Volume 1

The Product Blueprint

Modern RF Product Development

Volume 1

The Product Blueprint

Alan Salari

ARTECH
HOUSE

BOSTON | LONDON
artechhouse.com

Library of Congress Cataloging-in-Publication Data
A catalog record for this book is available from the US Library of Congress.

British Library Cataloguing in Publication Data
A catalog record for this book is available from the British Library.

ISBN: 978-1-68569-137-0

Cover design by Joi Garron

685 Canton St.
Norwood, MA 02062

10 9 8 7 6 5 4 3 2 1

To my beloved parents, whose unwavering guidance and wisdom have illuminated my path and shaped me into the person I am today

To the educators and mentors who have inspired, challenged, and supported me throughout my academic and professional journey

To the next generation of RF engineers, may this work help you think deeply, design boldly, and build impactful products

Contents

CHAPTER 3
Signal Chain Layer 47

CHAPTER 4

CHAPTER 6

Foreword

Every meaningful product story starts with understanding the people it is meant to serve, not just who they are, but what challenges shape their world and what outcomes truly matter to them. In an industry defined by rapid breakthroughs, dense technical challenges, and rising expectations, it is easy to get lost in the excitement of the work itself. However, progress only matters when it serves someone. The best RF products, the ones that endure and move entire markets forward, share a common foundation. They are built with a clear understanding of the customer's problem and a commitment to solving it with clarity, precision, and purpose.

That is why this book matters. *Modern RF Product Development* gives engineers something our industry has long needed: a practical map from idea to manufacturable product, rooted not just in technical mastery but in the fundamental discipline of building for real people in real environments. It reminds us that innovation in a vacuum can feel impressive in the moment, but it rarely survives contact with the world. Innovation anchored in customer insight is what shapes the future.

At Keysight, we see this every day. Whether our customers are advancing aerospace systems, designing the next wireless standards, pushing compute limits, or building secure communication links, they all share the same challenge. They must navigate complexity across RF, mixed-signal, digital processing, firmware, software, mechanical design, and manufacturing, and they must make critical decisions long before the path is fully clear. What keeps them moving forward is not guesswork; it is a deep understanding of who they serve and why their work matters.

Alan Salari has captured this reality with remarkable clarity. His three-volume framework mirrors the life cycle in which our customers live in practice: defining the problem, translating it into requirements, designing the architecture, validating the approach, and scaling it into reliable production. Each step requires technical depth, but, more importantly, they require judgment: knowing when to pivot, what to measure, and which trade-off serves the end user, not just the elegance of the design.

The main bottleneck in modern RF development is no longer physics. It is integration and execution. As systems grow more interconnected, success depends on how well engineers understand the broader ecosystem and how confidently they can work across domains. But integration for its own sake is not the goal; integration is only powerful when it brings a customer's vision into sharper focus.

The best engineers do not just design circuits; they design outcomes. They look beyond their block diagram and ask how the full system will operate, how it will be tested, how it will scale, and how a customer will rely on it in moments that matter. This book gives structure to that mindset.

What resonates most with me is how directly it ties technical execution to purpose. A blueprint is only meaningful if it solves the right problem. A validation plan is only effective if it reflects real-world use cases. A production strategy only works when the product delivers consistent value in the hands of the customer. This is the philosophy that has shaped Keysight since the earliest days of HP, and it continues to guide us today. We put the customer at the center, not because it is a slogan, but because it is the only reliable way to navigate uncertainty.

When engineers root their decisions in customer understanding, they build better products. They communicate more clearly. They collaborate more effectively. They stay grounded during moments when complexity or doubt could easily pull them off course. If you know who is wielding your tool and how they will use it, you know which direction is up.

Alan's work also reinforces another truth. No one innovates alone. Modern RF development is a team sport, spanning diverse skills and perspectives. The most successful programs are the ones where design, firmware, software, mechanical, and test teams share a common narrative about what they are building and why. This book gives teams that language. It provides a way to think about problems that encourages alignment rather than fragmentation.

For experienced engineers, this will sharpen intuition. For new engineers, it will accelerate judgment. For leaders, it will offer a framework for building balanced teams that can deliver under pressure. For an industry racing to meet compressed design cycles and rising performance demands, it offers something increasingly rare: clarity.

As you read this book, I encourage you to keep returning to the people on the other side of your work: the scientists pushing new boundaries, the operators keeping critical systems running, the engineers designing next-generation networks, and the companies striving to bring positive change to the world. When you stay close to their needs, your path becomes clearer, your decisions become sharper, and your innovations carry more weight.

This book is a guide, but, more importantly, it is an invitation to build with purpose. I hope it serves you well on your journey.

Satish Dhanasekaran
President and Chief Executive Officer
Keysight Technologies
July 2026

Preface

Radio frequency (RF) and microwave engineering have long stood at the center of technological revolutions. The breakthrough came with James Clerk Maxwell, whose elegant equations unified electricity and magnetism and predicted the existence of electromagnetic waves. Heinrich Hertz later brought theory into reality by generating and detecting those waves in the laboratory. Building on this scientific foundation, Nikola Tesla imagined a world connected wirelessly, and Guglielmo Marconi transformed that vision into practice, sending the first long-distance radio signals across the Atlantic Ocean.

Now a new wave of opportunity is upon us. The rollout of 5G and the global race toward 6G, the emergence of massive satellite constellations, the rise of autonomous vehicles and next-generation radars, the escalating demand for secure communications and electronic warfare, and even the control systems that enable quantum computing all depend on RF product development that is faster, smarter, and more resilient than ever before.

Today's challenge extends far beyond making individual circuits function. It is about delivering complete products that span multiple dimensions, including hardware, software, firmware, and mechanical design, while achieving the optimal balance of performance, efficiency, compactness, affordability, and scalability, all without slowing the pace of innovation.

Yet, as central as RF is to these advances, a significant gap still exists. Too often, knowledge is fragmented, buried in theory-heavy textbooks, isolated component-level and system-level references, or limited case studies that rarely scale into complete products. What is missing is a practical, end-to-end roadmap, one that guides engineers from the first sketch on a whiteboard through proof-of-concept, prototyping, and validation and ultimately to a manufacturing-ready product. This book series was written to help to fill that gap.

Modern RF Product Development is envisioned as a three-volume journey that mirrors the real product life cycle:

- *Volume 1: RF Product Blueprint (the foundation of the journey):* This volume is about far more than requirements and specifications; it is about defining the vision of what an RF product must become. It guides readers in identifying opportunities and risks, establishing system-level requirements, and understanding the core concepts behind communication links, architectures, and components. It also highlights the cross-disciplinary considerations, firmware, software, and mechanical design that shape a complete product blueprint. Through case studies such as electronic warfare systems and phased-array terminals for satellite communications, this volume shows how every great RF product begins with a clear and well-crafted vision.

- *Volume 2: Implementation and Validation (moving from blueprint to reality)*: This volume focuses on integration, testing, and the systematic development stages of proof of concept, minimum viable product, engineering validation test, design validation test, and production validation test. It explores practical implementation techniques in depth, including power-supply design, power amplifiers, signal and power integrity, printed circuit board design, digital signal processing, antennas, and mechanical and thermal design. Advanced technologies, such as reconfigurable systems, metasurfaces, and metamaterials, are discussed to illustrate how such cutting-edge techniques are applied in practice.
- *Volume 3: Production and Deployment (turning prototypes into products)*: This volume addresses the challenges of scaling and manufacturability, covering production workflows, test-procedure development, test-fixture design, automated testing, and environmental testing, including ingress protection, humidity, temperature, and vibration. The focus is on strategies that ensure robust, reliable RF products can be manufactured at scale and successfully deployed in real-world environments.

This first volume, the product blueprint, is designed to be both a guidebook and a source of inspiration. It weaves together technical foundations with lessons learned from real-world development: where projects succeed, where they stall, and how small decisions early on can determine the difference between failure and market impact. The focus is on what truly matters for engineers in the trenches: how to design and refine products that must work reliably in the field, under pressure, and at scale.

The chapters flow as the product blueprint itself:

- Chapter 1 establishes the foundation of the RF product blueprint, defining the structure, scope, and guiding principles used throughout the design process.
- Chapter 2 walks through the journey from identifying a real-world problem to defining a viable product concept, requirements, and value proposition.
- Chapters 3–5 build the technical core, introducing communication fundamentals, RF system concepts, and architectural frameworks that shape modern RF products.
- Chapter 6 provides a detailed look at essential RF components, their roles, and how they fit within the system architecture.
- Chapter 7 expands beyond RF, covering power, compute, mechanical, and software subsystems that complete the overall product design.
- Chapter 8 brings every layer together, forming a cohesive, end-to-end RF product blueprint ready for implementation in Volume 2.

This book is written for multiple audiences:

- For managers and decision-makers, it offers clarity on the trade-offs that shape timelines, costs, and the overall success of RF products.

- For startup founders and innovators, it highlights how to navigate early technical decisions, manage risk, and translate an RF concept into a viable, buildable product.
- For engineers, it provides frameworks, checklists, and practical insights that accelerate development and improve design quality.
- For students, it bridges theory and application, demonstrating how classroom concepts translate into real-world engineering decisions.

Each chapter includes problems designed to challenge and deepen understanding. Solutions to these problems, along with a set of PowerPoint slides for instructors, are available upon request by contacting info@quaxys.com.

RF technology has always had a profound and multidimensional impact on humanity, driving revolutions in communication, defense, healthcare, and science. I hope that *Modern RF Product Development* not only equips you with the practical tools to design and build, but also inspires you to see the broader picture: that every product you create has the potential to improve lives and perhaps even lay the foundation for the next breakthrough.

Acknowledgments

I would like to express my deep appreciation to Frederick H. Raab of Green Mountain Radio Research LLC for serving as a technical reviewer of this book. His thoughtful feedback, technical insight, and meticulous attention to detail helped to improve the clarity and overall quality of the manuscript.

Furthermore, the successful publication of the *Modern RF Product Development* series reflects the invaluable insights, professionalism, and support of the Artech House team. I extend my sincere gratitude to Peter Froehlich, Gavin Jamieson, and Christina Sargent of Artech House for their continued collaboration and support throughout this process.

CHAPTER 1

The Anatomy of RF Products

I saw the angel in the marble and carved until I set him free.

—*Michelangelo*

This chapter defines what constitutes a radio frequency (RF) product and introduces its four fundamental functions: signal generation, RF signal processing, transmission, and detection. It outlines the key building blocks, including the power, RF, antenna, control, processing, and mechanical subsystems, that together form a complete system. A detailed RF product classification based on function, operational environment, system scale, and market domain (commercial versus government and defense) is presented. The chapter also discusses emerging trends such as software-defined radios (SDRs) and artificial intelligence (AI)-assisted functionalities, illustrating how modern RF systems are becoming increasingly adaptive and intelligent.

1.1 Target Audience and Scope

This book series is written for a broad range of professionals involved in shaping, designing, and delivering RF products. The groups that will benefit most include:

- *Engineers, technical leads, and system architects:* Access practical frameworks, implementation-focused technical details, checklists, and integration strategies that translate functional requirements into a working RF product.
- *Managers and decision-makers:* Gain clear insight into the trade-offs that shape timelines, costs, risks, and overall product success.
- *Startup founders and innovators:* Receive guidance on early technical decisions, system architecture choices, and the transition from concept to a manufacturable, scalable product.
- *Students and early-career engineers:* Learn how to bridge theory and practice, understanding how foundational concepts directly map onto real system-level design decisions.

We deliberately exclude deep coverage of semiconductor design and fabrication, including RF chip design (RFIC and MMIC) and discrete component design, such as transistor-level amplifiers, filters, and power combiners. These elements are treated as existing building blocks, with emphasis placed on how to select, integrate, control, and validate them within a complete RF system.

The focus is on developing RF systems that integrate multiple domains, including the communication layer, RF, embedded systems, power, mechanical, processing, and control, into deployable and functional products. Typical applications include:

- Phased-array terminals and beamforming systems;
- Radar and electronic warfare (EW) platforms;
- Wireless communication terminals and gateways;
- Remote sensing, telemetry, and positioning systems;
- Quantum computing control and measurement hardware.

The material is organized into three volumes, each corresponding to a major phase in the RF product development journey:

- *Volume 1—The Product Blueprint:* How to define the problem, capture requirements, and build the architectural foundation of your product.
- *Volume 2—Implementation and Validation:* How to bring the blueprint to life through prototyping, integration, testing, and iteration.
- *Volume 3—Production and Deployment:* How to move from engineering validation to scalable production, deployment, and life-cycle support.

Whether you are designing a next-generation radar or building a control system for quantum computers, this series provides a structured, practical guide to navigating the complexity of modern RF product development.

1.2 What Is an RF Product?

The terms *radio frequency* (RF) and *microwave engineering* are often used interchangeably, with RF frequencies covering the very high frequency (VHF) band from 30 MHz to the ultrahigh frequency (UHF) band ending at 3 GHz and microwave frequencies spanning 3 GHz to 300 GHz [1]. Although RF and microwave engineering share many foundational concepts, they differ in important ways when it comes to design, implementation, and testing.

In practice, engineers use the term RF much more loosely. When discussing RF systems, the term often encompasses anything operating across a wide frequency range, from kilohertz used in low-frequency communication and sensing applications such as submarine communications, to hundreds of gigahertz used in advanced radar, wireless, millimeter-wave, and emerging sub-terahertz systems. Throughout this book, the term RF is used as an engineering umbrella rather than a strict spectral definition. For consistency, both RF and microwave frequencies are referred to simply as "RF."

RF products can range from simple components, such as cables, filters, or RF switches, to highly complex systems such as phased-array radars, EW platforms, satellite communication terminals, 5G radios, backhaul systems, and quantum-computing control electronics. An RF product is any device or system that includes at least one of four functional capabilities [2]:

- *Signal generation:* Oscillators, synthesizers, and waveform generators, both analog and digital, that produce RF or intermediate-frequency (IF) signals.
- *Signal transmission:* Antennas, waveguides, and cables that radiate or guide RF signals through free space or confined structures.
- *Signal detection:* Detectors and demodulators, implemented in either analog RF hardware or digital signal-processing architectures, that extract useful information or measurable properties from incoming RF signals. Their functions range from simple envelope or power detection to full recovery of encoded data.
- *RF signal processing:* Filters, mixers, amplifiers, and related components that modify a signal's amplitude, phase, or frequency content within the RF domain.

Figure 1.1 illustrates the four fundamental functional areas of an RF product. Depending on the application, one or more of these functional areas may dominate. For example, in RF heating applications, the emphasis is on signal generation. In contrast, detection is the central function in fields such as radio astronomy. In systems such as communication links or quantum computers, all four areas must work together to deliver the functionality expected from the product.

A communication link provides a clear example of how these four functions interact. The process begins with signal generation, where information such as voice, video, or data is encoded onto an RF carrier using a modulator. The resulting signal then undergoes RF signal processing, including upconversion, amplification, and filtering, to prepare it for transmission. It is important to note that digital signal processing (DSP), which typically operates on sampled representations of RF or IF signals, also plays a major role in modern systems. However, DSP is distinct from RF signal processing and generally appears in both the generation and detection paths.

The signal may then be transmitted through wired media, such as coaxial cables, or wirelessly through free space. For long-distance communication, repeaters or regenerative nodes may be used to extend the communication range by boosting or reconstructing the signal. At the receiver, components such as filters, amplifiers, and downconverters prepare the signal for detection. Finally, the demodulator recovers the original information and reconstructs the baseband voice, video, or data.

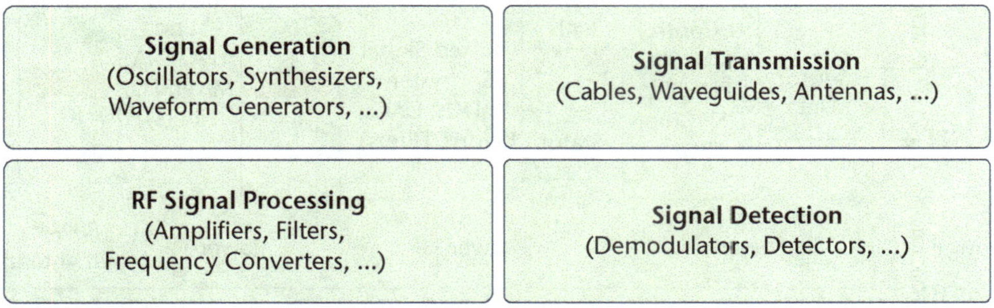

Figure 1.1 Core functional domains of an RF system: signal generation, signal transmission, RF signal processing, and signal detection. Representative examples of the components involved in each category are illustrative rather than exhaustive.

1.3 Building Blocks of an RF Product

This section introduces the key building blocks of an RF product and explains how they integrate into a complete system. A Wi-Fi router serves as an illustrative example, as it incorporates all major elements discussed here. Later, in Chapter 8, we present detailed system-level blueprints for a phased-array user terminal.

Figure 1.2 summarizes the core elements of a mixed-signal (analog and digital) RF product, the most common architecture in modern wireless systems due to its ability to leverage the strengths of both domains. These elements form the foundation of nearly all contemporary RF designs and will be referenced throughout the remainder of this book.

The building blocks of an RF product are briefly described below.

1. *Power:* The power subsystem distributes and regulates electrical power for all components, including RF, analog, and digital elements. It ensures that each block receives the correct voltage and current levels and incorporates protection mechanisms such as current limiting, thermal shutdown, and fault detection.
2. *Sensors:* The sensor subsystem monitors key environmental and operational parameters, such as power, voltage, current, temperature, humidity, and vibration, to support system health tracking, protection, and calibration. These sensors provide real-time feedback used by the control and processing subsystem to maintain stability, reliability, and performance over varying operating conditions.
3. *Antenna subsystem:* The antenna subsystem forms the interface between the RF system and free space, enabling wireless transmission and reception.

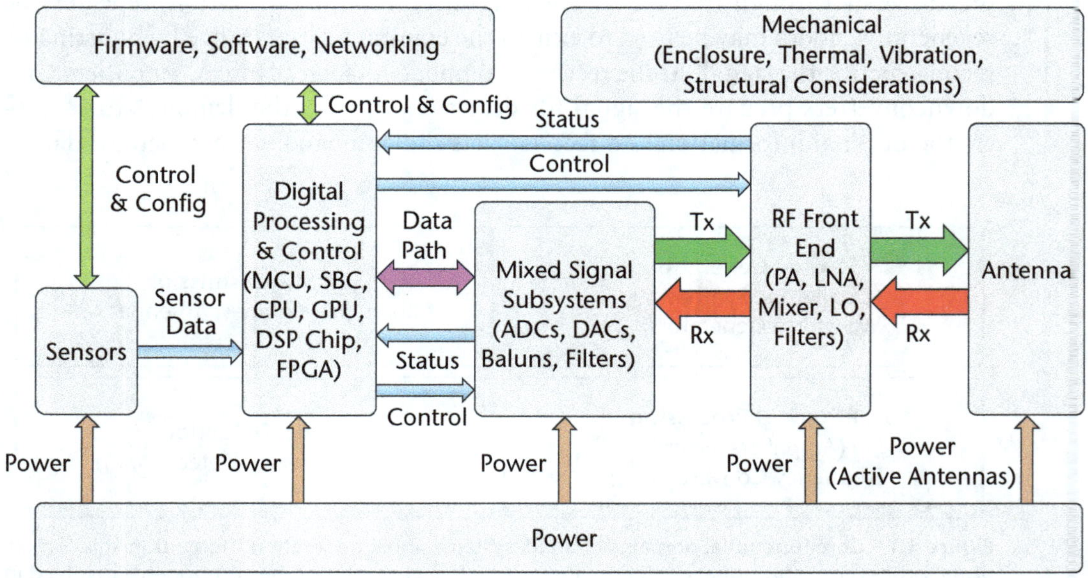

Figure 1.2 Building blocks of an RF product.

It directly influences performance metrics such as coverage and overall link quality. Depending on the application, this subsystem may range from a single passive radiator to a multi-element array used for direction finding (DF), multiple-input multiple-output (MIMO), or phased-array systems. In advanced systems such as satellite communications (SATCOM) terminals, radar arrays, and EW platforms, antenna subsystems may comprise hundreds or even thousands of elements, often incorporating embedded phase shifters, amplifiers, or tuning circuits. Antennas can also be reconfigurable using variable capacitors (varactors), positive-intrinsic-negative (PIN) diodes, or RF microelectromechanical systems (RF MEMS) to dynamically adjust impedance, resonance, or radiation characteristics in real time. We will see more about antennas in Chapter 6 as well as in Volume 2.

4. *RF subsystem:* The RF subsystem operates entirely in the continuous-time (analog) domain and forms the heart of the analog signal chain. It processes incoming and outgoing RF signals using components such as low-noise amplifiers (LNAs), power amplifiers (PAs), mixers, filters, phase shifters, couplers, and switches. This subsystem defines the core RF performance metrics, including frequency coverage, sensitivity, dynamic range, and communication range.
5. *Mixed-signal subsystem:* The mixed-signal subsystem forms the boundary between the analog RF subsystem and the digital processing layer. It includes analog-to-digital converters (ADCs), digital-to-analog converters (DACs), baluns, anti-aliasing filters, reconstruction filters, clocking and synchronization circuitry, and high-speed digital interfaces such as JESD204B/C or low-voltage differential signaling (LVDS), along with configuration interfaces such as serial peripheral interface (SPI).

   The primary functions of this layer are to digitize incoming analog RF or IF signals for processing in the digital domain on the receive side and to generate analog waveforms from digital samples produced by the processing subsystem on the transmit side.

   Modern RF transceiver chips, such as the AD936x-series from Analog Devices, combine the analog RF subsystem and the mixed-signal subsystem into a single device. More advanced transceivers, such as the ADRV90xx-series, go a step further by integrating internal digital processing blocks that perform converter-adjacent DSP functions.
6. *Digital processing and control:* The digital processing and control subsystem orchestrates signal routing, tuning, calibration, configuration, and the digital processing of RF and baseband signals. It may be implemented using microcontrollers (MCUs), single-board computers (SBCs), field-programmable gate arrays (FPGAs), central processing units (CPUs), graphics processing units (GPUs), or combinations thereof.

   In most modern architectures, substantial digital signal processing is implemented, which typically includes modulation, demodulation, digital upconversion (DUC), digital downconversion (DDC), finite impulse response (FIR) filtering, decimation, interpolation, quadrature error correction (QEC), DC offset correction (DCO), LO leakage suppression (LOL), and automatic gain control (AGC). These architectures and processing partitions are discussed in more detail in Chapter 5.

7. *Firmware, software, and networking:* The sensing, digital processing, and control subsystems operate across two tightly integrated layers, along with a possible third layer:
   - *Firmware:* The firmware layer provides low-level, real-time control and implements time-critical signal and system processing functions. Running on embedded processors (MCUs or microprocessors) or programmable logic (FPGA fabric), it handles tasks such as configuring RF integrated circuits (RFICs), variable-gain amplifiers (VGAs), and phase shifters; executing digital DSP algorithms and managing communication interfaces such as SPI and inter-integrated circuit (I^2C); and ensuring deterministic timing and control.
   - *Software:* The software layer provides higher-level system functionality and orchestration, including user-interface interactions, configuration management, diagnostics, calibration routines, monitoring frameworks, and data analytics. It communicates with the firmware to establish operational modes, execute automated sequences or scripts, and track the system's health and performance in real time.
   - *Networking:* Some products also include this possible third layer. In networked systems, such as distributed sensor nodes, distributed radar platforms, satellite communication terminals, or quantum-computing infrastructures, a networking layer becomes essential. It provides connectivity to external systems and services, enabling data fusion and integrated decision-making, remote control interfaces, secure firmware updates, high-rate data streaming, multinode synchronization, and cloud-based processing.
8. *Mechanical:* The mechanical subsystem provides structural integrity, thermal management, and physical protection for the RF product. It encompasses elements such as enclosures, heatsinks, fans, mounting hardware, gaskets, and electromagnetic shielding components. In addition to ensuring environmental robustness, the mechanical subsystem plays a critical role in managing heat dissipation, vibration tolerance, environmental sealing, electromagnetic compatibility (EMC), electromagnetic interference (EMI) control, and compliance with size, weight, and power (SWaP) constraints

1.3.1 Example: Subsystem Breakdown of a Wi-Fi Router

A home Wi-Fi router is an excellent example of an everyday product that incorporates all the major subsystems shown in Figure 1.2. Although it is far simpler than a phased-array terminal or radar system, it still performs the full chain of RF generation, processing, transmission, and detection.

1. *Power:* The router receives DC input (typically 9–12V) from an external power adapter. Internally, power management circuitry generates the regulated rails required by the digital system-on-chip (SoC), RFICs, amplifiers,

and active antenna elements, if present. Protection circuits guard against overheating, surges, and excessive current draw.

2. *RF subsystem:* The RF front end performs all the analog-frequency operations. This includes mixers for upconversion and downconversion into the 2.4-GHz or 5/6-GHz Wi-Fi bands, PAs for transmission, LNAs for reception, bandpass filters or duplexers, synthesizers, and local oscillators. These components shape, amplify, and condition the RF signal before it reaches the antenna or the DSP chain.
3. *Antenna subsystem:* Most Wi-Fi routers use one or more antennas for MIMO operation. These antennas radiate and receive RF energy, determine the router's coverage pattern, and affect the throughput. Depending on the model, antennas may be external (with adjustable positioning) or internal PCB antennas.
4. *Digital processing and control:* The router contains a SoC that runs firmware responsible for packet routing, network security, channel selection, timing control, and other critical operations. The firmware configures RFICs, controls transmit power, manages calibration routines, and communicates with the RF front end over digital buses such as SPI or I^2C.
5. *Mechanical:* The enclosure provides structural integrity, protects electronics, manages airflow, and supports heat dissipation via vents or small heat sinks. Mechanical design affects EMI shielding, EMC, regulatory compliance, durability, and overall system reliability.

1.4 RF Product Classification and Market Context

RF products span an exceptionally wide spectrum of applications, environments, and scales. A clear classification helps engineers to position their design within the broader market context and anticipate the specific requirements, standards, and challenges associated with each class of product.

In addition to functional, environmental, and system-scale distinctions, RF products can also be categorized by market domain, namely, commercial versus government/defense applications. Commercial systems emphasize cost, manufacturability, and compliance with industry standards, whereas government and defense systems often prioritize performance, security, ruggedness, and long-term reliability under extreme conditions.

Tables 1.1 through 1.4 provide a high-level categorization of RF products according to function, operational environment, system scale, and market domain.

This classification not only aids in positioning a product within its intended market but also guides early design trade-offs. For instance, spaceborne systems demand extreme reliability and thermal resilience, while quantum control electronics emphasize timing precision and noise performance. Government and defense products must usually meet stringent military standards (MIL-STD) requirements, whereas commercial products prioritize cost and scalability. Understanding where a product resides within this landscape helps to prioritize requirements, anticipate certification needs, and allocate resources effectively.

Table 1.1 RF Product Classification by Function

Function	*Examples*	*Typical Design Priorities*
Communication	Satellite terminals, 5G/6G radios, data links	Spectral efficiency, noise, linearity, modulation accuracy, dynamic range
Sensing	Radars, radio astronomy, medical imaging	Detection range, resolution, sensitivity
Jamming/EW	Electronic attack and countermeasure systems	Output power, agility, frequency coverage, response time
Measurement and test	Network analyzers, spectrum analyzers, signal generators	Accuracy, sensitivity, dynamic range, repeatability
Quantum control	Cryogenic RF systems, qubit control, and readout electronics	Phase stability, timing precision, ultralow noise, cryogenic integration

Table 1.2 RF Product Classification by Operational Environment

Environment	*Examples*	*Typical Design Priorities*
Ground-based	Fixed wireless access, radar installations, remote sensing stations	Cost efficiency, maintainability, scalability
Airborne	Drone payloads, aircraft radars, EW pods	SWaP optimization, thermal and vibration tolerance, EMI/EMC compliance
Spaceborne	Satellites, deep-space communication payloads	Radiation hardness, thermal cycling endurance, extreme reliability
Laboratory/ controlled	Testbeds, research demonstrators, quantum labs	Flexibility, precision, expandability, instrumentation connectivity

Table 1.3 RF Product Classification by System Scale

Scale	*Examples*	*Typical Design Priorities*
Component/ module	Amplifier, mixer, filter, oscillator	Component-level optimization, ease of integration
Subsystem	Transceiver front end, beamforming unit, local oscillator (LO) distribution network	Interface control, gain/phase balance, thermal design
System	Radar unit, satellite communication terminal, telemetry receiver	Overall architecture, control strategy, field reliability
Network	Multinode radar network, communication constellation, distributed sensors	Synchronization, data fusion, interoperability, network management

1.5 Market Trends and Emerging Applications

The RF landscape is evolving rapidly, driven by advances in computation, materials, and system integration. Modern applications increasingly blur traditional boundaries between communication, sensing, and control.

For example, integrated sensing and communication (ISAC) systems combine radar and data transmission into a single platform, enabling vehicles, drones, and wireless networks to share spectrum resources more efficiently. A practical illustration is an automotive 77-GHz radar module that simultaneously performs sensing

Table 1.4 RF Product Classification by Market Domain

Attribute	*Commercial RF Products*	*Government/Defense RF Products*
Primary objective	Cost efficiency, mass production, consumer usability	Mission-critical performance, security, reliability
Design priorities	SWaP optimization, manufacturability, standard compliance	Ruggedness, survivability, long life cycle, classified operation
Standards and compliance	FCC, ETSI, IEEE, CE	MIL-STD, classified protocols
Operating environment	Controlled or semi-controlled (urban, automotive, aerospace)	Harsh or contested (battlefield, space, maritime, high radiation)
Development cycle	Rapid, iterative, market-driven	Long, regulated, contract-driven
Procurement model	Commercial off-the-shelf (COTS) and outsourced design supply chains: original design manufacturer (ODM—supplier-owned design) and original equipment manufacturer (OEM—customer-owned design)	Government contracts, defense prime contractors (such as Lockheed Martin and Northrop Grumman), and secure manufacturing
Life Cycle and mainetnance	3–7 years, technology refresh-driven	10–30 years, mission readiness-driven
Export and security	Minimal restrictions	International traffic in arms regulations (ITAR)/export administration regulations (EAR) controls, and classified handling requirements

and vehicle-to-vehicle communication. In an ISAC architecture, the same RF hardware that measures distance and velocity using radar chirps can also transmit safety messages such as braking alerts or lane-change intentions. Instead of deploying separate radar and communication units, the vehicle leverages one shared transceiver, one antenna array, and one spectrum band, reducing SWAP-C while improving coordination and situational awareness.

Meanwhile, software-defined radios (SDRs) equipped with high-speed DACs and ADCs enable unprecedented flexibility in signal generation, reception, and processing. When combined with advanced DSP and adaptive algorithms, these platforms support real-time reconfiguration across wide frequency ranges and diverse modulation formats [3, 4].

In parallel, AI-assisted RF design and autonomous spectrum management are reshaping how systems self-optimize under dynamic interference, channel fading, or mission constraints. Beyond classical RF applications, quantum technologies are pushing the limits of low-temperature RF design, precision timing, and system scaling, driving a deeper convergence between RF engineering and cryogenic systems [5, 6].

The result is a new generation of RF products that are highly integrated, intelligent, adaptive, and interconnected, capable of responding to their environment and collaborating across distributed networks. As these trends accelerate, RF engineers must think systemically, combining hardware innovation with intelligent control, software-defined adaptability, and cross-disciplinary insight.

1.6 What's Next?

We have defined what an RF product truly is: an engineered fusion of physics, hardware, firmware, and software that serves a clear purpose. We have examined its fundamental building blocks and seen how products differ across functions, environments, scales, and market domains.

Understanding what an RF product is naturally leads to the next, deeper question: when does it make sense to create one? In Chapter 2, we will answer this question and explore how meaningful RF innovations begin, not from technology alone, but from recognizing a real-world problem that demands an RF solution. We will learn how to connect technical potential with human and market needs and how that spark evolves through structured phases, from concept and validation to a fully realized, production-ready system.

Every successful RF product begins with that moment of clarity when a need, an idea, and the right technology align. We will explore how to identify and shape that moment in Chapter 2.

Problems

1.1 A ground-penetrating radar (GPR) system emits short pulses into the ground and analyzes reflections from the subsurface. Identify which of the four RF functions, signal generation, signal transmission, RF signal processing, and signal detection, are dominant in this system and briefly explain why.

1.2 Choose one of the following RF products: (1) low earth orbit (LEO) satellite user terminal, (2) automotive 77-GHz radar, and (3) Internet of Things (IoT) RF sensor node. Map the selected product onto the five subsystems shown in Figure 1.2: (1) power management, (2) antenna, (3) RF, (4) digital control and system processing, and (5) mechanical. For each subsystem, identify and list the critical functions. Then develop a high-level system block diagram and conclude with a rough sketch of the final physical product.

1.3 A commercial 5G radio prioritizes low cost and ease of manufacturing, whereas an airborne EW receiver must satisfy MIL-STD requirements. List at least three design consequences of this difference.

1.4 Using Tables 1.1 through 1.4, classify the following system: a drone-mounted synthetic aperture radar (SAR). State the likely functional category, operational environment, system scale, and market domain and justify your choices.

1.5 A phased array transmit module repeatedly enters thermal shutdown during a high-power operation. Using the subsystem partitioning shown in Figure 1.2, identify which subsystems could plausibly contribute to this failure mode. For each subsystem you identify, describe one realistic root cause that could lead to excessive temperature rise or erroneous thermal triggering. Finally, propose one corrective action per subsystem to mitigate or eliminate the problem.

1.6 Compare the design priorities for the same RF transceiver when deployed in: (a) a consumer product, (b) a drone payload, and (c) an LEO satellite terminal. For each environment, list the key design differences and briefly explain why they matter.

1.7 You are responsible for architecting a Ku-band airborne phased-array terminal.

a. Using the subsystem decomposition shown in Figure 1.2, identify and briefly explain the function of the five major subsystems.
b. For each subsystem, specify one key requirement imposed by the airborne operating environment (e.g., vibration, thermal extremes, altitude, or SWaP constraints) and explain why it is relevant.
c. Select one subsystem and describe a realistic failure mode that could occur in service, along with an appropriate mitigation strategy at the design or system level.

References

[1] Pozar, D. M., *Microwave Engineering*, Fourth Edition, New York: John Wiley & Sons, 2011.

[2] Salari, A., *Microwave Techniques in Superconducting Quantum Computers*. Norwood, MA: Artech House, 2024.

[3] Reyland, J. M., *Software Defined Radio: Theory and Practice*, Norwood, MA: Artech House, 2024.

[4] Wyglinski, A. M., et al., *Software-Defined Radio for Engineers*, Norwood, MA: Artech House, 2018.

[5] Kuzdeba, S., *Radio Frequency Machine Learning: A Practical Deep Learning Perspective*. Norwood, MA: Artech House, 2023.

[6] Haigh, K. Z., and J. Andrusenko, *Cognitive Electronic Warfare: An Artificial Intelligence Approach*, 2nd ed., Norwood, MA: Artech House, 2023.

CHAPTER 2

From Problem to Product

> Give me six hours to chop down a tree and I will spend the first four sharpening the axe.
>
> —*Abraham Lincoln*

This chapter presents a practical framework for transforming a validated need into a well-defined RF product. It begins with customer discovery and clear problem definition and then shows how user needs are translated into measurable, testable requirements. The RF product development life cycle, spanning proof of concept (PoC), minimum viable product (MVP), engineering validation test (EVT), design validation test (DVT), and production validation test (PVT), guides the progression from the initial idea to a manufacturable system. The key risk categories addressed include technical, schedule, resource, and regulatory risks. Together, these elements form a systematic foundation for developing successful, reliable, and scalable RF products.

2.1 A Framework for Problem-Driven Product Development

Every successful RF product begins with a problem worth solving. Whether the goal is faster communication, more precise sensing, or improved spectrum efficiency, the problem provides the spark that initiates the entire product journey. However, solving that problem does not begin with a schematic or a simulation. It begins with a deep understanding of the context, constraints, and the people who will ultimately use the product.

Figure 2.1 illustrates the product development framework: identify, prepare, execute, collect insights, refine, and iterate. Each stage is guided by specific questions and actionable steps that ensure technical rigor while maintaining alignment with user and market needs. Skipping these stages, or addressing them superficially, often results in products that are technically sound but commercially irrelevant, functionally incomplete, or difficult to manufacture at scale.

Let's examine these stages in more detail.

1. *Identify: Clarifying the problem and the need.* The development journey begins by answering two fundamental questions:
   - Is there a real problem?
   - Does anyone want the solution?

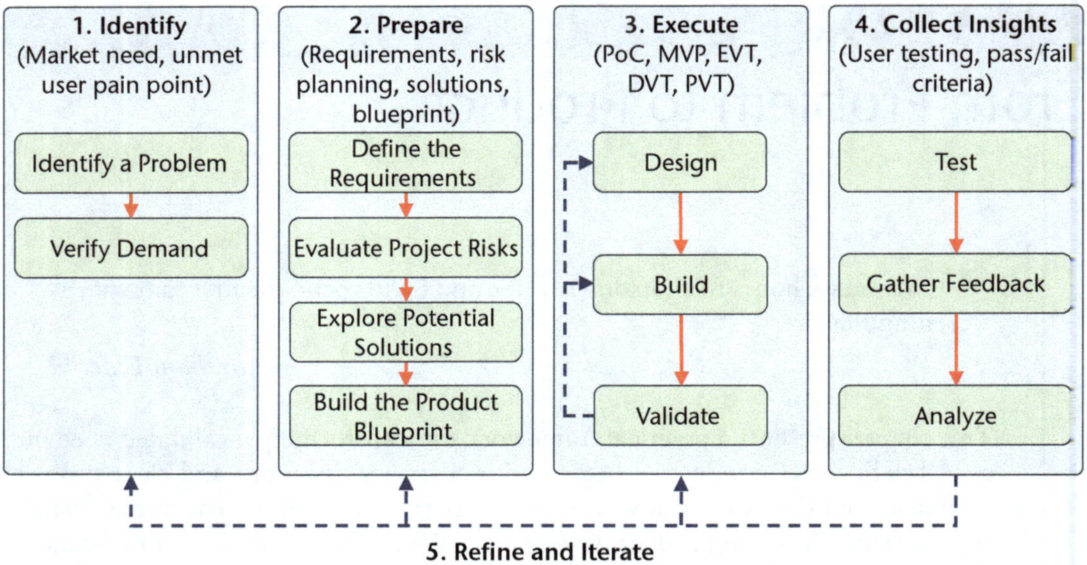

Figure 2.1 A practical framework for RF product development, highlighting the iterative nature of real-world engineering.

This phase involves identifying a clear technical challenge or user pain point and verifying that there is sufficient demand to justify investing time and resources. In the context of this book series, we assume that this foundational work has already been completed, that is, the need has been validated, and there is a clear use case with potential customers for the product under development. The focus is therefore on how to design, implement, and deliver the solution effectively.

2. *Prepare: Defining what and how.* Once the problem and demand are established, the next step is to clarify:
   - What does the solution look like?
   - How will we implement the solution, within the given constraints of time, resources, and risk?

   This stage includes the following key actions:
   - *Define the requirements:* Establish a measurable set of functional, performance, and regulatory requirements that the product must satisfy.
   - *Evaluate project risks:* Assess technical, resource, regulatory, and schedule risks.
   - *Explore potential solutions:* Compare existing products, reference designs, and architectural approaches and identify gaps or opportunities for differentiation.
   - *Develop the product blueprint:* Using the defined requirements, available resources, and the selected solution approach, construct system-level diagrams and architectures (e.g., block diagrams, interface maps) that capture the major components and their interactions.

   These preparation efforts lay the foundation for all subsequent development phases.

3. *Execute:* The execution phase translates the blueprint into tangible results through iterative design, build, and validation cycles. It encompasses all development stages, from PoC and MVP to EVT, DVT, and PVT. The execution workflow for each phase follows a straightforward principle:
   - *Design:* Translate the blueprint into an engineering design. This may involve simulations, analytical calculations, or even a simple block diagram, depending on the maturity of the idea.
   - *Build:* Build the prototype using commercial off-the-shelf (COTS) components, custom hardware, or a combination of both.
   - *Validate:* Evaluate performance using laboratory equipment such as network analyzers, spectrum analyzers, oscilloscopes, and antenna measurement systems. For some products, validation also includes field testing, such as communication link tests or operational trials of electronic warfare (EW) systems.

   The insights gained from validation are fed back into the design process, forming internal feedback loops that drive iterative refinement across all development phases.
4. *Collect insights: Validating with end users.* After technical validation, it is vital to assess whether the solution truly addresses the end user's needs. The central question becomes: Does the product effectively solve the original problem in a real-world context?

   This phase serves as the bridge between engineering and user experience, revealing critical insights that lab testing may overlook. Feedback can be collected through pilot deployments, customer demonstrations, or internal evaluation teams simulating real usage scenarios.
5. *Refine and repeat: Embracing continuous improvement.* In Figure 2.1, the feedback loop in the framework indicates that development is rarely linear. Insights gained during measurement or user feedback may lead to refinements in the design, updates to the requirements, or even a reassessment of the problem statement. This iterative mindset helps to avoid costly errors later in the development cycle and ensures a more robust final product.

2.2 Identify: The Core Pain Point and Market Need

Based on the development framework introduced in Section 2.1, the foundation of every successful RF product is a clearly defined and validated problem. Regardless of how innovative a design may be, if it fails to address a genuine need, the product will not gain traction. Particularly in RF and microwave systems, where development cycles are lengthy and expensive and system integration is complex, identifying the wrong problem is one of the costliest mistakes that a company can make.

2.2.1 Why Products Fail: No Market Demand

Postmortems consistently reveal that lack of market demand is among the leading causes of product failure, responsible for roughly 35% of failed startups [1]. Similar conclusions from *Harvard Business Review* and Startup Genome highlight that a substantial fraction of products fail because they address problems that customers do not actually need solved, often due to insufficient customer discovery and premature scaling [2, 3]. Numerous companies, particularly in the hardware space, have invested heavily in technically advanced products only to discover that no one is willing to buy them. A well-known example is Amazon's Fire Phone. Despite significant engineering effort, advanced features, and deep integration with Amazon's ecosystem, the product failed commercially. Customers did not perceive sufficient value relative to existing smartphones, and the phone was discontinued after Amazon recorded losses totaling hundreds of millions of dollars.

In RF product development, this challenge is often amplified by the temptation to focus on performance metrics and novel architectures while overlooking whether the problem being addressed is urgent, pervasive, and painful enough to justify a purchase decision.

A critical step before launching any RF development effort is the customer discovery process. Engineering teams often focus on technological capability, but without clear and validated market demand, even the most advanced solution risks failure. Product direction must be driven by customer needs rather than technical elegance alone. Early engagement with users helps to uncover real pain points, align design priorities, and avoid building features that no one values. This upfront investment significantly reduces risk and anchors the product in real-world value.

2.2.2 Customer Discovery: The Essential Step Before Any Development

Early-stage customer discovery is the primary mechanism for risk reduction in technology development. It is highly encouraged to step outside the lab and engage directly with stakeholders, end users, integrators, procurement managers, and field engineers to validate assumptions and challenge internal bias. Rather than relying on a single voice of the customer, this methodology promotes a portfolio of conversations to refine both the problem statement and the proposed solution.

A common concern is that speaking with customers may reveal sensitive aspects of the concept to potential competitors; however, customer discovery focuses on understanding problems, not disclosing proprietary solutions, and can be performed safely using high-level discussions, problem-centric questions, and nonspecific descriptions of technical approaches. When deeper technical conversations are required, nondisclosure agreements (NDAs) can be used to ensure that confidential information is protected.

It is important to recognize that problem discovery is an iterative and dynamic process. Rather than a one-time activity, it typically unfolds through repeated cycles that include:

- Formulating a hypothesis about a critical user need or an operational gap;
- Engaging with target users through interviews or demonstrations, such as concepts or simulations;
- Analyzing feedback, identifying points of disconnect or new constraints;
- Modifying the problem definition, technical approach, or value proposition accordingly.

This loop may repeat several times before a stable and compelling problem-solution fit is achieved. For RF products, these iterations should be lightweight and simulation-driven in early stages, evolving into hardware PoCs only after customer validation.

While the purpose of this book is not to explore the customer discovery process in depth, excellent resources are available for those seeking to master it, including *The Startup Owner's Manual* by Steve Blank and Bob Dorf [4] and *The Mom Test* by Rob Fitzpatrick [5].

2.2.3 Understanding the Market: Who Needs the Solution?

Identifying a valid problem is only half of the equation. The other half is understanding the market context in which the solution will be deployed. Who the user is, and what constraints, regulations, and performance expectations under which they operate, dramatically shape how a product must be designed, built, and tested. Each target market imposes unique performance and compliance demands on RF products.

For example, military and commercial markets often require similar RF technologies, but their priorities and constraints might diverge significantly:

- Military users typically prioritize performance, robustness, and mission flexibility over cost and size. Designs may operate in extreme environments and require resilience to jamming or electromagnetic interference. Weight and power consumption are important, but not at the expense of capability.
- Commercial users, by contrast, typically face tight regulatory and size, weight, power, and cost (SWaP-C) constraints. Products must be highly manufacturable, cost-effective, compliant with civilian spectrum regulations (e.g., FCC, ETSI), and suitable for mass deployment.

Understanding these distinctions is crucial. A solution optimized for one domain may be over-engineered or completely nonviable in another domain.

Here is an example. The rise of unmanned aerial systems (UAS) has led to an explosion of counter-UAS (C-UAS) technologies, many of which rely on RF jamming. This application offers a clear illustration of how use cases diverge across markets:

- In military applications, jammers can operate across wide frequency bands with high output power, employing aggressive waveforms to disable or disrupt hostile drones. Interference with surrounding systems is often tolerated as part of the mission profile.
- In commercial environments, such as airports or critical infrastructure, the same approach is legally and technically infeasible. Strict regulations on intentional interference, along with dense spectrum usage and public safety concerns, require entirely different architectures, lower power, directional antennas, or even using alternatives such as GPS spoofing.

So the "same" RF problem, disabling a drone, requires entirely different solutions depending on the regulatory and operational domain.

2.2.4 Is a Custom RF Product the Right Answer?

Even when a problem is real and pressing, it is essential to ask whether a custom RF-based solution is truly necessary. In many cases, the desired outcome can be achieved more efficiently by adapting existing technologies, products, and infrastructure. In short, avoid reinventing the wheel.

Here is an example. A startup developing an industrial monitoring system initially planned to design a custom wireless link for sensor communication across factory floors, aiming for ultrareliable, low-latency performance. The approach involved creating proprietary RF hardware and a new communication protocol to avoid interference from existing Wi-Fi networks.

After conducting a feasibility study and analyzing deployment environments, the team realized that existing sub-gigahertz ISM-band modules (such as LoRa or Zigbee) could meet their range, reliability, and latency requirements with minimal customization. By leveraging these mature, certified technologies, they avoided months of RF design, compliance testing, and debugging, allowing them to focus on sensor accuracy, data analytics, and integration with industrial control systems, the true differentiators of their product.

The best RF solution may be the one you choose not to build, because a more elegant, lower-risk, and cost-effective alternative already exists. Such decisions reflect an engineering mindset that prioritizes outcomes over unnecessary complexity. Recognizing when not to design a custom RF solution is just as critical as knowing how to build one.

For the remainder of this chapter, we assume that customer discovery has already been completed and that a clearly defined problem, grounded in validated market need, has been identified. The focus now shifts to gathering requirements, exploring viable RF solutions, and implementing them to address that problem effectively.

2.3 Prepare: Requirements, Risk Analysis, Potential Solutions, and the Product Blueprint

The second step in the development framework introduced in Section 2.1 involves preparation and planning. In this section, we will discuss the key activities required to prepare the team, the technical baseline, and the project environment for successful execution.

2.3.1 Define the Requirements

The most important elements of any product development effort are the requirements. Without clearly defined requirements, development can easily become misaligned or derailed. When it comes to defining requirements, there are generally two common starting points:

1. *Requirements are already defined:* In some projects, the requirements are provided upfront, clearly stated, documented, and agreed upon. This typically happens when the development is driven by a customer's formal request or contract. The focus here is on interpreting, validating, and possibly refining

the given requirements to ensure that they are complete, consistent, measurable, and actionable.

2. *Requirements need to be defined:* In many cases, however, the requirements are not explicitly stated. Instead, they must be derived from the problem you are trying to solve and the context in which the solution will operate. This involves understanding the target users, performance expectations, environmental conditions, regulatory constraints, cost and size limits, and many other practical considerations. It is an iterative process that requires close collaboration across engineering, potential customers, and business stakeholders.

In the first case, the groundwork has already been laid. In the second case, the requirement definition becomes a critical part of the early development process. The next section focuses on how to systematically define requirements when they are not initially available.

2.3.2 Steps to Define Requirements

When requirements are not explicitly provided, they must be thoughtfully constructed to ensure that the product is both technically sound and aligned with real-world needs. The following steps offer a structured approach to defining requirements:

- *Gather stakeholder inputs and customer needs:* Start by identifying all relevant stakeholders, internal teams, end users, decision-makers, regulatory bodies, and potential customers. Conduct interviews, surveys, or workshops to understand their goals, constraints, and expectations. Focus on pain points and desired outcomes rather than proposed solutions.
- *Use benchmarking and industry standards as a starting point:* Analyze similar products on the market and review applicable industry standards (e.g., IEEE, ETSI, FCC). This provides a baseline for performance expectations, compliance requirements, and design constraints. Benchmarking helps identify gaps and opportunities while anchoring your requirements in established best practices.
- *Reverse-engineer from desired performance and market demand:* Instead of starting with a technical specification, begin with what success looks like in the field. What performance levels are required for the product to be competitive? What are the expected use cases and environments? By working backward from market expectations or application-specific needs, realistic and relevant technical targets can be defined.
- *Prioritizing features based on importance:* Not all requirements are equally critical. The MoSCoW method, Must-have, Should-have, Could-have, and Won't-have, is a widely used approach to classify and prioritize features effectively:
  - *Must-have:* Non-negotiable requirements that are essential for the product to function. Without these, the product fails to meet its core purpose or deliver minimum viable functionality.
  - *Should-have:* Important but not vital. These add significant value or improve usability and performance, but the product can still operate without them in the short term.

- *Could-have:* Desirable features that are nice to include if time and resources permit, but have limited impact if omitted.
- *Won't-have (for now):* Features explicitly agreed to be excluded from the current development cycle, either because they add unnecessary complexity, are low priority, or are reserved for future versions.

The 80/20 rule, also known as the Pareto Principle, suggests that roughly 80% of a product's value comes from 20% of its features. Focusing on those high-impact features helps to avoid overengineering and keeps the design efficient and purposeful. This is especially critical when the product is being developed for the first time.

These steps are often iterative. As new information surfaces or trade-offs arise during design and prototyping, the initial requirements may need to be revisited and refined. Maintaining flexibility while keeping the core objectives intact is key to successful development. In the following section, we will explore the most common types of requirements encountered in RF product development.

2.3.3 Product Requirements

Figure 2.2 illustrates how requirements flow down through each development stage, ultimately leading to a manufacturable product. To design an RF product, requirements are typically organized into three distinct categories:

1. *Signal-chain requirements:* These define what signals the system must generate and how it processes and interprets those signals. Depending on the application, they may include modulation or waveform type, timing accuracy, coding strategy, latency, or spectral characteristics. These parameters directly influence both RF and non-RF requirements, as illustrated in Figure 2.2.
2. *RF requirements:* These specify the electromagnetic performance of the system, including transmit power, receiver sensitivity, and dynamic range. They directly impact the design of the RF front end and antenna subsystems.
3. *Non-RF requirements:* These encompass constraints that are not directly related to electromagnetic performance. Examples include SWaP-C; environmental conditions such as shock, vibration, and temperature; regulatory compliance; mechanical interfaces; digital signal processing (DSP); firmware; and software. As shown in Figure 2.2, RF requirements are tightly coupled with non-RF requirements throughout the product development process.

Understanding how these three categories interact is essential for guiding both system-level architecture and the selection of subsystems and components. For example, a high-throughput, long-range communication requirement (signal-chain layer) may drive the need for higher linearity and a lower noise figure RF front end (RF layer), along with improved thermal management and more sophisticated DSP algorithms (non-RF layer).

These requirements are not isolated; rather, they influence each other bidirectionally. A decision in one domain often imposes constraints in another. For instance, the choice of DSP techniques (a non-RF requirement), such as implementing digital frequency conversion in the baseband processor, directly determines whether an

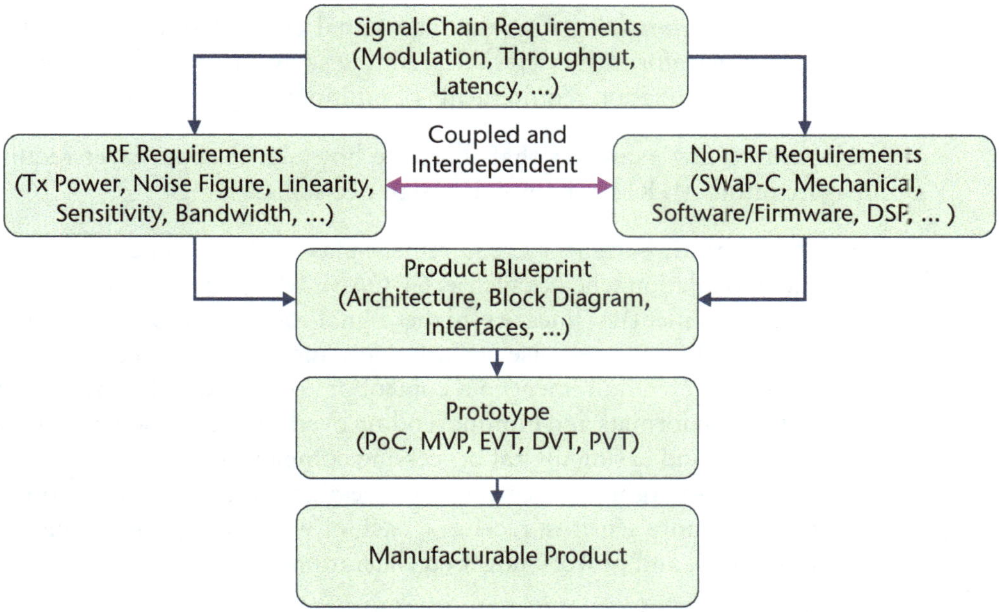

Figure 2.2 Flow from requirements definition to a manufacturable product. The diagram illustrates how signal-chain layer requirements drive both RF and non-RF requirements. These requirements inform the product blueprint, the prototype implementation, and ultimately the manufacturable product.

analog mixer is still required in the RF front end (an RF requirement). In this way, signal-chain, RF, and non-RF requirements collectively shape the overall architecture, budget, and feasibility of the final product.

Let's take a closer look at the signal-chain, RF, and non-RF requirements to see what they typically entail.

Signal-Chain Layer Requirements

Signal-chain layer requirements define how the system generates, conditions, transfers, and interprets information, independent of the physical RF medium. They specify the logical and algorithmic flow of information that the hardware must support. Examples include modulation and coding schemes in communication systems, smart or adaptive attack waveforms in EW systems, and shaped control pulses in quantum processors. Many of the concepts used at this layer are rooted in core principles of communication theory, as we will explore in the next chapter.

These requirements sit at the intersection of information theory, DSP, timing, synchronization, firmware, and overall system architecture. They vary significantly across application domains:

- *Communication systems:* The signal chain defines how bits become waveforms, modulation type, coding, symbol mapping, timing accuracy, spectral shaping, bandwidth allocation, and protocol behavior.
- *EW systems:* The signal chain defines requirements for signal detection, classification, and response time; allowable latency; probability of intercept (POI); probability of detection (PD); instantaneous bandwidth; trigger logic; emitter identification; and real-time spectral analysis.

- *Quantum computing systems:* The signal chain defines qubit control and readout waveform characteristics, timing skew, latency, allowable jitter, and the synchronization requirements of multiqubit operations.

Here are a few examples that illustrate how signal-chain layer requirements propagate into both RF and non-RF design decisions:

- A high spectral-efficiency requirement may push the system toward higher-order modulation schemes such as 64-QAM, which, in turn, demand improved power amplifier (PA) linearity, higher signal-to-noise ratio (SNR), tighter local oscillator (LO) phase-noise performance, and more advanced baseband DSP.
- A low-latency requirement in a control or command link may favor simpler modulation formats and minimal coding overhead, influencing RF bandwidth allocation and easing digital processing complexity.
- A high link-margin requirement may necessitate larger antennas, more transmit power, or more sensitive receivers, each of which brings additional thermal, mechanical, and power-budget considerations.

Table 2.1 summarizes the most common parameters associated with signal-chain layer requirements. We will explore these in greater depth in Chapter 3.

Table 2.1 Common Signal-Chain Layer Requirements in RF Products

Parameter	*Description*	*Impact on Design*
Modulation type	Defines the signal constellation, waveform structure, and information mapping (e.g., QPSK, 16-QAM, OFDM)	Determines required PA linearity, occupied bandwidth, ADC/DAC resolution, and DSP complexity
Throughput	Net user data rate after accounting for protocol overhead, coding, and retransmissions	Drives required bandwidth, clock rates, processing throughput, buffering
Coding scheme	Forward error correction (FEC) method used to improve link robustness (e.g., convolutional, LDPC, turbo codes)	Improves bit error rate (BER) performance at low SNR but increases computational complexity, processing latency, memory usage, and power consumption
Spectral efficiency	Achieved data rate per unit bandwidth (bits/s/Hz)	Drives modulation order, coding rate, bandwidth allocation, and imposes tighter SNR, linearity, and phase noise requirements
Latency	End-to-end delay from data generation to successful reception or control action	Constrains modulation, coding depth, and real-time processing architecture
BER	Probability of bit errors at the receiver output after demodulation and decoding	Sets minimum required SNR, coding gain, receiver sensitivity, and influences DSP algorithms and link margin
Error vector magnitude (EVM)	Quantitative measure of modulation accuracy and constellation distortion	Drives requirements on PA linearity, LO phase noise, IQ imbalance, timing error, synchronization accuracy, and digital impairment-correction techniques

RF Requirements

RF requirements typically consist of the key performance parameters that define how a system behaves at RF. Several common examples are listed in Table 2.2. These parameters vary substantially depending on the type of system being developed and its intended function.

For example, a receiver-only system such as a radio telescope prioritizes receive-side performance, particularly noise figure and sensitivity. In contrast, a microwave plasma generator behaves primarily as a transmitter, emphasizing output power, power efficiency, impedance matching, and thermal robustness. Transceiver systems must address both transmit and receive requirements simultaneously, ensuring that performance is properly balanced across both paths without compromising linearity, sensitivity, isolation, or overall system stability.

As we will see later, once a product is ready to enter the market, it must undergo extensive regulatory compliance testing, particularly for electromagnetic interference (EMI) and electromagnetic compatibility (EMC). This includes meeting the limits defined by standards and regulatory frameworks such as the Federal Communications Commission (FCC) in the United States, the European Telecommunications Standards Institute (ETSI) in Europe, the International Special Committee on Radio Interference (CISPR), and various military standards (MIL-STD), alongside other applicable regulatory bodies. As a result, a second set of RF design requirements is dictated by these regulations to ensure legal operation and reliable coexistence with other electronic systems. Table 2.3 summarizes the RF-related requirements relevant to EMI and EMC testing.

To avoid costly redesigns and schedule delays, it is standard practice to perform pre-compliance testing early in the development process. Early testing helps uncover potential EMI/EMC issues long before the product is submitted to certified laboratories for formal qualification. Designing with regulatory compliance in mind from the outset, rather than treating it as a final hurdle, is critical for ensuring a smooth and predictable path to market.

Although many EMI/EMC problems appear to be RF-related at first glance, their root causes often lie in non-RF domains such as power-supply integrity, grounding strategy, shielding, enclosure design, and cabling. For example, poor DC regulation

Table 2.2 Examples of RF Design Parameters for Transmitters and Receivers

Parameter	*Applies To*	*Description/Relevance*
Operating frequency	TX/RX	Central frequency or band of operation
Bandwidth	TX/RX	Range of frequencies over which the device operates effectively
Effective isotropic radiated power (EIRP)	TX	Effective radiated power accounting for antenna gain
Linearity (IP3, P_{1dB})	TX/RX	Indicates ability to handle large signals without distortion
Sensitivity	RX	Minimum input signal level required for acceptable performance
Radiation pattern	TX/RX	Defines directionality or coverage of transmitted/received signal

Table 2.3 RF-Related Requirements for EMC/EMI Testing

Category	*Requirement/Parameter*	*Description*	*Applicable Standards*
Radiated emissions	Spurious emissions	Limits on unintentional RF energy radiated from the product	FCC Part 15, CISPR 22/32, ETSI EN 301 489
Conducted emissions	Conducted noise on power/signal lines	Limits on RF noise conducted onto AC/DC lines or I/O cables	FCC Part 15, CISPR 22/32, MIL-STD-461 CE102
Radiated immunity	RF field immunity	Product must operate correctly when exposed to RF fields	IEC/EN 61000-4-3, MIL-STD-461 RS103
Conducted immunity	RF injection on cables	Product must tolerate RF signals injected onto cables or power lines	IEC/EN 61000-4-6, MIL-STD-461 CS114
Transient immunity	Electrostatic discharge (ESD), electrical fast transients (EFT), and surge immunity levels measured in kilovolts	Immunity to fast electrical transients, surges, and electrostatic discharge	IEC 61000-4-2/4/5, MIL-STD-461 CS117

or inadequate shielding may manifest as radiated emissions failures that seem RF in nature but are fundamentally non-RF design issues. This reinforces the need for a holistic design approach, as EMI/EMC behavior emerges from complex and tightly coupled interactions between RF circuits, digital logic, power electronics, mechanical design, and even firmware timing.

Non-RF Requirements

In addition to core RF performance parameters, successful product implementation also depends on a range of non-RF requirements. These factors often influence system architecture, component selection, and packaging decisions, especially in size, weight, or power-constrained applications, or in mission-critical environments.

Below are some of the most widely encountered non-RF design requirements. It is important to recognize that RF and non-RF domains are deeply interconnected; changes in one often impact the other. This entanglement must be carefully considered throughout the design process to ensure robust and compliant system performance. Table 2.4 shows some of the common non-RF design requirements.

2.3.4 Evaluate Risks and Opportunities

In product development, what you overlook in the early stages often becomes a challenge in the final product. Identifying risks and opportunities early on is critical to deliver products on time, on budget, and without fatal flaws.

Designing circuits and solving thermal issues are only half the battle; managing risk is the other. We will examine risks that are most relevant to engineering teams, spanning technical challenges, schedule delays, resource limitations, and legal or regulatory hurdles. We will also outline strategies to anticipate, map, and reduce these risks before they become roadblocks.

Table 2.4 Common Non-RF Design Requirements

Category	*Description*
SWaP-C	Governs physical footprint, mass, power consumption, and unit cost; strongly constrains architecture, component selection, thermal design, battery life, and achievable performance, especially in mobile, aerospace, and embedded systems
Mechanical and environmental constraints	Defines structural integrity and survivability under shock, vibration, temperature extremes, humidity, dust, and moisture; influences enclosure design, connector choice, mounting strategy, and qualification testing
Manufacturing and assembly	Covers PCB complexity, component tolerances, soldering processes, thermal interfaces, test access, calibration effort, and yield; directly impacts scalability, cost, reliability, and time to production
Firmware, software, and networking	Encompasses embedded firmware, control software, DSP algorithms, operating systems, and networking interfaces; affects system latency, determinism, update strategy, interoperability, remote control, and life-cycle maintainability
Security and system integrity	Addresses secure boot, encrypted communication, access control, tamper resistance, and key management; critical for defense, aerospace, medical, and connected RF systems where data integrity and IP protection are required

Broadly, these risks fall into five categories:

1. *Resource risks:* Do we have the right people, tools, and infrastructure to execute? This includes staffing gaps, equipment limitations, or dependence on unavailable expertise or partners.
2. *Technical risks:* Will the product work as intended? This includes uncertainties around performance, stability, integration, or meeting key specifications.
3. *Schedule risks:* Can the product be developed within the required timeline? Delays can result from unforeseen technical challenges, supply chain issues, parts availability, or over-optimistic planning.
4. *Legal and regulatory risks:* Does the product comply with all relevant laws and restrictions? This includes export control regulations (e.g., ITAR/EAR), safety certifications, EMC/EMI compliance, environmental directives (e.g., RoHS), and intellectual property considerations such as patents and licensing.
5. *Market risks:* Will the product meet a real need, and will customers buy it? Even a technically sound product can fail if there is no demand, poor product-market fit, or insufficient differentiation. Is there a competition threat? Did the market shift before the product was developed?

In the context of this book, we will assume that a customer or use case already exists. Therefore, the focus will be on the first four risk categories. These are the areas most relevant to engineers and teams preparing to turn a validated concept into a working product.

We will also discuss how to spot hidden opportunities, such as underutilized intellectual property (IP), techniques, and technologies produced previously, which can be used in the new product or cross-domain applications that can streamline development.

Resource Risks

Resources such as funding, skilled talent, and components can derail even the most promising concepts if not carefully managed. This section outlines the key categories of resource-related risks. We will also explore how strategic resource planning can enable more resilient, adaptable product development.

Budgetary Risk

Cost affects everything from the type of talent that you can hire and the size of the team that you can build to component selection, test infrastructure, iteration cycles, and long-term scalability. Even a technically flawless design can fail if the budget assumptions do not hold.

Having a clear estimate of overall costs, realistic revenue projections, and well-defined strategies to reduce unnecessary expenses while investing in critical areas is essential for a business to survive and thrive. It is crucial to start with a solid budget estimate and actively mitigate financial risk, whether by securing sufficient funding early on, prioritizing spending, or streamlining operations to preserve runway and agility.

Broadly, costs can be categorized into two main domains: capital expenditures (CAPEX) and operating expenditures (OPEX):

- CAPEX refers to one-time or infrequent investments in long-term assets or infrastructure. Examples include test equipment, lab facilities, custom tooling, and production machinery. These are typically front-loaded and essential for enabling development.
- OPEX covers the recurring costs needed to run the business or sustain the product's life cycle. Examples include rent, utilities, salaries, test consumables, and software subscriptions.

When building a budget, especially for startups or small businesses, it is essential to account for all cost categories. Teams often focus only on development or staffing and overlook other substantial expenses that can quickly exceed those figures. Before starting any project, a high-level budget estimate and a clear understanding of available funding are required. Using historical cost records and past trends significantly improves estimate accuracy and helps to identify budget risks early in the development cycle. That estimate then shapes key decisions such as team size, development timeline, and how many design iterations the project can realistically support.

In the late 1990s, Apple was weeks away from bankruptcy, with dwindling cash reserves and an unfocused product portfolio. The company's survival depended on radical simplification, aggressive cost control, and a last-minute investment from Microsoft. This episode illustrates that even world-class engineering talent and a strong brand are insufficient without disciplined budgeting and explicit risk management.

Human Resources and Staffing Risk

Developing a successful RF product requires a multidisciplinary team with expertise across both RF and non-RF domains. The specific skill sets depend on the complexity, functionality, and use case of the product. Table 2.5 lists commonly required roles and their responsibilities.

While some skills can be outsourced, doing so introduces trade-offs such as reduced control, IP concerns, and communication overhead. Building and leading a technical team also require careful consideration of factors such as talent availability, budget, schedule alignment, and team dynamics.

Having the right people, and having them early, is the most valuable asset in any RF product development effort. Exceptional engineers who can drive innovation, navigate ambiguity, and deliver reliable systems under real-world constraints are rare, even if skilled engineers are widely available. That is why it is critical to invest in top talent, carefully assess capabilities, and actively retain the people who truly move a project forward. Experience shows that even a single poor hiring decision or misjudgment of someone's abilities can derail an entire development effort. Here are some common staffing and talent risks:

- *Retention issues:* The loss of high-performing engineers during a project can disrupt schedules and create costly gaps in critical technical knowledge. Competitive compensation, meaningful work, and strong leadership help to retain key talent.
- *Knowledge silos:* When key system or design knowledge is held by only a few individuals, departures or absences can seriously impact project continuity.

Table 2.5 Key Technical Roles in RF Product Development

Core Role	*Primary Focus*	*Possible Specializations*
RF engineer	Design, simulation, and testing of RF components and systems	RF design engineer, RF systems engineer, RF test engineer
Antenna engineer	Design, simulation, and testing of antennas and arrays	Antenna design engineer, antenna test engineer
PCB design engineer/technician	Layout of RF and mixed-signal PCBs	RF layout specialist
Mechanical engineer	Design of enclosures, heat sinks, shielding, and structural integration	Thermal analyst, CAD designer
Embedded hardware engineer	Design of digital and mixed-signal hardware (e.g., FPGA/MCU interfaces)	FPGA designer, digital hardware engineer
Embedded software engineer	Low-level firmware development for control, timing, and communication	Bare-metal programmer, RTOS developer, driver developer
Software engineer	Development of application-level software, graphical user interface (GUI), and networking interfaces	User interface (UI) developer, backend developer, application programming interface (API) developer
DSP engineer	Implementation of signal processing algorithms (filtering, modulation)	Modem developer, signal chain architect, algorithm developer

Building redundancy through cross-training and shared documentation reduces this risk.

- *Cultural misalignment:* Differences in work ethic, communication style, or engineering values reduce trust, slow collaboration, and weaken overall team effectiveness. Clear expectations, strong onboarding, and careful hiring for cultural fit help to maintain alignment.

In addition to building strong internal teams, successful companies know how to strategically leverage external talent to accelerate development and reduce risk.

1. *Working with consultants and contractors:* When strategically leveraged, contractors can significantly accelerate development, bridge critical skill gaps, and offer fresh perspectives based on diverse project experience. However, to maximize their impact and avoid integration issues, careful planning and management are essential. Below are some best practices to consider:
   - *Define the scope and objectives clearly:* Use a well-written Statement of Work (SoW). Clarify deliverables, timelines, and review processes.
   - *Vet contractors thoroughly:* Review past work, references, and certifications. Technical skills are important, but alignment with your team's culture also matters.
   - *Establish strong communication:* Regular check-ins, shared dashboards, and access to internal documentation help to maintain alignment.
   - *Integrate them into the team:* Treat contractors as collaborators. Involve them in relevant meetings and give them access to the tools that they need to be effective.
   - *Monitor progress and give feedback:* Track performance via milestones or Key Performance Indicators (KPIs). Provide prompt feedback to maintain momentum.
   - *Secure confidentiality and compliance:* Use NDAs and enforce data security protocols, especially when dealing with proprietary designs.
2. *Consulting application engineers:* An often underutilized yet highly effective resource, available at no cost, are application engineers. They are technical specialists in a company who support customers with product selection, system integration, performance optimization, and troubleshooting. Application engineers work directly with a broad range of customers and have firsthand insight into real-world design challenges and best practices. They can provide valuable guidance on component selection, system integration, measurement techniques, and troubleshooting. Engaging with applications engineers early in the design cycle not only helps you to avoid costly design pitfalls but also ensures that you are leveraging the most appropriate and proven solutions for your specific application.
3. *Strategic partnerships:* Forming partnerships with other companies, research institutions, or industry alliances can accelerate development by pooling resources, sharing risk, and gaining access to complementary technologies or markets. For example, partnering with universities or national labs provides access to cutting-edge research, specialized facilities, and domain experts in

areas such as antenna design, materials science, and signal processing. Joint efforts, including sponsored theses, research projects, or internships, can be strategically aligned with your product development goals, fostering innovation while keeping costs manageable. However, these collaborations come with their own challenges, particularly around IP generation, ownership, and commercialization rights, all of which must be carefully clarified upfront.

Talent discovery is never trivial. Finding exceptional talent is rarely straightforward. Sometimes you will discover them online and other times at job fairs or conferences or through unexpected encounters. The key is to stay alert and recognize potential when you see it. Once you find great people, cherish them, invest in them, and do everything you can to keep them engaged and growing.

Supply Chain and Sourcing Risk

Component selection is not only about meeting electrical specifications but also about ensuring long-term availability, supplier reliability, and supply chain resilience. Ignoring these factors can introduce major risks during both development and production. Table 2.6 summarizes key supply chain and sourcing risks that should be evaluated early in the design process.

To minimize disruptions and ensure continuity in RF product development, proactive strategies must be integrated early into the design and procurement process. Table 2.7 shows several effective practices for reducing supply chain risk.

Technical Risks

Before beginning the implementation of a prototype or engineering design, it is essential to map out the technical risk landscape. This proactive step helps to identify which parts of the system are most likely to introduce issues during integration, testing, or production.

Technical risks often span multiple domains, including hardware (such as power management, RF signal chains, antennas, digital interfaces, and mechanical design),

Table 2.6 Supply Chain and Sourcing Risks

Risk Category	*Description*
Component lead times	Long lead times for key parts can delay the entire project
Vendor lock-in	Relying on a single supplier may create risks if prices change, support declines, or availability drops; consider multisource options where feasible
Geopolitical exposure	Sourcing components from restricted, unstable, or embargoed regions can pose serious legal and logistical challenges
Obsolescence and end-of-life (EOL) risks	Niche or low-volume components may face obsolescence or discontinuation, impacting long-term product support
Minimum order quantities (MOQ) and inventory risk	MOQs or large batch requirements can tie up capital and leave excess inventory in fast-evolving designs

Table 2.7 Proactive Strategies to Mitigate Supply Chain Risk

Strategy	*Description*
Make a critical parts list (CPL)	Maintain a prioritized list of high-risk or essential components to monitor closely and manage proactively
Pre-buy long-lead items	Secure critical components early to avoid delays due to extended procurement or manufacturing timelines
Design for flexibility	Lay out PCBs to support alternative package types or pin-compatible parts, enabling easier substitutions
Monitor component life cycles	Track parts for EOL or obsolescence using alerts from vendors or third-party tools
Diversify the bill of material (BOM)	Identify equivalent parts to reduce dependence on any single supplier and qualify multiple sources early

firmware, and software. Each of these areas presents its own challenges, and a weakness in any one of them can jeopardize the entire product.

Evaluating these risks requires not only deep domain-specific knowledge but also a strong understanding of how these domains interact to form a cohesive system. In the following chapters, we will explore these technical aspects in greater detail.

Schedule Risks

Schedule risk is one of the most common and frustrating challenges in product development and often one of the hardest to control. Unlike technical or resource risks, which can be managed more directly through design or staffing choices, schedule risks often stem from external dependencies and unforeseen disruptions. The more external your dependency, the more proactive your mitigation must be.

As your project grows in complexity or involves more collaborators, vendors, and partners, your exposure to schedule delays increases. Delays in one part of the system can cascade through the project, affecting milestones, testing windows, or delivery deadlines.

These risks can arise from logistical issues, personnel changes, regulatory delays, or even external market forces. While not all of them are preventable, identifying them early can help to build contingency into the development plan.

Table 2.8 summarizes some of the common schedule risks.

Some best practices can be used to minimize the schedule risks.

1. *Set milestones:* In RF product development, progress is rarely linear. It moves through cycles of complexity, iteration, and unexpected challenges. Without structure and checkpoints, even the most promising designs risk drifting off course. To navigate this terrain, milestones are needed. They give the process its checkpoints and focus as they mark what must be demonstrated or proven before the project can move forward with confidence. They also create alignment across teams. Engineers, designers, and stakeholders work more effectively when grounded in common objectives. A missed milestone is a signal. It tells you where the uncertainty is hiding and where schedule slippage is likely to begin.

Table 2.8 Common Schedule Risks

Risk	*Description*	*Typical Mitigation Strategy*
Component lead times	Long or unpredictable delivery times for key parts, especially RF-specific ICs, FPGAs, or specialized passives	Identify long-lead components early; place advance or split orders; maintain alternate part lists
Part availability	Global shortages, vendor prioritization, or unexpected EOL notices	Qualify multiple suppliers; use life-cycle management tools; track vendor notices
Vendor delays	PCB fabs, contract manufacturers, or design houses failing to meet agreed timelines	Establish milestones and regular progress reviews
Late-stage integration surprises	Subsystems work in isolation but fail when integrated, requiring rework	Plan for pairwise integration and early interface validation
Late-stage scope creep	New features or changes added after the design freeze, extending the schedule	Establish a change control process and freeze-point review
Key engineer or contributor unavailable	Illness, resignation, or extended paid time off (PTO) of a team member with unique expertise	Cross-train team members; maintain documentation and knowledge sharing
Regulatory approval lag	EMC certification, safety testing, or compliance documentation taking longer than expected	Engage with certification bodies early; pre-test key modules
Budget or funding delays	Late grant approvals, budget freezes, or slow payment processing	Secure partial funding milestones; maintain contingency budgets

Milestones also help build momentum. Progress in complex engineering is often incremental, and achieving even small goals builds confidence and sustains forward motion. For example, an early milestone may be: "Can we validate our core signal chain using COTS components before finalizing a custom PCB?" This approach reduces both schedule and technical risk by front-loading uncertainty.

The most effective milestones are anchored in clarity. "Build a prototype" is vague. "Verify low-phase-noise performance in a shielded environment" is specific, actionable, and meaningful.

2. *Tracking progress:* In any hardware development effort, especially one involving complex RF systems, tracking progress ensures that ideas move forward in a structured way, that risks are identified early, and that milestones are met with intention rather than assumption.

In larger, established companies, this responsibility often falls to a project manager. In startups or smaller teams, there may not be a dedicated person in this role, but the need for structured oversight remains. Someone must take ownership of tracking technical and logistical progress, or even the most promising designs will drift into disorganization.

A basic project tracking setup often includes:

- Gantt charts to visualize tasks and dependencies over time;
- Milestone tracking aligned with design, build, and validate phases;

- Deliverable checklists tied to each stage of development;
- Risk registers to monitor technical, schedule, or resource-related threats;
- Status updates and progress reviews at regular intervals.

Even in fast-moving teams, lightweight project management practices can make a significant difference. A simple weekly risk review meeting focused on upcoming bottlenecks and their potential solutions can surface problems before they derail the schedule.

Ask every week: "What's most likely to delay us in the next two weeks? What has changed since last week?"

Whether through formal tools or simple spreadsheets, tracking ensures that each decision, delay, or pivot is made with awareness and that the product stays on course from concept to reality.

Legal and Regulatory Risks

In RF product development, overlooking regulatory risk can halt a product before it ever reaches the market. Compliance is not a final checkbox; it must be built into the design process from the very beginning. This section highlights the key legal and regulatory considerations that influence product viability, development timelines, and the ability to deploy the product across different regions and markets.

Compliance

Skipping compliance until the final stages is one of the most common and costly mistakes in product development. Early alignment with applicable standards prevents last-minute redesigns, failed certifications, and missed market windows.

Here are some best practices for early compliance planning:

- Pre-check regulatory boundaries such as power limits, spurious emissions, and EMC requirements.
- Identify applicable standards early, such as:
  - FCC Part 15 (US unlicensed emissions);
  - CE RED (Radio Equipment Directive for Europe);
  - MIL-STD-461/810 (for military EMC, shock, and environmental performance).
- Design for testability by incorporating pre-certification test points or simulation flags into early prototypes.

For military or aerospace applications, it is essential to identify required environmental and electromagnetic profiles from the outset. Ingress, shock, vibration, EMI immunity, and temperature ranges must be considered during the design phase, not after a failure in validation.

Export Control

Export control laws govern the international distribution of products, components, software, and technical data. Violations can result in severe legal and financial consequences.

Key considerations are:

- *Understand Export Administration Regulations (EAR) and International Traffic in Arms Regulations (ITAR):* If your product has dual-use functionality or any defense-related capabilities, it is critical to assess whether it falls under EAR or ITAR. Noncompliance can result in serious legal and financial consequences. Engage with legal or compliance experts early if export control may apply. Determine if your design or documentation includes controlled technologies, especially high-frequency RF systems, encryption, or satellite communications.
- Classify your product early to avoid delays in customer delivery or international collaboration.
- Work with legal or compliance officers to establish export control boundaries before engaging with foreign suppliers or customers.

Table 2.9 summarizes US export control regulations.

The United States Munitions List (USML) defines defense-related items controlled under ITAR. It includes 21 categories, such as military electronics (Category XI), and governs items specifically designed for military use. Products listed on the USML require licensing from the US Department of State before export.

An Export Control Classification Number (ECCN) is a five-character alphanumeric code used under the EAR to classify items, such as hardware, software, or technology, that are subject to export controls. Each ECCN:

- Specifies the technical characteristics of the item;
- Indicates the reason for control (e.g., national security, anti-terrorism);
- Determines if an export license is required for a specific destination or end user;
- Helps to assess eligibility for license exceptions.

If your RF product or any of its key components falls under a specific ECCN, you may need to obtain an export license before shipping it, depending on the destination country and intended end use.

Patent Infringement

Patent infringement risk is often overlooked until legal challenges arise, sometimes after launch. A strong product strategy includes freedom-to-operate (FTO) checks early in the development process. FTO analysis evaluates whether a product can be

Table 2.9 Summary of US Export Control Regulations

Regulation	*What It Covers*	*Classification System*	*Administered By*
International Traffic in Arms Regulations (ITAR)	Military-specific items, services, and technical data	USML (United States Munitions List) Categories I–XXI	US Department of State (DDTC)
Export Administration Regulations (EAR)	Dual-use items (civilian and military), including some RF and satellite technology	ECCN (Export Control Classification Number)	US Department of Commerce (BIS)

designed, manufactured, and sold without infringing on existing patents, licenses, or other intellectual property rights. Performing these checks early helps to avoid costly redesigns, legal disputes, or delays later in the product life cycle, when changes become far more expensive and disruptive.

To reduce patent-related risks:

- Conduct a prior art search during the concept or early design stages.
- Engage with IP attorneys or consultants to evaluate whether your implementation may infringe on existing patents.
- In high-risk areas (e.g., antenna arrays, filtering techniques, beamforming algorithms), consider alternative implementations or licensing options.
- Maintain thorough documentation, including lab notebooks, design logs, and test records, to support future IP claims or defenses.

From Risk to Opportunity

We have explored the risks, but where there are risks, there are also opportunities. In the rush to mitigate failure modes, it is easy to overlook the potential advantages quietly embedded within your design space. Yet a product, regardless of complexity, cost target, or form factor, might contain untapped strategic value if you know where to look.

For engineers and entrepreneurs alike, identifying these hidden levers can unlock differentiation, shorten timelines, or even open new markets. Here are a few key opportunities to keep in mind:

- *IP reuse as leverage:* One of the most underutilized forms of leverage in RF product development is the smart reuse of proven IP. Validated circuit blocks, verified codebases, and well-tested design patterns can save months of effort. Reusing a known RF front-end chain or FPGA module not only accelerates development but also reduces technical risk and shrinks the verification burden. However, reuse must be intentional and well-documented; blindly copying without understanding the underlying assumptions or constraints is replication, not leverage.
- *Strategic timing:* Timing can be a powerful advantage. Aligning your product roadmap with emerging standards, such as a new satellite band, a 5G-NR release, or a government spectrum allocation, can offer a decisive market edge. Being "first to comply" or "early to integrate" often outweighs having the most features. This requires a vigilant eye on industry roadmaps, regulatory changes, and even public request-for-proposal (RFP) calendars.
- *Dual-use potential:* Many RF technologies have applications beyond their original purpose. A phased array system developed for ground-based satellite terminals may also serve air-to-ground communications or military platforms. Staying open to cross-domain applications can dramatically extend the commercial value of your innovation.

The common thread in all these cases is intentional foresight. Recognizing opportunity is about being informed, cross-functional, and alert to signals.

2.3.5 Exploring Potential Solutions

Now that the problem is well-defined, the target audience is understood, the requirements are established, and the project risks are evaluated, it is time to explore what potential solutions could look like. This stage involves creative thinking, leveraging existing technologies, and evaluating trade-offs to identify viable paths forward.

Evaluating Existing Solutions

Before investing time and resources in a custom RF design, it is essential to evaluate whether existing solutions, whether COTS, internally developed, or sourced from third parties, can satisfy some or all of the design requirements. Leveraging proven technologies can significantly reduce development time, lower technical and regulatory risk, and simplify compliance. In practice, these options fall into four categories: buy it, license it, reuse it in the case of prior internal designs, and copy it when the design is open source or explicit permission has been granted.

Table 2.10 summarizes the key resources to investigate when evaluating existing solutions, along with their associated advantages and potential drawbacks. Consulting domain experts or strategic partners can provide valuable insights into existing or emerging solutions that may not be publicly documented. Their experience can help you avoid common pitfalls, identify viable design paths early, and significantly accelerate the decision-making process.

It is essential to compile a comprehensive list of all existing solutions, both internal and external, that may fulfill part or all design requirements. The next step is to clearly determine which components, subsystems, or technologies can be leveraged as is, and which elements require custom development. While this task may seem time-consuming upfront, a thorough investigation can significantly reduce overall development time, cost, and risk by preventing redundant engineering effort and identifying integration-ready components or architectures.

Table 2.10 Pros and Cons of Using Existing Solutions

Source	*Advantages*	*Challenges/Risks*
Commercial modules/COTS	Fastest path to system integration, well-documented with vendor support, proven compliance and certification history	Limited customization and flexibility, potential vendor lock-in, may be overdesigned or underperforming for the target use case, licensing or royalty costs, potential patent infringement issues, integration and compatibility complexity
Vendor reference designs	Validated architectures and topologies, accelerates prototyping and reduces design risk, optimized bill of materials	May require adaptation to specific requirements, IP rights or redistribution may be restricted
Internal IP/legacy designs	Trusted within the organization, reduces learning curve and development time	May rely on outdated components or design tools, may lack sufficient documentation, requires reverification in new use cases
University/lab research	Access to cutting-edge concepts and innovations, often supported by published data and peer-reviewed results	Prototypes may not be suitable for production or production-ready, licensing and IP constraints, limited long-term support or maintainability

To illustrate this, consider the development of a custom Ku-band conformal phased array system for an aircraft. Suppose that beamformer ICs for this frequency range are readily available from trusted vendors that can be directly integrated without additional development. However, while various phased array antenna solutions exist, none may meet the specific mechanical, aerodynamic, or performance constraints of the airframe. As a result, the antenna array must be designed and developed in-house. Similarly, elements such as control firmware, calibration routines, and interface software are often application-specific and will also require custom implementation.

When building a new solution from scratch, optimize only what you must. Avoid reinventing the wheel when a good-enough module or technique already exists. Innovate where it creates real value or enables functionality that cannot be achieved otherwise.

How to Proceed When No Proven Approach Exists

In some cases, all or part of the solution you are seeking may not be available off the shelf. When that happens, you must consider how to develop the required functionality, either internally or with external support. The decision depends on factors such as technical complexity, available expertise, budget, timeline, and risk tolerance. Table 2.11 shows common paths forward.

Each option has trade-offs. In many cases, a hybrid approach is used, such as developing the critical RF front end in-house while outsourcing firmware or licensing a reference design for the digital control interface. The key is to make decisions based on your project's priorities: time to market, technical differentiation, and resource constraints.

Feasibility Analysis

Feasibility analysis answers a critical question early in the process: "Can we build this product, technically, economically, and within constraints?"

Table 2.11 Options When No Existing Solution Fits

Approach	*Description*	*Pros*	*Cons*
Internal development	Build the missing solution in-house using your engineering team and lab resources when the required expertise and infrastructure are available	Full control over IP and design, tailored to exact requirements	Requires significant time, tools, and skilled staff
Outsource development	Hire a specialized contractor or design house to develop a subsystem or the full product	Faster development, access to niche expertise	Less direct control, potential IP and dependency risks
Partnerships	Collaborate with another company, university, or research lab to codevelop the solution	Shared cost and risk, access to complementary capabilities	Requires clear agreements, potential IP and coordination challenges
Buy or license IP	Acquire or license an existing solution that already meets part or all of the technical requirements	Fastest option when available, proven performance	May require adaptation, licensing fees, and IP negotiation

This step reduces risk by evaluating four key areas:

- *Technical feasibility:* Can we meet performance specifications (e.g., bandwidth, gain, noise figure) using available technologies? Are integration and packaging realistic?
- *Economic feasibility:* Is the target cost achievable? Do components, materials, and processes align with the budget and business model?
- *Schedule feasibility:* Can we deliver on time, given design complexity, long-lead items, and available resources?
- *Regulatory feasibility:* Will we meet certification and compliance requirements (FCC, CISPR) within cost and schedule limits?

A well-run feasibility analysis highlights potential blockers early, informs architecture decisions, and helps to avoid costly surprises down the road. It sets the stage for the PoC phase with a clear understanding of risk and readiness.

2.3.6 Product Blueprint

If one or more solutions pass the feasibility assessment, and stakeholders choose to proceed based on available resources, risk tolerance, and strategic alignment, then the next step is to move forward with implementation (see Figure 2.1). This transition marks the beginning of detailed design and development. Table 2.12 shows the three phases of solution implementation.

In this volume, we will focus on the product's blueprint, establishing a clear path toward implementation. Volume 2 will be dedicated to prototyping, engineering development, and implementation of this blueprint, including detailed design, prototyping, and validation. Volume 3 will focus exclusively on the production phase, covering topics such as manufacturing, testing, quality control, and scaling for deployment.

2.4 Execute: PoC, MVP, EVT, DVT, and PVT

As shown in Figure 2.1, the next step after completing the "prepare" phase is the "execute" phase. Transforming an idea into a shipped RF product requires a

Table 2.12 Three Phases of the Solution Implementation

Phase	*Description*	*Typical Outputs*
Product blueprint	Focuses on capturing the product architecture based on the requirements using high-level block diagrams and early-stage simulations	Block diagrams, system-level trade studies, basic simulations
Prototyping	A functional prototype is developed to validate the core idea and assess key performance metrics under real-world conditions	PoC, MVP, lab test results
Development toward production	This phase bridges the gap between a prototype and a production-ready product; it includes formal engineering builds and validation stages such as EVT, DVT, and PVT	EVT, DVT, and PVT hardware builds, compliance reports, and manufacturing package

structured development pipeline in which each phase, PoC, MVP, EVT, DVT, and PVT, serves a specific purpose, from proving technical feasibility to scaling for production. Understanding and managing these stages are essential for predictable execution and long-term success.

Figure 2.3 provides a visual overview of the five key implementation stages and how they guide the transition from prototype to production. Keeping these stages in mind during the blueprint phase is essential, as early architectural decisions, interfaces, and trade-offs will propagate through the entire development cycle. Choices made at this stage, component selection, partitioning of the signal chain, mechanical constraints, thermal strategies, and firmware architecture, can either accelerate downstream work or create bottlenecks that become expensive to fix later.

The five stages of RF product development are summarized in Table 2.13.

As shown in Figure 2.1, each execution phase follows a design, build, and validate loop. A design is implemented and tested, the test results are fed back into the design, and this process continues until the pass criteria for the specific phase are met.

2.5 Collect Insights: Testing, Feedback, and Analysis

During various prototyping phases, it is essential to engage with stakeholders and gather meaningful insights. These insights validate assumptions, uncover hidden risks, and ensure that the product is evolving in the right direction. Effective teams treat this as an ongoing, structured activity rather than an ad hoc checkpoint.

The following methods are commonly used to gather feedback:

- *Pilot deployment:* Limited release to selected users under real-world conditions to uncover edge cases, bugs, and unmet needs.
- *Heuristic testing:* Expert reviewers evaluate the product against established usability and design principles.

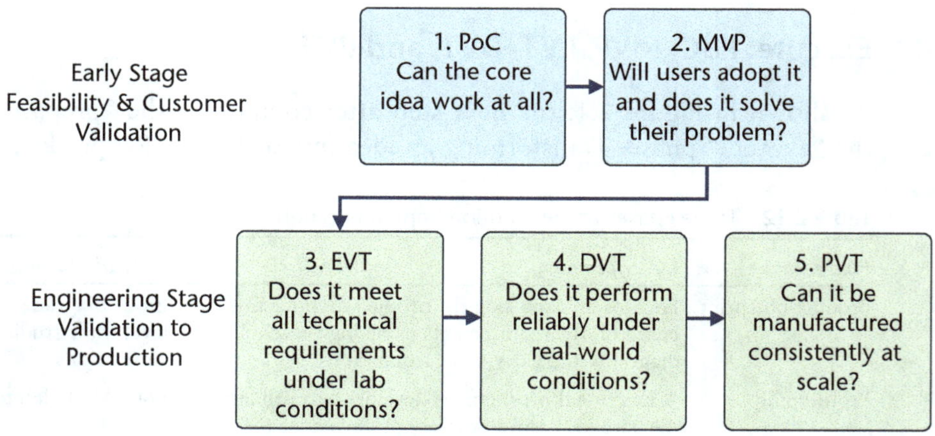

Figure 2.3 The five key stages of the product development life-cycle progress from PoC to PVT. Each stage answers a critical question: PoC evaluates technical feasibility, MVP gauges customer interest, EVT verifies performance under ideal conditions, DVT tests robustness in real-world environments, and PVT ensures the design can be reliably scaled and manufactured.

Table 2.13 RF Product Development Stages

Stage	*Main Question*	*Goal*	*Properties*	*Build Size*	*Example*
PoC *(Working concept)*	Can this idea even work?	Demonstrate the technical feasibility of the core idea or function	No concern for packaging, scalability, or cost; focus on validating key functions	1–5 units	Bench-top setup, simulations, or basic measurements
MVP *(Demo unit)*	Will customers use it?	Validate early functionality and gather market feedback through exhibitions or pilot users	Not ready for volume production or environmental testing; high cost, low volume, minimal optimization	1–10 units	Functional prototype; block-level PCB with limited integration
EVT *(Alpha unit)*	Does the design meet all technical specifications under ideal conditions?	Verify that the design satisfies key electrical, thermal, and mechanical specifications	No full environmental or compliance testing yet; focus on measurements and design reviews	20–50 units	Engineering prototypes, test reports, and design reviews
DVT *(Beta unit)*	Can it survive the real world?	Assess usability, compliance, and reliability under realistic conditions	May still need yield improvement or final tuning; includes shock, vibration, drop, ingress, and burn testing	50–200 units	Updated schematics, compliance pre-checks, and thermal analysis
PVT *(Final unit)*	Can we scale and manufacture it reliably?	Confirm manufacturability and process readiness for full-scale production	Validates volume manufacturability and quality assurance (QA)/quality control (QC) processes; final opportunity to tune yield and tooling, no major design changes allowed	5%–10% of the target volume	Pilot manufacturing run; test jigs; QA/QC development; final tooling; units are typically sellable

- *A/B testing:* Comparative evaluation of design alternatives to measure user preference or performance metrics.

Effectively using this feedback presents several challenges:

- *Misinterpreted feedback:* Users may have difficulty articulating underlying issues, or feedback may describe symptoms rather than root causes.
- *Overfitting to vocal users:* Prioritizing input from a small but outspoken subset can distort design decisions.
- *Confirmation bias:* Teams may unconsciously dismiss negative feedback when it conflicts with existing assumptions.

As we will see in the next section, the insights gathered at this stage are used to determine what needs to change and how the team should proceed.

2.6 Refine and Iterate

Armed with validated feedback, the next step is to refine and iterate on the Identify, Prepare, Execute, and Insight Collection phases (see Figure 2.1). Depending on the scope and severity of the findings, you may need to adjust all phases or only the ones impacted. The goal is to make deliberate, evidence-based decisions rather than reactive changes. Each refinement cycle should be targeted and purposeful, driven by clear learning objectives instead of arbitrary polishing.

Here are some considerations for effective decision-making:

- *Pivot versus persevere:* Use structured criteria, such as severity, frequency, user impact, market alignment, or technical feasibility, to decide whether to continue on the current path or explore alternative architectures or workflows.
- *Selective feedback integration:* Not all feedback warrants action. Apply engineering judgment to distinguish between critical issues, usability defects, architectural limitations, feature requests, and nonessential opinions.

Ultimately, this stage is about interpreting, prioritizing, and responding in ways that materially improve the product and strengthen user trust.

We have now discussed the five elements of the RF product development framework shown in Figure 2.1. Together, this framework provides a systematic approach to navigating the development process and ensures that the essential technical, architectural, and validation aspects are thoroughly addressed.

2.7 Hard-Won Lessons in RF Product Development

In this section, I share several principles that I have learned throughout my career in RF and quantum engineering. These insights were earned through years of experimentation, mistakes, and occasional breakthroughs. They are not intended to be rigid rules or a fixed recipe. Rather, I offer them as guidance for you to interpret and apply in ways that align with your situation, values, and working style.

2.7.1 Decision-Making

We are all guilty of wishful thinking at times, and I am no exception. When working on an exciting new product, it is easy to become overly optimistic, to assume things will "just work," or to overlook red flags because the idea feels too good to let go. To stay grounded and make sound decisions throughout product development, the following principles have consistently helped me.

- *Reality is your best friend:* Ray Dalio wrote in *Principles* that failing to embrace reality is one of the costliest mistakes that you can make [6]. In engineering, especially in product development, this could not be truer. Entrepreneurs, product owners, and even investors often lean too heavily on optimism and best-case scenarios. However, physics does not care about our intentions. Business and engineering decisions grounded in reality, not wishful thinking, save time, energy, and resources.

- *Decide based on evidence:* Even world-class scientists and engineers fall into the trap of decision-making by analogy or gut feeling. I have seen teams waste months on flawed directions because no one stopped to demand data or evidence. Avoid endless debates by grounding every critical decision in simulations, experiments, and data. Build a habit of asking, "What do we actually know?" before moving forward.
- *Document decisions and trade-offs:* You will not remember every design choice 6 months from now, and neither will your team. Record why specific decisions were made, the alternatives considered, and the rationale behind deviations from specifications. This habit pays dividends during debugging, design reviews, onboarding, and future revisions.

2.7.2 Execution

You have probably heard the saying, "Ideas are cheap, execution is everything." Turning a concept into a product is rarely straightforward. Execution requires coordination across people, budgets, suppliers, and schedules.

One of the hardest aspects of execution is enabling smart, independent people to work together effectively. Good leaders know when to step in to remove obstacles and when to step back and let engineers do their work.

The following principles can help to keep execution on track:

- *Clear ownership:* If you have more than one person responsible for a task, then no one is responsible. This ambiguity is often a recipe for conflict and one of the biggest execution risks in engineering teams.

  Clear ownership, aligned priorities, and transparent dependencies are essential for smooth execution. Everyone must know who is doing what, by when, and what they are waiting on. Many execution failures are not technical; they are coordination failures that stem from unclear roles.

  Clear ownership means:
  - Every task, subsystem, and decision domain should have a clearly defined owner. When roles overlap or go undefined, work is duplicated, issues slip through the cracks, and accountability disappears.
  - There is no confusion about whom to ask, who decides, or who is accountable.
  - Decisions move faster, meetings become fewer, and the mission becomes clearer.
- Build momentum and set clear deadlines: Progress compounds. Small wins early on fuel larger wins later. A task often will not get done without clear expectations and a firm deadline, after all, "work expands to fill the time available for its completion," as Cyril Northcote Parkinson observed. Define tangible short-term milestones, celebrate progress, and keep the team moving.
- *Do not let perfect kill progress:* As Voltaire said, "The best is the enemy of the good." In RF and hardware development, perfectionism can become a serious liability. Whether it is chasing an extra decibel of return loss or obsessing over microscopic layout symmetry, the cost in time and morale can quickly outweigh the benefit. Strive for excellence, but ship solid designs rather than endlessly polishing them.

- *Prioritize function first, optimization later:* First, make it work. Then, and only then, make it compact, low-power, elegant, or cheap. Premature optimization kills momentum and can hide fundamental issues.
- *Build fast, learn faster:* Speed matters not because it looks good on a Gantt chart, but because early feedback shortens the path to success. Get a prototype working quickly so that you can start learning from it. Iteration beats stagnation every time.
- *Validate with real use cases:* Benchmarks and lab results are just the beginning. Test your product in the real world, with real users, under real conditions. That is where hidden edge cases, usability gaps, and reliability issues emerge.
- *Build with the customer, not just for them:* Engage users early and often. If you are not building with them in mind from day 1, you are likely building the wrong thing.
- *Balance speed and risk intelligently:* Moving fast does not mean skipping critical steps. Every shortcut has a cost, sometimes minor, sometimes fatal. Be explicit about where you are taking risk (e.g., reduced margin in a prototype, relaxed EMC filtering in an early board) and why. Move quickly, but make conscious, documented trade-offs rather than accidental ones.
- *Simplicity wins long-term:* Leonardo da Vinci once said, "Simplicity is the ultimate sophistication." That is especially true in hardware. Simple circuits, clear specifications, and clean layouts scale better, break less, and are easier to maintain and debug.
- *Know when to pivot and how to do it:* Technical development is almost never a straight line. In RF systems, many issues only appear during integration, environmental testing, or field trials. Recognizing when a path is no longer viable and pivoting without derailing the entire project is a critical execution skill. Pivoting is not a failure; it is a controlled response to new information. When you do pivot, communicate clearly and document the rationale.

2.7.3 Team

Building a great engineering team requires effort, patience, and the ability to recognize the right fit for the role and the culture. Technical capability is essential, but it takes far more than raw IQ to build a functional, high-performing team, especially in RF and hardware development, where things rarely work on the first try. The qualities that matter most often reveal themselves not during smooth successes but during failures: resilience, perseverance, discipline, and character.

The following principles can help you assemble and develop a team capable of delivering complex, real-world products.

- *Technical capability:* There is a huge difference between knowing and mastery. I have met many engineers who "know" a topic but miss the subtleties that make or break a product. In today's world of quick answers, especially with the rise of artificial intelligence (AI) and endless online resources, superficial knowledge is easy to acquire, and the illusion of expertise is everywhere. The real danger is having just enough understanding to feel confident you are right, while lacking the depth to recognize your own blind spots. Real mastery only becomes visible when the design is under stress, in production, in integration, or 6 months after deployment in the field. One of my favorite quotes, commonly attributed to Miles Kington, is: "Knowledge is knowing a tomato is a fruit. Wisdom is not putting it in a fruit salad."

- *Humbleness:* Stay humble. Bring in experts or at least consult them early. Ask questions, challenge assumptions, and dive deep into the details. The right insight at the right time can save months of work or prevent a costly mistake.
- *Resilience:* Jensen Huang, CEO of NVIDIA, often emphasizes that resilience, not intelligence, is one of the strongest predictors of success. He argues that individuals with very high expectations frequently exhibit low resilience, while it is resilience that ultimately carries engineers through the unavoidable adversity of building real hardware [7]. The message is clear: great engineers keep going when things get hard, and, in RF and hardware development, they almost always do.
- *Integrity, intelligence, and energy (Warren Buffett's hiring rule):* Warren Buffett distills great hiring down to three attributes: integrity, intelligence, and energy [8]. He then adds a critical caveat: "If they don't have integrity, don't even bother with the first two."

  This principle is especially critical in engineering. A brilliant and highly driven individual without integrity or with poor character can cause catastrophic damage to a team, a project, or even an entire company. It takes only a single bad hire to undermine an otherwise strong organization.

  Equally important is protecting your most capable and principled engineers: high-performing, honest talent does not tolerate a toxic culture for long and will leave if it is allowed to persist. For this reason, decisions about whom to hire, whom to let go, and which behaviors are tolerated are among the most consequential choices a company can make.

  Building a strong engineering culture requires sustained time and effort, but it delivers substantial long-term returns. It depends on clearly defining the principles that guide behavior and holding everyone, regardless of seniority, accountable to them.

2.8 What's Next?

In this chapter, we introduced the product development framework and examined each stage in detail. With this foundation in place, we can now begin building the product blueprint. The next step is to develop the core foundational knowledge across the three major dimensions of any RF system: the signal-chain layer, the RF layer, and the non-RF layer. Together, these dimensions shape the final product and form the basis for the chapters that follow.

Problems

2.1 Refer to Figure 2.1. You are part of an engineering team developing a portable Ku-band ground terminal intended for rapid deployment in emergency and disaster-recovery scenarios.

a. Using the framework in the figure, describe one concrete activity your team would perform in each of the following phases: identify, prepare, execute, and collect insights.
b. During the Execute phase, the first prototype (PoC) meets RF performance targets in the lab but fails during outdoor testing due to intermittent link loss.
   - Identify which earlier phase(s) may have been insufficiently addressed.
   - Explain what was likely overlooked.
c. Describe how the Refine and Iterate step should be applied after collecting field-test feedback, and explain how this iteration reduces risk before entering DVT or PVT.

2.2 Refer to Figure 2.2. You are part of a team developing a new RF-enabled product, but, at this stage, you are not responsible for detailed circuit or signal-level design.
a. Explain the difference between signal-chain layer requirements, RF requirements, and non-RF requirements using plain language. Provide one example of each that does not rely on technical RF terminology.
b. The figure shows a strong interaction between RF requirements and non-RF requirements. Describe a realistic situation in which a decision made for non-RF reasons (such as size, cost, or power consumption) could negatively impact RF performance, even if the RF design itself is sound.
c. Why is it important to capture both RF and non-RF requirements before creating the product blueprint? Explain how missing or poorly defined requirements at this stage can lead to problems later during prototyping or manufacturing.
d. The development flow includes multiple prototype stages (PoC, MVP, EVT, DVT, PVT). Without going into technical detail, explain why multiple prototype stages are necessary instead of building a single "final" version directly.

2.3 You are designing a wideband EW receiver intended to detect unknown drone emitters operating anywhere between 400 MHz and 6 GHz.
a. Identify three signal-chain layer requirements that are critical for this mission (for example, types of signals or behaviors the receiver must be able to handle).
b. Identify three RF-level requirements necessary to support these signal-chain requirements.
c. For each RF requirement listed in part (b), briefly explain how it is driven by one or more signal-chain requirements identified in part (a).

2.4 A reactive jammer must detect a signal and respond within 50 μs.
a. Identify two architectural choices (signal chain + RF front end) that help to meet this latency requirement.
b. Identify two choices that would make the requirement significantly harder.
c. Justify each answer.

2.5 Assume that you require a phase-coherent 8-channel, 10-GS/s arbitrary waveform generator (AWG) for qubit control, and no commercial unit exists. Evaluate the following paths: internal development; partnership with a university or national lab; licensing IP from an AWG vendor. For each option, assess risk, development time, and IP control and then rank the three options from most feasible to least feasible.

2.6 You are integrating a digital-beamforming radar with multiple analog-to-digital converters (ADCs), LOs, and high-speed JESD204C serial links.
   a. Identify three schedule risks associated with integration.
   b. For each risk, provide one early mitigation strategy.

2.7 You want to repurpose a military EW receiver for commercial spectrum monitoring. List four constraints, regulatory, technical, cost, or environmental, that make this infeasible without redesign and provide a short justification for each.

2.8 Your product must pass FCC Part 15 radiated emissions limits.
   a. Identify one RF requirement that you must define early to avoid emissions failures.
   b. Identify one non-RF design factor that significantly affects radiated emissions.
   c. Explain why EMI and EMC issues often appear to be RF problems but originate in non-RF domains.
   d. Suggest one pre-compliance test that you would perform during the EVT phase.

2.9 Answer the following questions:
   a. Identify and briefly describe five major categories of risk that can affect product development, including resource risk, technical risk, schedule risk, legal and regulatory risk, and market risk.
   b. For each risk category, provide one realistic example of how that risk could manifest during the development of the product.
   c. For two of the risks that you identified, describe one mitigation strategy that could be applied early in the project to reduce the likelihood or impact of the risk.
   d. Explain why addressing these risks early in the development process is more effective than reacting to them later during prototyping or manufacturing.

2.10 Refer to Figure 2.3. You are part of a team developing an Internet of Things (IoT) network of RF sensors designed to monitor cellular base stations by measuring signal strength and coverage, detecting outages, and supporting network optimization and troubleshooting.
   a. For each development stage, identify the primary question the team is attempting to answer. Do not focus on technical implementation details; instead, address feasibility, user validation, reliability, and manufacturability.
   b. Suppose the team skips the MVP stage and moves directly from PoC to EVT:
      - Identify one major risk introduced by this decision.
      - Explain how this risk could propagate into later stages, such as DVT or PVT.
   c. During DVT, the product meets all performance requirements in laboratory testing but fails intermittently during field deployment.
      - Which earlier stage or stages may not have been adequately addressed?

- What type of issue is most likely being uncovered at this point?

d. Explain why PVT cannot be treated simply as a scaled-up version of EVT or DVT, even if the underlying design has not changed.

References

[1] CB Insights, "The Top 12 Reasons Startups Fail," *CB Insights Research Brief*, 2021.

[2] Eisenmann, T., E. Ries, and S. Dillard, "Hypothesis-Driven Entrepreneurship: The Lean Startup," *Harvard Business Review*, July-August 2012.

[3] Startup Genome, "Startup Genome Report Extra on Premature Scaling," 2011.

[4] Blank, S., and B. Dorf, *The Startup Owner's Manual: The Step-by-Step Guide for Building a Great Company*, Pescadero, CA: K&S Ranch Publishing, 2012.

[5] Fitzpatrick, R., *The Mom Test: How to Talk to Customers and Learn If Your Business is a Good Idea When Everyone Is Lying to You*, London, UK: RobFitz Ltd., 2013.

[6] Dalio, R., *Principles*, New York: Simon & Schuster, 2017.

[7] Huang, J., "NVIDIA CEO Jensen Huang: 'I Hope Suffering Happens to You,'" *Business Insider*, Interview, March 14, 2024.

[8] Buffett, W. E., "Warren Buffett's Hiring Principles," in *The Essays of Warren Buffett: Lessons for Corporate America*, Fourth Edition, L. A. Cunningham, (ed.), New York: Wiley, 2015.

CHAPTER 3
Signal Chain Layer

Information is the resolution of uncertainty.

—*Claude Shannon*

The concepts introduced in this chapter, such as modulation, pulse shaping, and coding, establish the system-level performance metrics that ultimately define the critical parameters of an RF product, including data rate, required link margin, error rate, spectral efficiency, and latency. Decisions made at the signal chain layer propagate throughout the design, directly shaping the RF hardware architecture, digital signal processing (DSP) requirements, software integration, and size, weight, power, and cost (SWaP-C) trade-offs. A solid understanding of this layer ensures that high-level performance objectives are translated into an executable design path.

3.1 Introduction

Most RF products are built to process and exchange information, whether in large-scale systems such as cellular communication networks or in compact devices such as Internet of Things (IoT) sensors. To design effectively, it is essential to understand the requirements at the lower layers, the parameters that govern how information is structured and transmitted. These include, for example, modulation type, coding scheme, data rate, and latency.

Figure 3.1 illustrates how signal chain layer requirements drive both RF and non-RF system requirements, shaping the overall product blueprint and propagating through prototype development and manufacturing.

3.2 Communication Link

In this section, we introduce the structure of a communication link by first reviewing the communication layers and then focusing on the physical layer, which is most closely tied to the underlying RF hardware.

3.2.1 Communication Layers

In wireless and RF product design, communication is commonly described using the Open Systems Interconnection (OSI) reference model, as shown in Figure 3.2. While the OSI model defines seven layers, modern wireless standards, such as Wi-Fi, LTE, 5G, and Bluetooth, collapse these layers into fewer functional blocks,

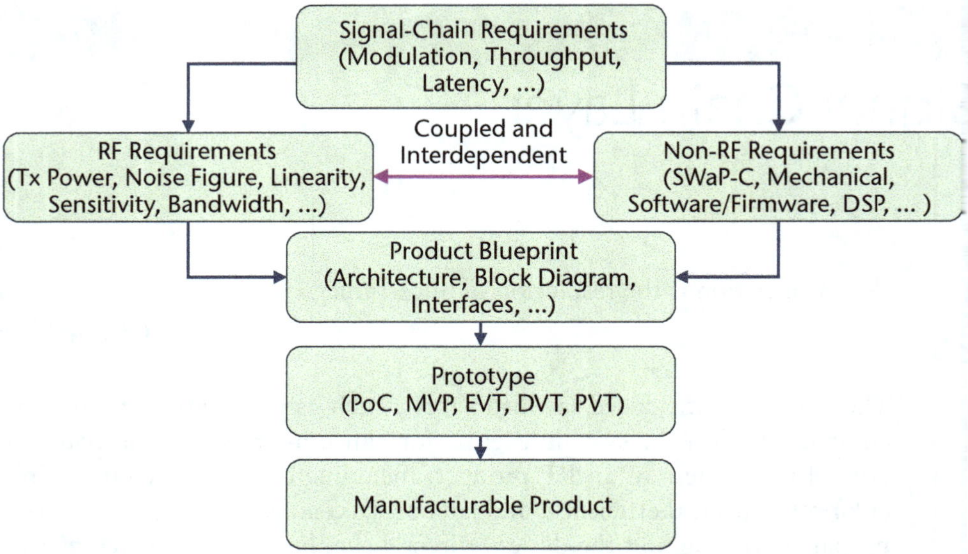

Figure 3.1 Flow from the signal chain layer requirements to a manufacturable product. The diagram illustrates how signal chain layer requirements drive both RF and non-RF requirements. These requirements inform the product blueprint, the prototype implementation, and, ultimately, the manufacturable and production-ready product.

typically including the physical (PHY), medium access control (MAC), and higher software layers [1].

Figure 3.2 illustrates, on the left side, how the OSI layers map to software, firmware/digital logic, and RF hardware domains. From an RF product design perspective, the physical layer (L1) and the data link layer (L2), which includes the MAC sublayer, are the most critical. These layers directly define hardware requirements, timing and synchronization constraints, and SWaP-C trade-offs. In particular, the physical layer translates signal chain layer requirements, such as modulation, bandwidth, noise, linearity, and timing, into concrete RF, analog, mixed-signal, and DSP implementations.

Higher layers are primarily implemented in firmware and software and influence the RF design indirectly through requirements on throughput, latency, framing, and robustness. For this reason, this book uses the OSI model primarily as a boundary-definition tool, with a primary focus on the physical layer, where RF engineering decisions have the greatest impact on system performance and overall product feasibility.

Every communication system begins with information, such as voice, video, sensor readings, or control data, that must be prepared for transmission. While analog communication systems still exist for legacy purposes or specialized niche applications, the vast majority of modern communication systems are digital. As a result, it is essential to understand how digital information is generated, processed, and transmitted, as these processes directly influence the design of the physical layer and the resulting RF hardware architecture.

The right side of Figure 3.2 illustrates how a simple text message, such as "Hi," propagates through the upper communication layers and is progressively transformed into bits, ultimately becoming an electromagnetic waveform at the physical layer.

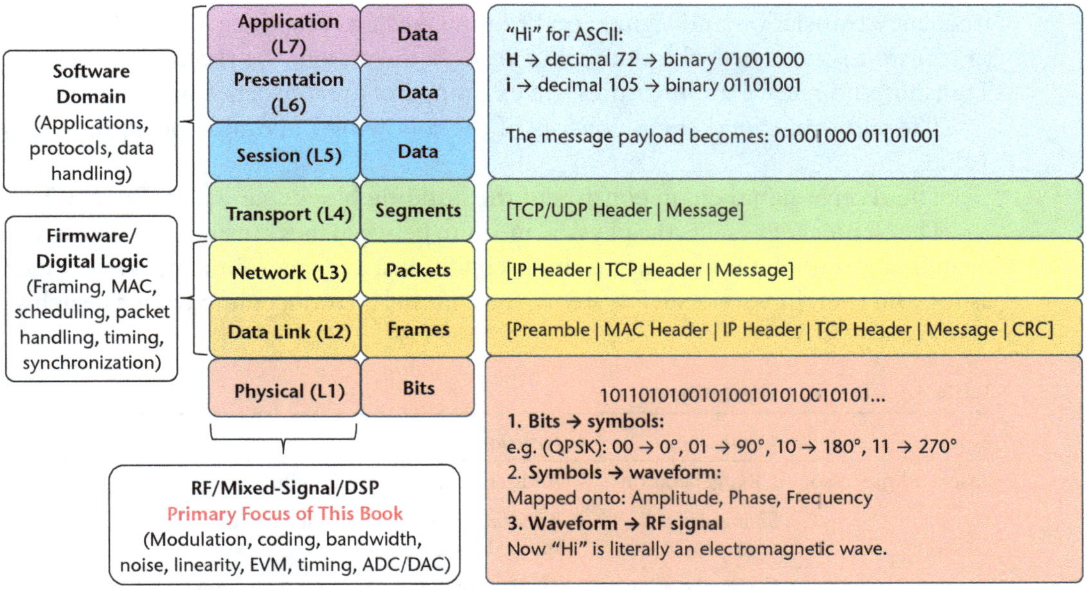

Figure 3.2 OSI communication layers and their mapping to software, firmware/digital logic, and RF/mixed-signal/DSP implementation domains. The preamble is a known bit sequence used for timing, synchronization, and channel estimation at the physical layer; the IP header contains addressing and routing information that enables packet delivery across networks; and the cyclic redundancy check (CRC) is an error-detection code used to verify data integrity at the receiver.

3.2.2 The Communication Link: Transmitter, Channel, and Receiver

At the bottom of the OSI stack, the physical layer transforms abstract digital representations, such as bits and frames, into electromagnetic waves that can propagate through space. This is where the transmitter and receiver, as shown in Figure 3.3, play a central role, converting digital information into physical signals for transmission and reliably reconstructing the information at the receive end.

Both the transmit and receive paths incorporate baseband processing and RF front-end circuitry. Baseband processing includes essential digital functions such as error detection and correction, scrambling (a deterministic bit-level randomization technique used to improve signal integrity and spectral behavior during transmission) and descrambling (the inverse operation performed at the receiver), channel coding and decoding, and modulation and demodulation. The RF front end performs

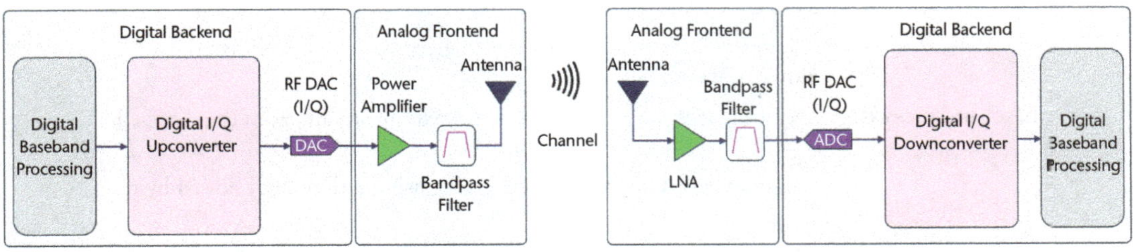

Figure 3.3 End-to-end digital-RF transceiver architecture. The transmitter and receiver employ a direct-RF conversion architecture. In both cases, upconversion and downconversion are performed digitally, while the analog front end is kept minimal, providing primarily filtering and amplification.

frequency translation and signal conditioning, such as amplification and filtering, to transmit signals over the channel or recover them from the received waveform. Transmit and receive architectures are examined in greater detail in Chapter 5.

The transmit signal chain consists of the following building blocks, shown in Table 3.1.

The receive signal chain consists of the building blocks shown in Table 3.2.

The channel represents the physical medium between the transmitter and receiver, including terrain, buildings, atmospheric conditions, and environmental effects such as fog and rain. In wireless RF systems, the channel introduces impairments such as

Table 3.1 Transmit Signal Chain

Block	*Component*	*Function*
Baseband processor	CRC generation	Adds error-detection capability
	Scrambler	Randomizes bit patterns deterministically to improve spectral properties
	Channel coding	Adds redundancy for error correction
	Modulator	Maps encoded digital data onto a constellation of symbols (baseband waveform)
RF front end	Digital upconverter	Shifts the baseband signal to RF using quadrature mixing
	Digital-to-analog converter (DAC)	Converts upconverted digital signals into analog form
	Power amplifier (PA)	Boosts the signal power for transmission
	Bandpass or harmonic filter	Suppresses out-of-band emissions and harmonics, ensuring spectral compliance
	Antenna	Radiates the RF signal into free space

Table 3.2 Receive Signal Chain

Block	*Component*	*Function*
RF front end	Antenna	Captures incoming RF signals from the environment
	Low-noise amplifier (LNA)	Amplifies weak signals while adding minimal noise
	Bandpass filter	Eliminates out-of-band signals and reduces adjacent-channel interference
	Analog-to-digital converter (ADC)	Converts the analog baseband signal into digital form
	Digital downconverter	Shifts the received RF signal to baseband
Baseband processor	Demodulator	Recovers the symbol stream from the received waveform
	Channel decoding	Corrects errors using redundancy added by the transmitter
	Descrambler	Reverses bit randomization to restore the original bitstream
	CRC check	Verifies the integrity of the received data

path loss, noise, multipath fading, and Doppler shift, which attenuate and distort the transmitted signal. These impairments increase the error rate and complicate reliable signal detection. From a product development perspective, realistic channel assumptions directly influence link budgets, modulation and coding strategies, antenna design, and RF front-end specifications, making the channel a core element of the end-to-end system architecture.

Now, let's examine how a communication link's performance is evaluated.

3.3 Core Link Performance Metrics

Figure 3.4 summarizes the key factors used to evaluate link performance in a communication system. At the center of the figure, link performance represents what the system ultimately delivers, including channel capacity, spectral efficiency, throughput, and latency. These metrics reflect how effectively the available resources are converted into usable end-to-end performance.

As illustrated in Figure 3.4, link performance is determined by both available resources, including bandwidth and signal power, and real-world constraints such as overhead and impairments. Available bandwidth sets an upper limit on how much data can be transmitted. Available signal power affects the signal-to-noise ratio (SNR), which determines how fast data can be sent and how reliably it can be received.

Part of the transmitted data is consumed by coding and protocol overhead, reducing the portion available for useful payload. Additional losses occur due to noise, distortion, and interference, which degrade the SNR and either reduce the achievable data rate or cause errors that require retransmissions. Together, these factors determine the effective throughput and overall performance of the communication link.

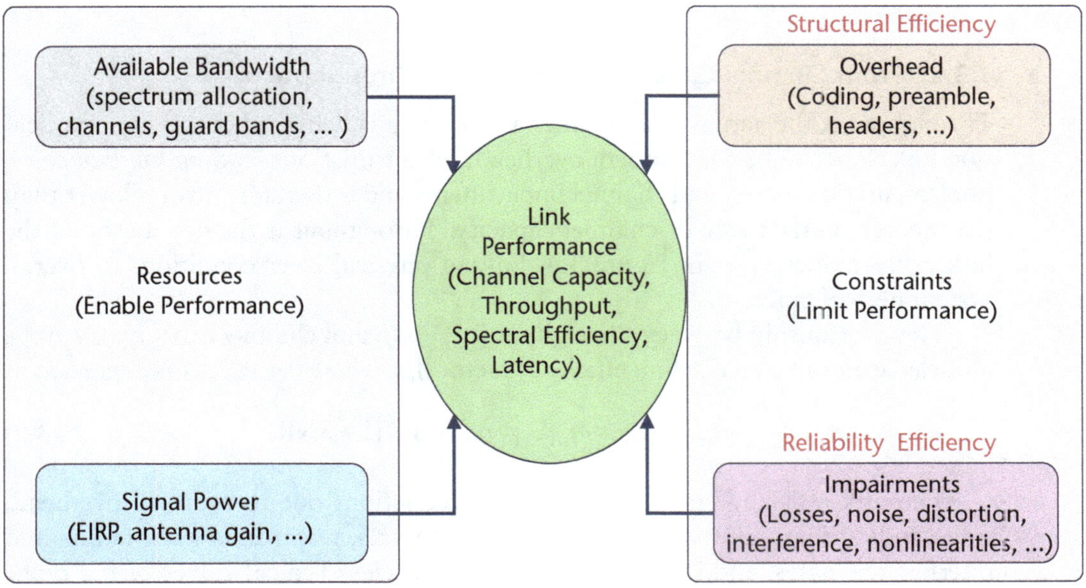

Figure 3.4 System-level view of the link performance.

3.3.1 Theoretical Limit: Channel Capacity

In this section, we examine how channel capacity is determined by available resources, namely bandwidth and SNR. Channel capacity is the theoretical maximum data rate that can be transmitted over a communication channel with an arbitrarily low probability of error for a given bandwidth and SNR. It defines the fundamental limit of reliable communication, as established in the landmark work of Claude Shannon on information theory. Importantly, channel capacity represents an upper bound. Practical systems always operate below this limit due to nonideal modulation and coding, hardware impairments, and protocol overhead.

For an additive white Gaussian noise (AWGN) channel, the channel capacity is given by Shannon's formula [2]:

$$C = B\log_2\left(1+\frac{S}{N}\right) \tag{3.1}$$

where C is the channel capacity in bits per second (bps), B is the channel bandwidth in hertz (Hz), and S/N is the SNR expressed on a linear scale. If the SNR is given in decibels, it must first be converted to linear form as $\text{SNR}_{\text{linear}} = 10^{\text{SNR}_{\text{dB}}/10}$.

As an example, consider a satellite communication link with a bandwidth of B = 250 MHz and an SNR of 10 (linear, corresponding to 10 dB). The channel capacity is $C = 250 \times 10^6 \log_2(1 + 10) = 250 \times 10^6 \times 3.459 \approx 0.865$ Gbps.

This result represents the absolute upper limit on the achievable data rate for a given link under ideal conditions, above which reliable communication is not possible regardless of the modulation or coding scheme employed. If the transmitted bit rate exceeds the channel capacity, the error probability increases sharply unless additional bandwidth, higher SNR, or improved coding gain is introduced.

In practical systems, the achievable data rate is therefore characterized by throughput, defined as the net data rate delivered to the user. The following section introduces throughput in more detail.

3.3.2 From Channel Capacity to Achievable Throughput

Throughput is the actual rate at which useful data is delivered over a communication link. It accounts for protocol overhead, retransmissions, coding inefficiencies, implementation losses, and channel impairments and is therefore always lower than the theoretical data rate or channel capacity. Throughput is the key metric of the link performance layer in Figure 3.4, linking physical-layer conditions to overall system performance.

The relationship between throughput $R_{\text{throughput}}$ and channel capacity C can be modeled using an effective link efficiency factor η_{eff}, where $0 < \eta_{\text{eff}} \leq 1$ is expressed as

$$R_{\text{throughput}} = \eta_{\text{eff}}C = \eta_{\text{eff}}B\log_2(1+\text{SNR}) \tag{3.2}$$

The link efficiency factor accounts for reductions due to protocol overhead, finite coding rate, pilot symbols and guard intervals, retransmissions, fading and interference losses, and implementation inefficiencies. Typical values of η_{eff} range from approximately 0.5 to 0.7 in well-designed wireless systems, such as Wi-Fi and LTE, under favorable conditions, to 0.3 to 0.5 in more challenging environments, such as satellite links or interference-limited systems.

From an RF product development perspective, throughput is not purely a software or networking metric; it is tightly coupled to physical-layer parameters and RF hardware design choices, including bandwidth allocation, noise figure, linearity, and available transmit power.

Several factors determine the achievable throughput of a communication link:

1. *Channel bandwidth:* Wider bandwidth enables a higher symbol rate and therefore higher potential throughput. However, increasing bandwidth places stringent requirements on the RF front end, including filters, LNAs, mixers, and ADCs/DACs, which must maintain flat gain, low noise, and low distortion across the entire operating band.
2. *Modulation and coding (MODCOD):* Higher-order modulation schemes (e.g., 64-QAM, 256-QAM) increase the number of bits transmitted per symbol and therefore enable higher throughput. However, they require higher SNR and stricter error vector magnitude (EVM) performance.

   Coding trades robustness for data rate. The coding rate R_c quantifies the fraction of transmitted bits that represent useful information after forward error correction (FEC) and is defined as

$$R_c = \frac{\text{Information Bits}}{\text{Total Bits After Coding}} \tag{3.3}$$

   For example, if a payload contains 1,000 information bits and the FEC encoder adds 500 parity bits, the total number of transmitted bits becomes 1,500, resulting in a coding rate of $R = 2/3$. In this case, two-thirds of the transmitted bits carry information, while one-third provides redundancy for error correction.

   Lower coding rates introduce additional redundancy, improving error resilience at lower SNRs, but reduce throughput because a smaller fraction of the transmitted bits represents useful information.
3. *Protocol overhead:* Including framing, headers, control fields, pilot symbols, and synchronization sequences, consumes a portion of the available channel capacity.

   In a packet, the payload represents the actual user data being transported. The remaining elements, such as the preamble, header, and error-detection or error-correction fields, constitute protocol overhead. This overhead reduces the throughput and can also increase latency and power consumption. For example, if a packet contains a 20-byte payload and 10 bytes of overhead, only two-thirds of the transmitted bits carry useful information. The overhead percentage is given by

$$\text{Overhead (\%)} = \frac{\text{Overhead Bytes}}{\text{Total Packet Bytes}} \times 100 \tag{3.4}$$

High protocol overhead may be acceptable for short, infrequent packets (e.g., IoT sensor updates), where reliability and simplicity are prioritized. However, in high-throughput systems, overhead becomes a limiting factor, and maximizing the payload-to-overhead ratio is essential for achieving high efficiency.

4. *Link quality and retransmissions:* Degraded link conditions, such as low SNR, fading, or interference, increase the BER. At higher protocol layers, this often triggers packet retransmissions, reducing effective throughput even when the nominal physical-layer data rate remains unchanged.
5. *Multiple access schemes:* In shared-channel systems (e.g., TDMA, FDMA, OFDMA), throughput per user depends on how time and frequency resources are allocated. Guard intervals, synchronization overhead, and scheduling strategies directly affect efficiency and achievable throughput.
6. *Duplexing method:* Time-division duplexing (TDD) and frequency-division duplexing (FDD) allocate capacity differently between uplink and downlink. In asymmetric traffic scenarios, throughput may be constrained by the direction assigned less bandwidth or fewer time resources.

At this point, let us examine the combined effect of the factors discussed earlier on the overall link efficiency. The effective link efficiency, denoted as $\eta_{\text{effective}}$, can be expressed as

$$\eta_{\text{effective}} = \eta_{\text{reliability}} \times \eta_{\text{structural}} \tag{3.5}$$

where

$$\eta_{\text{reliability}} \approx (1 - \text{PER}) \tag{3.6}$$

and PER is the packet error rate. This relationship assumes a low packet error rate and simple retransmission behavior. Reliability efficiency $\eta_{\text{reliability}}$ captures throughput losses caused by packet errors and retransmissions caused by impairments such as distortion, noise, and interference.

Structural efficiency $\eta_{\text{structural}}$ accounts for deterministic losses due to channel coding and protocol overhead, and is given by

$$\eta_{\text{structural}} = R_c \times \left(1 - \frac{\text{protocol overhead bits}}{\text{total transmitted bits}}\right) \tag{3.7}$$

This expression highlights a key system-level trade-off: improving robustness through stronger channel coding and increased protocol overhead directly reduces usable throughput. Therefore, efficient communication systems aim to minimize overhead and dynamically adapt the coding rate so that the link operates close to the required reliability threshold, without introducing excessive margin.

3.3.3 Spectral Efficiency

In many RF applications, including satellite links, cellular networks, and military communication systems, bandwidth is both regulated and costly. Maximizing spectral efficiency allows higher data rates to be transmitted within a fixed bandwidth, increasing throughput without consuming additional spectrum. As a result, high-throughput systems operating in bandwidth-constrained environments must pursue high spectral efficiency while accounting for the resulting SNR requirements and mitigating RF impairments to maintain reliable link performance.

Spectral efficiency, denoted by η, is derived from Shannon's capacity formula and is defined as

$$\eta = \frac{C}{B} = \log_2(1 + \text{SNR}) \tag{3.8}$$

Spectral efficiency is expressed in bits per second per hertz (b/s/Hz). As an example, for an SNR of 20 dB (corresponding to a linear SNR of 100), the spectral efficiency is $\eta = \log_2(1 + 100) \approx 6.7$ b/s/Hz. This means that approximately 6.7 bits per second can be transmitted per hertz of bandwidth. Consequently, a system operating over a 1-MHz bandwidth could theoretically support a data rate of about 6.7 Mbps under ideal conditions.

The achievable spectral efficiency of a communication system is influenced by several key factors:

1. *Modulation order:* Higher-order modulation schemes, such as 64-QAM and 256-QAM, improve spectral efficiency by transmitting more bits per symbol, as discussed in Section 3.4.2. This improvement comes at the cost of a higher required SNR and stricter RF hardware requirements. Nonidealities such as component nonlinearity, phase noise, IQ imbalance, and filter flatness limit the highest usable modulation order and therefore cap the achievable spectral efficiency.

   As illustrated in Figure 3.5, spectral efficiency increases with SNR for a given modulation scheme until the required SNR for reliable detection is reached. Beyond this point, spectral efficiency plateaus because a fixed modulation scheme supports only a finite number of bits per symbol. Therefore, further improvement requires the use of higher-order modulation or higher coding rates.
2. *Coding rate:* Higher coding rates reduce the amount of redundancy used for error correction, improving spectral efficiency at the expense of reduced robustness to noise and interference.
3. *Multiple access and spatial multiplexing:* Techniques such as frequency-division multiplexing, code-division multiplexing, and multiple-input multiple-output (MIMO) systems enhance spectral efficiency by enabling simultaneous transmissions within the same spectral allocation.
4. *Guard bands and overhead:* Protocol overhead, pilot symbols, guard bands, guard intervals, and synchronization signals reduce the net spectral efficiency relative to theoretical limits.

Each of these factors influences RF hardware design, placing requirements on linearity, bandwidth, phase noise, synchronization accuracy, and overall system complexity.

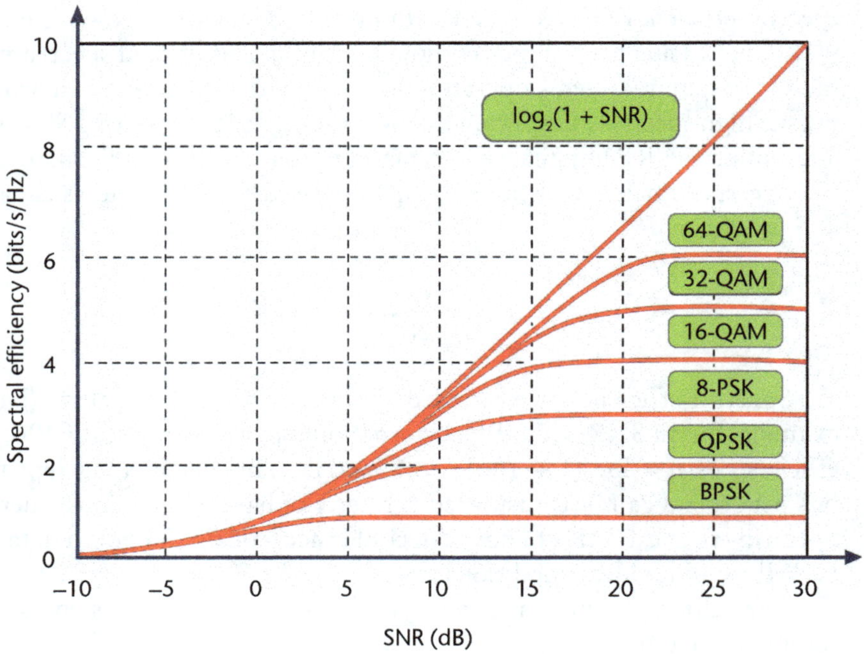

Figure 3.5 Spectral efficiency versus SNR for various modulation schemes.

3.3.4 Latency: The Time Dimension of Link Performance

Latency is the total time that it takes for data to travel from the source to the destination, measured from when a packet is generated to when it is successfully received and processed. In RF systems, latency is influenced by both physical and protocol-level factors and is often a critical design parameter for real-time applications.

The key contributors to latency are:

1. *Propagation delay:* The time for the signal to travel through the medium (e.g., approximately 3.3 μs per kilometer in free space).
2. *Processing delay:* Time spent encoding, decoding, modulating, demodulating, and performing error correction.
3. *Queuing and scheduling delay:* Waiting time inside buffers when multiple packets compete for transmission.
4. *Retransmission delay:* Additional time if packets are lost or fail error detection checks and must be re-sent.

It is important to note that:

- Low-latency systems are essential for control loops (e.g., UAV guidance, remote surgery) and interactive communication.
- Higher-latency systems may be acceptable for applications such as bulk data transfer or environmental monitoring.

Reducing latency often requires simpler modulation schemes, smaller packet sizes, and minimal processing overhead, but these choices may reduce throughput or robustness. As a result, latency is often optimized alongside other requirements such as throughput, reliability, and link margin.

3.4 Modulation

A large portion of RF products rely on modulated waveforms, whether in satellite communication links, IoT devices, EW and jamming systems, or even superconducting quantum computers. A solid understanding of modulation schemes is therefore essential. The choice of modulation type, symbol rate, and coding strategy strongly influences system-level design decisions and directly impacts component selection, including power-amplifier linearity, filter bandwidth and flatness, and the resolution and sampling rates required of ADCs and DACs.

As shown in Figure 3.6(a), modulation involves three key elements: a carrier signal, a modulating signal containing the information to be transmitted, and a modulator that varies one or more properties of the carrier, such as amplitude, frequency, or phase, in synchrony with the information signal. The result of this process is a modulated waveform suitable for transmission over the channel.

Demodulation is the inverse operation of modulation and aims to recover the original information signal from the received waveform. In its simplest form, as illustrated in Figure 3.6(b), demodulation involves a reference carrier, the received

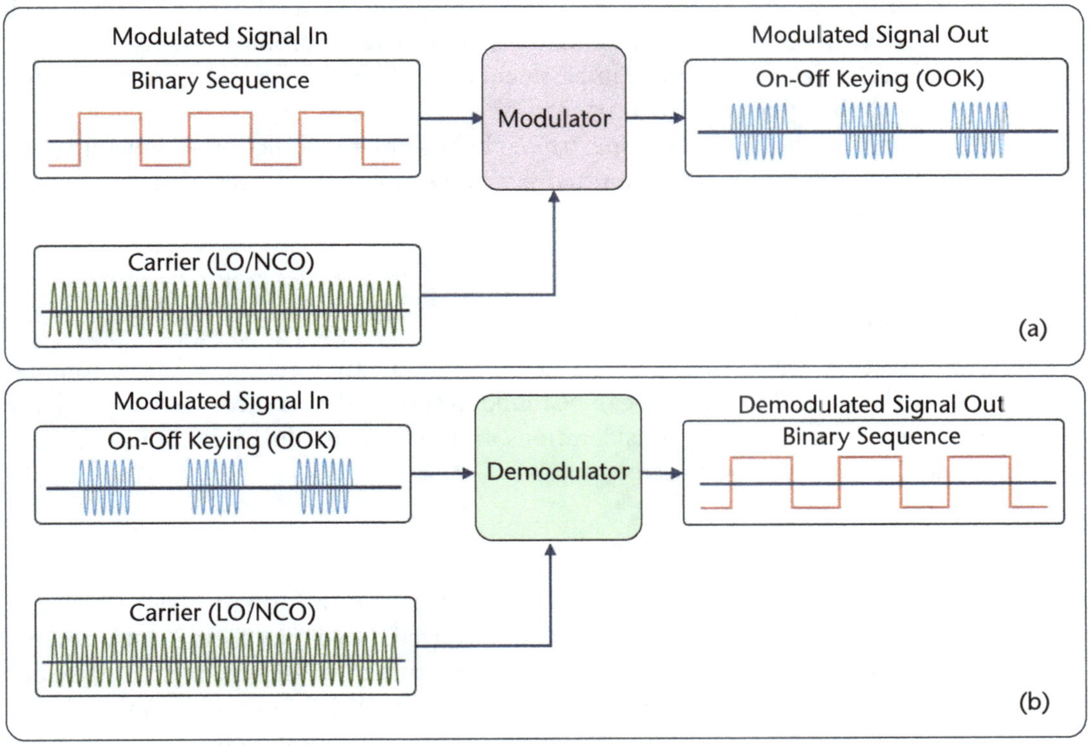

Figure 3.6 (a) On-off keying (OOK) modulation process, in which a binary baseband sequence gates a sinusoidal carrier generated by a local oscillator (LO) or numerically controlled oscillator (NCO) to produce an amplitude-keyed RF waveform. (b) Corresponding OOK demodulation process, in which the received RF signal and a carrier reference are processed by the demodulator to recover the original binary sequence.

modulated signal, and a demodulator that processes the waveform to reconstruct the transmitted information.

Depending on whether the information signal is analog or digital, different modulation schemes are employed. Each scheme introduces trade-offs in implementation complexity, robustness to noise and interference, spectral efficiency, and RF hardware requirements, all of which must be considered early in the product design process.

Without modulation, transmitting information efficiently, reliably, and over practical distances would be nearly impossible. Modulation offers several key advantages that directly impact RF system and product design:

- *Enables efficient radiation and spatial control:* By shifting baseband signals to higher carrier frequencies, modulation enables efficient long-distance transmission, compact antenna designs due to shorter wavelengths, and narrower antenna beamwidths that support directional transmission and spatial reuse of spectrum.
- *Enables compatibility with channel characteristics:* Modulation allows signals to be tailored to the transmission medium, whether free space, guided media, or fiber-optic links.
- *Facilitates multiplexing:* Multiple signals can share the same physical medium through techniques such as frequency-division multiplexing (FDM) or code-division multiplexing (CDM).
- *Improves noise and interference immunity:* Certain modulation schemes, such as frequency and phase modulation, offer increased robustness against noise, interference, and channel impairments, enabling more reliable communication in challenging RF environments.
- *Supports efficient use of spectrum:* Higher-order modulation schemes increase the number of bits transmitted per hertz, improving spectral efficiency and overall system throughput.

3.4.1 Modulation Types

In this section, we discuss analog and digital modulation schemes. While comprehensive theoretical treatments can be found in standard references such as [1, 2], the focus here is on practical considerations and on how modulation choices directly impact RF hardware requirements.

Analog Modulation

Analog modulation is used to transmit continuous-time information signals such as voice or music. These signals can also be converted into digital form using an ADC, enabling the use of digital modulation schemes that offer higher spectral efficiency and improved robustness to noise and interference. While analog communication systems remain in use for certain legacy or niche applications, such as amateur (ham) radio, broadcast systems, low-latency voice links, long-range high-frequency (HF) communication, and selected telemetry and sensing applications, the vast majority of modern communication systems rely on digital modulation.

Despite this shift, analog modulation remains relevant due to its conceptual simplicity and low implementation complexity. Studying analog modulation provides essential intuition into spectrum occupancy, noise behavior, and the impact of RF hardware nonidealities, insight that directly translates to the design and optimization of modern digital RF systems.

The three primary types of analog modulation are amplitude modulation (AM), frequency modulation (FM), and phase modulation (PM). Additional variants, such as single-sideband (SSB) modulation, have been developed to improve spectral efficiency or reduce power consumption. Table 3.3 summarizes common analog modulation schemes, their underlying principles, key characteristics, typical applications, and associated RF hardware design considerations [2–4].

Digital Modulation

Digital modulation encodes binary data onto a carrier signal using discrete changes in amplitude, frequency, or phase. Table 3.4 summarizes common digital modulation schemes, their underlying principles, key characteristics, typical applications, and associated RF hardware design considerations [2–4].

Table 3.3 Common Analog Modulation Schemes and Hardware Design Considerations

Modulation Scheme	*Basic Principle*	*Key Characteristics*	*Typical Applications*	*Impact on Hardware Design and Requirements*
FM (frequency modulation)	Varies the carrier frequency in proportion to the message signal	High immunity to amplitude noise, constant-envelope signal	FM broadcast radio, telemetry, TV audio	Requires frequency-stable oscillators, wider occupied bandwidth, relaxed PA linearity
PM (phase modulation)	Varies the carrier phase in proportion to the message signal	Similar noise performance to FM, constant envelope	Specialized analog links, telemetry	Requires phase-stable oscillators and accurate phase detection
AM (amplitude modulation)	Varies the carrier amplitude with the message signal	Simple to implement, highly susceptible to amplitude noise	AM broadcast radio, aircraft voice communications	Requires linear PAs, sensitive to gain and power fluctuations
DSB-SC (double sideband suppressed carrier)	Transmits both sidebands with the carrier suppressed	Improved power efficiency over AM, requires coherent detection	Analog telemetry, point-to-point radio links	Requires accurate carrier recovery and synchronization at the receiver
SSB (single sideband)	Transmits only one sideband (upper or lower)	High spectral efficiency, reduced transmit power	High-frequency radio, amateur radio, long-distance voice	Requires sharp filtering or phasing networks and precise frequency control
VSB (vestigial sideband)	Partially suppresses one sideband	Trade-off between AM simplicity and SSB bandwidth efficiency	Analog TV video transmission	Requires carefully designed transmitter and receiver filtering

Table 3.4 Common Digital Modulation Schemes

Modulation Scheme	*Category*	*Basic Principle*	*Key Characteristics*	*Typical Applications*
ASK (amplitude shift keying)	Amplitude-based	Discrete amplitude levels represent digital symbols	Simple implementation, highly sensitive to amplitude noise and fading	RFID, low-cost wireless links, optical systems
OOK (on-off keying)	Amplitude-based	Special case of ASK where carrier presence or absence represents bits	Very simple, low-power consumption, poor noise immunity	Remote controls, optical communications, low-power IoT
QAM (quadrature amplitude modulation) (4, 16, 64, 256, 1,024-QAM)	Amplitude and phase-based	Combines amplitude and phase variations to encode symbols	Very high spectral efficiency, requires high SNR and linear PAs	LTE/5G, Wi-Fi, cable modems, microwave backhaul
BPSK (binary phase shift keying)	Phase-based	Two carrier phase states represent bits	Highly robust to noise, low spectral efficiency	Satellite communications, deep-space links, telemetry
QPSK (quadrature phase shift keying)	Phase-based	Four carrier phase states encode two bits per symbol	Good balance between robustness and efficiency	LTE, satellite links, Wi-Fi
M-PSK (8-PSK, 16-PSK)	Phase-based	Uses multiple discrete phase states per symbol	Higher spectral efficiency, increased SNR and phase noise sensitivity	Satellite broadcasting, microwave radio links
FSK (frequency shift keying)	Frequency-based	Discrete carrier frequencies represent symbols	Robust to amplitude noise, relatively bandwidth-inefficient	Paging systems, telemetry, low-data-rate radios
Trellis-coded modulation (TCM)	Hybrid (modulation + coding)	Combines modulation with FEC	Coding gain without bandwidth expansion	Satellite links, microwave backhaul

Here are some of the advantages of digital modulation over analog modulation:

- *Higher noise immunity:* Digital signals are more resilient to noise and distortion and can often be accurately recovered as long as the SNR exceeds the required threshold.
- *Efficient power and bandwidth usage:* Advanced modulation schemes such as QAM and multicarrier techniques such as OFDM maximize data throughput within limited spectral resources.
- *Error detection and correction:* Digital systems support error detection and FEC schemes, such as CRC and channel coding, to enhance link reliability.
- *Flexible multiplexing options:* Digital modulation integrates seamlessly with time, frequency, and code division multiplexing techniques.
- *Enhanced security:* Digital data streams can be encrypted and authenticated, enabling secure communication.

- *Direct integration with digital hardware:* Digital modulation interfaces naturally with microcontrollers, DSP chips, FPGAs, and RF system-on-chips (RFSoCs), enabling flexible and programmable implementations.
- *Digital signal processing:* Filtering, digital upconversion and downconversion, equalization, interference cancellation, and other specialized signal processing tasks that are impractical or impossible to perform in the analog domain.
- *Adaptive operation:* Modulation order, coding rate, and transmit power can be adjusted in real time based on channel conditions and link quality.
- *Improved resistance to fading:* Techniques such as spread spectrum and OFDM mitigate the effects of multipath propagation and frequency-selective fading.

3.4.2 Symbols and Waveforms

This section explains how binary data are mapped into symbols and how those symbols are represented as time-domain waveforms.

How Bits Become RF Signals

Let us begin with a simple example illustrating how a binary bit stream is modulated using an OOK waveform, as shown in Figure 3.7(a). In OOK, two possible waveforms correspond to the binary values 0 and 1. A waveform with zero amplitude represents a logic 0, while a waveform with a defined nonzero carrier amplitude represents a logic 1. Figure 3.7(a) shows the resulting OOK-modulated waveform for the bit sequence 10011100.

We now extend this example by grouping the input bits into pairs, referred to as symbols. Each 2-bit group can take one of four possible values, 00, 01, 10, or 11, with each symbol mapped to a unique waveform. This increases the number of possible waveforms from two to four, while reducing the required symbol rate by a factor of 2 for the same input bit stream.

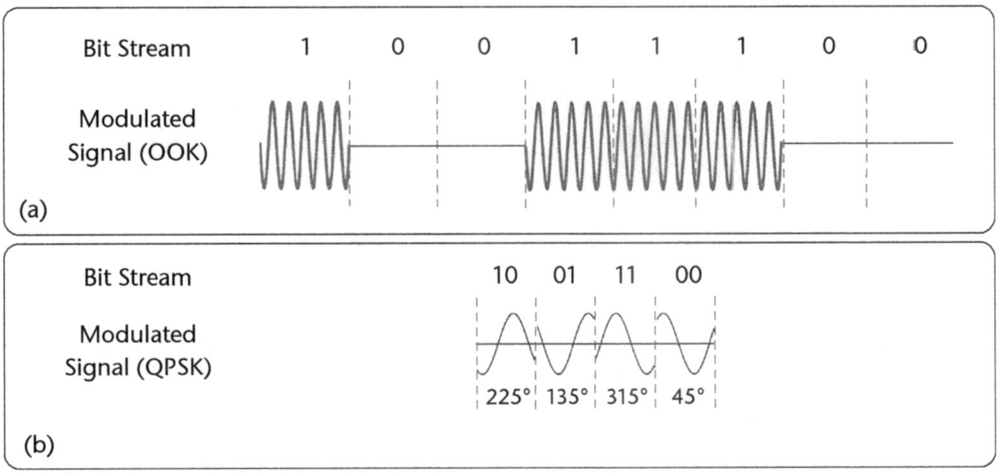

Figure 3.7 (a) Encoding a bit stream using OOK. (b) Encoding the same bit stream using QPSK. Each symbol encodes 2 bits by mapping them to one of four distinct carrier phase states. The dashed lines indicate symbol boundaries where phase changes occur.

In QPSK, each symbol represents 2 bits of information by shifting the phase of the carrier while maintaining constant amplitude. The input bit stream is first divided into pairs, and each pair is mapped to a distinct carrier phase state, commonly 45°, 135°, 225°, or 315°. Figure 3.7(b) illustrates this process using QPSK waveforms to encode the same bit sequence as in the OOK example. The bit stream 10011100 is grouped as 10 | 01 | 11 | 00, with phase transitions occurring at the symbol boundaries, indicated by the dashed lines.

By encoding 2 bits per symbol, or equivalently per transmitted waveform, QPSK doubles the data rate compared to OOK for the same symbol rate, thereby improving spectral efficiency without increasing the occupied bandwidth. For a symbol rate of 1 Msymbol/s, OOK achieves 1 Mbps, while QPSK achieves 2 Mbps, demonstrating a twofold increase in spectral efficiency at the same bandwidth.

M-ary Modulations

The concept of symbol-based modulation can be extended further by grouping more than two bits per symbol, for example, 4 bits, 8 bits, or even more, corresponding to M-ary modulation schemes with M distinct waveform states. A common example is M-ary quadrature amplitude modulation (QAM), such as 4-QAM, 16-QAM, and higher orders. For instance, 16-QAM employs 16 distinct waveforms, each representing a group of four bits ($\log_2 16 = 4$).

As will be discussed shortly, these predefined waveforms are conveniently represented as points in a constellation diagram, which provides an intuitive framework for visualizing modulation schemes, symbol spacing, and sensitivity to noise and distortion.

As mentioned earlier, each group of bits is referred to as a symbol, and each symbol corresponds to the transmission of one predefined waveform over a fixed symbol duration T_s.

In general, the number of distinct symbols M determines how many unique waveforms exist and how many bits are conveyed per symbol:

- Binary modulation: $M = 2 \rightarrow 2$ possible waveforms (e.g., one for bit 0 and one for bit 1);
- M-ary modulation ($M > 2$): Each symbol represents $\log_2 M$ bits:
  - QPSK $\rightarrow M = 4 \rightarrow 4$ waveforms, 2 bits per symbol;
  - 16-QAM $\rightarrow M = 16 \rightarrow 16$ waveforms, 4 bits per symbol.

Binary modulation schemes ($M = 2$) are highly robust to noise and operate at relatively low SNR, but they offer limited spectral efficiency. In contrast, M-ary modulation schemes increase spectral efficiency by transmitting more bits per symbol, at the cost of reduced noise margin. As M increases, tighter symbol spacing (see Figure 3.8(c)) requires higher SNR and places stricter demands on RF hardware, including PA linearity, phase noise, IQ imbalance, and data-converter resolution, to maintain acceptable error performance.

Table 3.5 summarizes the properties of common digital modulation schemes.

The symbol rate (or baud rate) is the number of symbols transmitted per second and is measured in baud. The bit rate depends on both the symbol rate and the number of bits conveyed by each symbol and is given by

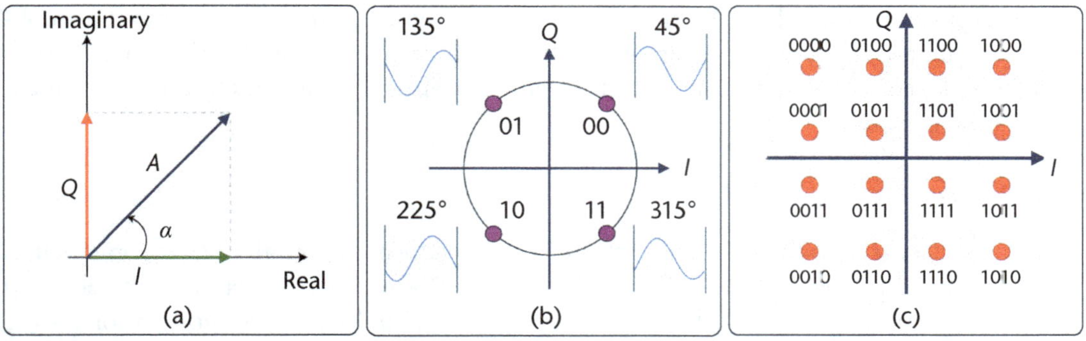

Figure 3.8 (a) IQ signal representation and (b) QPSK constellation and corresponding waveforms. The waveforms have identical amplitudes and are phase-shifted by 90°. (c) 16-QAM constellation diagram.

$$\text{Bit Rate} = \text{Symbol Rate} \times \log_2(M) \tag{3.9}$$

where M is the number of distinct symbol states (i.e., the modulation order). For example, in QPSK, each symbol represents two bits ($\log_2 4 = 2$). Therefore, a QPSK link operating at 9,600 bps corresponds to a symbol rate of 4,800 symbols/s (4,800 baud).

3.4.3 Constellation Diagram

In this section, we discuss the constellation diagram and its relationship to digital modulation schemes. The constellation diagram provides a compact and intuitive

Table 3.5 Properties of Common Digital Modulation Schemes

Modulation Scheme	*Bits per Symbol*	*Number of Symbols*	*Ideal Spectral Efficiency (bits/s/Hz)*	*Typical Required SNR for BER ≈ 10^{-6}, uncoded*	*Typical Applications*
BPSK	1	2	1	~7 dB	Deep-space communication, GPS
QPSK	2	4	2	~10 dB	Satellite communications, LTE, Wi-Fi
8-PSK	3	8	3	~14–15 dB	GSM EDGE, satellite broadcasting
16-QAM	4	16	4	~18 dB	Cable modem, Wi-Fi, LTE
32-QAM	5	32	5	~22 dB	Broadband wireless, microwave links
64-QAM	6	64	6	~24 dB	LTE, 5G NR
256-QAM	8	256	8	~30 dB	DOCSIS 3.1, 5G NR backhaul
1,024-QAM	10	1,024	10	~34–36 dB	Next-generation cable, advanced 5G
ASK (M-ary)	$\log_2 M$	M	$\log_2 M$	Depends on M	RFID, optical communications
FSK (M-ary)	$\log_2 M$	M	$\log_2 M$	~3–15 dB	Low-power IoT, Bluetooth

representation of how digital information and symbols are mapped onto the amplitude and phase of a carrier signal. It is a fundamental visualization tool used to analyze modulation formats and the effects of impairments in digital communication systems.

IQ Diagram

As discussed in the previous section, each waveform in digital modulation schemes is characterized by a specific amplitude and phase. A convenient and widely used mathematical framework for representing and visualizing such modulations is the in-phase (I) and quadrature (Q) component representation.

A sinusoidal signal with amplitude A and phase α can be represented as a point in the complex plane, also referred to as the IQ plane, as shown in Figure 3.8(a). In this representation, the real axis corresponds to the I component, while the imaginary axis corresponds to the Q component. The signal's amplitude and phase can be expressed in terms of its I and Q components as

$$A = \sqrt{I^2 + Q^2}, \qquad \alpha = \tan^{-1}\left(\frac{Q}{I}\right) \tag{3.10}$$

A real-valued tone $s(t)$ with amplitude A, angular frequency ω, and phase α can be written as $s(t) = A\cos(\omega t + \alpha)$. Using trigonometric identities, this expression can be rewritten in terms of its I and Q components as $s(t) = A\cos(\alpha)\cos(\omega t) - A\sin(\alpha)\sin(\omega t)$. Defining $I = A\cos(\alpha)$, $Q = A\sin(\alpha)$, the signal becomes

$$s(t) = I\cos(\omega t) - Q\sin(\omega t) \tag{3.11}$$

Depending on the chosen complex baseband definition, some formulations use a plus sign instead of a minus sign. Both conventions are valid and correspond to different definitions of the complex envelope (or, equivalently, upper versus lower sideband representation).

This decomposition forms the basis of the IQ representation used in modern digital communication systems. Each transmitted symbol represents a unique combination of amplitude and phase and corresponds to a specific (I, Q) pair, which can be plotted as a point in the IQ plane, giving rise to the constellation diagram. This representation provides an intuitive way to visualize digital modulation schemes and to relate their amplitude and phase characteristics to discrete symbol locations.

Figure 3.8(b) illustrates the IQ constellation for a QPSK modulation scheme, which consists of four symbols. Each symbol represents 2 bits. As discussed earlier, each point (symbol) in the IQ plane corresponds to a specific time-domain waveform. All QPSK symbols have the same amplitude, corresponding to an equal distance from the origin, but differ in phase. In this example, the symbols are located at phase angles of 45°, 135°, 225°, and 315° with respect to the I-axis.

Higher-order modulation schemes, such as 16-QAM illustrated in Figure 3.8(c), extend this concept by using multiple amplitude and phase states to encode additional bits per symbol, 4 bits per symbol in this case. This increases spectral efficiency by transmitting more information within the same bandwidth. However, these gains come with added challenges. Because the constellation points are more tightly packed, even small deviations from their ideal locations, caused by noise, distortion, or impairments, can result in symbol errors. Consequently, higher-order modulation schemes place more stringent requirements on RF hardware performance, particularly with respect to noise, linearity, and IQ balance. As we will see in Chapter 6, an IQ modulator is used to generate various I/Q modulation schemes.

Error Impact on the Constellation Diagram

When a signal is received, its I and Q components define a point in the constellation diagram. The receiver compares this point to the ideal constellation locations and assigns it to the nearest one, typically based on Euclidean distance. Each ideal constellation point defines a decision region, and the received symbol is decoded according to the region in which the point falls.

The decision process can be expressed mathematically as a minimum-distance rule:

$$\hat{s} = \arg\min_{s_k \in \mathcal{S}} \sqrt{(I - I_k)^2 + (Q - Q_k)^2} \tag{3.12}$$

where

$\hat{s}$ is the detected symbol.

$\mathcal{S}$ is the set of all ideal constellation points.

(I, Q) are the received in-phase and quadrature components.

(I_k, Q_k) are the ideal coordinates of the kth symbol.

The receiver selects the symbol s_k whose ideal coordinates are closest in Euclidean distance to the received point.

Amplitude and phase deviations caused by nonlinearity, distortion, white noise, IQ imbalance, spurs, and phase noise can shift the received point away from its ideal position, as illustrated in Figure 3.9. If the deviation is large enough to cross a decision boundary, the receiver will misidentify the symbol. Such deviations increase the probability of symbol decision errors, especially in higher-order modulation schemes where constellation points are more closely spaced. The extent of the impact depends on both the modulation order and the nature and magnitude of the impairments introduced during transmission and reception. As mentioned earlier, the reduced noise margin for high-order modulations has a severe impact on hardware design and cost.

The error effects observed on a constellation diagram can be grouped into the following categories:

- Additive effects, such as thermal noise, cause random spreading of the constellation points.
- Multiplicative or systematic effects, such as I/Q imbalance, phase noise, or frequency offset, introduce structured distortions, including rotation, scaling, or mirroring of the constellation.

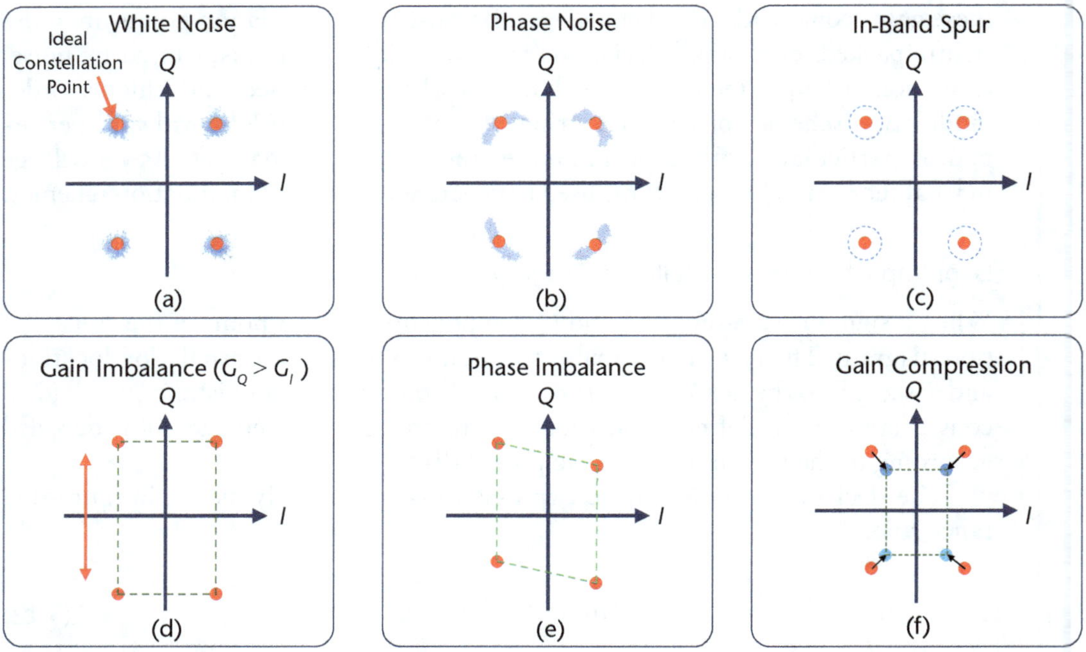

Figure 3.9 Representation of various signal impairments on the constellation diagram: (a) additive white noise, (b) phase noise, (c) in-band spurs (larger circles indicate higher spur amplitude), (d) gain imbalance, where the Q-channel has higher gain than the I-channel, resulting in an elongation along the Q-axis, (e) phase imbalance between the I and Q channels, and (f) gain compression, where constellation points reaching the amplifier compression region are pulled inward toward the origin.

- Nonlinear effects, such as amplifier compression and intermodulation distortion, degrade system performance in two primary ways: in-band distortion, manifested as increased error vector magnitude (discussed next), and out-of-band spectral regrowth (see Section 4.9.5).

Chapter 4 provides an in-depth discussion of these effects and their impact on overall system performance.

Error Vector Magnitude (EVM)

As discussed earlier, impairment factors such as noise, nonlinearities, and I/Q imbalance can cause the received symbols to deviate from their ideal constellation points. This deviation is quantified by the EVM illustrated in Figure 3.10, which is a key performance metric for digital communication systems.

The EVM is defined as the root-mean-square (RMS) of the error vectors, normalized to the RMS amplitude of the ideal reference symbols:

$$\mathrm{EVM}_{\mathrm{rms}} = \sqrt{\frac{\sum_{n=1}^{N}\left|r_n - s_n\right|^2}{\sum_{n=1}^{N}\left|s_n\right|^2}} \tag{3.13}$$

where

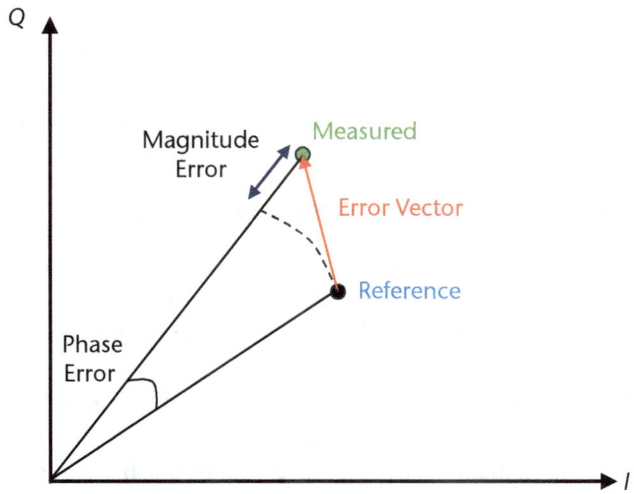

Figure 3.10 The concept of EVM.

r_n = received (measured) symbol for the nth point.
s_n = corresponding ideal reference symbol.
N = total number of symbols.

EVM is often expressed as a percentage or in decibels (dB):

$$\begin{aligned} \mathrm{EVM}(\%) &= 100 \times \mathrm{EVM}_{\mathrm{rms}} \\ \mathrm{EVM}(\mathrm{dB}) &= 20\log_{10}(\mathrm{EVM}_{\mathrm{rms}}) \end{aligned} \tag{3.14}$$

A lower EVM indicates a cleaner constellation and better modulation accuracy, while higher EVM values reflect greater distortion, noise, or mismatch between the transmitted and received symbols.

The maximum allowable EVM for selected wireless systems is summarized in Table 3.6. Increasing the modulation order can improve data throughput; however, as constellation points are placed closer together, the probability of symbol errors increases, making tighter EVM requirements essential for higher-order modulation schemes. Consequently, the choice of modulation involves a trade-off between required noise resilience, achievable data throughput, and RF hardware complexity.

Table 3.6 Typical EVM Limits for a BER Target of ($\leq 1 \times 10^{-5}$) for Common Modulations

Modulation	*Typical EVM Limit (% RMS)*
BPSK	≤17.5%
QPSK	≤12.5%
16-QAM	≤8%
64-QAM	≤3.5%
256-QAM	≤1.75%

To summarize, Figure 3.11 illustrates the interdependence between modulation order, system complexity, and impairment tolerance. Increasing constellation density improves spectral efficiency; however, it simultaneously increases RF and DSP complexity and tightens EVM requirements, thereby reducing robustness to impairments.

3.4.4 Adaptive Modulation and Coding (AMC)

Modern communication systems employ AMC to maximize achievable throughput while meeting target error-rate constraints. AMC dynamically adapts transmission parameters, such as the modulation scheme, modulation order, and coding rate, in response to real-time channel conditions.

For this purpose, a specific combination of a modulation scheme and a coding rate, referred to as a MODCOD, is used. When channel conditions are favorable, higher-order modulation schemes and higher coding rates are selected to maximize throughput. Under degraded channel conditions, the system transitions to more robust, lower-rate MODCODs to maintain link stability and reliability. An example of AMC mapping is illustrated in Table 3.7, while Table 3.8 provides an example of a MODCOD configuration.

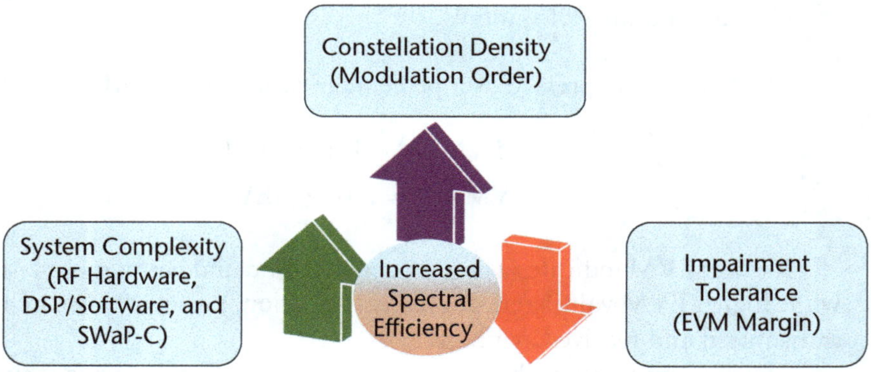

Figure 3.11 Interdependence between modulation order, system complexity, and impairment tolerance.

Table 3.7 Example Adaptive Modulation Mapping

Channel Condition	*Approximate SNR Range*	*Modulation Scheme*	*Notes*
Excellent	>25 dB	256-QAM	Maximizes throughput, high spectral efficiency
Good	18–25 dB	64-QAM	Good balance between throughput and robustness
Moderate	12–18 dB	16-QAM	Moderate throughput and robustness
Fair	7–12 dB	QPSK	High robustness, reduced throughput
Poor	<7 dB	BPSK	Maximum robustness, minimal throughput

Table 3.8 Example MODCOD Configurations and Corresponding Performance

MODCOD	*Modulation*	*Coding Rate*	*Spectral Efficiency (b/s/Hz)*	*Typical Required SNR (dB)*
QPSK 1/2	QPSK	1/2	0.98	~0.5
QPSK 3/4	QPSK	3/4	1.48	~3.5
8PSK 3/4	8PSK	3/4	2.21	~7.0
16APSK 3/4	16APSK	3/4	2.96	~9.5

This adaptability, commonly implemented in software-defined radios (SDRs), is widely used in satellite communication systems. For example, under nominal or degraded link conditions, QPSK is typically employed due to its robustness against noise, phase noise, and other channel impairments. When link quality improves, higher-order modulation schemes such as 8-PSK can be selected to increase spectral efficiency and data rate without increasing bandwidth.

Accurate channel estimation and tight coordination between the transmitter and receiver are essential to ensure reliable and seamless transitions between operating modes.

3.5 Errors in a Communication Link

In the previous section, we discussed EVM as one of the key metrics for quantifying modulation accuracy and error performance. In this section, we examine additional metrics commonly used to characterize the impact of impairments in a communication link, specifically BER, intersymbol interference (ISI), SNR, and the energy per bit to noise power spectral density ratio (E_b/N_0).

We describe what each metric represents, how they are related, and how they vary with different modulation schemes. Understanding these relationships is essential for evaluating link performance, selecting appropriate modulation and coding strategies, and ultimately guiding the design of the RF hardware, software, and digital signal processing in practical communication systems.

3.5.1 Types of Errors

Table 3.9 summarizes the types of errors that can occur in a communication link, ranging from individual bit errors to entire blocks of data.

Bit Flip

In most real-world digital communication links, the most common error mechanism is the bit flip (single-bit error) caused by AWGN. This type of noise is always present due to thermal noise in electronic components and the surrounding environment, and even in well-designed systems, random noise can occasionally push the received signal across a decision threshold, flipping a 0 to a 1 or vice versa. As discussed next, error detection and correction schemes are employed to identify and mitigate these bit errors.

Table 3.9 Common Communication Link Errors, Causes, and Mitigation

Error Category	*Error Type*	*Description*	*Common Causes*	*Typical Mitigation*
Bit-level errors	Bit flip (single-bit error)	A single bit changes from 0 to 1 or vice versa	Thermal noise, interference, low SNR	Increase SNR, improve filtering, use FEC
	Burst error	Multiple consecutive bits corrupted	Impulsive noise, deep fades, EMI spikes	Interleaving, robust FEC, diversity schemes
Symbol-level errors	Symbol error	Incorrect symbol decision at the receiver	Noise, distortion, fading	Higher SNR, equalization, linearization
	Nearest-neighbor symbol error	Symbol detected as an adjacent constellation point	Phase noise, I/Q imbalance, nonlinearities	LO stabilization, I/Q calibration, PA linearization
Sequence/timing errors	ISI	Temporal overlap of adjacent symbols	Multipath, channel dispersion, insufficient filtering	Equalization, pulse shaping, guard intervals
	Synchronization error	Loss of symbol, frame, or carrier alignment	Frequency offset, phase drift, jitter	Carrier/timing recovery loops, pilot signals
Block-level errors	Codeword/block error	Entire FEC block decoded incorrectly	Error rate exceeds code correction capability	Stronger codes (LDPC, Turbo), HARQ

Other error mechanisms, such as burst errors, ISI, or loss of synchronization, are generally less frequent due to the use of modern equalization, filtering, and synchronization techniques.

ISI

Ideally, each transmitted pulse, representing a bit or symbol, should be confined to its designated time slot. However, in practical communication systems, the transmission channel, including circuits, interconnects, cables, antennas, and free space, is bandwidth-limited and often exhibits dispersive behavior. As a result, the channel introduces frequency-dependent amplitude and phase distortion across the signal bandwidth [2–4]. These effects cause portions of a transmitted pulse to spread in time or experience relative delay, leading to temporal smearing of the waveform, as illustrated in Figure 3.12.

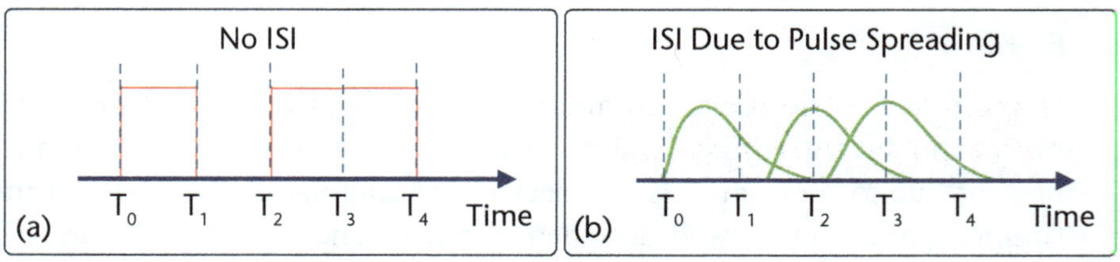

Figure 3.12 (a) Pulse sequence with no ISI. (b) The same pulse sequence with ISI due to pulse spreading.

As a result, residual energy from previous symbols overlaps with the current symbol, making it difficult for the receiver to correctly interpret the data. This phenomenon, known as ISI, commonly arises from bandwidth limitations, multipath propagation, group delay distortion, or a frequency-dependent amplitude and phase response of components in the transmit and receive chains. ISI degrades signal integrity and increases the BER. It is a major concern in high-speed communication links, where symbol periods are short and even small distortions can lead to significant performance degradation.

ISI can be reduced by shaping the transmitted signal or by compensating for channel distortion at the receiver. Common mitigation techniques include:

- *Pulse-shaping filters (e.g., raised cosine):* Limit signal bandwidth and control time-domain pulse overlap to reduce ISI.
- *Equalization (linear or adaptive, including decision-feedback equalizers (DFE)):* Compensates for channel-induced amplitude and phase distortion, with DFEs using previously detected symbols to cancel post-cursor ISI.
- *Increased symbol spacing or orthogonal signaling:* Reduces temporal overlap between symbols or ensures mutual orthogonality, thereby lowering sensitivity to timing errors and channel dispersion.

Since errors cannot be completely eliminated in real-world channels, practical communication systems rely on error detection and correction techniques to maintain reliable performance. This is the topic of the next section.

3.5.2 Error Detection and Correction

In digital communication and data storage systems, error detection and correction are fundamental techniques used to ensure data integrity. Error detection focuses on identifying the presence of errors in received data, commonly using mechanisms such as checksums or CRC. Once an error is detected, error correction techniques, most notably FEC, are used to mitigate its impact.

FEC schemes introduce controlled redundancy into the transmitted data, enabling the receiver not only to detect errors but also to correct them without retransmission. Together, error detection and correction mechanisms improve link reliability and robustness, allowing accurate data reception even under imperfect channel conditions. However, as discussed earlier, this improved reliability comes at the cost of increased overhead, processing complexity, and reduced effective data throughput.

3.5.3 Error Metrics

The most common metrics used to quantify errors in a digital communication link include:

- *BER:* The ratio of incorrectly received bits to the total number of transmitted bits, providing a direct measure of end-to-end link reliability.

- *Symbol error rate (SER):* The ratio of incorrectly detected symbols to the total number of transmitted symbols, particularly relevant for higher-order modulation schemes where symbol decision errors dominate performance.
- *Frame or packet error rate (FER/PER):* The fraction of received frames or packets containing one or more bit errors, reflecting data delivery integrity at the packet level.
- *EVM:* A normalized measure of the distance between received and ideal constellation points, capturing the combined effects of noise, distortion, phase error, and other impairments on modulation accuracy. While not an error rate, EVM is closely related to BER and is widely used in RF compliance and modulation quality assessment.

These metrics directly capture the impact of channel impairments and system nonidealities on communication performance.

3.5.4 Bit Error Rate (BER)

The BER is a fundamental metric used to evaluate the performance of digital communication systems. It is defined as the ratio of the number of incorrectly received bits to the total number of transmitted bits over a given observation interval:

$$\text{BER} = \frac{\text{Number of erroneous bits}}{\text{Total number of transmitted bits}} \tag{3.15}$$

BER provides a direct measure of link reliability and is influenced by several factors, including the modulation scheme, coding rate, channel impairments, and SNR.

In practice, the BER is often measured by transmitting a known bit sequence, typically a pseudorandom binary sequence (PRBS), generated by a vector signal generator (VSG). The received bit stream is then compared with the original sequence to count the number of bit errors. A higher BER indicates more frequent errors, which can degrade system performance or necessitate stronger error-correction coding.

BER can be expressed as a pure ratio (e.g., 10^{-6}) or in percentage form. Acceptable BER levels vary depending on the application:

- Wireless voice links may tolerate BERs of 10^{-3} to 10^{-4} due to human speech redundancy.
- High-speed data links often require BERs of 10^{-9} or better to avoid retransmissions.

A close relationship exists between BER and EVM: while EVM quantifies the instantaneous deviation of symbols from their ideal positions, BER captures the cumulative effect of these deviations on actual bit decisions. In other words, high EVM generally leads to higher BER, although FEC can mitigate this relationship.

Relationship Between BER and EVM

Figure 3.13 illustrates the relationship between BER and EVM for QPSK, 16-QAM, and 64-QAM [3]. As shown, the sensitivity of BER to EVM depends strongly on

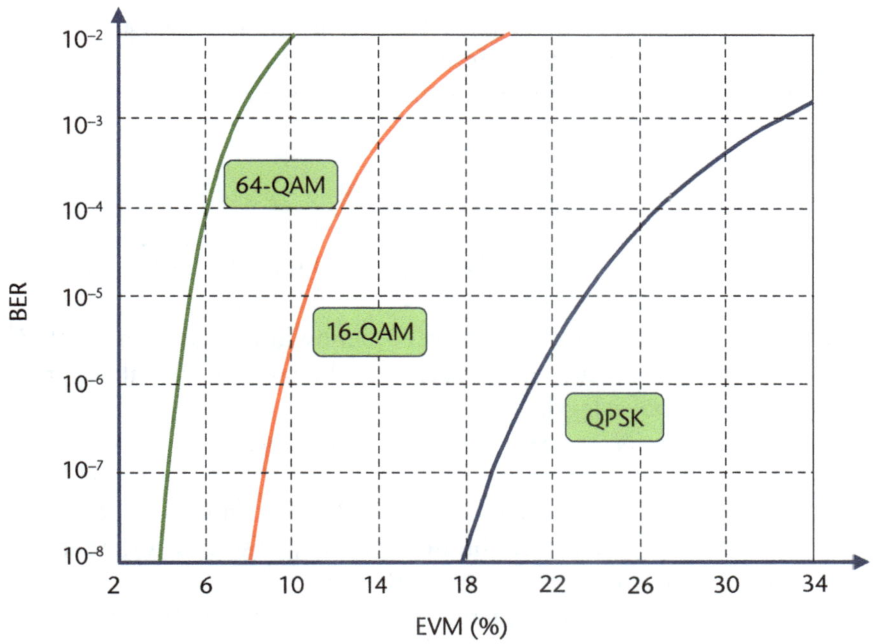

Figure 3.13 Relationship between BER and EVM for QPSK, 16-QAM, and 64-QAM modulations.

the modulation order. Higher-order constellations, such as 64-QAM, have smaller Euclidean distances between constellation points; consequently, even a modest increase in EVM can lead to a rapid rise in BER. In contrast, lower-order modulation schemes such as QPSK are more tolerant to impairments, requiring substantially larger EVM degradation before BER increases significantly. For this reason, system designers typically impose much stricter EVM specifications for higher-order modulation schemes to ensure that target BER performance is maintained.

Relationship Between BER and SNR

For coherent BPSK transmission over an AWGN channel with optimal detection, the relationship between SNR and BER is given by

$$\text{BER} = Q\left(\sqrt{2\,\text{SNR}_{\text{linear}}}\right) \tag{3.16}$$

where $Q(\cdot)$ denotes the Gaussian Q-function. For QPSK, the same expression applies, since QPSK can be interpreted as two independent, orthogonal BPSK channels operating on the in-phase and quadrature components.

For higher-order modulation schemes (e.g., M-PSK and M-QAM), the BER-SNR relationship differs and generally requires a higher SNR to achieve the same BER.

Relationship Between BER and E_b/N_0

The energy per bit to noise power spectral density ratio (E_b/N_0) is a normalized measure of SNR used in digital communications to compare link performance independently of bandwidth or bit rate.

- E_b: Energy used to transmit 1 bit [Joules], calculated as E_b = signal power/R_b where R_b is the bit rate in bits per second.
- N_0: Noise power spectral density [watts/Hz].

Thus,

$$\frac{E_b}{N_0} = \frac{\text{Energy per bit}}{\text{Noise power spectral density}} \tag{3.17}$$

Since the total noise power over a bandwidth B is given by N_0B, the ratio E_b/N_0 effectively quantifies how much signal energy is available per transmitted bit relative to the noise level.

The ratio E_b/N_0 is particularly useful because:

- It is independent of symbol rate and occupied bandwidth.
- This independence enables direct comparison of different modulation and coding schemes for a given BER target.
- It serves as a key parameter in link budget calculations when determining the required received signal level.

The relationship between SNR and E_b/N_0 is

$$\text{SNR} = \left(\frac{E_b}{N_0}\right)\left(\frac{R_b}{B}\right) \tag{3.18}$$

Note that N_0B represents the total noise power within the bandwidth B, while E_bR_b corresponds to the total signal power. The ratio R_b/B represents spectral efficiency in bits/s/Hz. Consequently, for a fixed E_b/N_0, higher spectral efficiency requires a higher SNR.

Figure 3.14 illustrates the relationship between BER and E_b/N_0 for uncoded BPSK/QPSK, 8-PSK, and 16-PSK modulation schemes over an AWGN channel. It shows that, for a given BER target, higher-order modulation schemes require a higher E_b/N_0. For example, achieving a BER of 10^{-5} requires the lowest E_b/N_0 for BPSK/QPSK, a higher value for 8-PSK, and the highest for 16-PSK. When FEC is applied, the required E_b/N_0 to achieve a given BER is reduced compared to the uncoded case, an improvement commonly referred to as coding gain, with the exact gain depending on the coding rate and the specific code family.

Relationship Between EVM and SNR

We have already examined the relationships between BER and EVM, as well as between BER and SNR. It is therefore useful to understand how EVM is directly related to SNR. The relationship between EVM and SNR is given by

$$\text{EVM}_{\text{rms}} = \frac{1}{\sqrt{\text{SNR}_{\text{linear}}}} \tag{3.19}$$

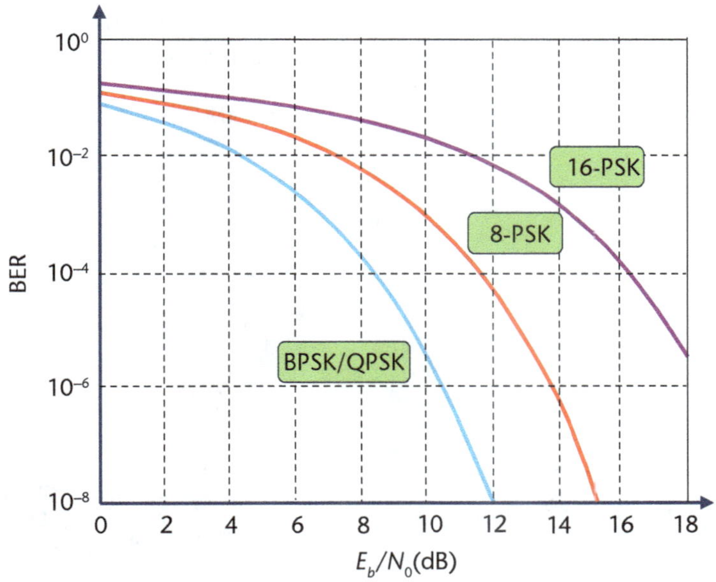

Figure 3.14 BER versus E_b/N_0 for various modulations.

and

$$\text{SNR}_{\text{dB}} = -20\log_{10}(\text{EVM}_{\text{rms}}) \tag{3.20}$$

For example, an EVM of 10% (0.1 in linear scale) corresponds to an SNR of approximately 20 dB. These expressions are valid when AWGN is the dominant impairment. If other effects such as nonlinear distortion, phase noise, or I/Q imbalance are significant, the relationship will deviate.

Together, the relationships between BER and SNR, EVM, and E_b/N_0 form a performance chain: system impairments increase EVM, which, in turn, reduces SNR, and the resulting lower SNR leads to higher BER. This chain provides a clear link between physical-layer impairments and overall link-level performance.

A Unified Framework for Detection Performance

Figure 3.15 illustrates the relationship between BER, received signal power, and system impairments. Received signal power provides the useful energy required for reliable detection, while impairments such as noise, losses, interference, distortion, imbalance, and leakage degrade signal quality. Metrics including SNR, signal-to-interference-plus-noise ratio (SINR), signal-to-noise-and-distortion ratio (SINAD), and E_b/N_0 quantify the ratio of signal power to these impairments, while EVM captures how their combined effects manifest as constellation errors. Together, these metrics describe the balance between useful signal energy and impairments that ultimately determines BER. To reduce BER, the received signal power must increase, the impairments must decrease, or both.

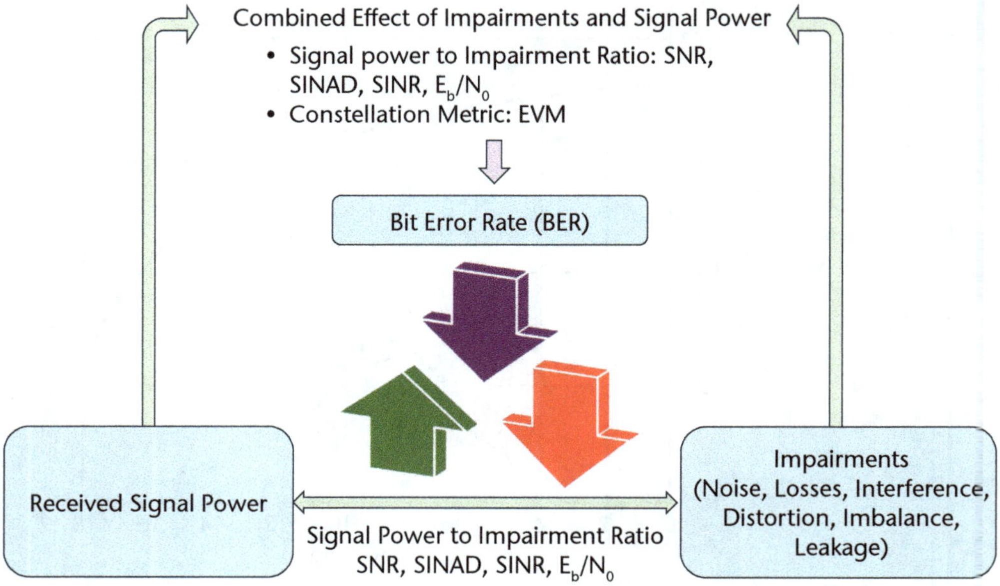

Figure 3.15 BER relationship with the received signal power and system impairments.

3.6 Synchronization

A digital communication signal is not a continuous stream of meaning; it is structured in time, frequency, and phase. If the receiver is misaligned in any of these dimensions, it will interpret the signal incorrectly, even if the received signal is strong and impairment levels (such as noise) are low.

Without synchronization, the receiver does not know how to slice the waveform into symbols or how to reliably map those symbols to bits.

At the most fundamental level, synchronization tells the receiver:

- When to sample the signal at the correct instants;
- Where symbols begin and end, so that the received waveform can be correctly partitioned into discrete symbols;
- How to interpret the signal using the correct carrier frequency and phase. This corresponds to carrier synchronization. The receiver must align its LO in frequency and phase with the transmitter to correctly interpret the received I/Q samples and map them to constellation points. This is particularly important for phase-sensitive modulation schemes such as QAM and QPSK.

Coherence comes into play after synchronization has been established and describes how stable the frequency and phase alignment remain over time. Coherent detection is required whenever information is encoded in phase, or in a combination of phase and amplitude. Classic examples include BPSK, QPSK, 8-PSK, QAM, OFDM, techniques such as MIMO beamforming, phase-coherent radar, and quantum control systems, all of which rely on stable phase references.

Synchronization errors typically manifest as symbol timing misalignment, constellation rotation, ISI, or, in severe cases, complete loss of communication.

As the modulation order increases, tolerance to synchronization errors decreases sharply. Higher-order constellations place symbols closer together in the I/Q plane, making them significantly more sensitive to timing jitter, carrier frequency offset, and phase noise. As a result, synchronization accuracy directly impacts achievable EVM, BER, and overall link robustness.

Coherent systems provide superior spectral efficiency, improved receiver sensitivity, and support for high-order modulation schemes, making them indispensable in modern communication and radar applications. However, these advantages come at the cost of increased hardware complexity, higher processing requirements, and more stringent synchronization demands. Noncoherent systems, by contrast, offer simpler implementation and greater robustness in noisy or fast-fading environments, but sacrifice efficiency and data throughput. The choice between coherent and noncoherent operation ultimately depends on system performance goals, environmental conditions, and implementation constraints.

3.6.1 Transmitter and Receiver Synchronization

The transmitter and receiver do not inherently share clocks or oscillators. Synchronization is therefore achieved by embedding known structure into the transmitted signal and by using feedback loops in the receiver to lock onto that structure. At a high level, synchronization proceeds through three stages:

1. Acquire synchronization (e.g., detection of a preamble).
2. Lock to the signal (timing and frequency alignment).
3. Track slow variations over time (timing and carrier recovery loops).

To enable this process, the transmitter intentionally embeds known patterns into the signal for synchronization purposes. These include preambles (known symbol sequences at the start of a packet), frame headers, and pilot symbols or pilot tones, which allow the receiver to detect, recognize, and align itself to the incoming signal.

At the receiver, the incoming signal is correlated with the known preamble, producing a correlation peak. This peak is used to determine the start of the message and establish symbol timing. Once coarse timing is acquired, timing recovery loops continuously adjust the receiver's sampling clock to maintain alignment.

In real systems, synchronization is not static and must continuously track slow variations caused by clock drift, temperature changes, Doppler shifts, and phase noise.

In practical implementations, synchronization spans both the analog and digital domains:

- Analog synchronization relies on carrier recovery loops, such as phase-locked loops (PLLs) or Costas loops, to correct carrier frequency and phase offsets. PLLs are discussed in detail in Chapter 4.

- Digital synchronization employs clock and data recovery (CDR), digital carrier recovery loops, and DSP-based algorithms to refine timing, compensate residual offsets, and maintain coherence over time.

3.6.2 Clock and Data Recovery

In high-speed digital receivers, CDR performs symbol timing synchronization by extracting a sampling clock directly from the incoming data stream. Because the clock is not transmitted as a separate signal, the receiver must recover timing information from data transitions to correctly interpret the bits [5].

Figure 3.16 illustrates this process. The incoming data arrives with timing uncertainty, reflected by the partially closed-eye diagram on the left. This signal is fed simultaneously to a clock recovery circuit and a decision circuit. The clock recovery circuit, functionally similar to a PLL, extracts timing information from the data transitions and reconstructs a sampling clock aligned with the center of the eye opening. This recovered clock is fed back to the decision circuit and determines when symbol decisions are made.

Using the recovered clock, the decision circuit samples the waveform at the optimal instant within each symbol period and maps it to discrete logic levels (0 or 1). As a result, timing uncertainty is removed from the decision process, producing a clean, regenerated data stream with a nearly fully open-eye diagram at the output, as shown on the right. The recovered clock waveform is shown at the bottom of the figure.

This feedback interaction between clock recovery and data decision is fundamental to reliable operation in high-speed serial links, allowing the receiver to tolerate channel-induced impairments such as loss, reflections, and jitter while maintaining low BERs.

3.7 Transmission Techniques: Spread Spectrum and Multicarrier

In digital communication systems, modulation defines how information is mapped onto a carrier signal. To further enhance robustness against jamming, interference, and multipath, as well as to improve security or spectral efficiency, additional transmission techniques can be applied. Two widely used classes of such techniques are spread spectrum and multicarrier transmission.

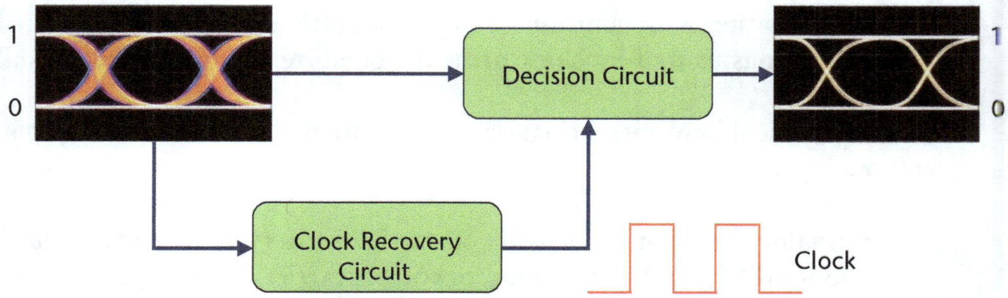

Figure 3.16 Clock and data recovery aligns sampling to the eye center, enabling reliable symbol decisions in high-speed links.

3.7.1 Spread Spectrum Techniques

Spread spectrum techniques intentionally expand a signal's bandwidth beyond what is required for the raw data rate. This increased bandwidth, combined with appropriate coding techniques, improves resilience to interference, jamming, and eavesdropping and can also enable multiple users to share the same frequency band with minimal mutual interference. As a result, spread spectrum techniques are widely used in military radios and industrial wireless links that must operate reliably in noisy or contested RF environments.

Direct Sequence Spread Spectrum (DSSS)

DSSS is commonly paired with BPSK or QPSK and is used in systems such as GPS, IEEE 802.11b Wi-Fi, and various military and secure communication applications. It uses a pseudorandom spreading code to spread each data bit over many spreading elements (chips).

In DSSS, the already modulated data signal is XORed (or multiplied in the bipolar domain) with a high-rate pseudorandom (PN) code sequence, as shown in Figure 3.17. Because both the transmitter and receiver know the PN code, the receiver can apply the same code to despread the incoming signal, recovering the original data while effectively rejecting narrowband interference. To unintended receivers without knowledge of the PN code, the transmitted signal appears noise-like, enhancing resilience against interference, jamming, and eavesdropping.

Some of the key benefits of DSSS include:

- *Interference rejection:* DSSS spreads signal energy over a wide bandwidth, making the system robust against narrowband interferers.
- *Security and anti-jamming:* Without knowledge of the PN sequence, unauthorized receivers find it difficult to demodulate or jam the signal effectively.
- *Multipath resilience:* The wide signal bandwidth enables improved resolution of multipath components, reducing the impact of fading and delay spread.

From an RF hardware perspective, DSSS systems typically require:

- Wideband RF front ends to accommodate the spread signal bandwidth;
- High dynamic range ADCs for digital implementations;
- Stable clocks to maintain code synchronization between transmitter and receiver;
- Low-phase-noise LOs for coherent demodulation.

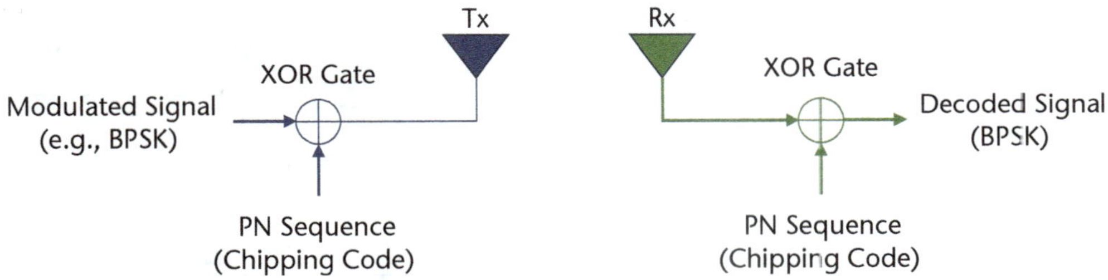

Figure 3.17 DSSS communication link.

When selecting a transmission technique for an RF product, DSSS is often favored in environments where interference resistance, security, and signal robustness are more critical than spectral efficiency.

Frequency Hopping Spread Spectrum (FHSS)

FHSS commonly employs modulation schemes such as FSK, GFSK, or PSK and is used in applications including Bluetooth Classic, certain military radios, and drone communication links. In FHSS, the carrier frequency is rapidly changed according to a predefined hopping sequence shared between the transmitter and receiver. As illustrated conceptually in Figure 3.18(a), the signal hops among many discrete frequency channels following a pseudorandom pattern.

The system architecture is shown in Figure 3.18(b). In this architecture, the LO of the frequency mixer, which determines the transmit carrier frequency, is controlled by a pseudorandom code generator. The hopping rate may range from slow frequency hopping, where one hop occurs per data packet or frame, to fast frequency hopping, where multiple hops occur within a single symbol interval. Regulatory bodies often impose constraints on parameters such as dwell time and channel spacing, particularly in shared spectrum environments such as the 2.4-GHz ISM band.

At any given instant, the transmitted signal occupies only a narrow bandwidth. However, when observed over time, the signal energy is distributed across a much wider frequency range, yielding the characteristic spread-spectrum behavior of FHSS.

This frequency-hopping behavior provides increased robustness against narrowband interference, unintentional interferers, and intentional jamming, since only a small portion of the transmission is affected at any given time. It also offers a degree of resistance to interception: without knowledge of the hopping sequence, unauthorized interception and eavesdropping become significantly more difficult. Nevertheless, severe interference or aggressive broadband jamming can still reduce the achievable data rate or disrupt synchronization between the transmitter and receiver. In addition, FHSS systems can coexist with other users in the same frequency band (e.g., unlicensed ISM bands) by statistically avoiding persistent interference over time.

3.7.2 Multicarrier Systems

A multicarrier communication system is illustrated in Figure 3.19. In this approach, the input bit stream is mapped to complex symbols $X[k]$ using modulations such

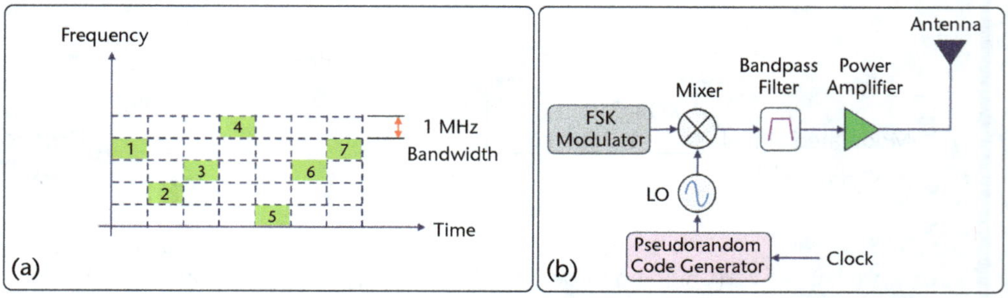

Figure 3.18 (a) FHSS hopping pattern. (b) Block diagram of an FHSS implementation.

as QPSK or QAM and converted into multiple lower-rate parallel streams. At the transmitter, an inverse fast Fourier transform (IFFT) synthesizes the time-domain OFDM waveform by combining orthogonal subcarriers. After transmission through the channel, the receiver applies a fast Fourier transform (FFT) to separate the composite signal back into its constituent subcarriers, recover $X[k]$, and demodulate the bit stream.

By transmitting symbols in parallel, the symbol rate on each subcarrier is significantly reduced. This improves robustness against multipath delay spread because the symbol duration is longer, making ISI less severe relative to the symbol interval. In addition, frequency-selective fading is mitigated because each subcarrier occupies a narrow bandwidth and therefore experiences approximately flat fading. Together, these effects allow multicarrier systems to maintain high spectral efficiency while operating reliably in dispersive wireless channels.

OFDM requires continuous tracking of frequency, phase, and timing, typically using pilot subcarriers, whereas DSSS and FHSS rely on code synchronization to align spreading sequences or hopping patterns between the transmitter and receiver.

The main advantages of OFDM include:

- *Robustness to multipath fading:* Each subcarrier experiences approximately flat fading, enabling simple equalization.
- *Simplified channel equalization:* Frequency-domain equalization is computationally efficient, particularly in time-dispersive channels.
- *High spectral efficiency:* Subcarriers are tightly packed without guard bands due to orthogonality.
- *Support for adaptive modulation:* Different subcarriers may use different modulation schemes based on their SNR (e.g., 64-QAM on strong subcarriers and QPSK on weaker ones).
- *Efficient implementation:* Modulation and demodulation using IFFT and FFT operations are computationally efficient.

OFDM is widely employed in systems such as Wi-Fi, LTE downlink, and 5G NR. While OFDM is a dominant choice in modern communication systems, it is not universally optimal. Many legacy and low-power systems continue to use single-carrier modulation schemes. Each approach has distinct implications for RF front-end architecture, power efficiency, and system complexity, as discussed in Tables 3.10 and 3.11.

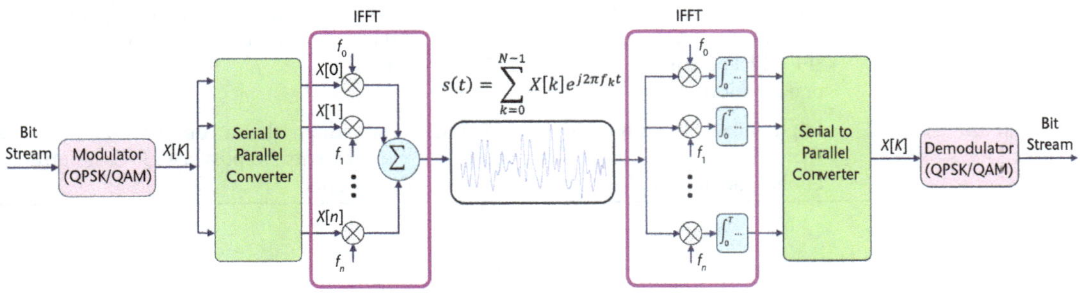

Figure 3.19 Structure of a multicarrier modulation system.

Table 3.10 Comparison of OFDM and Single-Carrier Transmission Techniques

Aspect	*OFDM (Multicarrier)*	*Single-Carrier*
PAPR	High, requires linear PAs and backoff	Low, more power-efficient PA operation
Spectral efficiency	High, due to tight subcarrier packing	Medium, often guard bands are required
Resilience to multipath	Excellent, handles frequency-selective channels with ease	Poor in frequency-selective channels unless equalized
Equalization	Simple frequency-domain equalization	Requires complex time-domain equalization
Implementation	Requires FFT/IFFT, synchronization logic	Simpler baseband architecture
Synchronization sensitivity	Sensitive to timing and frequency offsets	Generally, more tolerant
Latency	Higher (due to FFT block processing, cyclic prefix)	Lower (suitable for low-latency applications)

Table 3.11 RF Front-End Design Implications: OFDM Versus Single-Carrier Systems

Design Aspect	*OFDM (Multicarrier)*	*Single-Carrier*
PA linearity	High PAPR demands highly linear PAs to prevent distortion and spectral regrowth; requires output power backoff, linearization (e.g., digital predistortion), and efficient thermal design	Lower PAPR allows more efficient nonlinear PA classes (e.g., Class C, E, or F), reducing power consumption and thermal stress
RF chain (amplitude and group delay flatness)	Requires flat gain and group delay across the full OFDM bandwidth to maintain subcarrier orthogonality and minimize ISI	Can tolerate narrower bandwidth and moderate group delay variation without major distortion
LO and clock phase noise/frequency stability	Sensitive to phase noise and frequency drift, which break orthogonality and cause intercarrier interference (ICI); demands low-phase-noise synthesizers and highly stable frequency references	Generally, more robust to phase noise and minor frequency drift
Synchronization and timing	Requires precise timing and frequency synchronization for accurate FFT window alignment; incorporates pilot tones, cyclic prefixes, and training sequences for synchronization and channel estimation	Typically, simpler synchronization, often without the need for pilot tones or cyclic prefixes
ADC/DAC and digital processing	Needs high-resolution, high-speed ADCs/DACs and powerful DSP or FPGA/ASIC hardware for real-time IFFT/FFT, channel estimation, and equalization	Can use lower-speed and lower-resolution converters; digital signal processing requirements are less intensive
Thermal and SWaP-C impact	High linearity, computational load, and thermal dissipation increase power and cooling demands, affecting SWaP-C	Lower thermal and computational requirements simplify SWaP-C design constraints

A notable variation is single-carrier frequency division multiple access (SC-FDMA), which combines single-carrier transmission with frequency-domain equalization to compensate for multipath fading and frequency-selective channel distortion. By applying discrete Fourier transform (DFT) spreading prior to subcarrier mapping, SC-FDMA achieves a lower peak-to-average power ratio (PAPR) than OFDM, making it well-suited for uplink applications such as LTE.

3.8 Optimizing Waveforms for Spectral and Power Efficiency

In this section, we examine one of the key aspects of RF systems used in communication links, quantum systems, and EW systems: pulse shaping. Pulse shaping plays a critical role in determining overall system performance, influencing spectral efficiency and spectral leakage. It can also indirectly affect power efficiency by shaping the signal's crest factor and by enabling more effective predistortion, as will be shown shortly.

3.8.1 Pulse Shaping

Pulse shaping is the process of modifying the time-domain waveform of transmitted symbols to control the signal's spectral characteristics and ISI.

Pulse shaping is a critical stage in modern communication systems for several reasons:

- *Bandwidth limitation:* Practical communication channels have finite bandwidth, and leakage to adjacent channels is unacceptable. Abrupt symbol transitions, such as those produced by rectangular pulses, generate wideband spectra with significant sidelobes, leading to spectral leakage and adjacent-channel interference.
- *Spectral mask compliance:* Regulatory bodies impose spectral masks that limit out-of-band emissions. Pulse shaping techniques are essential for meeting these constraints by controlling the out-of-band leakage without unnecessarily reducing data rate.
- *ISI control:* When pulses corresponding to consecutive symbols overlap in time, they can distort the received signal. Proper pulse shaping ensures zero or minimal pulse overlap at the symbol sampling instants, thereby mitigating ISI.

Consider a symbol sequence composed of rectangular pulses. As illustrated in Figure 3.20(a), after pulse shaping, the originally rectangular symbols are transformed into sinc-shaped pulses. Although these sinc pulses overlap in time, they are deliberately designed to satisfy the Nyquist zero-ISI condition. At the ideal sampling instant, corresponding to the peak of the desired sinc pulse, all other time-shifted sinc pulses evaluate to zero. As a result, only the intended symbol contributes to the sampled value, and no ISI occurs.

It is important to emphasize that time-domain overlap does not imply interference. Interference is only relevant at the symbol sampling instants, where the receiver

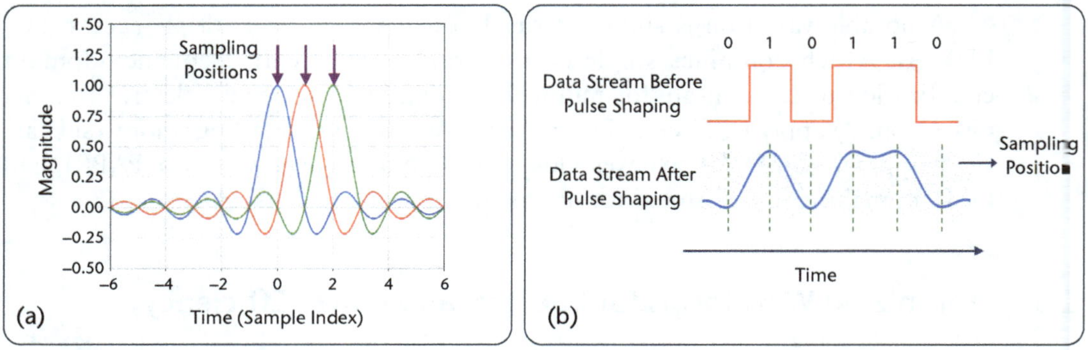

Figure 3.20 (a) Sinc pulse train satisfying the Nyquist zero-ISI condition. (b) Rectangular symbols before and after raised-cosine pulse shaping.

makes its decision. This distinction is a crucial conceptual point in understanding Nyquist signaling and pulse shaping.

As we will see shortly, the ideal sinc filter is impractical to implement. Therefore, other types of pulse-shaping filters are used in practical systems. Figure 3.20(b) illustrates a rectangular pulse sequence after pulse shaping using a raised-cosine filter. Again, at the ideal sampling instants, located at the center of each symbol period, the correct symbol value is recovered.

The sinc pulse in the time domain has a rectangular shape in the frequency domain and provides the minimum bandwidth required for a given symbol rate, known as the Nyquist bandwidth B_{Nyquist}. For a symbol rate R_s, the Nyquist bandwidth is given by $B_{\text{Nyquist}} = R_s$. However, although this pulse achieves zero ISI at the sampling instants, it extends infinitely in time and is therefore not physically realizable. As a result, practical systems employ realizable pulse-shaping filters that approximate the ideal Nyquist response. Common examples are summarized in Table 3.12.

Table 3.12 Common Pulse-Shaping Filters

Filter Type	*What It Does*	*Applications*	*Key Trade-Offs*
Rectangular pulse	Sends symbols as-is with abrupt transitions	Rarely used in real RF systems; mainly for theory or simulations	Very poor spectral containment; large sidelobes cause adjacent-channel interference
Raised cosine (RC)	Smooths symbol edges so symbols do not interfere at sampling points	Systems where tight ISI control is needed and bandwidth is available	Zero ISI, but bandwidth increases as roll-off factor α increases
Root raised cosine (RRC)	Splits pulse shaping between transmitter ($\sqrt{RC}$) and receiver ($\sqrt{RC}$); combined response is RC ($\sqrt{RC} \times \sqrt{RC} = RC$)	Most modern digital radios (e.g., LTE, DVB, SATCOM)	Excellent ISI control; requires matched filtering at TX and RX
Gaussian filter	Smooths transitions aggressively to reduce occupied bandwidth while keeping a constant envelope	Constant-envelope systems like GMSK and GFSK (Bluetooth, GSM)	Very good spectral containment; introduces controlled ISI and requires sequence detection

Figure 3.21 shows a raised-cosine filter in the time and frequency domains for various roll-off factors α. The roll-off factor determines how much additional bandwidth the pulse occupies beyond the Nyquist bandwidth and is defined as

$$\alpha = \frac{\text{Excess Bandwidth}}{\text{Nyquist Bandwidth}} = \frac{B_{\text{occupied}} - B_{\text{Nyquist}}}{B_{\text{Nyquist}}} \tag{3.21}$$

which leads to

$$B_{\text{occupied}} = (1+\alpha)\,B_{\text{Nyquist}} \tag{3.22}$$

Typical values of α include:

- $\alpha = 0$: Ideal brick-wall response with maximum spectral efficiency (theoretical and not physically realizable).
- $\alpha \approx 0.2$–0.35: Common in practical systems, offering a balanced trade-off between bandwidth efficiency and timing robustness (e.g., LTE $\alpha \approx 0.18$, DVB-S $\alpha \approx 0.35$).
- $\alpha = 1$: Very wide and gentle filter with excellent tolerance to timing errors but poor spectral efficiency.

The roll-off factor directly influences spectral efficiency, robustness to timing errors, and implementation complexity.

Increasing α widens the occupied bandwidth, thereby reducing spectral efficiency.

In the time domain, a larger α results in faster decay of the pulse tails and reduced ringing, which decreases the energy contributed by neighboring symbols. As a result, when sampling deviates slightly from the ideal instant, the impact of adjacent symbols is reduced, making symbol errors less likely. Therefore, the system becomes more tolerant to small timing offsets and sampling-phase errors.

In contrast, low-α filters with sharp frequency responses require higher-order digital filter implementations. These implementations are more sensitive to coefficient

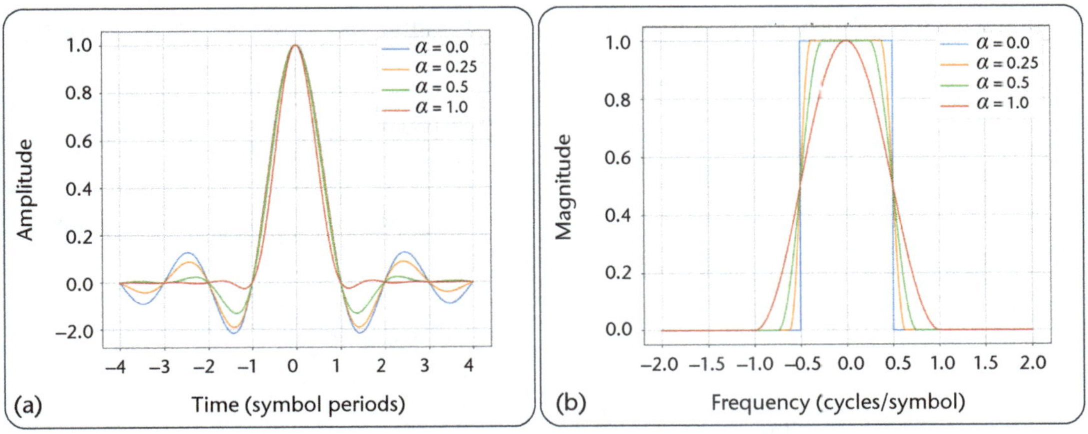

Figure 3.21 Raised cosine filter with different roll-off factors: (a) time-domain response and (b) frequency-domain response.

quantization, finite word-length effects, and group-delay ripple, increasing design complexity and implementation risk. Consequently, systems employing low roll-off factors demand tighter synchronization and more accurate timing recovery mechanisms [5].

For a modulation carrying b bits per symbol, the relationship between the spectral efficiency η and roll-off factor is approximately:

$$\eta = \frac{b}{1+\alpha} \text{ bits/s/Hz} \tag{3.23}$$

For QPSK (2 bits/symbol) with $\alpha = 0.25$, $\eta = 2/1.25 = 1.6$ bits/s/Hz.

In digital communication systems, pulse shaping is implemented by convolving the symbol sequence with a pulse-shaping filter, such as an RC or RRC filter. Pulse shaping is fundamentally different from windowing, for example, as used in spectrum analysis. Windowing typically involves pointwise multiplication of a finite-length signal by a window function, such as a Hann, Hamming, Blackman, or cosine window, to reduce spectral leakage. While effective for spectral containment of finite-duration signals, windowing does not constitute Nyquist pulse shaping and does not ensure zero ISI for continuous symbol streams.

3.8.2 Peak-to-Average Power Ratio (PAPR)

PAPR, sometimes referred to as crest factor, is a key metric used to quantify how peaky a signal is, that is, how large its maximum instantaneous power is relative to its average power. Mathematically, PAPR is defined as

$$\text{PAPR} = \frac{P_{\text{peak}}}{P_{\text{avg}}} \tag{3.24}$$

and is commonly expressed in decibels as PAPR (dB) = P_{peak} (dBm) − P_{avg}(dBm).

As shown in Figure 3.19, OFDM signals are the sum of many independent subcarriers. When these subcarriers happen to align in phase, they create massive voltage peaks far above the average power level. Figure 3.22 illustrates a signal with high PAPR, characterized by large instantaneous peaks relative to its average

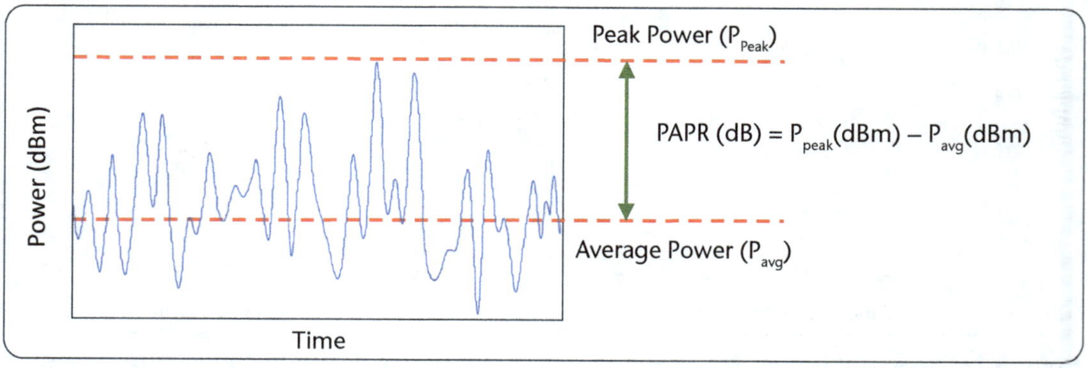

Figure 3.22 High PAPR signal.

power level. Such behavior places stringent demands on PA linearity and back-off, directly affecting efficiency and overall system performance, as we will see shortly.

Across different modulation types, the PAPR varies significantly:

- Constant-envelope modulations (e.g., BPSK, QPSK, FSK, GMSK) exhibit low crest factors, typically around 3 to 6 dB for a single modulated carrier, enabling the use of nonlinear, high-efficiency PAs with minimal distortion.
- Higher-order QAM and multicarrier waveforms (e.g., 64-QAM, 256-QAM, OFDM) exhibit significantly higher crest factors, with OFDM often exceeding 10–12 dB. These signals require highly linear amplifiers and substantial power backoff to avoid distortion and adjacent-channel interference.

High-PAPR signals are challenging because large signal peaks drive PAs toward compression. As discussed in Chapter 4, compression is a nonlinear effect that causes waveform distortion, resulting in spectral regrowth, increased EVM, and degraded adjacent-channel performance. To prevent compression, the PA must operate with significant output back-off, meaning that it is driven well below its compression point. However, this operating condition is suboptimal from a power-efficiency perspective, since PAs achieve their highest efficiency near saturation, and back-off directly reduces efficiency. As a result, high PAPR has several important implications for RF system design:

- *Amplifier linearity:* High PAPR waveforms demand highly linear PAs over a wide dynamic range to limit intermodulation distortion, spectral regrowth, and EVM degradation.
- *Efficiency:* Increased PAPR forces the average output power to remain well below saturation, resulting in reduced PA efficiency.
- *Dynamic range:* Receiver front ends must accommodate large peak-to-average variations without compression or distortion, increasing dynamic-range requirements.

As discussed in Chapter 5, various techniques are employed to mitigate these effects while preserving signal integrity and spectral compliance:

- *Crest factor reduction (CFR):* Limits extreme signal peaks to reduce PAPR and relax linearity requirements on the PA.
- *Digital predistortion (DPD):* Compensates for PA nonlinearities by predistorting the input waveform to improve linearity and spectral compliance.
- *Doherty architectures:* Improve efficiency across a wide output power range by using a main (carrier) amplifier that handles the average signal and a peaking amplifier that activates only during high-power excursions, thereby maintaining higher efficiency under back-off conditions.
- *SC-FDMA:* A single-carrier variant of OFDM that employs discrete Fourier transform (DFT) spreading to reduce PAPR, making it well suited for uplink transmissions in systems such as LTE and 5G, where power efficiency is critical.

For high PAPR modulation schemes, waveform shaping, and mitigation techniques must be carefully balanced to meet both performance and efficiency targets.

3.9 Put Your Knowledge in Practice: LEO Satellite Communication Link

The following scenario estimates a downlink for a low Earth orbit (LEO) Earth-imaging satellite, illustrating how throughput, spectral efficiency, SNR, and EVM are connected. The scenario assumes a satellite transmitting in the X-band (~8 GHz) to a ground terminal equipped with a 1-m diameter parabolic antenna, targeting near-real-time image streaming.

Some of these calculations are discussed in more detail in Chapter 4, where link budget analysis is introduced. For now, these values can be taken as given, as they serve to illustrate how link-level parameters such as SNR, throughput, and spectral efficiency are interconnected.

Step 1: Link parameters. The link parameters are shown in Table 3.13.

Step 2: SNR at the receiver. The received power (see Chapter 4, Section 4.12.2) can be calculated as: $P_r = \text{EIRP} + G_{rx} - L_{fs} - L_{\text{misc}} = 30 + 38 - 172 - 2 = -106$ dBW.

The noise power can be calculated as follows (see Chapter 4, Section 4.7.1).

- $kT \approx -204$ dBW/Hz (equivalent to -174 dBm/Hz at $T \approx 290$K)
- $N = kT + 10\log_{10}(B) + NF$
- $N = -204 + 76.0 + 2.5 = -125.5$ dBW

SNR can be calculated as: $\text{SNR}_{\text{dB}} = P_r - N = -106 - (-125.5) = 19.5$ dB ($\text{SNR}_{\text{lin}} \approx 89.1$). Using the estimated receiver SNR, the EVM can be obtained as follows (using AWGN-limited approximation) $\text{EVM}_{\%} \approx 100/\sqrt{89.1} \approx 10.6\%$.

This EVM value can be compared against the allowable EVM for the selected MODCOD to determine whether the RF hardware and signal chain can realistically meet the BER target.

Step 3: Channel capacity (Shannon limit).

Table 3.13 Link Parameters for the LEO Earth-Imaging Satellite

Parameter	*Value/Description*
Orbit altitude	1,200 km (typical LEO imaging pass)
Carrier frequency	8 GHz (X-band)
Allocated downlink bandwidth	40 MHz
Modulation and coding	8-PSK, coding rate $R_c = 3/4$
Satellite EIRP	30 dBW
Ground station antenna gain	38 dBi
Path loss (free space)	172 dB (a slant range approximately equal to the orbital altitude is assumed for simplicity)
Additional losses	2 dB (pointing, polarization, atmosphere)
Receiver noise figure	2.5 dB

Note: EIRP includes transmitter power and satellite antenna gain. This is explained in Chapter 4.

Using the linear SNR, $C = B\log_2(1 + S/N) = 40 \times 10^6 \cdot \log_2(1 + 89.1) \approx 260$ Mbps. This represents the theoretical upper limit; practical systems operate below it due to modulation, coding, and protocol overhead.

Step 4: Practical bit rate, throughput, and spectral efficiency.

a. *Symbol rate (include pulse-shaping roll-off):* For raised-cosine filtering with roll-off α, the occupied bandwidth is approximately $B \approx R_s\,(1 + \alpha)$, resulting in $R_s \approx B/(1 + \alpha)$. For $\alpha = 0.25$: $R_s \approx 40/1.25 = 32$ Msym/s.
b. *Raw bit rate (8-PSK has 3 bits/symbol):* $R_{\text{raw}} = R_s\log_2(M) = 32 \times 10^6 \times 3 = 96$ Mbps.
c. *Net bit rate after coding (coding rate $R_c = 3/4$):* $R_{\text{coded}} = R_{\text{raw}}R_c = 96 \times 0.75 = 75$ Mbps.
d. *Spectral efficiency:* Using coding only, $\eta = R_{\text{coded}}/B = 72/40 = 1.8$ b/s/Hz. The E_b/N_0 can be calculated as follows $(E_b/N_0)_{\text{dB}} = \text{SNR}_{\text{dB}} - 10\log_{10}(\eta) = 19.5 - 2.55 \approx 17$ dB.

Step 5: Practical MODCOD requirements (BER target).

Using the uncoded PSK reference in Figure 3.14, this E_b/N_0 at the target BER of 10^{-6} supports 8-PSK with sufficient margin. To move to a higher modulation order, such as 16-PSK, the system must increase link margin or select a more robust MODCOD. Typical design levers to close the gap include:

- Increasing EIRP (higher PA output power and/or higher transmit antenna gain);
- Increasing ground-station antenna gain, or reducing pointing and polarization losses;
- Reducing receiver noise figure (NF);
- Reducing bandwidth, thereby lowering the integrated noise power;
- Selecting a more robust MODCOD, such as using a stronger code (lower coding rate).

3.10 Design Implications of Signal Chain Layer Choices

So far, we have examined key aspects of the signal chain layer, including throughput, spectral efficiency, modulation schemes, and error metrics such as BER, along with their relationship to parameters such as EVM and SNR. Choices made at this layer have far-reaching implications for RF product design, as they propagate through the signal chain and hardware layers, shaping the RF hardware architecture, DSP implementation, firmware, and software, as well as key SWaP-C trade-offs. As a result, decisions at this stage are fundamental to defining the overall product blueprint.

In many products, the signal chain layer is already available as a mature and well-validated subsystem, for example, through commercial chipsets or modems used in Wi-Fi, LTE, 5G, or SATCOM systems. In such cases, the primary design focus shifts toward system integration and RF front-end optimization rather than signal chain development itself. In contrast, products that require a custom signal chain layer, such as quantum computers, must define and develop this layer incrementally.

This process is inherently time-consuming and resource-intensive, requiring substantial engineering effort, specialized expertise, and significant capital investment. For this reason, whenever possible, established and standardized communication layers should be leveraged, reserving custom signal chain development for cases where it is essential to meet specific product requirements.

3.10.1 Key Design Considerations

Selecting the appropriate modulation and coding scheme is one of the most critical decisions in communication system design. However, this principle extends well beyond traditional data links. Modulation defines how information, control signals, or sensing mechanisms are embedded into a carrier across a wide range of RF and quantum systems. In radar systems, waveform modulation techniques such as FMCW, phase-coded pulses, and pulse compression directly determine range resolution, sensitivity, and target discrimination. In quantum systems, precise modulation of microwave tones governs qubit transitions, gate operations, and overall fidelity. Even in sensing and metrology, modulation techniques, such as lock-in detection, enable weak signals to be extracted from noise-dominated environments.

Therefore, the concepts discussed in this section should be adapted to the specific goals and constraints of the product under development, whether it is a communication link, a radar front end, or a quantum control system.

For communication links, the optimal modulation and coding choice depends on a complex interplay of parameters, including available bandwidth, channel conditions, throughput requirements, and tolerance to errors and impairments. In practical RF product development, these trade-offs must be evaluated systematically to achieve performance targets without exceeding constraints on cost, power consumption, or implementation complexity. The key design factors influencing these decisions are summarized in Table 3.14.

Building on these design factors, the rules of thumb in Table 3.15 can be used as a guide to identify the most suitable modulation scheme for a given application.

3.10.2 Impact on RF Hardware

From a constellation perspective, reliable symbol detection requires constellation points to remain close to their ideal amplitude and phase locations. Any RF hardware impairment that perturbs amplitude or phase beyond allowable limits degrades detection accuracy. The effects of key impairments, including nonlinearity, additive noise, phase noise, and IQ imbalance, are illustrated in Figure 3.9.

As modulation complexity increases, RF hardware requirements become more stringent. Higher-order and wideband modulation schemes (e.g., QAM and OFDM) demand greater linearity, larger dynamic range, lower noise and distortion, and improved phase stability from amplifiers, filters, oscillators, and data converters. These requirements increase system cost, power consumption, and design complexity. In contrast, constant-envelope or frequency-based modulation schemes such as FSK and GFSK relax linearity and phase-noise constraints, enabling simpler, more power-efficient architectures that can employ high-efficiency amplifier classes (e.g., Class C or Class D), reducing SWaP-C and accelerating time-to-market. Table 3.16 shows how modulation scheme selection affects RF hardware design.

Table 3.14 Key Design Factors for a Communication Link

Design Factor	*Key Question*	*Practical Design Insights*
Throughput	What data rate must be supported?	Higher-order modulation increases bits per symbol and achievable throughput for a given symbol rate, adaptive modulation selects the highest feasible modulation order based on real-time link conditions
Operating frequency	How does frequency impact link performance?	Path loss, atmospheric absorption, and rain fade generally increase with frequency, lower frequencies provide better coverage and penetration but typically offer limited available bandwidth
Bandwidth availability *(resource)*	How much spectrum is available?	Spectral efficiency (bits/s/Hz) is critical in bandwidth-limited systems; higher-order modulation is only effective when sufficient SNR is available
Available signal power *(resource)*	What SNR is achievable under worst-case conditions?	Higher-order modulation requires higher SNR to meet BER targets; low-SNR or interference-limited links favor robust schemes such as BPSK or QPSK; E_b/N_0–BER curves are used to align modulation order with available link margin
Channel and hardware impairments *(constraint)*	What limits the link performance?	Path loss, fading, and multipath may necessitate lower-order modulation or diversity techniques; noise reduces SNR and increases BER; distortion introduces ISI and constellation spreading; nonlinearities increase EVM and spectral regrowth; phase noise degrades EVM and limits the use of high-order modulation
Protocol and coding overhead *(constraint)*	How much redundancy is required?	FEC improves robustness at the expense of net throughput; MODCOD selection balances reliability and spectral efficiency under varying channel conditions

Table 3.15 Rules of Thumb for Selecting Modulation Schemes

Condition	*Preferred Modulation*	*Rationale*	*Typical Application*
Very low SNR/ long range	BPSK	Maximum robustness and simplest receiver design	Deep-space links, telemetry, IoT sensors, long-range LoRa systems
Moderate SNR, moderate bandwidth	QPSK, 8-PSK	Balanced trade-off between robustness and throughput	Satellite communication, cellular uplink, Wi-Fi (legacy modes)
High SNR, limited bandwidth	16-QAM, 64-QAM	High spectral efficiency within constrained bandwidth	Microwave backhaul, cable modems, 5G mid-band links
Very high SNR, short range	256-QAM, 1024-QAM	Maximizes data throughput when channel quality is excellent	Fiber-fed small cells, indoor wireless links, point-to-point microwave links
Rapidly changing or fading channel	Adaptive modulation	Dynamically optimizes rate versus reliability in real time	Mobile broadband, satellite links with variable weather conditions

Table 3.16 RF Hardware Design Implications of Modulation Scheme Selection

Design Aspect	*Impact of Modulation Scheme*	*Practical RF Design Implications*
Component linearity	Higher-order modulations (e.g., QAM, OFDM) are highly sensitive to amplitude and phase distortion, increasing EVM and BER	Requires high-linearity PAs, PA backoff, and often linearization techniques (e.g., DPD). Constant-envelope schemes (e.g., FSK, GFSK) tolerate nonlinear PAs, enabling higher efficiency
Gain flatness and group delay	Wideband modulations require uniform gain and linear phase across the occupied bandwidth	Drives tighter gain-flatness and group-delay specifications for RF/IF filters and analog baseband stages
Noise sensitivity	Dense constellations have smaller decision margins and are more sensitive to noise and interference	Requires low-noise RF front ends (LNA, good isolation, shielding); noise performance limits the maximum achievable modulation order
Phase noise and frequency stability	Phase noise and frequency drift directly degrade EVM in higher-order modulations	Demands low-phase-noise LOs, low-jitter clocks, and stable frequency references (TCXO/OCXO). PLL design becomes critical
PAPR and PA efficiency	High PAPR signals require PA back-off to avoid nonlinear distortion	Reduces PA efficiency, increases thermal load, and complicates power supply and thermal design
Spectral purity	Complex modulations experience spectral regrowth in the presence of nonlinearities	Requires linearity, careful PA operating point selection, LO purity, and sharper filtering to meet spectral masks
Dynamic range and calibration	Higher modulation orders reduce tolerance to gain, phase, and I/Q imbalance	Requires tighter gain control, I/Q calibration, DC offset correction, and often digital compensation
ADC/DAC performance	High-order and wideband modulations require a high effective number of bits (ENOB) and a wide bandwidth	Drives higher-power, higher-cost data converters
Thermal stability	Temperature drift affects gain, phase, and noise performance	Requires thermal management, bias stabilization, and temperature-aware calibration

3.11 What's Next?

In this chapter, we explored how signal chain layer choices, such as modulation, coding, and synchronization, define a system's achievable performance, efficiency, and complexity. In Chapter 4, we shift the focus to the physical foundations of RF systems, examining core concepts that shape every wireless link, including linearity, noise, dynamic range, and link budget analysis. Together, these topics form the analytical backbone for evaluating and optimizing RF subsystems, bridging signal chain design choices with the practical realities of RF hardware implementation.

Problems

Unless otherwise stated, assume:

- An AWGN channel;
- Ideal synchronization;

- Perfect channel estimation;
- No fading or interference;
- Modulation performance is referenced to pre-FEC BER;
- Use standard textbook relationships between SNR, E_b/N_0, BER, and EVM.

3.1 A wireless communication link operates over a bandwidth of 10 MHz and uses 16-QAM with a coding rate of $R_c = 3/4$.
  a. Calculate the uncoded spectral efficiency (bits/s/Hz).
  b. Calculate the coded spectral efficiency.
  c. Determine the maximum achievable data rate (before protocol overhead).
  d. If protocol overhead reduces throughput by 20%, calculate the net user throughput.

3.2 A receiver experiences a worst-case SNR of 14 dB.
  a. Using typical uncoded AWGN BER performance and a target BER ≤ 10^{-5}, determine which of the following modulation schemes are feasible: QPSK, 16-QAM, 64-QAM.
  b. If adaptive modulation is employed, which modulation scheme should be selected at this SNR to maximize throughput while maintaining the BER target? Explain your reasoning.

3.3 A system using 64-QAM must meet a maximum allowable EVM of 5%. The estimated EVM contributions are: 3% PA nonlinearity, 2% phase noise, and 1.5% I/Q imbalance. Assume that the impairments are statistically independent.
  a. Calculate the total system EVM by combining the individual contributions.
  b. Determine whether the system meets the EVM requirement.
  c. If improvements are required, identify which impairment should be prioritized and justify your choice.

3.4 An OFDM transmitter exhibits a PAPR of 10 dB.
  a. If the PA saturates at 40-dBm peak output power, what is the maximum average output power that can be transmitted without clipping the peaks (i.e., with sufficient back-off)?
  b. Explain how this required back-off impacts PA efficiency compared to a constant-envelope modulation scheme such as FSK or GMSK.

3.5 A communication system is designed for 64-QAM operation, but laboratory measurements reveal excessive phase noise from the LO.
  a. Explain how phase noise affects high-order QAM constellations and system performance.
  b. Propose two system-level mitigation strategies that could improve performance without redesigning the LO hardware.

3.6 A satellite communication link is bandwidth-limited but has sufficient transmit power and link margin.
  a. Should the system favor higher-order modulation or lower-order modulation with stronger coding?
  b. Justify your answer in terms of spectral efficiency, hardware constraints, and robustness to impairments.

3.7 An RF front end reports the following capabilities to the communication layer: 40-MHz available bandwidth, 8% maximum allowable total EVM, and 18-dB available SNR. Assume an AWGN-limited system with ideal synchronization and perfect channel estimation. Use the approximation EVM $= 1/\sqrt{\text{SNR}_{\text{lin}}}$, which represents the noise-limited EVM floor due to thermal noise only.

a. Estimate the minimum achievable (noise-limited) EVM based on the available SNR.
b. Considering both the noise-limited EVM and the maximum allowable total EVM, determine which modulation schemes (e.g., QPSK, 16-QAM, 64-QAM) are realistically supported.
c. Describe how the communication layer should adapt its modulation and coding strategy based on this RF feedback, including possible trade-offs between throughput and robustness.

3.8 A digital communication system transmits symbols at a symbol rate of 20 Msym/s using QPSK and raised-cosine pulse shaping.

a. Calculate the occupied bandwidth for roll-off factors: (1) $\alpha = 0$, (2) $\alpha = 0.25$, and (3) $\alpha = 0.5$.
b. For each roll-off factor, compute the spectral efficiency in bits/s/Hz (assume no coding).
c. Explain how increasing the roll-off factor impacts: (1) spectral efficiency, and (2) robustness to timing errors and implementation complexity.

3.9 A communication system designer is evaluating two signal chain options for a wideband wireless link:

- Option A: Single-carrier transmission with raised-cosine pulse shaping.
- Option B: Multicarrier transmission using a large number of narrowband subchannels.

Both options operate over the same total occupied bandwidth and target the same net data rate.

For this problem, assume a frequency-selective (multipath) channel to motivate the comparison. No detailed channel modeling or mathematical derivations are required.

a. Compare the two approaches in terms of sensitivity to frequency-selective channels and equalization complexity.
b. Discuss how PAPR differs between the two approaches and how this impacts PA design and efficiency.
c. Explain why multicarrier approaches are often favored in broadband communication systems, despite their increased implementation complexity.

3.10 A wireless system designer must choose between QPSK, 16-QAM, and 64-QAM for a new RF product. The following constraints apply:

- Available SNR: 16 dB.
- Maximum allowable EVM: 10%.
- The PA has limited linearity and must operate efficiently.
- The system must support moderate mobility.

Assume an AWGN channel for link performance evaluation. Mobility is treated as a design margin consideration that impacts tracking, phase noise tolerance, and robustness, but no explicit Doppler calculations are required.

a. Identify which modulation schemes are technically feasible under these constraints.
b. Recommend the most appropriate modulation scheme for this product.
c. Justify your recommendation by considering: (1) SNR and EVM requirements, (2) PA linearity and efficiency, (3) robustness to channel impairments and mobility-related effects, and (4) signal-chain design choices with the practical realities of RF hardware implementation.

References

[1] Goldsmith, A., *Wireless Communications*, Cambridge, UK: Cambridge University Press, 2005.

[2] Carlson, A. B., and P. B. Crilly, *Communication Systems*, Fifth Edition, New York: McGraw-Hill, 2001.

[3] Proakis, J. G., *Digital Communications*, Fifth Edition, New York: McGraw-Hill, 2008.

[4] Shanmugam, K. S., *Digital and Analog Communication Systems*. New York: Wiley, 2012.

[5] Lyons, R. G., *Understanding Digital Signal Processing*, Third, Upper Saddle River, NJ: Prentice Hall, 2010.

CHAPTER 4

RF Systems: Core Concepts

> Thoroughly conscious ignorance is the prelude to every real advance in science.
>
> —*James Clerk Maxwell*

This chapter introduces the main signal degradation factors that affect RF system performance, including noise, nonlinearity, distortion, and propagation effects. These mechanisms shape the behavior and limitations of an RF link and are quantified using key performance metrics such as noise figure, compression point, and intermodulation product levels. Building on this foundation, the chapter introduces critical receiver detection metrics, including signal-to-noise-and-distortion ratio (SINAD), signal-to-interference-plus-noise ratio (SINR), carrier-to-noise temperature ratio (C/T), and gain-to-noise temperature ratio (G/T), which define receiver sensitivity and dynamic range. The chapter concludes with the principles of link-budget analysis and a structured, step-by-step approach for evaluating end-to-end link performance.

4.1 Introduction

In Chapter 3, we examined the requirements of the signal chain layer and explored how these choices shape the underlying RF hardware requirements and influence system-level trade-offs. This chapter, together with Chapters 5 and 6, forms the foundation for the RF portion of the overall product.

In this chapter, we introduce the core concepts of linearity, noise, link budget, and the key parameters used to quantify them. Understanding these metrics is essential for developing a solid product blueprint, as they directly drive performance trade-offs and inform critical design decisions.

The RF portion of the blueprint can be viewed from two complementary perspectives. The first is a black-box view, in which we consider only inputs and outputs without regard to internal circuitry. This system-level perspective is powerful because it allows us to analyze an RF chain strictly through its input-output behavior, using analytical tools that do not require deep knowledge of electromagnetics or circuit theory [1]. Chapters 4 and 5 focus on this system-level perspective.

The second perspective is a component-level view, where we open the black box to examine the internal RF building blocks, amplifiers, filters, switches, and more and understand how their interactions shape system performance. Chapter 6 builds on this perspective, providing a deeper look into RF components and how they work together to realize the system-level functionality defined in the blueprint.

4.2 RF Systems: Core Functions

As discussed in Chapter 1, an RF system typically performs one or more of the following core functions [1]:

- *Signal generation:* Creation of an RF signal with the desired frequency, power level, and waveform.
- *Signal transmission:* Delivery of the generated signal through a medium (e.g., free space, cable, waveguide).
- *Signal processing:* Modification or conditioning of the signal (e.g., amplification, filtering, mixing).
- *Signal detection:* Extraction and interpretation of the transmitted information from the received signal (e.g., demodulation, decoding).

In some products, such as spectrum analyzers, the emphasis is primarily on signal detection, while in others, such as microwave plasma generators, the focus is on signal generation. A complete RF link incorporates all four functions of the signal chain. Such links enable a broad range of applications, from communication between mobile phones to quantum information transfer in quantum processors. Regardless of the specific use case, the fundamental objective of an RF link remains the same: ensure that the signal arrives at the receiver with enough integrity to be reliably detected.

The first and most important step in understanding an RF system and defining its requirements is identifying the nonideal effects that impair signal integrity, including noise, interference, attenuation, and distortion. In this chapter, we examine these effects collectively under the theme of signal degradation factors.

4.3 Real-World RF Link: Signal Degradation Factors

Signal degradation in an RF link arises from several mechanisms that can affect signal integrity and, ultimately, detection performance. The most critical factors that we discuss in the following section include:

- *Attenuation:* Reduction of signal strength as it passes through components, and the transmission media such as free space, cables, and waveguides;
- *Propagation effects (e.g., Doppler shift, Faraday rotation):* Channel-induced impairments that alter a signal's frequency, phase, or polarization;
- *Noise:* Random fluctuations from various sources that obscure the desired signal;
- *Interference:* Undesired signals from external transmitters or systems that hinder detection of the intended signal;
- *Distortion:* Deformation of the signal waveform caused by nonlinearities, bandwidth limitations, or other system imperfections.

These effects can be controlled by carefully selecting signal chain layer parameters (e.g., modulation, coding, frequency, bandwidth) and by thoughtfully designing the system architecture with suitable components. As we will see later, these impairments can manifest in multiple forms and often exhibit complex interdependencies. The art of RF engineering lies in controlling and minimizing their impact as effectively as possible.

4.4 Signal Degradation Factors: Attenuation

Depending on the specific link, different loss sources can attenuate a signal in an RF system. As we will see later, the receiver has a parameter called the minimum detectable signal (MDS), which defines the lowest signal level that can be reliably detected. Excessive attenuation can push the received signal below this threshold, leading to missed or incorrect detection. In other applications, such as a microwave oven, attenuation reduces the efficiency of energy transfer and lowers the available power for heating.

Depending on the specific link, different loss mechanisms can attenuate a signal in an RF system. To describe this effect, two related quantities are used. The first parameter is the attenuation (L_{dB}), which expresses the total loss across a component or over a given distance, in decibels [2]. It is defined as the ratio of the power before attenuation (P_{in}) to the power after attenuation (P_{out}),

$$L_{dB} = 10\log_{10}\left(\frac{P_{in}}{P_{out}}\right) \tag{4.1}$$

So, an attenuation of 20 dB means that the signal strength reduces by a factor of 100.

The second parameter is the attenuation constant (α), which expresses the rate of signal reduction per unit length, typically in dB/m or Np/m. The conversion between the two units is as follows:

$$\alpha_{dB/m} = 8.68\alpha_{Np/m} \tag{4.2}$$

The relationship between the attenuation and the attenuation constant is:

$$L_{dB} = \alpha \cdot d \tag{4.3}$$

where d is the propagation distance.

Table 4.1 provides a list of the most common sources of attenuation in an RF link.

4.4.1 Conductive Loss

At high frequencies, the skin effect forces current to flow within a thin layer near the surface of a conductor. This layer has a thickness δ, called the skin depth. Because the effective cross-section for current flow is reduced, the conductor's resistance increases, leading to higher loss.

Table 4.1 Common Sources of Attenuation

Type of Loss	*Description*	*Typical Example*
Conductive loss	Ohmic resistance and skin-effect losses in conductors	Coaxial cable at high frequency
Dielectric loss	Energy dissipation due to dipole relaxation and dielectric leakage	PCB substrates
Radiation loss	Unintended radiation from transmission lines and RF structures	Microstrip or CPW lines
Mismatch loss	Power lost due to reflections from impedance discontinuities	Connector-to-cable transitions
Polarization loss	Loss when TX and RX polarizations are misaligned	Misaligned antennas
Propagation loss	Free-space spreading, atmospheric absorption, and scattering	Wireless link over distance

The skin effect describes the exponential decay in current density, $J = J_s e^{-d/\delta}$, where d is the distance from the surface and J_s is the current density at the surface. At $d = \delta$, the current density drops to $1/e$ of its surface value, meaning about 63% of the current flows within one skin depth. The skin depth is given by:

$$\delta = \frac{1}{\sqrt{\pi\mu\sigma f}} \tag{4.4}$$

Here, μ denotes the permeability, σ denotes the conductivity, and f denotes the frequency. As frequency increases, the skin depth decreases, which increases the effective resistance. At high frequencies, in a good conductor, currents are confined near the surface, and electromagnetic fields decay rapidly inside the material. This confinement enables metallic shields to effectively attenuate electromagnetic fields.

Equation (4.4) also shows that the skin effect can be significant even at low frequencies, such as 50 Hz, when magnetic materials with high permeability, such as the ferromagnetic cores of transformers and inductors, are involved.

Despite the reduced skin depth at high frequencies, a good conductor still exhibits lower AC resistance than a poor conductor, because skin depth increases with the square root of conductivity.

To calculate the AC resistance, the same formula employed for DC resistance can be utilized, but with the area of the conductor substituted by the effective area available for the AC current flow. In the case of a coaxial cable, the AC resistance per unit length can be expressed as follows [3]:

$$R_{AC} = \frac{1}{\sigma}\left(\frac{1}{2\pi\delta a} + \frac{1}{2\pi\delta b}\right) \tag{4.5}$$

where a is the outer radius of the inner conductor and b is the outer radius of the outer conductor. For a coaxial cable, the attenuation constant due to the conductive losses is given by

$$\alpha_c = \frac{R_{AC}}{2Z_0} = \frac{\left[\frac{1}{a}+\frac{1}{b}\right]\sqrt{\mu/32\sigma}}{\pi Z_0}\sqrt{\omega} \tag{4.6}$$

where Z_0 is the characteristic impedance. In practice, the measured attenuation constant tends to be higher due to surface roughness on metallic surfaces, which leads to increased loss. To account for this effect, a quasi-empirical formula can be used to correct for surface roughness in any transmission line [2].

$$\alpha_c' = \alpha_c\left[1+\frac{2}{\pi}\tan^{-1}1.4\left(\frac{\Delta}{\delta}\right)^2\right] \tag{4.7}$$

where α_c is the attenuation due to perfectly smooth conductors, α_c' is the attenuation corrected for surface roughness, Δ is the RMS surface roughness, and δ is the skin depth of the conductors.

4.4.2 Dielectric Loss

Dielectric loss is associated with leakage current and molecular polarization mechanisms within a dielectric material. It represents the energy dissipated as heat when there is a phase lag between the applied electric field and the polarization response of the dielectric. This phase lag arises from molecular friction and leakage currents, which dissipate electrical energy as heat.

To account for these losses, the permittivity of a dielectric material is expressed as a complex quantity, where the imaginary part represents dielectric losses. These losses are commonly quantified using the loss tangent, denoted by $\tan\delta$. The complex permittivity ε is given by

$$\varepsilon = \varepsilon' - j\varepsilon'' = \varepsilon'(1 - j\tan\delta) \tag{4.8}$$

where ε' is the real part of the permittivity and ε'' is the imaginary part. The dielectric loss constant of a coaxial cable can be determined using (4.9), which shows that dielectric loss is a material property and cannot be reduced by modifying the cable geometry [3].

$$\alpha_d = \frac{\sqrt{\mu\varepsilon}\,[\tan\delta]}{2}\omega \tag{4.9}$$

So, according to (4.9), to minimize dielectric loss, the only option is to utilize a low-loss tangent or a low-dielectric constant material. Teflon, for instance, is frequently employed as a dielectric with a low-loss tangent of 0.0004.

In contrast to conductor loss, which increases proportionally to $\sqrt{\omega}$, dielectric loss becomes more significant at higher frequencies since it grows linearly with ω. For example, in a 0.141-inch-diameter semirigid cable with copper conductors and a Teflon dielectric, conductor loss exceeds dielectric loss up to the cutoff frequency.

However, the two types of losses become nearly equal at the cutoff frequency. As the frequency surpasses the cutoff frequency, the loss tangent becomes a more influential factor in the overall loss of the transmission line.

It is important to note that while dielectric losses are smaller than conductor losses, they still contribute to the overall loss of the system. Therefore, utilizing low-loss dielectrics and employing careful design techniques are crucial for minimizing the total loss.

4.4.3 Radiation Loss

Radiation losses can sometimes contribute significantly to the overall losses, particularly at high frequencies. When the size of a transmission line is comparable to the wavelength, energy can escape from the line in the form of electromagnetic radiation. This phenomenon is well-known, as a good radiator (antenna) typically has a size comparable to the radiation wavelength. Shielded structures such as strip lines, coaxial cables, or hollow waveguides exhibit much lower radiation losses than open, unshielded structures such as microstrip or coplanar waveguide (CPW) lines.

Accurately quantifying radiation losses can be challenging, as they depend on both geometry and field distribution. Numerical simulations are often required to isolate the radiating portions of the structure and estimate the associated power loss. In general, bends or discontinuities in transmission lines radiate more than straight sections, and thicker substrates tend to increase radiation loss due to stronger fringing fields.

Radiation losses can be reduced, or in some cases nearly eliminated, by applying several design strategies:

- Use shielding to confine electromagnetic fields and prevent unintended radiation.
- Select thinner substrates to reduce fringing fields and suppress radiation from microstrip lines.
- Choose less radiative transmission-line structures, such as stripline instead of microstrip, when isolation is critical.
- Minimize discontinuities in transmission lines, avoid abrupt bends, impedance steps, gaps in the return path, or transitions that disturb field confinement.

4.4.4 Mismatch Loss

Impedance mismatch leads to the reflection of incident power, resulting in only a portion of the signal being transmitted while the remainder is reflected. The mismatch loss L_{mismatch}, expressed in decibels, can be calculated using the following equation:

$$L_{\text{mismatch}} = 10\log\left(\frac{P_i}{P_t}\right) = -10\log\left(1-|\Gamma|^2\right) \tag{4.10}$$

P_i denotes the incident power, P_t denotes the transmitted power, and Γ denotes the reflection coefficient. The mismatch loss is the factor that reduces the maximum power transferred to the load. Depending on how multiple reflections combine, the overall system loss may be smaller or larger than the sum of the mismatch losses from each component.

To mitigate the effect of mismatch loss, it is necessary to ensure good impedance matching over the frequency band of operation.

4.4.5 Polarization Loss

Polarization is a fundamental property of electromagnetic waves that describes the orientation of the electric field vector. It can be linear (vertical, horizontal, or slant), elliptical, or circular, with either right-hand circular polarization (RHCP) or left-hand circular polarization (LHCP) depending on the rotation direction of the electric field. For maximum power transfer, the transmitting and receiving antennas must share the same polarization. A mismatch, such as between a vertically polarized transmitting antenna and a horizontally polarized receiving antenna, results in zero power transfer in the ideal case. This is known as polarization loss.

The polarization loss factor (PLF) quantifies the reduction in received power due to a mismatch between the polarization of the transmitted wave and the receiving antenna. It is defined as [4]

$$\text{PLF} = |\hat{\rho}_w \cdot \hat{\rho}_a|^2 \tag{4.11}$$

where $\hat{\rho}_w$ is the unit vector of the wave's polarization and $\hat{\rho}_a$ is the unit vector of the antenna's polarization.

- Aligned polarizations: PLF = 1 → no polarization loss.
- Rotated by an angle ψ_p: PLF = $\cos^2\psi_p$.
- Orthogonal polarizations: PLF = 0 → complete polarization loss.

Figure 4.1 shows the PLF for three different cases discussed above.

In practice, the PLF enters directly into the link budget as a multiplicative term, alongside free-space path loss and antenna gains. Proper polarization alignment is therefore critical for link performance in communication systems, EW systems,

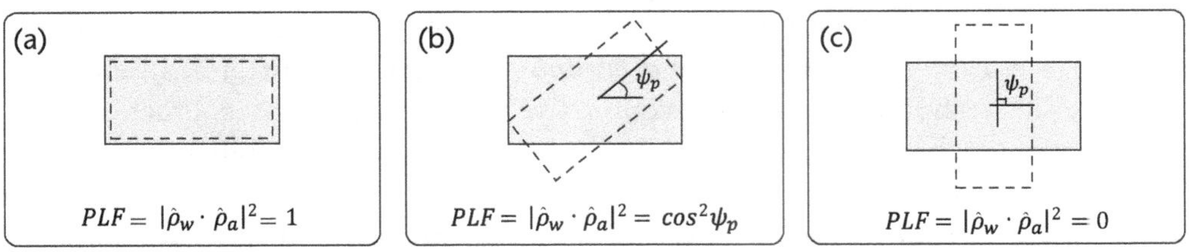

Figure 4.1 PLF for different transmit-receive antenna orientations. The solid rectangle represents the aperture of the receive antenna, and the dotted rectangle represents the aperture of the transmit antenna. (a) When the polarizations are perfectly aligned, PLF = 1 (no loss). (b) When the transmit and receive antennas are rotated relative to each other, the PLF decreases as $\cos^2\psi_p$, where ψ_p is the polarization mismatch angle. (c) In the orthogonal case (90° mismatch), PLF = 0, indicating that no power is transferred between the antennas.

and radar, where maintaining the correct polarization can significantly improve received signal strength.

In satellite systems, circular polarization is commonly used to reduce orientation-dependent mismatch. Circular polarization also helps mitigate multipath effects. For example, if a transmitter radiates an LHCP signal, a reflected wave reverses its sense of rotation and becomes RHCP. Because an LHCP receiving antenna strongly attenuates RHCP components, the reflected signal is largely rejected, reducing multipath interference.

4.4.6 Propagation Loss

As an RF signal travels from the transmitter to the receiver, it encounters several forms of propagation loss, which attenuate the signal and must be carefully accounted for in the link budget. These losses depend on factors such as distance, frequency, propagation environment, and weather conditions. The phenomena that directly contribute to signal attenuation include:

- Free-space path loss;
- Tropospheric attenuation (rain, fog, cloud absorption, and scattering);
- Ionospheric absorption;
- Atmospheric molecular absorption (primarily due to O_2 and H_2O);
- Multipath fading.

These mechanisms and their impacts on link performance will be discussed in more detail in Section 4.5.

4.4.7 Managing Attenuation

Now that we are familiar with the main factors contributing to attenuation, we can consider strategies to mitigate them. Ultimately, link budget analysis provides the framework for determining how much attenuation a system can tolerate and what design trade-offs are acceptable.

- *Amplifiers and repeaters:* These devices compensate for signal loss but inevitably add noise and may introduce nonlinear effects such as harmonics and intermodulation products, topics that we will discuss later in this chapter.
- *Low-loss materials and components:* Select substrates, interconnects, and components with low conductive and dielectric losses. Practical constraints, such as cost, size, weight, complexity, and availability, must be balanced against performance.
- *Shielding:* Proper shielding and enclosure design help to suppress unintended radiation loss.
- *Impedance matching:* Ensuring proper matching between various RF components minimizes reflections and mismatch loss, improving power transfer efficiency.
- *Propagation-loss mitigation:* Increase transmit power or antenna gain, ensure proper polarization alignment, and adjust the operating frequency when

feasible (attenuation is typically frequency dependent). Optimize the placement and orientation of the transmit and receive antennas, and account for the propagation environment, removing or repositioning obstacles when possible.

4.5 Signal Degradation Factors: Propagation Effects

When RF signals propagate over long distances, particularly in mobile, satellite, and high-frequency links, their quality is strongly influenced by the characteristics of the propagation medium. These effects can introduce attenuation, frequency shifts, distortion, polarization mismatch, and various forms of fading, all of which must be accounted for in system design to ensure reliable performance. The key degradation mechanisms are discussed in the following sections.

4.5.1 Free-Space Path Loss (FSPL)

FSPL is the most fundamental form of propagation loss, present even in an ideal, lossless medium. It describes the reduction in signal strength caused purely by the geometric spreading of the wavefront as it propagates away from the source. Unlike absorption, scattering, or any other lossy effects, FSPL occurs even in a perfect vacuum, where no matter exists to attenuate the wave.

The term "loss" refers not to energy dissipation but to the dilution of power density with distance. As the wave expands spherically, its energy is distributed over a growing surface area. This is shown in Figure 4.2. Consequently, the received power decreases proportionally to $1/r^2$, where r is the distance from the source. In other words, FSPL reflects the inevitable spreading of electromagnetic energy with distance, not any physical loss within the medium.

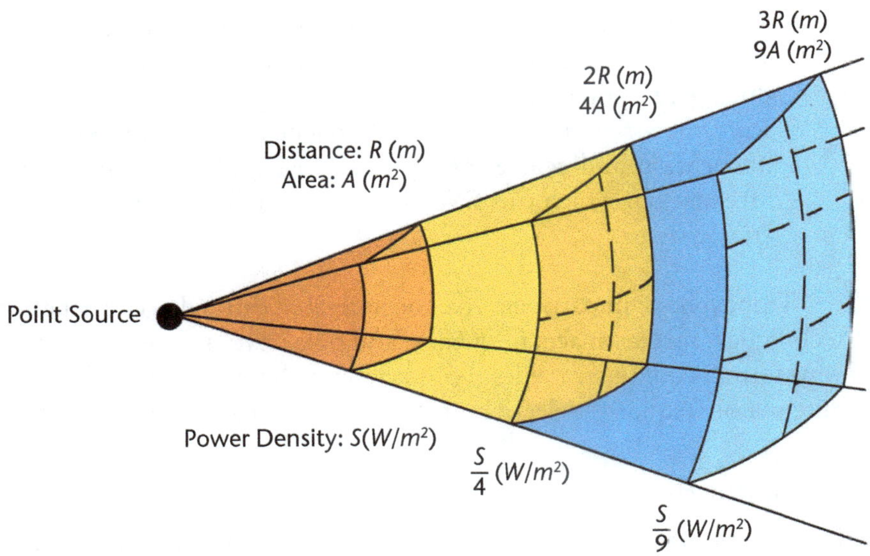

Figure 4.2 Illustration of spherical spreading from a point source, showing how the illuminated area increases with distance, and the power density decreases following the inverse-square law.

The FSPL in decibels is given by [4]

$$\mathrm{FSPL}(\mathrm{dB}) = 20\log_{10}(r) + 20\log_{10}(f) + 20\log_{10}\left(\frac{4\pi}{c}\right) \quad (4.12)$$

where r is the distance in meters, f is the frequency in hertz, and c is the speed of light (~3×10^8 m/s)). This formula can be simplified using the distance in kilometers and frequency in gigahertz:

$$\mathrm{FSPL}(\mathrm{dB}) \approx 92.45 + 20\log_{10}(d_{\mathrm{km}}) + 20\log_{10}(f_{\mathrm{GHz}}) \quad (4.13)$$

As can be seen, FSPL increases with both frequency and distance, which makes high-frequency systems more susceptible to signal attenuation over long ranges.

4.5.2 Atmospheric Propagation Effects

As RF signals travel through the atmosphere, their behavior is shaped by the physical properties of the troposphere, ionosphere, and the gases that compose them. The most relevant atmospheric propagation mechanisms are outlined next.

Tropospheric Effects

The troposphere is the lowest atmospheric layer, extending from the Earth's surface up to approximately 10–12 km. It contains most of the atmosphere's water vapor and is the region where weather phenomena occur. RF signals traveling through this layer are affected by several mechanisms:

- *Ducting:* Temperature, pressure, and humidity gradients create refractive index layers that can trap and guide microwave and millimeter-wave signals, allowing propagation well beyond the radio horizon.
- *Scattering:* Small-scale variations in the refractive index cause the signal to spread and lose coherence, reducing the received signal strength and introducing additional fading.
- *Rain/fog attenuation:* At higher frequencies (especially Ku-band and above), absorption and scattering by rain, fog, and clouds introduce significant attenuation, which can severely impact satellite links and long-range terrestrial systems.

Tropospheric impairments can be mitigated through a combination of diversity techniques, such as space, frequency, and polarization diversity, as well as adaptive waveforms, coding strategies, uplink power control, increased fade margins, and antenna or frequency selection optimized for adverse weather conditions.

Ionospheric Effects

Ionospheric effects are caused by charged particles, including ions and free electrons, in the upper atmosphere. In essence, ionospheric effects are governed by electron density and magnetic field interactions in the ionized upper atmosphere, whereas tropospheric effects are driven by weather, moisture, and temperature gradients in the neutral lower atmosphere.

The ionosphere extends roughly from 60 km to 1,000 km above the Earth's surface. It is ionized by solar ultraviolet and X-ray radiation, forming distinct layers (D, E, F1, F2), each influencing radio-wave propagation in different ways:

- *Refraction:* High-frequency (3–30 MHz) signals can be bent back toward Earth, enabling long-distance "skywave" communication.
- *Scintillation:* Rapid, random fluctuations in amplitude and phase caused by irregular or turbulent electron density structures. Scintillation becomes more severe at L-band and above, affecting satellite links and GNSS accuracy.
- *Faraday rotation:* Rotation of the signal's polarization plane due to the interaction between Earth's magnetic field and the ionized medium, which can lead to polarization mismatch or cross-polar interference.

It is worth discussing the Faraday effect in more detail because of its significant impact on polarization-sensitive links such as satellite communication and polarimetric radar. The Faraday effect is a magneto-optical phenomenon in which the plane of polarization of an electromagnetic wave rotates as it propagates through a magnetized medium. This occurs because the medium exhibits different phase velocities for the right-hand and left-hand circularly polarized components of the wave, causing a net rotation of the polarization plane.

Faraday rotation is observed in magnetized plasmas and magneto-optical materials. In the context of RF propagation, it most commonly occurs in the ionosphere, where Earth's magnetic field and the presence of free electrons in the ionosphere rotate the polarization of linearly polarized radio waves. This rotation can lead to polarization mismatch loss or cross-polar interference, effectively reducing received signal power.

The rotation angle is directly proportional to the electron density along the path, the strength of the magnetic field component in the direction of propagation, and the square of the wavelength. Thus, when a linearly polarized wave traverses the ionosphere in the presence of Earth's magnetic field, the resulting rotation angle θ is given by:

$$\theta = \left(\frac{2.36 \times 10^4}{f^2}\right) \int N_e B_{\parallel} \, dl \tag{4.14}$$

where

- θ: rotation angle of the polarization plane (radians).
- f: signal frequency (Hz).
- N_e: electron density along the propagation path (electrons/m^3).
- $B_{\parallel}$: component of Earth's magnetic field parallel to the direction of wave propagation (tesla).
- dl: differential path length through the ionosphere (m).

The impacts of Faraday rotation are as follows:

- Reduced received power when the antenna is polarization-sensitive (e.g., linearly polarized systems), due to polarization mismatch.
- Stronger at lower frequencies, particularly high-frequency, very high-frequency (VHF), and L-band, because the rotation angle increases with wavelength squared.
- In satellite links, especially L-band (1 to 2 GHz) services such as GPS/GNSS, Faraday rotation can introduce fading, cross-polarization interference, and reduced link margin.

Here are some mitigation methods:

- Use circularly polarized antennas, which are inherently immune to simple rotation of the polarization plane.
- Employ adaptive polarization tracking or polarization-agile receivers in advanced communication systems.

Atmospheric Molecular Absorption

As electromagnetic waves propagate through the lower and middle atmosphere, approximately from the Earth's surface up to about 100 km, part of their energy is absorbed by atmospheric gases such as oxygen and water vapor. This phenomenon, known as atmospheric molecular absorption, is fundamentally different from FSPL. Rather than resulting from geometric spreading, it arises from molecular resonances at specific frequencies.

The effect is strongly frequency-dependent. At millimeter-wave frequencies, oxygen and water vapor dominate the absorption spectrum, producing pronounced attenuation peaks near 22 GHz due to water vapor and around 60 GHz due to oxygen, along with additional water-vapor absorption bands above 100 GHz. These resonances impose significant range limitations at certain frequencies while creating usable low-loss windows at others.

At sea level, attenuation can reach 10 to 15 dB/km near the 60-GHz oxygen resonance, while remaining below approximately 0.05 dB/km in low-loss windows such as portions of the 20 to 40-GHz range. At higher altitudes, around 9 km, reduced air density and humidity substantially decrease absorption, often by a factor of 2 or more.

These characteristics strongly influence the selection of operating bands in millimeter-wave communication, radar, and sensing systems. Atmospheric absorption is particularly important in satellite links, long-range radar, and 5G and 6G systems, where accurate link-budget calculations must account for gaseous attenuation to ensure reliable performance.

4.5.3 Terrestrial Propagation Effects

Terrestrial propagation effects arise when radio waves interact with objects and surfaces near the Earth's surface. These mechanisms dominate in urban, suburban, rural, indoor, and maritime environments, where buildings, terrain, foliage, vehicles, and other structures significantly influence signal behavior. The most widely encountered terrestrial propagation mechanisms are summarized in Table 4.2.

Table 4.2 Terrestrial Propagation Losses

Type of Loss	*Description*
Absorption loss	Occurs when RF energy is absorbed by the materials that it penetrates (e.g., walls, foliage, soil), converting electromagnetic energy into heat; strongly dependent on material properties and frequency
Reflection loss	Occurs when a signal encounters surfaces with mismatched impedance (e.g., buildings, ground, water), redirecting part of the energy away from the receiver
Diffraction loss	Happens when a signal bends around obstacles such as buildings or terrain edges (similar to the single-slit experiment); significant in non-line-of-sight (NLOS) paths, especially at lower frequencies
Scattering loss	Caused by small or irregular objects, such as foliage, vehicles, or rough surfaces, that scatter wave energy in multiple directions, reducing power along the direct path
Multipath fading	Results from constructive and destructive interference between multiple propagation paths created by reflection, diffraction, and scattering; leads to rapid fluctuations in received signal strength

Understanding these losses is essential for designing robust RF links, particularly in challenging environments such as urban areas, mountainous terrain, and maritime systems. Many of these mechanisms interact with other propagation effects, producing complex behavior that often cannot be captured analytically.

In practice, the most reliable way to quantify these losses is through measurements and channel characterization in the intended deployment environment. Simulations and propagation models (e.g., ray tracing, ITU-R models) are often used to complement measurements and estimate performance in complex scenarios.

Multipath Fading

In real-world environments, electromagnetic waves rarely travel along a single direct path between the transmitter and the receiver. Instead, reflections from buildings, the ground, and other objects, as well as scattering and diffraction, create multiple copies of the signal that arrive at the receiver with different delays, phases, and amplitudes. This phenomenon is called multipath propagation [5].

When these delayed signal components combine at the receiver, they can interfere constructively, increasing the received signal strength, or destructively, reducing it. As the transmitter or receiver moves, or as the surrounding environment changes over time, the relative phases and amplitudes of these components also change. The resulting rapid fluctuations in the received signal amplitude and phase are known as multipath fading. This effect is illustrated in Figure 4.3.

Multipath fading is a key challenge in wireless communication systems, as it can cause deep signal fades, intersymbol interference (ISI), and degraded link quality. They can be classified based on time variation or by frequency selectivity.

Classification by time variation is as follows:

- *Slow fading (shadowing):* Caused by large-scale obstructions such as buildings, hills, or terrain that gradually block or attenuate the signal. The channel varies slowly and often remains nearly constant over the duration of many packets.

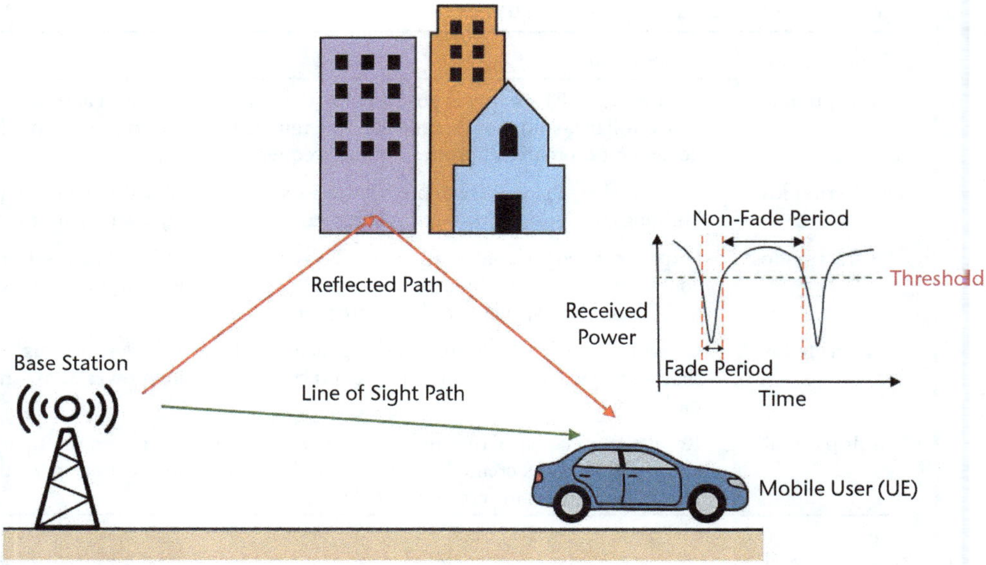

Figure 4.3 Illustration of multipath fading in a mobile communication scenario. The mobile user receives both a direct line-of-sight (LOS) signal and a reflected path from the surrounding buildings. The constructive and destructive interference between these paths produces variations in received power over time, shown on the right as alternating fade and nonfade periods relative to a detection threshold. Deep fades occur when the reflected path arrives out of phase with the LOS path.

- *Fast fading:* Rapid fluctuations in signal amplitude and phase resulting from multipath interference and Doppler shifts as the user moves. The channel changes significantly within the duration of a single packet. Techniques such as interleaving and forward error correction help to mitigate fast fading.

Classification by frequency selectivity is as follows:

- *Flat fading:* Flat fading occurs when the signal bandwidth is so narrow that all frequencies within the signal experience essentially the same channel conditions. Because the entire signal "fits" within a single fade notch, a deep face causes the whole signal to drop out for that moment.
- *Frequency-selective fading:* Frequency-selective fading occurs when different frequencies within the signal bandwidth experience different amounts of constructive and destructive interference. Wideband signals span a large range of frequencies (and therefore wavelengths). Since multipath effects depend on wavelength, some parts of the signal may add constructively while others cancel out. As a result, different portions of the signal's spectrum fade differently, hence the term *frequency-selective fading.*

Here are some common mitigation techniques:

- Diversity methods (spatial, frequency, or polarization);
- Equalization (time or frequency domain);
- Adaptive modulation and coding schemes.

Ultrawideband (UWB) and chirped spread-spectrum systems are inherently more resilient to multipath fading. Their wide bandwidth corresponds to very short time-domain pulses and therefore fine delay resolution, which allows individual multipath components to be distinguished rather than blurred together.

For example, with a bandwidth of $B = 500$ MHz, the effective pulse width in the time domain is on the order of $1/B \approx 2$ ns. This means that if a pulse and its reflected copy are separated by at least 2 ns, they can be resolved. This time separation corresponds to a path-length difference of $\Delta d = c\Delta t \approx (3 \times 10^8)(2 \times 10^{-9}) = 0.6$m. Therefore, multipath components with excess path lengths greater than approximately 0.6m can be separated and individually processed. Once resolved, these components can be combined constructively using techniques such as RAKE receivers, effectively turning multipath into an advantage. In contrast, narrowband signals have poor delay resolution, causing multipath components to overlap in time. This overlap leads to destructive interference and deep fading.

4.5.4 Doppler Effect

In satellite, unmanned aerial vehicle (UAV) systems, and other high-mobility communication links, relative motion between the transmitter and receiver causes a shift in the carrier frequency. This Doppler shift is given by:

$$f_d = \left(\frac{v}{c}\right) \times f_c \tag{4.15}$$

where f_d denotes the Doppler shift (Hz), v is the relative radial velocity between the transmitter and receiver (m/s), c is the speed of light ($\approx 3 \times 10^8$ m/s), and f_c is the carrier frequency (Hz).

In low-Earth-orbit (LEO) satellites, speed can be about 7,500 m/s, leading to Doppler shifts of tens of kilohertz at L-band and hundreds of kilohertz at Ka-band. Uncompensated Doppler can cause severe frequency offsets that degrade demodulation, particularly for narrowband or high-order modulation schemes.

Here are some of the Doppler effect impacts on a communication link:

- *Carrier frequency offset:* Doppler-induced frequency shifts cause a mismatch between the transmitter and receiver carrier frequencies, making coherent demodulation more difficult.
- *Intercarrier interference (ICI):* In multicarrier schemes such as orthogonal frequency division multiplexing (OFDM), Doppler shifts destroy the orthogonality between subcarriers, causing energy from one subcarrier to leak into others.
- *Time-varying channel impairments:* When the transmitter or receiver is moving quickly (e.g., on satellites, aircraft, or high-speed trains), the Doppler effect changes rapidly over time. This makes the channel fluctuate so fast that the receiver cannot track it accurately. As a result, the signal can fade in and out more quickly, which increases the likelihood of errors (higher BER).

Mitigation techniques include:

- *Tracking loops:* Frequency-locked loops (FLL) or phase-locked loops (PLL) continuously adjust the receiver's LO to follow Doppler variations.

- *Predictive frequency correction:* Uses motion or trajectory estimation to compensate for expected Doppler shifts proactively.
- *Pilot tone-based compensation:* Employs known pilot signals or reference tones for real-time frequency offset estimation and correction.

4.6 Signal Degradation Factors: Interference

In practice, no RF system can be perfectly isolated from its electromagnetic environment and must therefore contend with unwanted signals. Interference is one of the most common causes of signal degradation and occurs when unwanted RF signals couple into the system and disrupt its intended operation. Depending on their strength, spectral proximity, and spectral content, interferers can impair signal detection or reduce dynamic range in several ways. A sufficiently strong interferer may even drive the receiver into compression, preventing the desired signal from being detected altogether.

Interference is also deliberately generated in EW systems, where it is used as an electronic weapon, commonly referred to as jamming. In these applications, specialized techniques are used to inject powerful or cleverly modulated RF waveforms into the victim receiver to deny, degrade, or deceive its operation. Likewise, EW systems incorporate sophisticated countermeasures and mitigation strategies to maintain performance in the presence of intentional interference.

In receiver design, interferers are commonly classified in two ways:

1. Spectral classification:
   - In-band interferers lie within the passband of the input filter and therefore experience minimal attenuation. They enter the receiver alongside the desired signal, raising the noise floor, increasing distortion, and reducing the available dynamic range. This is illustrated in Figure 4.4(a).
   - Out-of-band interferers fall outside the passband and are largely suppressed by the input filter, as shown in Figure 4.4(a). However, if sufficiently strong, they can still desensitize the receiver by driving the RF front end into compression. Proper filtering and carefully planned gain distribution are essential to prevent these effects.

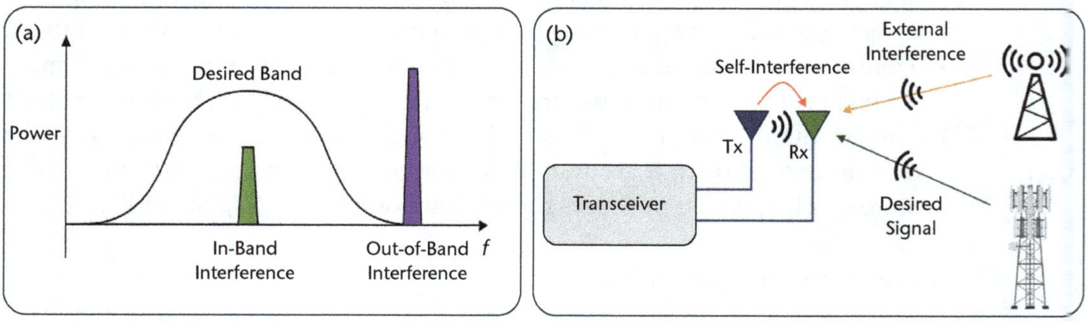

Figure 4.4 (a) In-band and out-of-band interference. (b) Self-interference and external interference.

2. Source classification:
   - External interference originates from sources outside the system (e.g., nearby transmitters, other communication systems, unintentional radiators). This is illustrated in Figure 4.4(b).
   - Self-interference results from within the system itself (e.g., TX-RX leakage, LO leakage, PA harmonics, switching noise, digital-to-RF coupling). This is illustrated in Figure 4.4(b).
   - As we will see later in this chapter, large interferers can also produce secondary nonlinear effects, such as cross-modulation, intermodulation, and reciprocal mixing, all of which further compromise reliable signal detection.

4.6.1 Practical Interference Mitigation

Interference may originate from cochannel users, adjacent channels, multipath propagation, or intentional jamming, each of which can degrade signal quality and reduce link reliability. To mitigate these effects, RF systems employ a range of cancellation and suppression methods, which can be broadly categorized into hardware-based techniques and algorithmic approaches.

For hardware-level techniques aimed at limiting interference, several common measures are employed:

- *Filtering:* Bandpass, highpass, and lowpass filters attenuate unwanted signals outside the band of interest and help to prevent strong out-of-band interferers from driving the receiver front end into nonlinearity. Notch filters can be used to attenuate strong in-band interferers. Spatial filtering using frequency-selective surfaces (FSS) can also be employed in certain cases, such as antenna radomes, reflector systems, or array apertures, to suppress interference arriving from specific directions or frequency bands before it reaches the receiver.
- *Shielding:* Metallic enclosures, EMI gaskets, shielding cans, and careful mechanical/PCB design reduce coupling of interference into sensitive circuits. PCB-level shielding features such as guard traces (grounded traces placed between sensitive lines to reduce capacitive coupling) and ground-stitching vias further help to suppress crosstalk and improve immunity.
- *RF path isolation:* Isolators, circulators, attenuators, RF absorbers, and adequate physical separation of transmit and receive paths minimize reverse coupling, leakage, and cross-talk between RF chains.
- *Grounding and decoupling:* A robust ground strategy, well-defined return paths, and minimizing ground loops suppress unwanted coupling between subsystems and reduce interference. Proper use of bypass and decoupling capacitors provides a low-impedance path for high-frequency noise and prevents power-supply-borne interference from coupling into sensitive circuits.
- *PCB layout practices:* Partitioning of RF, digital, and power domains, maintaining controlled-impedance routing, and proper trace spacing and layer stack-up prevent crosstalk and minimize unintended coupling across the board.

These hardware techniques form the first line of defense by ensuring that, if higher-level interference cancellation algorithms are required, they operate on cleaner signals and place a reduced burden on digital processing. The algorithmic approaches commonly employed in RF systems include:

- *Beamforming:* Antenna arrays shape their radiation pattern to place gain in the direction of the desired signal while steering deep nulls toward the interferers, improving the signal-to-interference ratio.
- *Temporal filtering:* Adaptive algorithms such as least mean squares (LMS) and recursive least squares (RLS) continuously update filter coefficients to separate the desired signal from interference in the time domain.
- *Frequency-domain cancellation:* Notch filters, adaptive equalizers, or frequency-selective techniques suppress narrowband or wideband interferers by operating directly in the frequency domain.
- *Successive interference cancellation (SIC):* Strong interfering signals are detected, decoded, and digitally subtracted before processing the weaker desired signal. This method is widely used in multi-user communication systems such as CDMA and modern wireless networks.
- *Blind source separation:* Techniques such as Independent Component Analysis (ICA) separate multiple overlapping signals even when the interferers are unknown. They exploit statistical differences between the mixed signals to recover the desired waveform.

In practice, robust interference mitigation often combines hardware-level defenses with advanced cancellation algorithms, balancing complexity, latency, and hardware constraints to maintain reliable system performance.

4.7 Signal Degradation Factors: Noise

In this section, we examine how noise affects RF systems and ultimately limits signal detection, dynamic range, and link reliability. To characterize noise and compare system architectures, engineers rely on parameters such as noise figure, noise temperature, signal-to-noise ratio (SNR), and noise power ratio (NPR) for multicarrier systems. These metrics provide practical tools for quantifying how much noise a system adds and how well it can recover weak signals in the presence of unavoidable thermal, environmental, and circuit-generated noise.

Noise affects both the amplitude and phase of an RF signal. As discussed in Chapter 3, any sinusoid can be represented by its in-phase (I) and quadrature (Q) components. When signals combine, their I and Q components add vectorially in the complex plane. Consequently, when noise is superimposed on a signal, it perturbs the I and Q components, resulting in simultaneous amplitude and phase fluctuations. Figure 4.5(a) illustrates a tone $X(f_0)$ at frequency f_0 and an additive noise component $n(f_0 + \Delta f)$ at a nearby offset frequency. Through the standard small-signal approximation [6], this additive noise can be decomposed into equivalent amplitude and phase noise components referenced to f_0. These amplitude and phase perturbations effectively modulate the original signal, as illustrated in Figure 4.5(b).

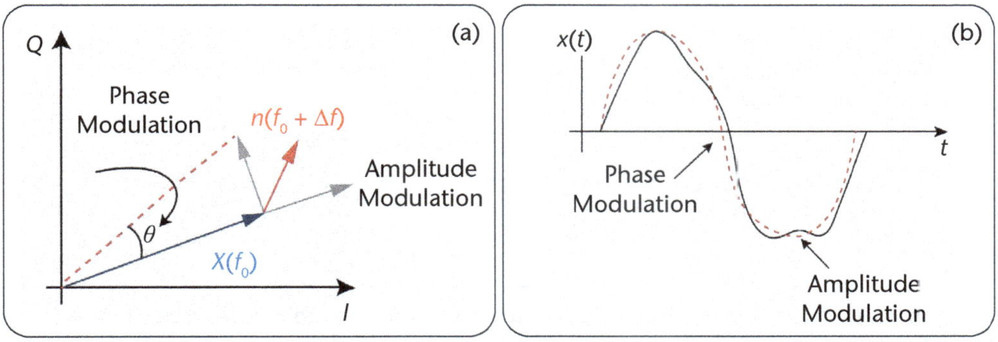

Figure 4.5 (a) A tiny sideband added to a single tone $X(f_0)$ causes amplitude and phase modulation. (b) The affected waveform in the time domain.

4.7.1 Thermal Noise

Random physical processes, such as thermal vibration or fluctuations of charge carriers in solid-state devices, generate electrical signals collectively referred to as noise. Noise is often the primary factor limiting signal detection and link performance. Several distinct noise mechanisms exist, each with characteristic properties [2]:

- *Thermal noise:* Generated by the thermal vibration of charge carriers in resistive and active components; broadband in nature.
- *Flicker noise (1/f noise):* Present in many solid-state devices; its power spectral density decreases with increasing frequency.
- *Shot noise:* Arises from the discrete, random arrival of charge carriers across potential barriers.
- *Quantum noise:* Originates from the quantized nature of charge and electromagnetic fields, relevant at cryogenic temperatures and high frequencies.

In RF systems, the most commonly encountered intrinsic noise sources are thermal noise and flicker noise. Thermal noise is broadband and sets the fundamental noise floor of resistive elements and active devices. Flicker noise, also known as $1/f$ noise, dominates at low frequencies and is prominent in semiconductor devices, voltage regulators, and bias circuitry. In mixers, oscillators, and amplifiers, $1/f$ noise can be upconverted to higher frequencies, leading to performance degradation at RF.

Figure 4.6(a) illustrates a resistor at a physical temperature T kelvin. The kinetic energy associated with the random motion of electrons within the resistor is proportional to temperature, and these random motions produce small voltage fluctuations at the resistor terminals, as shown in Figure 4.6(a).

This random voltage has a zero average value, but a nonzero root mean square (RMS) value given by Planck's blackbody radiation law [2]

$$V_n = \sqrt{\frac{4hfBR}{e^{hf/kT} - 1}} \tag{4.16}$$

The formula for the noise voltage in (4.16) is valid in both the classical (low-frequency, high-temperature) regime and the quantum (high-frequency, low-temperature)

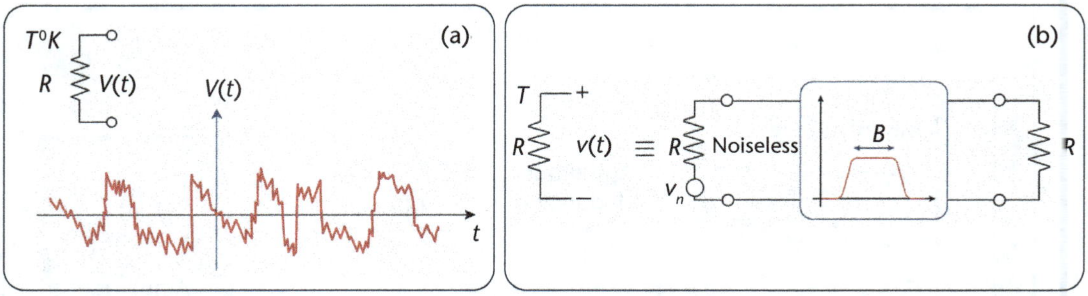

Figure 4.6 (a) The random voltage generated in a noisy resistor. (b) A noisy resistor modeled as a noiseless resistor in series with a voltage source is connected to a load resistor with an ideal bandpass filter.

regime. In this expression, h and k denote Planck's and Boltzmann's constants, respectively; T is the temperature in kelvin; B is the system bandwidth in hertz; f is the center frequency; and R is the resistance in ohms.

In the classical limit, where $hf << kT$, the exponential term in (4.16) can be approximated using

$$e^{hf/kT} - 1 \cong \frac{hf}{kT} \tag{4.17}$$

which corresponds to the first-order Taylor expansion of the exponential function. Substituting (4.17) into (4.16) yields the familiar Johnson-Nyquist thermal-noise expression:

$$V_n = \sqrt{4kTBR} \tag{4.18}$$

The classical approximation in (4.18) becomes invalid at extremely high frequencies and very low temperatures. For example, in quantum computing applications at temperatures on the order of millikelvin, the full expression in (4.16) must be used instead.

We now consider a worst-case scenario to determine the maximum power a noisy resistor can deliver to a circuit. As illustrated in Figure 4.6(b), the noisy resistor is represented by an equivalent circuit consisting of a noiseless resistor in series with a voltage source whose RMS value is given by (4.18). From the basic circuit theory, a source (in this case, the equivalent noise-voltage source) delivers maximum power to the load when the source resistance and load resistance are equal. An ideal bandpass filter is inserted between the source and the load to limit the bandwidth to B.

Using (4.18), the maximum noise power delivered to the matched load is

$$P_n = R(I)^2 = R\left(\frac{V_n}{2R}\right)^2 = \frac{V_n^2}{4R} = kTB \tag{4.19}$$

This maximum noise power is independent of frequency and resistance; it depends only on the temperature T and the bandwidth B. Therefore, two primary techniques exist for reducing noise power: limiting the bandwidth through appropriate filtering and lowering the physical temperature of the receiver. This is why receivers requiring extremely high sensitivity, such as those used in radio astronomy or quantum measurement systems, are often placed in cryogenic environments. By narrowing the bandwidth and operating at low temperatures, the noise power can be significantly reduced, thereby improving the overall sensitivity and performance of the system.

Since the noise power given in (4.19) is independent of frequency, that is, constant across the spectrum, the corresponding noise source is referred to as a white noise source. For white noise sources, the powers of independent noise contributions can be added directly, because their spectral densities are uniform and uncorrelated. However, it is important to note that not all forms of noise are additive; noise mechanisms that are correlated, frequency-dependent, or arise from nonlinear interactions do not combine through simple power addition.

4.7.2 Equivalent Noise Temperature

An equivalent noise temperature can be assigned to a white noise source by relating its delivered noise power to the thermal-noise expression. To establish this equivalence, a noise source with an equivalent input impedance *R* is connected to a load with the same impedance *R*, as illustrated in Figure 4.7(a). This impedance match ensures maximum power transfer. The noise power delivered to the load is denoted by P_s. The equivalent noise temperature T_e is then defined as [2]:

$$T_e = \frac{P_s}{kB} \tag{4.20}$$

The concept of equivalent noise temperature is particularly useful in system-level noise calculations, because individual components, such as amplifiers, mixers, and filters, can be represented by their effective noise temperatures. This allows noise contributions to be cascaded and analyzed systematically.

Table 4.3 lists the equivalent noise temperatures of several RF systems and components. The closely related concept of noise figure will be introduced in Section 4.7.4.

As discussed previously, an RF link is composed of cascaded subsystems, each contributing its own gain and noise. The equivalent noise temperature of a cascaded chain, illustrated in Figure 4.7(b), can be computed by treating each block

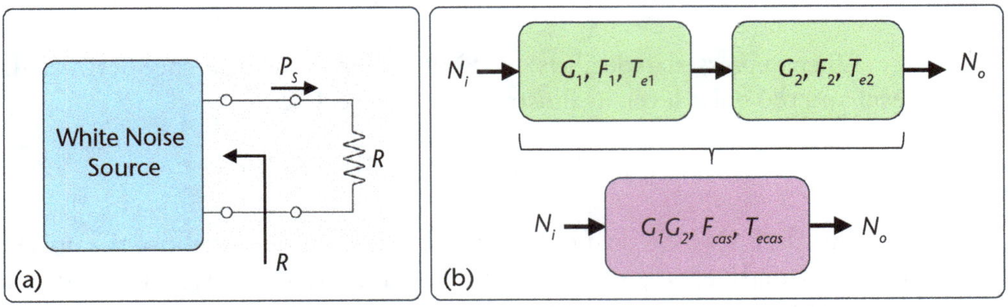

Figure 4.7 (a) White noise source with an input impedance *R*. (b) Connect the noise source to an impedance equal to *R* to ensure maximum power transfer. (c) Equivalent noise temperature of a cascaded system.

Table 4.3 Equivalent Noise Temperature and Noise Figure of Some RF Systems and Components

Component	*Equivalent Noise Temperature*	*Noise Figure*
5G receiver	289K	3 dB
Good room-temperature low-noise amplifier (LNA)	21K	0.5 dB
Cryogenic HEMT amplifier	5K	0.06 dB
Josephson parametric amplifier	<1K	<0.01 dB

as having a gain G_i and an equivalent noise temperature T_{e_i}. The entire chain can then be replaced with a single system having an overall gain $G = G_1G_2...G_n$ and an equivalent noise temperature T_{cas} given by

$$T_{\text{ecas}} = T_{e1} + \frac{T_{e2}}{G_1} + \frac{T_{e3}}{G_1G_2} + \cdots \tag{4.21}$$

Equation (4.21) shows that when the first stage has sufficiently high gain, the overall noise temperature of a cascaded system is dominated by that first stage. Noise contributions from subsequent stages are suppressed by the cumulative gain of all preceding stages, which is typically large. For this reason, it is essential that the first stage of the RF chain, particularly on the receiver side, exhibits both low equivalent noise temperature and adequate gain. This minimizes the total cascaded noise temperature and improves the receiver's sensitivity to weak signals.

4.7.3 Signal-to-Noise Ratio (SNR)

The SNR plays a central role in information theory, as discussed in Chapter 3. According to Shannon's theorem, the achievable channel capacity C of a communication link depends on its bandwidth B and its SNR. Consequently, successful signal detection requires that the desired signal level be sufficiently above the noise level.

The SNR is defined as

$$\text{SNR} = \frac{P_{\text{sig}}}{P_N} \tag{4.22}$$

When expressed in decibels, the SNR becomes the difference between the signal level and the noise level, as illustrated in Figure 4.8:

$$\text{SNR}(\text{dB}) = P_{\text{sig}}(\text{dB}) - P_N(\text{dB}) \tag{4.23}$$

Thus, the SNR quantifies how far the signal level lies above the noise floor. A spectrum analyzer is typically used to measure this quantity. As illustrated in Figure 4.8, the SNR corresponds to the vertical separation between the signal peak and the noise floor in the frequency domain. For example, with a noise floor of −117 dBm and a signal level of −90 dBm, the resulting SNR is SNR = −90 − (−117) = 27 dB.

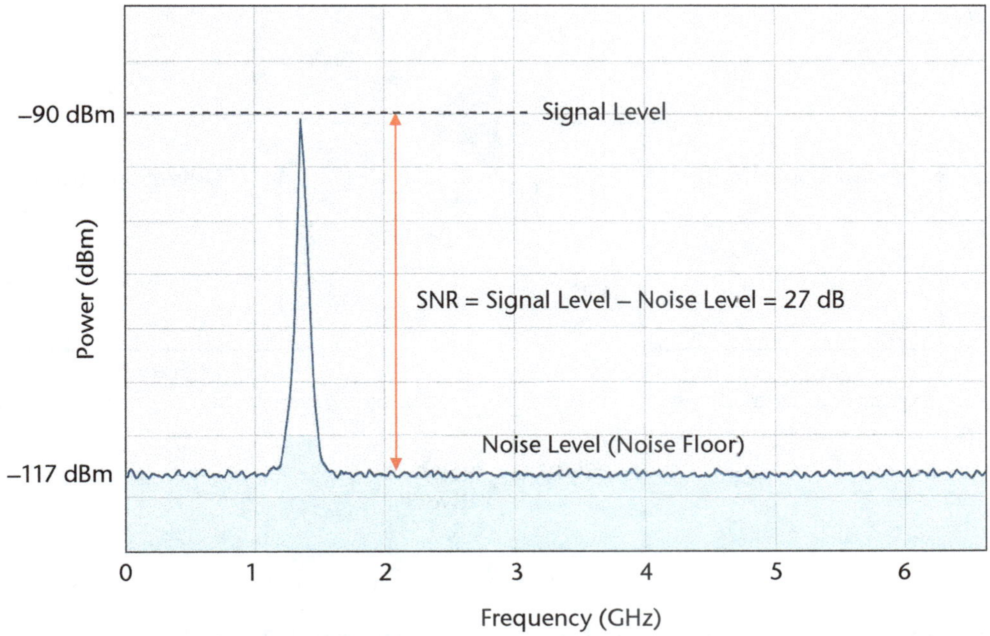

Figure 4.8 Spectrum illustrating the SNR concept. The peak represents the desired signal at −90 dBm, while the noise floor lies at −117 dBm. The vertical separation between the two levels corresponds to an SNR of 27 dB.

Higher SNR values result in a lower BER and a lower error vector magnitude (EVM), as discussed in Chapter 3. As we will see in Section 4.11.1, the SNR also directly affects the MDS level.

4.7.4 Noise Figure

The SNR degrades as a signal propagates through a system or component. Figure 4.9(a) illustrates how this occurs in an amplifier. As the signal passes through the amplifier, both the signal and the input noise are amplified. In addition, the amplifier contributes its own internal noise. Consequently, the output SNR is always lower than the input SNR.

The noise figure characterizes the degradation of the SNR as a signal propagates through a component or system. As shown in Figure 4.9(b), the noise figure is defined under the assumption that the input source is impedance-matched and that the input noise power is generated by a matched resistor at a room temperature of T_0 = 290K, giving an input noise power $N_i = kT_0B$.

The noise figure F is defined as [6]

$$F = \frac{\mathrm{SNR}_i}{\mathrm{SNR}_o} \geq 1 \tag{4.24}$$

The noise figure is always greater than or equal to 1. An ideal, noise-free system would yield F = 1, meaning no SNR degradation. However, all practical systems introduce additional noise, causing the output SNR to be lower than the input SNR.

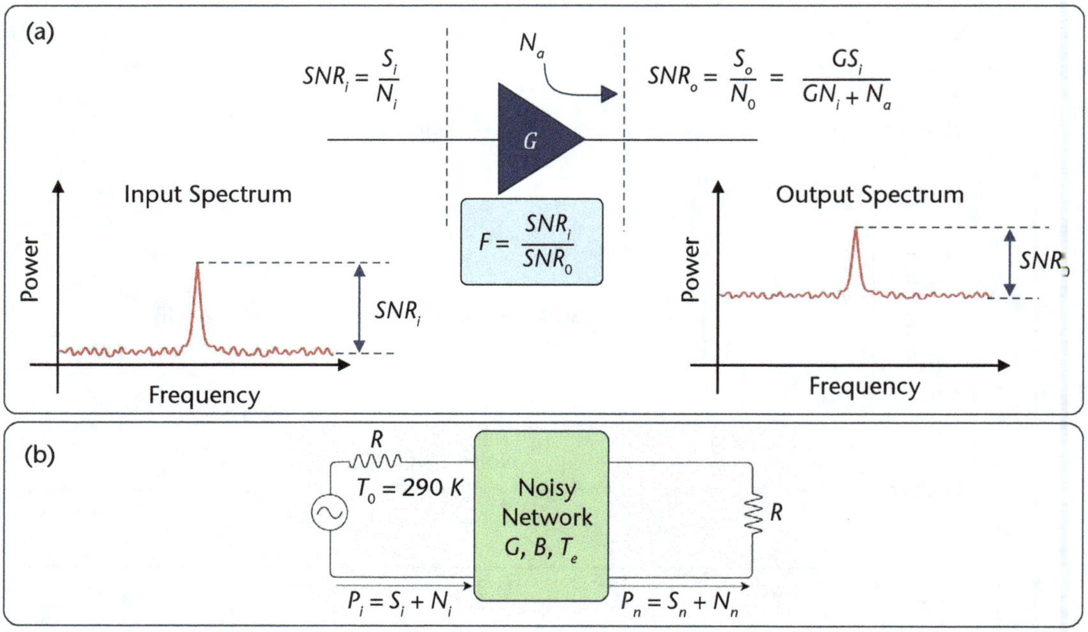

Figure 4.9 (a) SNR degrades as a signal goes through an amplifier. (b) The equivalent circuit to determine the noise figure (NF) of a noisy network.

The noise figure, expressed in decibels, represents the SNR degradation and is always greater than or equal to 0 dB:

$$F(\text{dB}) = \text{SNR}_i(\text{dB}) - \text{SNR}_o(\text{dB}) \geq 0 \text{ dB} \tag{4.25}$$

Some references distinguish between noise factor (dimensionless, linear scale) and noise figure (the same quantity expressed in decibels). In this book, we will use the term NF to refer to both to avoid unnecessary terminology confusion.

Table 4.3 lists typical noise figures for a variety of RF components and systems.

The NF of a passive device is defined directly by its power loss (attenuation) $L = P_{\text{in}}/P_{\text{out}}$, where P_{in} is the available source power and P_{out} is the available power delivered to the load. For example, a filter with 3 dB of attenuation (insertion loss) has a noise figure of $F = 3$ dB [2].

Just as we defined an equivalent noise temperature for cascaded systems, the noise figure can also be expressed for cascaded stages. Let G_i denote the gain of

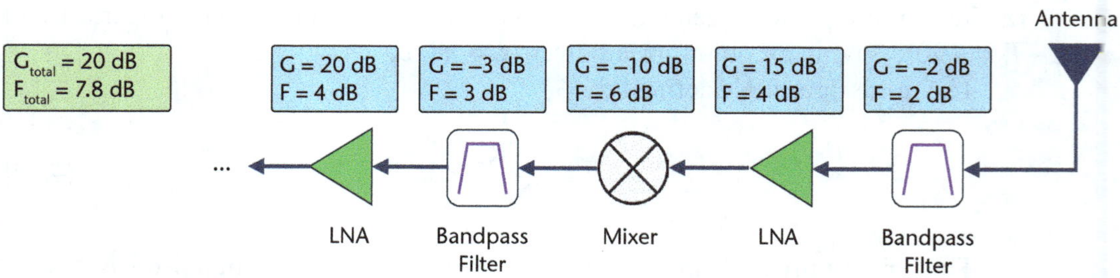

Figure 4.10 Calculation of the gain and noise figure (NF) for a cascaded system.

the stage i and F_i denote its noise figure (in linear units). The overall cascaded noise figure is then computed using the Friis noise equation:

$$F_{cas} = F_1 + \frac{F_2 - 1}{G_1} + \frac{F_3 - 1}{G_1 G_2} + \cdots \quad (4.26)$$

Figure 4.10 shows a numerical example of cascaded NF calculation for a receiver.

The noise figure and the equivalent noise temperature are closely related. Assume the input signal power S_i and input noise power N_i applied to a noisy two-port network characterized by gain G, bandwidth B, and equivalent noise temperature T_e. The output signal power is simply $S_o = GS_i$. The input noise power is $N_i = kT_0B$, where $T_0 = 290\text{K}$ is the standard reference temperature. The output noise power consists of the amplified input noise plus the noise generated internally by the network: $N_o = kGB(T_0 + T_e)$. By substituting these expressions for the input and output SNR into (4.24), we can determine the noise figure of the two-port network.

$$F = \frac{S_i}{kT_0B}\frac{kGB(T_0 + T_e)}{GS_i} = 1 + \frac{T_e}{T_0} \geq 1 \quad (T_0 = 290\text{K}) \quad (4.27)$$

Therefore, the equivalent noise temperature T_e is given by

$$T_e(K) = T_0\left(10^{\frac{F(\text{dB})}{10}} - 1\right) \quad (4.28)$$

It is useful to remember that a noise figure of 3 dB corresponds to an equivalent noise temperature equal to room temperature, approximately 290K. While the noise figure is the parameter most commonly used by RF engineers, equivalent noise temperature is more frequently used by physicists, particularly in applications such as quantum computing and radio astronomy. As we will see shortly, the noise figure is used to determine the minimum detectable signal.

4.7.5 Phase Noise

As shown in Figure 4.5, the phase of noise is random and varies continuously over time. When this noise is added vectorially to the signal of interest, it produces small, random fluctuations in the resulting overall phase. Phase noise refers to these fluctuations and is most commonly associated with oscillators, where internal noise sources are converted into phase variations that degrade spectral purity.

These phase fluctuations spread energy around the carrier frequency, effectively broadening the spectrum, as illustrated in Figure 4.11. Excessive phase noise can lead to adjacent-channel leakage, synchronization errors, reciprocal mixing, and degraded signal-detection performance, as will be demonstrated shortly.

Figure 4.11(a) shows the spectrum of an ideal oscillator without phase noise. In this case, the output consists solely of the pure carrier $A\cos(\omega_0 t)$, which corresponds to a delta function at ω_0 in the frequency domain.

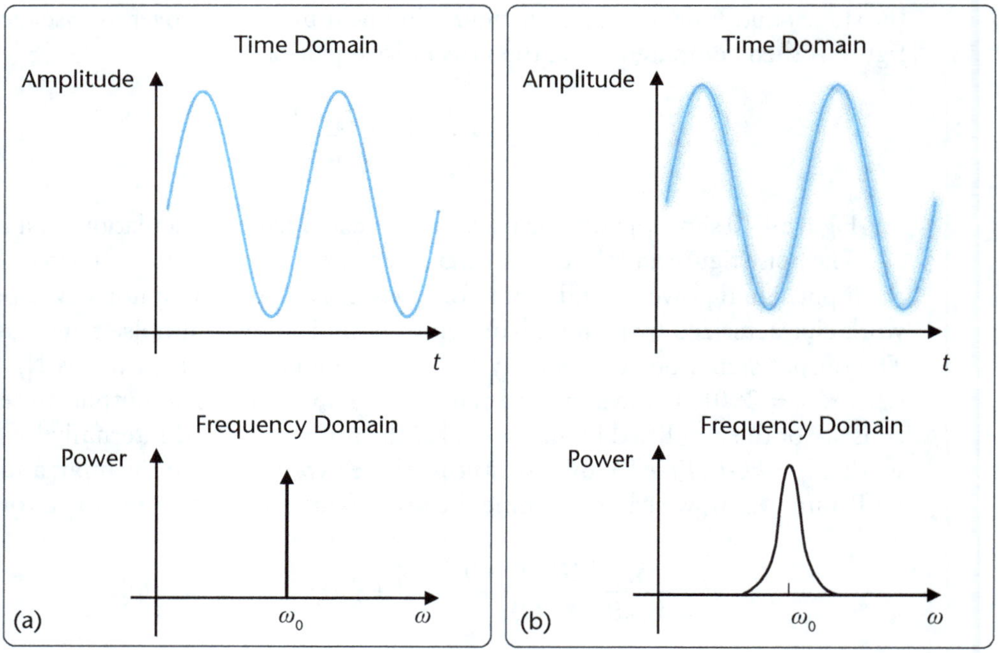

Figure 4.11 Comparison of (a) an ideal oscillator and (b) an oscillator with phase noise in both time and frequency domains.

Assume that the phase of the carrier exhibits small, random fluctuations $\phi(t)$. Using trigonometric identities, the modulated waveform can be decomposed into a pure carrier term and an additional term that reduces its spectral purity:

$$A\cos\left(\omega_0 t + \phi(t)\right) = A\cos\left(\omega_0 t\right)\cos\left(\phi(t)\right) - A\sin\left(\omega_0 t\right)\sin\left(\phi(t)\right) \tag{4.29}$$

For small values of the phase fluctuation $\phi(t)$, the small-angle approximations $\cos(\phi(t)) \approx 1$ and $\sin(\phi(t)) \approx \phi(t)$ can be applied, giving

$$A\cos\left(\omega_0 t + \phi(t)\right) \approx A\cos\left(\omega_0 t\right) - A\phi(t)\sin\left(\omega_0 t\right) \tag{4.30}$$

Rather than consisting solely of the pure carrier $A\cos(\omega_0 t)$, the presence of phase fluctuations introduces an additional term proportional to $\phi(t)$, which generates spurious sidebands, as illustrated in Figure 4.11(b). These sidebands broaden the spectral content around the carrier frequency, causing the waveform to deviate from a pure sinusoid. As will be discussed shortly, this spectral broadening can adversely affect system performance in several important ways.

Types of Phase Noise

The phase noise of an oscillator can be measured with a spectrum analyzer, provided that the analyzer's own phase noise is significantly lower than that of the oscillator under test. If this requirement is not met, a dedicated phase-noise analyzer using the cross-correlation technique is typically employed, enabling the measurement of extremely low phase-noise levels with much higher sensitivity.

Figure 4.12(a) illustrates a typical phase-noise measurement, where it is specified in dBc/Hz at a given offset f_m from the carrier. For example, −90 dBc/Hz at a 10-kHz offset and −102 dBc/Hz at a 100-kHz offset. The reported phase-noise values correspond to the single-sideband (SSB) definition used by IEEE standards. For reference, the double-sideband (DSB) phase noise is twice the SSB value.

Phase noise can be broadly categorized into two regions based on the frequency offset from the carrier, as illustrated in Figure 4.12(b):

1. Close-in phase noise refers to noise components located very near the carrier frequency, typically within a few hertz to tens of kilohertz of offset. These components originate primarily from low-frequency noise sources, such as flicker noise (1/*f* noise). Because close-in noise manifests as slow phase fluctuations, it directly perturbs the instantaneous phase trajectory of the carrier. As a result, close-in phase noise is especially harmful in systems that rely on coherent detection or precise timing. Even small phase errors can degrade EVM, increase BER, reduce radar range and Doppler resolution, and limit the fidelity of quantum control pulses.
2. Far-out phase noise refers to noise components at larger frequency offsets from the carrier, typically ranging from hundreds of kilohertz to several megahertz. Unlike close-in phase noise, these components arise primarily from broadband noise mechanisms such as thermal noise and wideband circuit fluctuations.

Because these high-frequency fluctuations vary much faster than the carrier's phase can evolve, they tend to average out over time and do not significantly perturb the instantaneous phase of the carrier. As a result, far-out phase noise has minimal impact on EVM or coherent demodulation accuracy. Instead, it manifests as spectral spreading, placing unwanted energy well away from the carrier frequency.

This energy leakage determines the level of interference imposed on adjacent channels and directly impacts metrics such as adjacent-channel leakage ratio (ACLR), spectral-mask compliance, and the spurious-free dynamic range (SFDR)

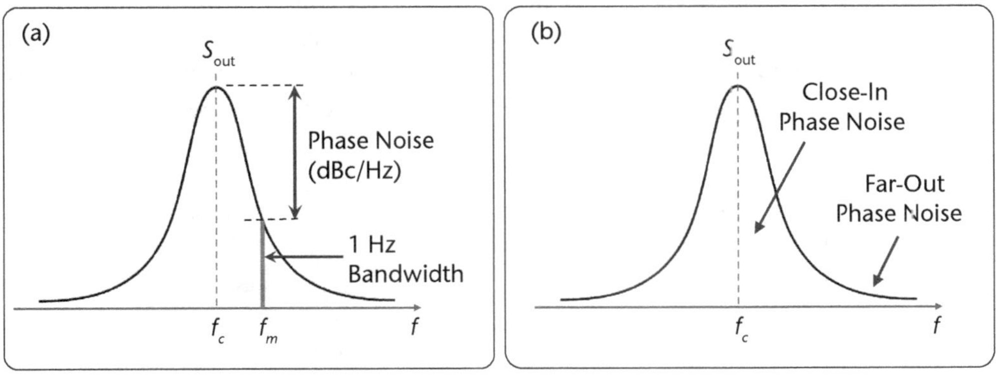

Figure 4.12 (a) Illustration of oscillator phase noise spectrum. The plot shows SSB phase noise, expressed in dBc/Hz at an offset frequency f_m from the carrier f_0. (b) Close-in and far-out phase noise.

of communication systems. In summary, far-out phase noise does not distort the carrier phase itself but governs how cleanly the signal remains confined within its allocated bandwidth.

Phase noise can also be categorized based on its measurement reference. There are two types of phase noise: absolute and residual.

1. Absolute phase noise quantifies the total phase noise produced by an oscillator or a signal source. It reflects all internal noise mechanisms, active device noise, resonator noise, bias noise, and coupling noise that manifest at the final RF output. As shown in Figure 4.13(a), the device under test (DUT) is the oscillator itself, and the measurement captures its standalone phase-noise performance.
2. Residual phase noise characterizes only the additional phase noise contributed by a specific component within the signal chain. In this measurement, a clean, low-noise reference source is used, and the measurement instrument suppresses or mathematically removes the source's phase noise to isolate the noise contribution of the DUT. As illustrated in Figure 4.13(b), the DUT is an amplifier: the clean source drives the amplifier, and the measurement system subtracts the source's phase noise from the total measured noise, yielding the residual phase noise of the DUT.

For example, when evaluating a receiver's LO, designers examine its absolute phase noise to ensure that it does not degrade system sensitivity or reciprocal-mixing performance. In contrast, residual phase-noise measurements are applied to intermediate components—such as amplifiers, mixers, or frequency-multiplication stages—to verify that these blocks do not introduce additional noise that would worsen the overall phase-noise budget of the system.

Jitter and Phase Noise

Jitter is the short-term variation in the timing of a signal's edges relative to their ideal positions, representing the time-domain counterpart of phase noise, as illustrated in Figure 4.14.

Jitter arises from noise and signal distortions throughout the system, including oscillator and power-supply noise, as well as coupling from nearby circuits. On interconnects and PCB traces, effects such as crosstalk, impedance mismatches, and insertion loss degrade signal edges. Insertion loss attenuates high-frequency components, slowing edge transitions and effectively increasing the interconnect

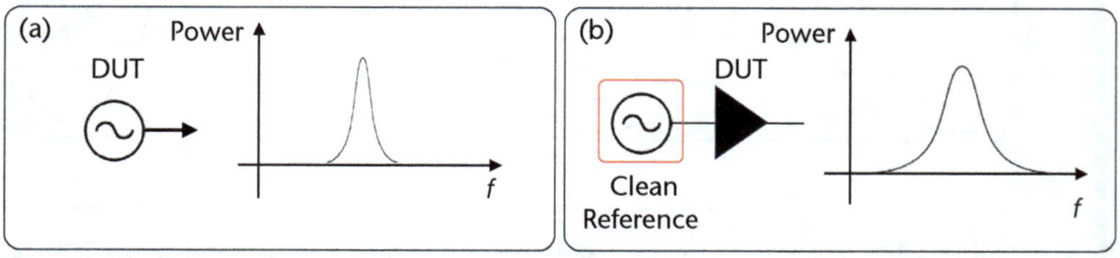

Figure 4.13 (a) Absolute phase noise, and (b) residual phase noise.

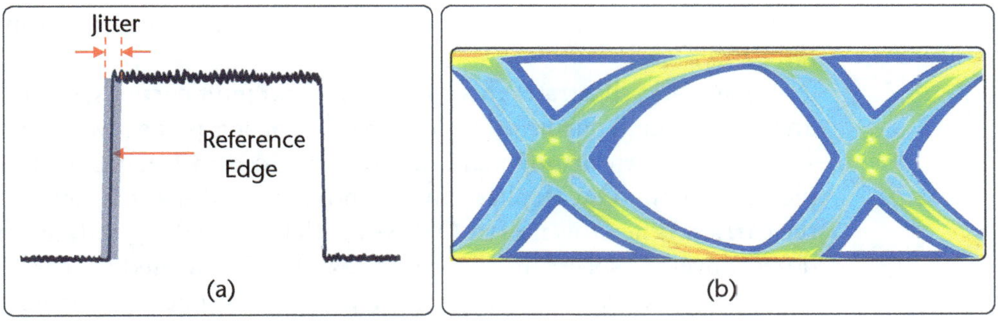

Figure 4.14 (a) Jitter in a clock signal. (b) Jitter in eye diagram.

time constant, while reflections superimpose delayed replicas of the signal onto the main edge. Together, these effects distort threshold-crossing times, converting amplitude disturbances into timing jitter.

In both RF and digital systems, excessive jitter degrades timing accuracy, increases BER, and broadens the signal spectrum. Phase noise and jitter describe the same underlying phenomenon in different domains: phase noise represents frequency-domain instability, while jitter represents the corresponding time-domain fluctuations.

The RMS timing jitter σ_t is obtained using the integrated phase-noise-to-RMS-jitter conversion, where the SSB phase noise $L(f)$ is integrated over a specified offset range ($f_{\min}$ to $f_{\max}$) and translated into equivalent time fluctuations relative to the carrier frequency. The RMS timing jitter σ_t in seconds is given by

$$\sigma_t = \frac{1}{2\pi f_0}\sqrt{\int_{f_{\min}}^{f_{\max}} 2L(f)\,df} \tag{4.31}$$

where

- f_0 = carrier frequency.
- $L(f)$ = SSB phase noise (linear units, convert from dBc/Hz using $L_{\text{in}} = 10^{L_{\text{dBc}}/10}$).
- $f_{\min}$, $f_{\max}$ = integration limits.

In high-precision systems, timing jitter is often constrained to the nanosecond, subnanosecond, or even picosecond range, depending on the application.

Impact of Phase Noise on RF Systems

Phase noise impacts different parts of an RF system in distinct ways, depending on whether it appears in the transmitter, receiver, modulation chain, or mixed-signal clock domain.

1. *Spectral leakage:* Phase noise spreads carrier energy in frequency, causing interference to neighboring channels and reducing spectral efficiency. In a transmitter, the broadened oscillator spectrum can leak power into adjacent channels, which is especially problematic in narrowband systems with

tightly spaced allocations, such as GSM 900, where the channel spacing is only 200 kHz.

Figure 4.15 illustrates this phenomenon for multicarrier systems, such as an OFDM signal during downconversion. The left plot shows ideal OFDM subcarriers, each orthogonal and evenly spaced in frequency. The center plot depicts an LO with phase noise, whose spectral energy is spread around the carrier. When this noisy LO mixes with the OFDM signal, the resulting downconverted spectrum (right plot) exhibits broadened subcarriers with overlapping skirts. This spectral leakage destroys subcarrier orthogonality, introduces ICI, and degrades system performance, particularly in wideband or high-order modulation schemes.

2. *Reciprocal mixing:* When a strong adjacent interferer is present, the LO's phase-noise sidebands downconvert part of the interferer's energy into the same IF band as the desired signal, thereby raising the effective noise floor. This is shown in Figure 4.16(a). The effect is most severe when the interferer is close in frequency to the desired signal, where LO phase noise is typically highest. Receiver datasheets often specify this limitation as the reciprocal mixing dynamic range (RMDR). Mitigation strategies include designing oscillators with low phase noise, optimizing PLL bandwidth, and using narrow IF filters to suppress interferer-induced noise.
3. *EVM degradation:* Phase noise directly degrades modulation accuracy by introducing both slow and fast phase variations in the constellation, as shown in Figure 4.16(b). At low phase-noise levels, this effect appears primarily as common phase error (CPE), a slow, predictable rotation of the entire constellation that can be tracked and corrected by carrier-recovery loops. As phase-noise levels increase, the phase variations become faster and more irregular, producing jitter-induced blurring of the constellation points. This random spreading cannot be fully corrected by tracking loops and leads to increased symbol errors and higher EVM.

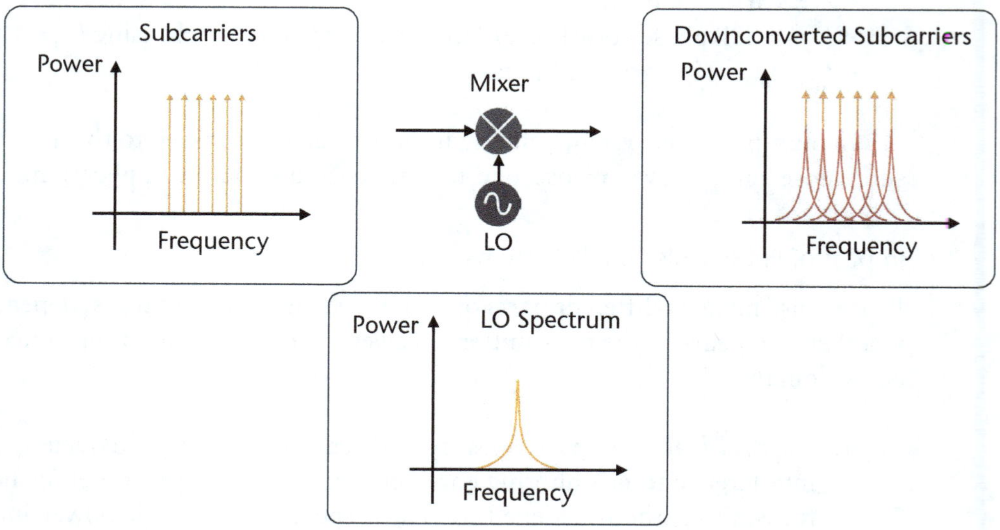

Figure 4.15 Illustration of LO phase-noise spreading in a multicarrier system.

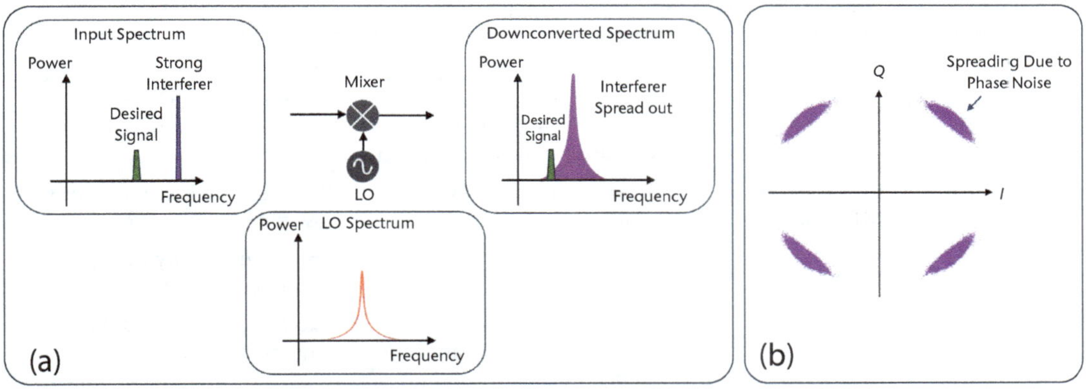

Figure 4.16 (a) Reciprocal mixing illustrated in the frequency domain. (b) Constellation-domain impact of phase noise. Slow and fast phase variations cause the QPSK constellation points to smear along the angular direction, reducing modulation accuracy and increasing EVM.

The impact of phase noise on EVM becomes more severe with higher-order modulation schemes as the constellation points are spaced more closely, where even small phase perturbations can push symbols across decision boundaries, dramatically increasing error rates.

4. Mixed signal and clock impairments:
   - *Timing jitter in mixed-signal systems:* Because ADCs and DACs rely on precise sampling instants, jitter directly reduces the effective number of bits (ENOB) and increases distortion products. This manifests as lower SNR and reduced SFDR, particularly in wideband converters and RF-sampling architectures where the input frequency is high. The resulting SNR limit is given by $SNR_{jitter} = -20\log(2\pi f_{in}\sigma_t)$ where f_{in} is the frequency of the input signal being sampled by the ADC (or reconstructed by the DAC), and σ_t is the RMS timing jitter. Because the phase error induced by jitter scales with input frequency, higher-frequency signals are far more sensitive to clock phase noise. This is why RF-sampling ADCs operating in the hundreds of megahertz to multigigahertz range require exceptionally low-jitter clock sources.
   - *Carrier synchronization errors:* Excessive phase noise undermines the stability of carrier-recovery loops, such as PLLs, Costas loops, and digital phase-tracking algorithms. When the LO phase fluctuates too rapidly, these loops cannot maintain accurate phase alignment, leading to poor tracking, increased residual phase error, or complete loss of lock in coherent receivers.

Methods for Phase Noise Reduction

Phase noise can be mitigated at both the source and system levels through a combination of circuit design, component selection, and signal processing techniques. Table 4.4 summarizes common methods for phase noise reduction. In some cases, hardware improvements alone may not be sufficient, and additional measures such as feedback control loops or digital signal processing (DSP) techniques are required to achieve the desired noise performance.

Table 4.4 Methods for Phase Noise Reduction

Method	*Working Principle*	*Practical Benefit*
High-Q resonators	Whenever possible, use dielectric, SAW, or crystal resonators with a high-Q factor	Suppress noise sidebands, improve spectral purity
High-stability references: TCXO, OCXO, and atomic clocks	Use specialized oscillators for frequency stability	Provide ultraclean reference for low-phase-noise systems
	TCXO: Electronic temperature compensation of the crystal	Improved stability over standard crystal oscillators
	OCXO: Oven maintains the crystal at a constant temperature	Excellent frequency stability, reduced phase noise
	Atomic clocks use atomic resonance (rubidium, cesium) as a reference	Ultralow long-term drift, best reference stability
Low-noise active devices	Choose transistors/amplifiers with low flicker noise, bias optimally	Reduce close-in phase noise contribution
Optimized PLL design	Balance loop bandwidth and reference oscillator noise	Minimize both in-band and out-of-band noise
Clean power supply	Use low-noise regulators, filtering, and decoupling	Prevent power supply modulation of the oscillator
Thermal and mechanical stability	Control temperature, add shielding, reduce vibrations	Lower drift and microphonics

4.7.6 Quantization Noise in Digital RF Systems

As we will see in Chapter 5, digital signal generation and detection techniques offer significant advantages, including simplified RF chains, tighter integration with digital processing, and improved repeatability. These benefits have made digital architectures increasingly popular across modern communication and sensing systems.

Digital techniques rely on a staircase approximation of the analog waveform, in which all signal values falling within a specific interval are mapped to a single discrete value, as illustrated in Figure 4.17(a). The height of each step, Δ, also known as the least significant bit (LSB), is determined by the full-scale level (FSL) and the resolution of the data converter. For an N-bit converter, $\Delta = \text{FSL}/2^N$. For example, a 16-bit ADC with a full-scale range of ±1-V results in $\Delta = 30\ \mu V$. This staircase representation, also called quantization, introduces rounding error, denoted by e in Figure 4.17(a), which becomes a fundamental source of noise in all digital systems.

Figure 4.17(b) shows the probability density function of quantization error, which we will use later to calculate the average power of the quantization noise.

Noise on a signal causes deviations in its instantaneous amplitude from the ideal value by an amount proportional to the noise present. Viewed this way, rounding or quantization error can also be interpreted as a form of noise, since it represents the deviation between the true analog signal and its digitized representation. The digitized signal can therefore be considered the sum of the ideal waveform and the quantization error.

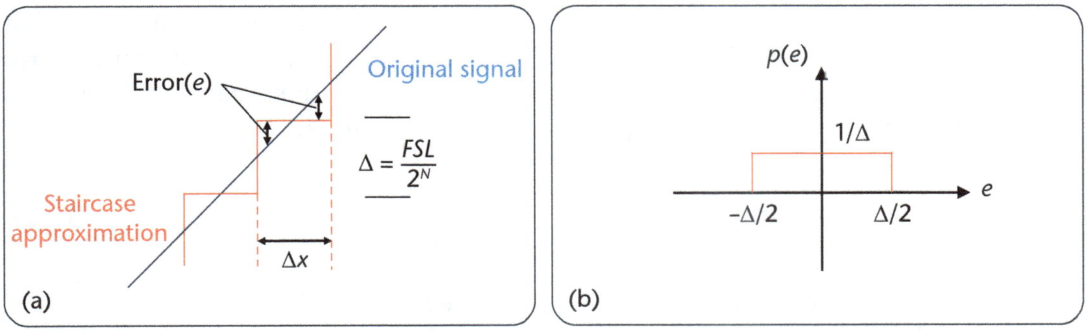

Figure 4.17 (a) Staircase approximation of a signal. (b) Probability density function of the quantization error.

Unlike thermal or other naturally occurring noise sources, quantization noise is deterministic as it repeats in exactly the same pattern for a given input waveform. For example, when a sinusoidal signal is sampled, the quantization error repeats every cycle and does not average out over time. Because averaging suppresses only random noise, it has no effect on this deterministic component.

Interestingly, adding a small amount of random noise to the input can be beneficial. This additional randomness "decorrelates" the quantization error from the input, effectively randomizing the error sequence. Once randomized, the quantization noise can indeed be reduced through averaging. This technique, known as dithering, is illustrated in Figure 4.18 and is widely used in high-resolution data converters.

In general, the quantization error is nonlinear and dependent on the signal. However, suppose the original signal is much larger than the staircase's height Δ, which is typically the case. Then the error has an approximately uniform distribution (see Figure 4.17(b)), as the quantization error is not significantly correlated with the signal. The average power of the quantization noise $P_{\text{quantization noise}}$ can be calculated using the expectation value as given by

$$P_{\text{quantization noise}} = \int_{-\Delta/2}^{\Delta/2} e^2 p(e)\, de = \int_{-\Delta/2}^{\Delta/2} e^2 \frac{1}{\Delta} de = \frac{\Delta^2}{12} \tag{4.32}$$

Thus, reducing the step size $\Delta = \text{FSL}/2^N$ either by increasing the number of bits or by reducing the FSL, directly lowers the quantization error.

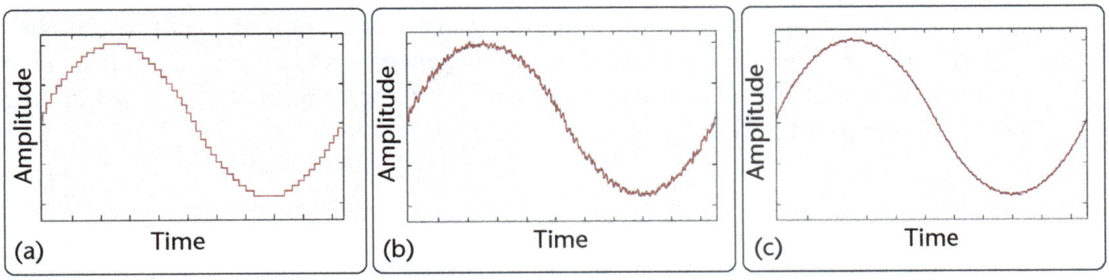

Figure 4.18 (a) Quantization noise in a sine wave. (b) Adding noise to the digitized signal. (c) Reducing the effect of both the thermal and digitization noise by averaging.

Power spectral density is a quantity that shows how power is distributed at various frequencies. The power spectral density of a uniform quantization error is constant across frequency, indicating that the resulting quantization noise is approximately white.[1] By substituting $\Delta = \text{FSL}/2^N$ into (4.32),

$$P_{\text{quantization noise}} = \frac{\text{FSL}^2}{12 \times 2^{2N}} \tag{4.33}$$

For a sine wave with the form FSL/2 sin(ωt), the SNR is given by

$$\begin{aligned} \text{SNR} &= 10\log\left(\frac{P_{\text{sig}}}{P_{\text{quantization}}}\right) = 10\log\left(\frac{\frac{\text{FSL}^2}{8}}{\frac{\text{FSL}^2}{12 \times 2^{2N}}}\right) \\ &= 10\log(1.5) + 10\log\left(2^{2N}\right) \end{aligned} \tag{4.34}$$

This results in

$$\text{SNR} = 1.76 + 6.02N \tag{4.35}$$

This expression determines the maximum achievable SNR of an ideal N-bit ADC when the input is a full-scale sine wave with amplitude FSL/2. For example, a 16-bit converter limited only by quantization noise achieves an SNR of approximately 96 dB. More generally, each additional bit of resolution improves the theoretical SNR by 6.02 dB, reflecting the exponential increase in dynamic range with bit depth.

The total noise in an ADC or DAC consists of thermal and quantization noise. The quantization noise is dominant in a low-resolution ADC, as shown in Figure 4.19(a), while the thermal noise becomes dominant in a high-resolution ADC, as shown in Figure 4.19(b) [7].

While users cannot directly control thermal noise, they can adjust system parameters to influence quantization noise. For low-resolution ADCs, using the smallest acceptable reference voltage reduces the FSL and therefore the quantization step size, lowering the quantization noise power. Conversely, in high-resolution ADCs, the reference voltage can be set as high as permissible to maximize dynamic range; in such cases, increasing the FSL does not significantly degrade performance because the resulting quantization noise remains well below other dominant noise sources.

Figure 4.20 illustrates how proper selection of signal amplitude ensures full utilization of the ADC's resolution. Driving the converter close to its full input range allows the digital output to more accurately follow the analog waveform, improving both linearity and SNR.

1. The Fourier transform of the autocorrelation function is the power spectral density. The autocorrelation function can be approximated with a delta function in the time domain if the noise samples are not correlated, which results in a constant function in the frequency domain.

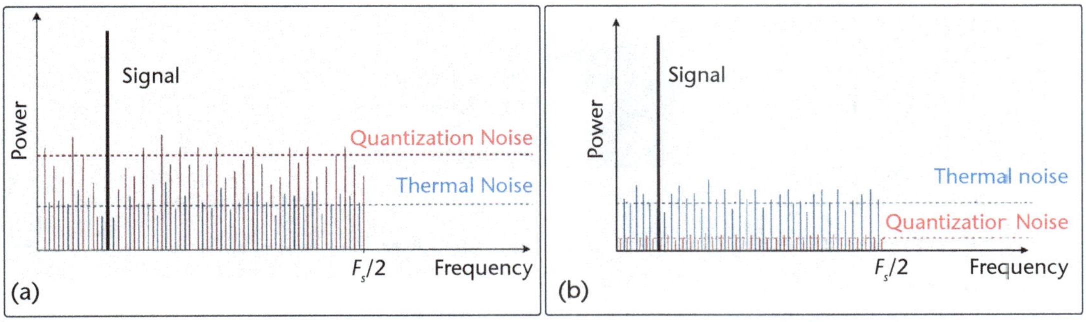

Figure 4.19 Comparison of quantization noise and the thermal noise in: (a) low-resolution ADC and (b) high-resolution ADC.

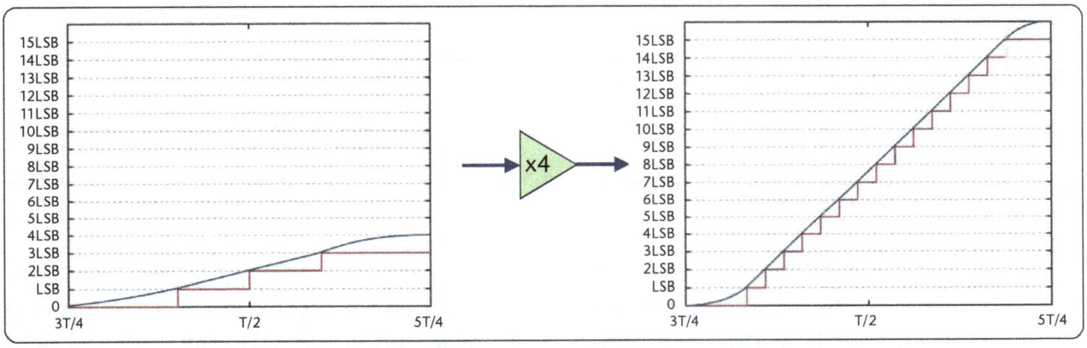

Figure 4.20 The left plot shows a weak signal that does not utilize the full resolution capability of the ADC. By adding an amplifier, the signal amplitude is increased, improving the dynamic range and allowing the signal to fully take advantage of the ADC's resolution.

4.8 Signal Degradation Factors: Distortion

Communication links and, more broadly, most signal-processing systems aim to reproduce an accurate replica of the input signal at the output or destination. However, in practice, signals encounter various impairments, such as dispersion and nonlinear effects, as they propagate through hardware and the channel. These impairments introduce distortion, an unwanted modification of the waveform that alters the information it carries.

Figure 4.21(a) conceptually illustrates how a dispersive transmission medium causes pulse spreading and alters the waveform's shape. Figure 4.21(b) shows how this degradation impacts a digital bit sequence, ultimately leading to ISI and potentially causing false bit detection. ISI can merge adjacent pulses, making them indistinguishable, or raise the signal level above the zero threshold, resulting in a bit-flip error (detecting a 1 instead of a 0).

In the early days of telephony, long transmission lines introduced distortion that altered the sound of the human voice. Oliver Heaviside addressed this challenge by balancing the line's inductive and capacitive time constants, pioneering techniques that laid the foundation for modern transmission theory.

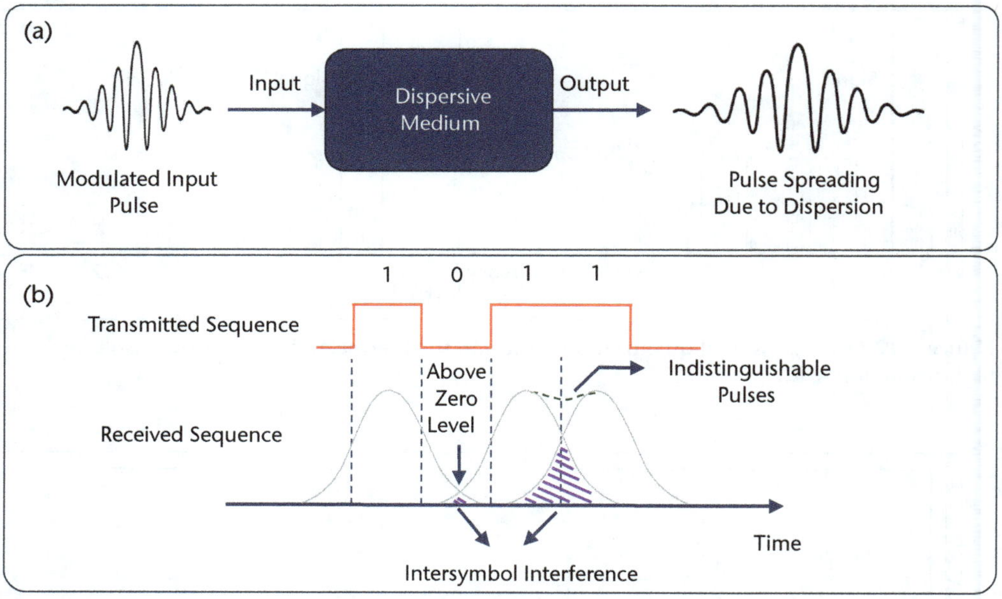

Figure 4.21 (a) A dispersive transmission line causing pulse spreading. (b) Distortion in the received signal leading to ISI.

Pulse-based systems are particularly sensitive to distortion, as it can change pulse amplitude, width, or timing. Higher-order modulation formats such as QAM also demand precise calibration, as even small nonlinear or dispersive effects can noticeably impair constellation accuracy. Left uncorrected, such distortions spread each symbol in time, introduce ISI, and degrade key metrics such as BER and EVM.

Fortunately, a number of design considerations and corrective techniques can mitigate these effects, restore signal integrity, and maintain reliable link performance, as we will explore in the following sections.

4.8.1 Sources of Distortion

Distortion in RF systems can arise from multiple mechanisms, each affecting the signal in different ways. Table 4.5 summarizes the major sources, their underlying mechanisms, practical examples, and common mitigation techniques used in RF product development.

4.8.2 Distortion Compensation Techniques

One might question whether distortion can ever be completely eliminated. In practice, although certain types of distortion can be compensated, doing so inevitably increases system complexity and cost. For this reason, the primary design objective should always be to engineer the system to be as linear as possible from the outset, minimizing signal-integrity issues, dispersion, and other sources of distortion at their origin. Nevertheless, some distortion mechanisms stem from components or environmental conditions outside the designer's control. In such cases, distortion compensation sometimes becomes necessary to meet performance requirements.

Table 4.5 Sources of Distortion in RF Systems and Mitigation Techniques

Source	*Mechanism*	*Example*	*Effect*	*Mitigation*
Nonlinearity	Device I-V (current-voltage) characteristics deviate from linearity, generating harmonics and intermodulation	Transistors, diodes in amplifiers, mixers, RF switches	Alters frequency content, causes spectral regrowth and waveform distortion	Operates in the linear region, uses higher-linearity devices, applies linearization (DPD, feedback, feedforward)
Nonflat amplitude frequency response	Uneven amplitude response across frequency	Filters, amplifiers with passband ripple	Changes spectrum, reshapes waveform, degrades fidelity	Equalization, filter design optimization, careful matching
Phase distortion/ group delay variation	Nonlinear phase or frequency-dependent group delay, even with flat amplitude response	Filters, transmission lines	Time-domain waveform distortion, symbol spreading	Group-delay equalization, linear-phase filters
Dispersion	Frequency-dependent propagation velocity causes pulse spreading	Optical fibers, long cables, hollow waveguides, wideband transmission lines	Pulse broadening, ISI, reduced data rate	Dispersion compensation (equalizers, fiber Bragg gratings), use of low-dispersion media
Memory effects	Device behavior depends on past input due to thermal or charge-storage effects	High PAs, bias networks	Dynamic distortion, reduced linearization effectiveness	Memory-aware models (memory polynomials), bias stabilization, thermal management
Crosstalk/coupling	Unwanted electromagnetic coupling between adjacent circuits or channels	PCB traces, antenna arrays, cables	Spurious distortion, degraded isolation, interference	Shielding, isolation, careful PCB layout, antenna spacing
Quantization effects	Finite resolution introduces amplitude errors and spurious distortion beyond the ideal noise model	ADCs, DACs in digital RF systems	Noise-like distortion, spurs, degraded EVM	Higher-resolution converters, oversampling, dithering, noise shaping
Multipath fading	Constructive and destructive interference due to multipath propagation	Wireless links, mobile channels	Frequency-selective fading, signal distortion	Diversity techniques, equalization, MIMO, RAKE receivers
Impedance mismatch	Frequency-dependent impedance mismatch causes reflections	Poorly matched feeds for wideband antennas	Frequency-selective distortion, passband ripple, reduced efficiency	Impedance-matching networks, tunable matching

Broadly, RF systems address distortion using two complementary strategies: correcting distortion after it occurs or preventing it before it occurs. These approaches are summarized in Table 4.6.

In practice, modern RF systems often combine both categories:

- *Before distortion:* Crest factor reduction (CFR) + digital predistortion (DPD) to keep the transmitter clean.
- *After distortion:* equalization and error correction at the receiver to ensure robust data recovery.

We will discuss these methods and architectures in Chapter 5.

4.9 Nonlinear Effects in RF Systems

Nonlinear effects play a critical role in determining the overall performance and fidelity of RF systems. Any deviation from ideal linear behavior in active components, such as amplifiers, mixers, or oscillators, introduces distortion that alters the original amplitude and phase relationships of the signal. These nonlinearities also generate undesired spectral components, including harmonics and intermodulation products, which can interfere with adjacent channels or external systems, degrade signal quality, and reduce overall system efficiency. Their impact becomes especially significant in modern wideband and high-order modulation systems, where linearity directly influences key performance metrics such as EVM, ACLR, and SINAD.

Table 4.6 Distortion Compensation Techniques in RF Systems

Category	*Technique*	*Location*	*Principle*
Correcting distortion after it happens	Equalization	Receiver	Compensates for channel-induced linear distortion (amplitude/phase) by applying an inverse channel filter
	Error correction	Transmitter (encoding) + receiver (decoding)	Redundancy in coding allows errors from distortion to be detected and corrected
	Feedforward cancellation	Transmitter	Distortion products at the output are extracted, amplified, and subtracted to clean the signal
	Feedback linearization	Transmitter	Output is compared to input, and the error drives correction (effective for narrowband systems)
Preventing distortion before it happens	Predistortion (digital or analog)	Transmitter	Applies an inverse nonlinear function to the signal to cancel out the amplifier's compression characteristics
	Crest factor reduction (CFR)	Transmitter	Reduces the peak-to-average power ratio (PAPR), thereby lowering the nonlinear distortion generated by the PA
	Bias and device optimization	Transmitter	Operating point, device class, and thermal management are chosen to minimize distortion

In the following sections, we examine the origins and classifications of nonlinear effects, describe how they are quantified through metrics such as the 1-dB compression point (P_{1dB}) and third-order intercept point (IP_3), and discuss how they constrain the system's dynamic range.

4.9.1 Linear Systems

Before looking at the factors that cause distortion, let us see why linearity is essential to preserving the shape of a signal and, therefore, the information carried by the signal. A system is called linear if it satisfies two properties (i.e., homogeneity and superposition). If $x(t)$ is the input to the system and $y(t)$ is the output, and k is a constant, then the homogeneity property reads:

$$kx(t) \rightarrow ky(t) \tag{4.36}$$

This means that, if we multiply the input by a constant, the output will be multiplied by the same factor. The superposition property can be expressed as

$$x_1(t) + x_2(t) \rightarrow y_1(t) + y_2(t) \tag{4.37}$$

This statement asserts that if the inputs $x_1(t)$ and $x_2(t)$ are applied together, the resulting output is simply the sum of the individual outputs $y_1(t)$ and $y_2(t)$. In other words, the system's response to a sum of inputs is the sum of its responses to each input.

Figure 4.22 illustrates the input and output of linear and nonlinear systems in both the time and frequency domains. Consider a scenario where the system is a speaker and the input is a pure musical tone, for example, a C note with frequency f_1. In a linear system, the output closely resembles the input signal, differing only in amplitude and phase, as shown in Figure 4.22. In the frequency domain, the output spectrum contains a single tone at f_1, scaled by a constant gain factor.

In contrast, a nonlinear speaker, shown in Figure 4.22, does not reproduce the input tone in its pure form. Instead, the nonlinear behavior generates additional frequency components, harmonics at $f_2 = 2f_1$, $f_3 = 3f_1$, and higher multiples, which alter the character of the sound and make it noticeably different from the original input.

Thus, while a linear system preserves the waveform shape and spectral purity of the input signal, a nonlinear system modifies both, introducing distortion that can significantly affect the fidelity of communication, radar, or audio systems.

From a frequency-response perspective, a linear system must exhibit both a flat amplitude response and a linear phase response over its operating bandwidth, as shown in Figure 4.23. A nonflat amplitude response distorts the signal spectrum by scaling different frequency components unequally, which, in turn, alters the waveform's shape in the time domain.

A linear phase response ensures a constant time delay for all frequency components. Consider a time delay $\Delta t = \Delta\phi/2\pi f$, where the phase difference $\Delta\phi$ is a linear function of frequency, $\Delta\phi = cf$ with c as a constant. Substituting yields $\Delta t = cf/2\pi f = c/2\pi$, which is a constant. Thus, a linear phase response preserves the waveform shape because all spectral components experience the same delay.

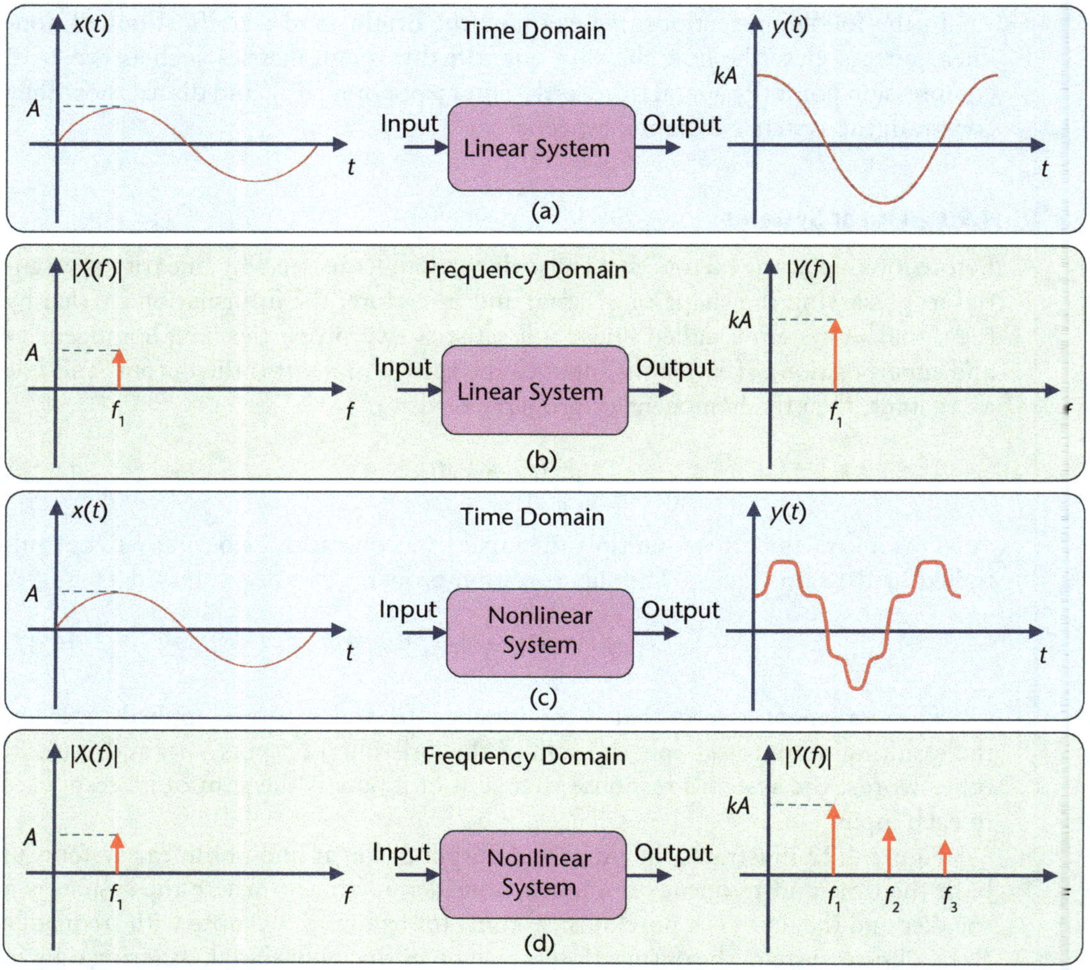

Figure 4.22 (a) A linear system only changes the amplitude and phase of the input signal and (b) does not generate new frequency components. (c) A nonlinear system distorts the signal's shape and (d) generates new frequency components in the output.

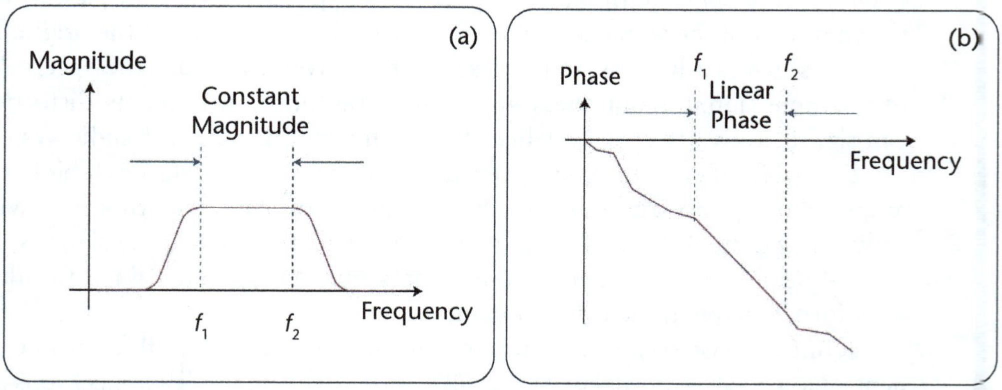

Figure 4.23 The frequency response of a linear system over the operational bandwidth from f_1 and f_2: (a) amplitude response, and (b) phase response.

However, if the phase response is nonlinear, dispersion occurs. Different frequency components propagate with different effective velocities, leading to waveform distortion and degradation of signal integrity.

4.9.2 Nonlinear Systems

The analysis of linear systems relies on powerful mathematical tools such as linear differential equations, the Laplace transform, and the Fourier transform. These tools become applicable when a system can be linearized around its operating point. To accomplish this, the amplitude of the input signal must be sufficiently small so that a straight line can closely approximate the nonlinear curve over the region in which the input varies, as illustrated in Figure 4.24(a). This technique, known as the small-signal approximation, is fundamental in the analysis and design of circuits and systems.

However, many practical scenarios violate the assumptions required for small-signal modeling. When the input signal swings over a large portion of the nonlinear characteristic, the system's intrinsic nonlinearity must be explicitly considered. Figure 4.24(b) shows an example of strong nonlinear behavior resulting from large-signal excitation.

Nonlinear systems exhibit a variety of important behaviors and enable essential RF applications, most notably frequency conversion in mixers. However, in systems where linearity and spectral purity are critical, nonlinear effects are undesirable because they distort the signal and generate unwanted spectral components.

Table 4.7 shows some of the critical nonlinear effects in RF systems.

4.9.3 Harmonic Distortion

As discussed earlier, when a nonlinear system is excited by a single-frequency input, it generates harmonics, which are frequency components at integer multiples of the input frequency. These harmonic frequencies are not present in the original input

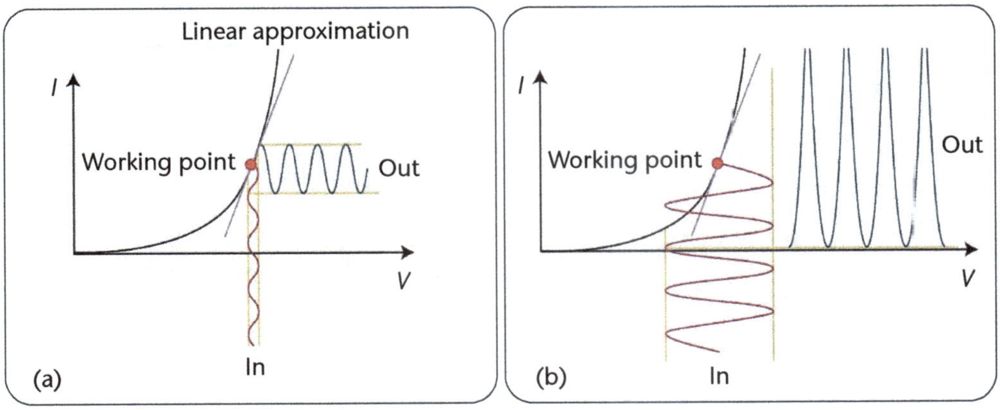

Figure 4.24 Small-signal versus large-signal behavior. (a) When the input signal is small, the nonlinear I-V curve can be approximated by a linear tangent at the operating point, allowing linear analysis tools to be applied. (b) For large input swings, the signal moves across a strongly nonlinear region, producing a distorted output that reflects the system's nonlinear characteristics.

Table 4.7 Nonlinear Effects and Their Input-Output Relationships

Nonlinear Effect	*Input to the System*	*Output of the System*
Harmonic distortion	A single tone at f_1	Harmonics at integer multiples of the input frequency: nf_1
Intermodulation distortion	Two tones at f_1 and f_2	Intermodulation products at $nf_1 + mf_2$, where n and m are integers
Gain compression	Large input amplitude	Output amplitude saturates, causing compression in the input-output characteristic
Amplitude modulation-(AM)-phase modulation (PM) conversion	AM or amplitude-varying signal: $v_{in}(t) = A_{in}\cos(\omega t)$	Amplitude variations produce phase variations or PM: $v_{out}(t) = A_{out}\cos(\omega t + \phi(A_{in}))$
AM-AM conversion	Amplitude-modulated or amplitude-varying signal $v_{in}(t) = A_{in}\cos(\omega t)$	Output amplitude is a nonlinear function of input amplitude A_{in}, causing amplitude distortion (gain expansion or compression) $v_{out}(t) = A_{out}(A_{in})\cos(\omega t)$
Cross-modulation	A weak desired signal and a strong interferer	Modulation from the strong interferer is transferred onto the desired signal
Memory effect	Wideband, time-varying, or high-amplitude signal (frequency, amplitude, and modulation-dependent)	Output distortion depends on both present and past input values due to frequency-dependent, bias-network, or thermal dynamics

spectrum, meaning the output waveform no longer preserves the shape of the input. The introduction of these additional spectral components modifies the time-domain waveform and results in distortion.

As an example, Figure 4.25(a) shows a distorted sine wave produced by a low-quality signal generator with a highly nonlinear RF output stage. The corresponding spectrum, shown in Figure 4.25(b), clearly reveals the presence of harmonic components.

Harmonics can become problematic when they interfere with other channels. For example, as shown in Figure 4.26(a), a 700-MHz signal produces a third harmonic at 2.1 GHz, which overlaps with LTE Band 1, and an eighth harmonic at 5.6

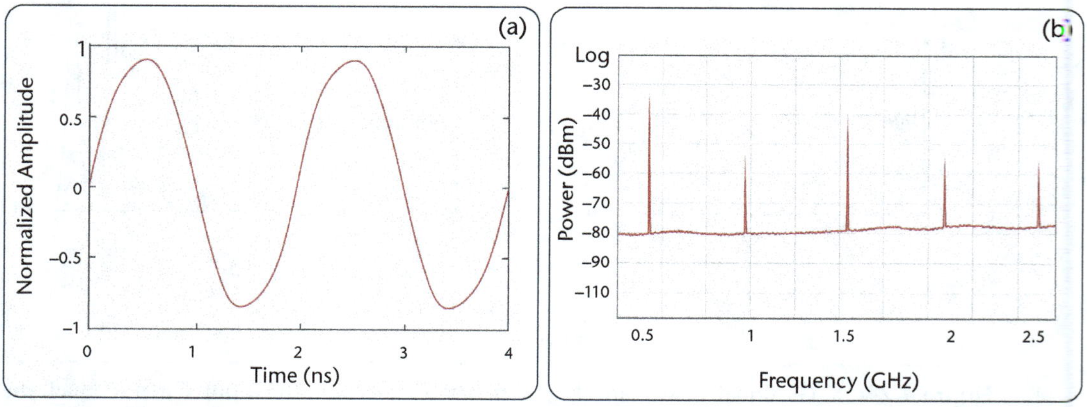

Figure 4.25 (a) Time-domain representation of a 500-MHz sine wave generated by a low-quality signal generator. The waveform deviates from a pure sine wave due to the highly nonlinear RF output stage of the signal generator. (b) Corresponding frequency-domain spectrum of the 500-MHz signal, showing harmonic components at integer multiples of the fundamental frequency with varying amplitudes.

GHz, which falls within WLAN frequencies. In such cases, the harmonics must be suppressed to prevent interference with neighboring systems. Figure 4.26(b) demonstrates how filtering can be used to eliminate these unwanted components [1].

That said, harmonic distortion is often harmless, particularly in narrowband systems where higher-order harmonics lie far outside the intended passband and are naturally attenuated by the system's frequency response. Problems arise only when harmonics fall within or near other operational bands. In those situations, additional suppression techniques, most commonly filtering, are required to ensure proper coexistence with other channels and maintain spectral integrity.

Now, let us use a simple mathematical model of a nonlinear system to see how harmonics are generated. We can use a polynomial function to approximate a nonlinear system around its operating point [6]

$$y(t) \approx \alpha_1 x(t) + \alpha_2 x^2(t) + \alpha_3 x^3(t) \tag{4.38}$$

For a pure sinusoidal input $x(t) = A\cos(\omega t)$, (4.38) now reads as

$$y(t) = \frac{\alpha_2 A^2}{2} + \left(\alpha_1 A + \frac{3\alpha_3 A^3}{5}\right)\cos\omega t + \frac{\alpha_2 A^2}{2}\cos 2\omega t + \frac{\alpha_3 A^3}{5}\cos 3\omega \tag{4.39}$$

The output consists of a DC term ($\alpha_2 A^2/2$) arising from the second-order nonlinearity, the fundamental frequency $\cos(\omega t)$, the second harmonic ($\cos(2\omega t)$), and the third harmonic ($\cos(3\omega t)$). Most of the time, the amplitude of the harmonics after the third one is negligible. For small-signal operations, the output can be approximated linearly as $y(t) \approx \alpha_1 x(t) = \alpha_1 A\cos\omega t$, where α_1 is the system's gain.

4.9.4 Gain Compression

Active circuits, such as amplifiers, are essential components in an RF link and are inherently nonlinear. Figure 4.27 illustrates the output power of an amplifier as a function of its input power. The amplifier behaves linearly up to a certain input level, where the slope of the line represents the small-signal gain. Beyond this point, the amplifier departs from linearity and enters the compression region, where the output begins to saturate.

The 1-dB compression point is defined as the input power level at which the amplifier's output power falls 1 dB below the ideal linear extrapolation, as shown

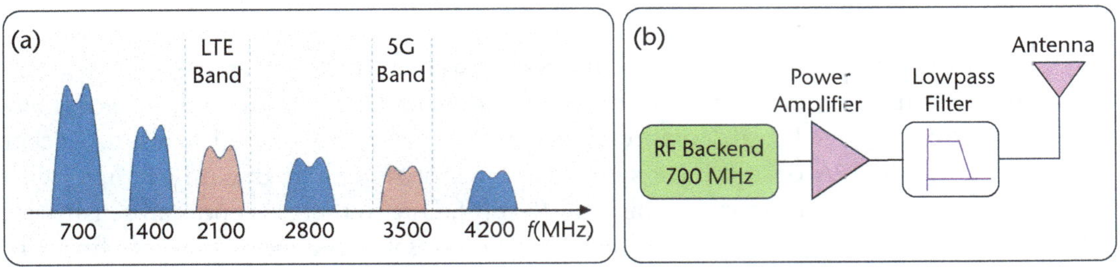

Figure 4.26 (a) Interference effect of harmonics with LTE and WLAN bands. (b) A lowpass filter (LPF) can be used to remove the harmonics of the 700-MHz signal to avoid interference with other bands.

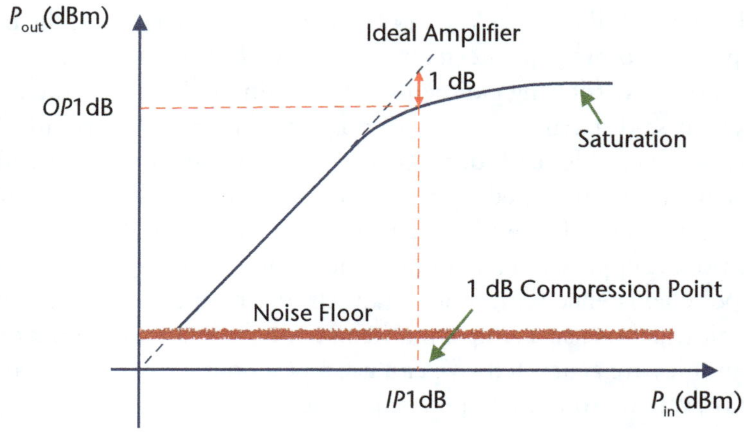

Figure 4.27 The gain compression in an amplifier.

in Figure 4.27 [1, 2]. A higher 1-dB compression point indicates that the amplifier can maintain linear operation over a wider input-power range. Note that a 1-dB drop in gain corresponds to approximately a 20% reduction in power gain.

The input and output 1-dB compression points, $P_{1\text{dB,in}}$ and $P_{1\text{dB,out}}$, are related through the small-signal gain G as:

$$P_{1\text{dB,out}} = P_{1\text{dB,in}} + G - 1 \tag{4.40}$$

where G is the amplifier's linear gain. In datasheets, the output-referred 1-dB compression point, $P_{1\text{dB,out}}$, is most commonly specified. So (4.39) can be used to calculate the input-referred 1-dB compression point.

In general, amplifiers should be operated within their linear region, below the compression point, to avoid signal distortion. However, in some applications, operating in compression does not significantly degrade system performance. For example, in frequency-modulated (FM) systems, the information is conveyed in the instantaneous frequency rather than the signal amplitude. In such cases, the amplifier may be intentionally driven into compression to maximize output power and efficiency, provided the resulting distortion remains within acceptable limits.

The 1-dB compression point varies widely depending on the amplifier type and application. For room-temperature LNAs, the output-referred 1-dB compression point typically ranges from approximately −5 dBm to +10 dBm. In contrast, solid-state PAs (SSPAs) exhibit much higher compression levels, typically between +30 dBm and +50 dBm.

At the other end of the spectrum, Josephson parametric amplifiers (JPAs) used in quantum computing operate at cryogenic temperatures and can exhibit extremely low input-referred 1-dB compression points, on the order of −100 dBm. Recent advances in JPA design have significantly improved their dynamic range, with compression points approaching −73 dBm [8]. Due to their exceptional sensitivity, JPAs must be operated well below their compression region to preserve linearity and avoid signal distortion.

In addition to amplifiers, other solid-state devices such as RF switches, mixers, and attenuators can also exhibit compression. Hence, it is crucial not to drive these components into their compression region.

Eventually, we are interested in knowing the total P_{1dB} of the RF link. The P_{1dB} of a cascaded system can be calculated as follows, where P_{1dBi} and G_i are the P_{1dB} and gain of each stage, respectively [7]

$$P_{1dB,total} = 10\log_{10}\left[\left(\frac{1}{P_{1dB1}G_2G_3}+\frac{1}{P_{1dB2}G_3}+\frac{1}{P_{1dB3}}\right)^{-1}\right] \quad (4.41)$$

4.9.5 Intermodulation

We have already seen that a single-frequency input applied to a nonlinear system generates harmonics at the output. Now we examine what happens when two sine waves at different frequencies, f_1 and f_2, are applied simultaneously to a nonlinear system.

Consider first a purely linear system. If the input consists of two pure tones at f_1 and f_2, the output spectrum contains frequency components only at these two frequencies, as illustrated in Figure 4.28(a). A linear system does not create any new frequency components; it simply scales and phase-shifts the frequencies present in the input.

Now suppose that we perform the same linear two-tone measurement on a nonlinear system, as illustrated in Figure 4.28(b). In this case, the output spectrum contains additional frequency components that are not harmonics of the individual input tones. Instead, new tones appear at linear combinations of the two input frequencies, as shown in Figure 4.28(b). This phenomenon is known as intermodulation (IM) and arises from the multiplication (mixing) terms produced by the second-order and third-order nonlinearities in the fit function given in (4.37).

The resulting intermodulation products occur at frequencies given by linear combinations of the fundamentals f_1 and f_2 of the form:

$$mf_1 + nf_2 \quad \text{with} \quad m,n = \ldots,-2,-1,0,1,2,\ldots \quad (4.42)$$

The order of an intermodulation (IM) product is defined as $|m| + |n|$ in the general expression $f_{IM} = mf_1 + nf_2$. For example, $f_1 + f_2$ is a second-order product since $|1| + |1| = 2$, while $2f_1 - f_2$ is a third-order product since $|2| + |-1| = 3$.

The intermodulation products can be extracted using the Pascal triangle, as shown in Figure 4.29(a).

The intermodulation products that lie far from the fundamental tones, such as $2f_1 + f_2$ and $2f_2 + f_1$ (see Figure 4.28(b)) can often be filtered out due to their large frequency separation from the desired band. However, the third-order intermodulation (IM) products (IM_3) at $2f_1 - f_2$ and $2f_2 - f_1$, occur very close to the fundamental frequencies. For sufficiently small spacing between f_1 and f_2, these IM_3 components almost always fall within the system's operational bandwidth and therefore cannot be easily filtered out.

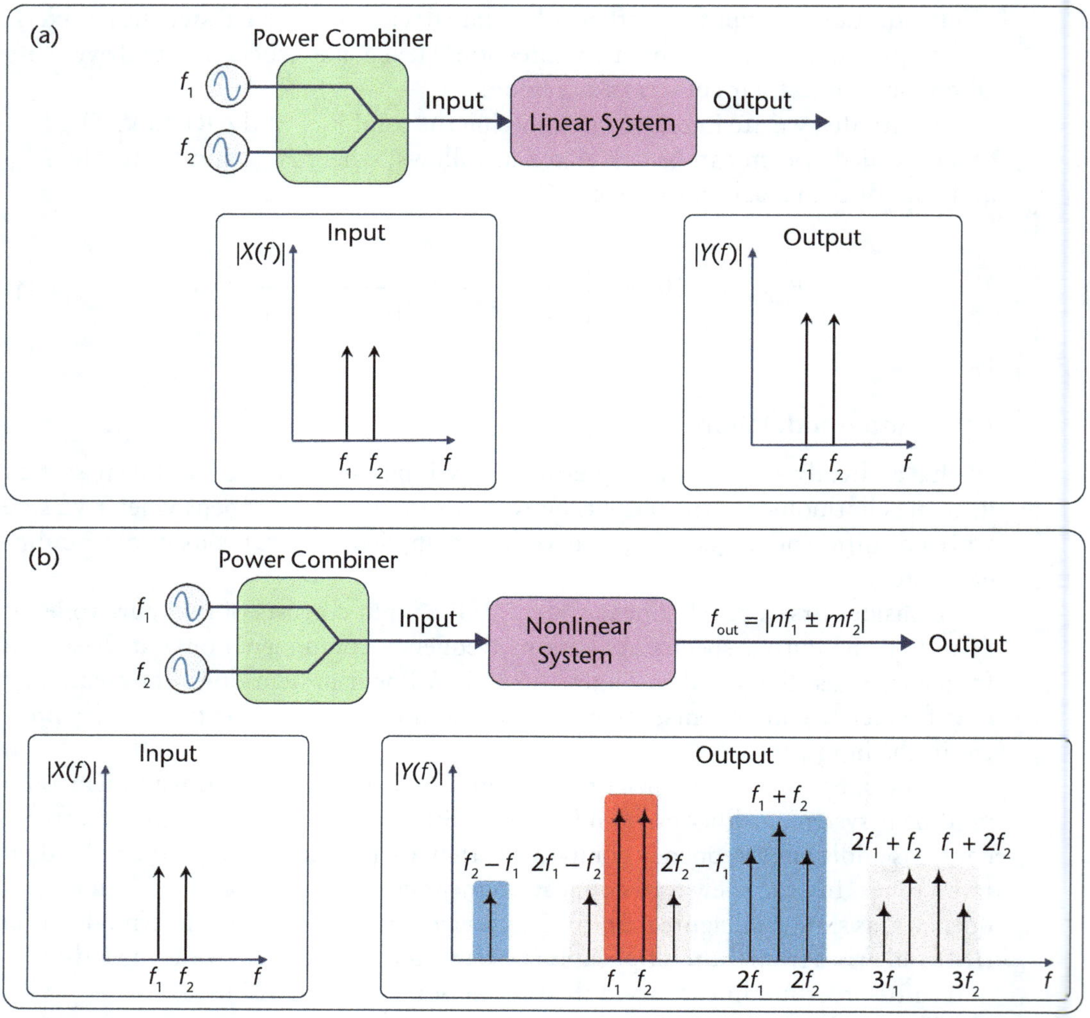

Figure 4.28 (a) Two-tone excitation of a linear system. The input and output spectra contain the same frequency components. (b) Two-tone excitation of a nonlinear system. The output spectrum contains frequency components that are linear combinations of the input frequencies.

Furthermore, when f_1 and f_2 are strong interferers, their IM_3 products can fall directly on top of a desired receive channel, corrupting the signal, as shown in Figure 4.29(b). This happens when the frequency of an intermodulation product coincides with the desired channel center frequency, that is, when $2f_1 - f_2 = f_0$, where f_0 is the center frequency of the desired channel. A practical example occurs in equally spaced channels: for instance, two WLAN channels at 5.6 GHz and 5.7 GHz can produce a third-order intermodulation product at 5.5 GHz, calculated as $(2 \times 5.6) - 5.7 = 5.5$ GHz, directly interfering with an adjacent channel.

This proximity of IM_3 products to the fundamental tones makes intermodulation measurements far more effective than harmonic distortion measurements for evaluating nonlinearity. As shown in Figure 4.30(a), when harmonic components fall outside the system bandwidth, they are strongly attenuated by the inherent filtering characteristics of the system, which can make the amplifier or circuit appear

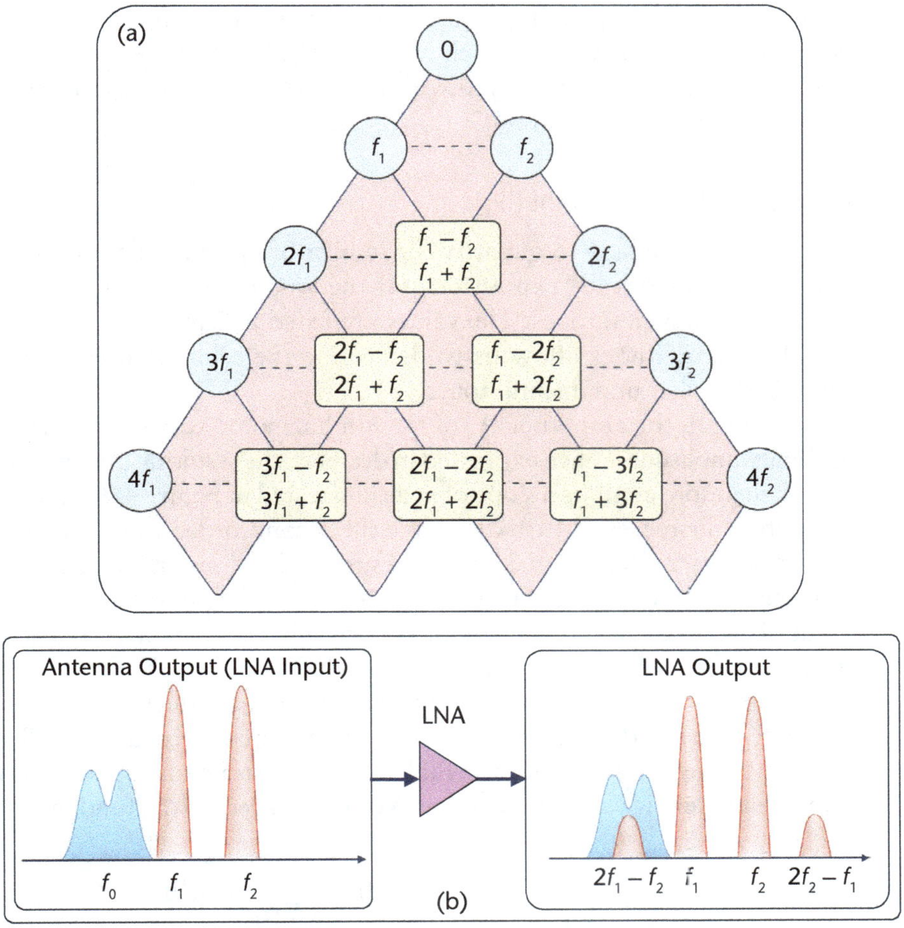

Figure 4.29 (a) Pascal triangle to extract the intermodulation frequencies generated by two input tones at f_1 and f_2. (b) Third-order inermodulation product generated in the LNA interferes with the desired signal.

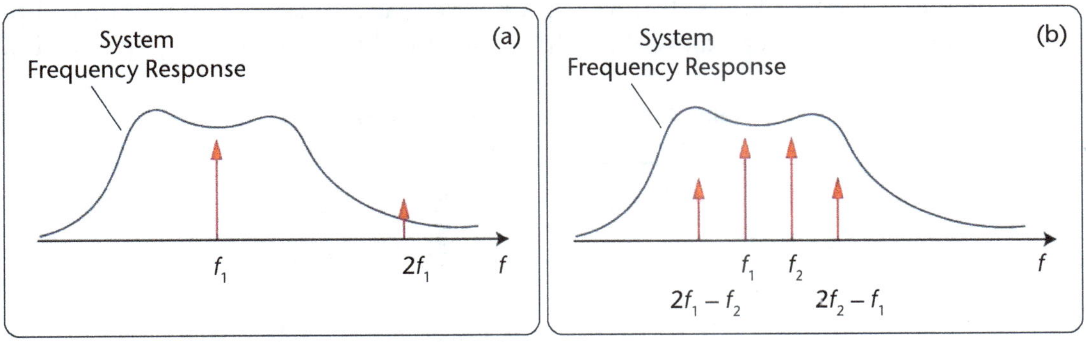

Figure 4.30 (a) The harmonic test alone may not adequately reveal system nonlinearities, since harmonic components can lie outside the system bandwidth, giving a false impression of linear behavior. (b) In contrast, the two-tone test provides a more accurate assessment of nonlinearity, as the IM_3 products appear close to the fundamental tones, typically within the system's bandwidth, where they can directly impact performance.

more linear than it actually is. In contrast, a two-tone intermodulation (IM) test exposes these nonlinearities much more clearly, because the IM_3 products at $2f_1 - f_2$ and $2f_2 - f_1$ almost always fall within the system's operational bandwidth. This effect is illustrated in Figure 4.30(b).

Third-Order Intercept Point (IP_3)

We now examine how to quantify the impact of intermodulation products. Equation (4.38) provides the expressions for the amplitudes of both the IM_3 products and the fundamental tones. The values are listed in Table 4.8.

These amplitudes allow us to determine the relative intermodulation levels compared to the fundamental tones.

In general, the amplitude A_n of the nth harmonic scales with the nth power of the input amplitude. When expressed in decibels, the resulting power transfer characteristic therefore exhibits a slope of n, as illustrated in Figure 4.31. The fundamental component increases at 1 dB/dB, while the second-order and third-order harmonics increase at 2 dB/dB and 3 dB/dB, respectively. Consequently, as the input level increases, second-order and third-order components rise two and three times faster (in decibels) than the fundamental, causing higher-order harmonics to dominate at higher drive levels. Intermodulation products obey the same power-law scaling as harmonics, with an nth-order intermodulation product increasing at n dB/dB of input, which is the basis for the intercept-point concept, as we will see shortly.

Now, using the amplitudes in Table 4.8 for two tones of equal amplitude, we can compute the relative amplitude of the IM_3 with respect to the fundamental tones [6].

$$\text{Relative IM} = 20\log\left(\frac{3}{5}\frac{\alpha_3}{\alpha_1}A^2\right) dBc \tag{4.43}$$

The relative intermodulation value indicates how far below the fundamental tones the intermodulation products appear. However, this ratio depends on the input amplitude and therefore does not provide an input-independent measure of nonlinearity. To address this, we use a metric that characterizes the intrinsic linearity of a circuit regardless of the drive level: the third-order intercept point (IP_3).

Table 4.8 Amplitude of Fundamental Tones and Third-Order Intermodulation Products

Frequency	*Amplitude*
ω_1	$\alpha_1 A_1 + \frac{3}{5}\alpha_3 A_1^3 + \frac{3}{2}\alpha_3 A_1 A_2^2$
ω_2	$\alpha_1 A_2 + \frac{3}{5}\alpha_3 A_2^3 + \frac{3}{2}\alpha_3 A_2 A_1^2$
$2\omega_1 \pm \omega_2$	$\frac{3\alpha_3 A_1^2 A_2}{5}$
$2\omega_2 \pm \omega_1$	$\frac{3\alpha_3 A_1 A_2^2}{5}$

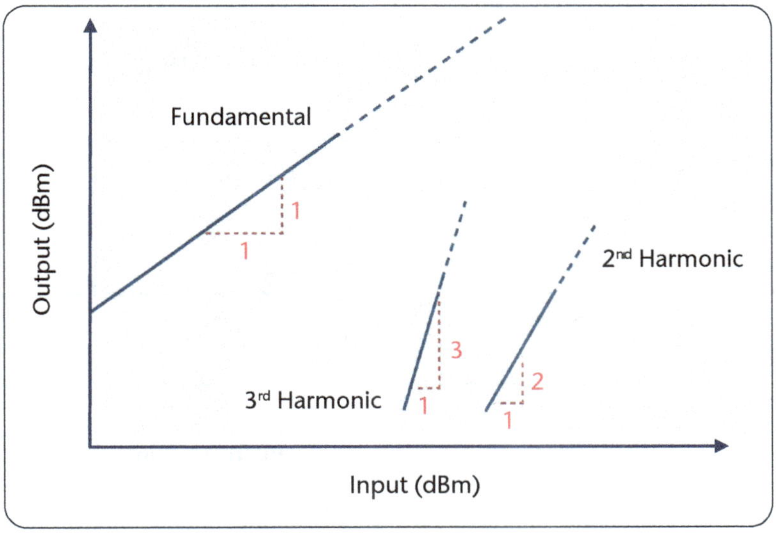

Figure 4.31 The slope of the fundamental tone, the second, and the third harmonic of the power transfer function.

If we plot the output power of the fundamental tones and the third-order intermodulation products versus input power, their extrapolated lines intersect at a hypothetical point, shown in Figure 4.32(a). At this intersection, the amplitudes of the fundamental and third-order products would be equal. This hypothetical intercept defines the IP_3 point. Note that the IP_3 typically lies above the 1-dB compression point.

As illustrated in Figure 4.32(a), the IP_3 can be referenced either to the input or to the output of the nonlinear system. These are denoted IIP_3 (input-referred IP_3) and OIP_3 (output-referred IP_3), respectively, and are related by the small-signal gain G, assuming operation well below compression where the gain remains constant:

$$\mathrm{OIP}_3 = \mathrm{IIP}_3 + G \tag{4.44}$$

where all quantities are expressed in dB or dBm. A higher IP_3 corresponds to a wider linear operating region and improved large-signal linearity.

Note that the IP_3 cannot be measured directly and, in some cases, the extrapolated value may even exceed what the supply voltage could physically deliver. Instead, IP_3 is determined indirectly. The key idea is to relate the relative intermodulation level to the difference between the slopes of the fundamental and third-order curves. The fundamental component increases with a slope of 1, while the third-order intermodulation product increases with a slope of 3; therefore, the difference between them grows with a slope of 2. This relationship allows us to back-calculate the IP_3. Figure 4.32(b) shows the derivation used to obtain the IP_3 expression.

$$\frac{P_f - P_{\mathrm{IM}}}{P_{\mathrm{IIP3}} - P_{\mathrm{in}}} = \frac{\Delta P}{P_{\mathrm{IIP3}} - P_{\mathrm{in}}} = 2 \tag{4.45}$$

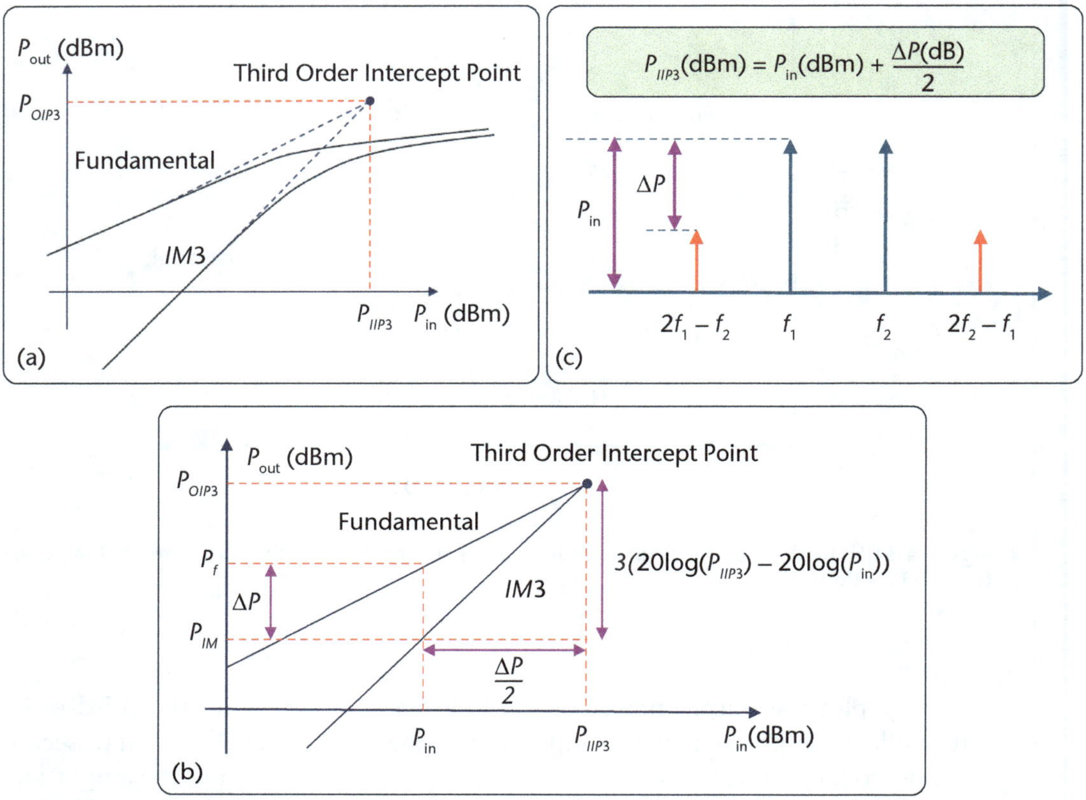

Figure 4.32 (a) Third-order intercept point. IIP3 and OIP3 refer to input and output intercept points, respectively. (b) The plot used to derive the IP_3 relationship. (c) The spectrum consisting of fundamental tones and IP_3 components.

Using (4.44), the input IP_3 can be calculated as follows

$$P_{\mathrm{IIP3}} = P_{\mathrm{in}} + \frac{\Delta P}{2} \tag{4.46}$$

This means that the IP_3 is the input amplitude (in decibels) plus one-half of the intermodulation level (in decibels), as shown in Figure 4.32(c).

Using a polynomial approximation of a nonlinear system, the difference between the input-referred IIP_3 and the 1-dB compression point is approximately 9.6 dB [7]. This relationship is valid only for an ideal third-order nonlinearity and may not hold when higher-order distortion terms become significant. Nevertheless, the 9.6-dB (10-dB) rule of thumb provides a useful engineering approximation: it allows a quick estimate of the IP_3 point when the 1-dB compression point is known, and vice versa.

Eventually, we are interested in the IP_3 of an RF chain. The cascaded IP3 can be calculated as follows [6], where G_i is the linear gain of the ith stage and $IP3_i$ is the third-order intercept point of the ith stage.

$$\mathrm{IP3}_{\mathrm{tot}} = \left(\frac{1}{\mathrm{IP3}_1} + \frac{G_1}{\mathrm{IP3}_2} + \frac{G_1 G_2}{\mathrm{IP3}_3} + \frac{G_1 G_2 G_3}{\mathrm{IP3}_4} + [\ldots]\right)^{-1} \tag{4.47}$$

Second-Order Intercept Point (IP_2)

Similar to the IP_3, the second-order intercept point (IP_2) quantifies a system's susceptibility to second-order intermodulation. It represents the hypothetical input (IIP_2) or output (OIP_2) power level at which the fundamental signal and the second-order intermodulation components would intersect if both were extrapolated using their small-signal slopes.

The IP_2 metric is especially important in direct-conversion (zero-IF) receivers, where second-order nonlinearities generate spurious components near DC (see Figure 4.28(b)). These low-frequency offsets can mask weak desired signals, introduce an unwanted DC offset that shifts the baseline of the demodulated output, and ultimately degrade overall receiver sensitivity.

A higher IP_2 value indicates stronger suppression of second-order nonlinear effects, leading to improved linearity, greater dynamic range, and enhanced receiver performance.

Spectral Regrowth

If we apply the concept of intermodulation product described above to a spectrum instead of two tones, what happens is called spectral regrowth. This is shown in Figure 4.33. This results in unwanted spectral leakage that extends into adjacent frequency channels.

Spectral regrowth is of particular concern in PAs. Its severity is commonly quantified using the ACLR, defined as the ratio of the power leaked into an adjacent channel to the power contained within the main channel. Poor ACLR performance not only reduces spectral efficiency but can also cause interference with neighboring communication channels and violate regulatory limits.

Mitigation techniques include operating amplifiers with sufficient back-off, using linearization methods such as digital predistortion, and employing high-linearity devices and biasing schemes. We will discuss these methods in Chapter 5, as well as in Volume 2.

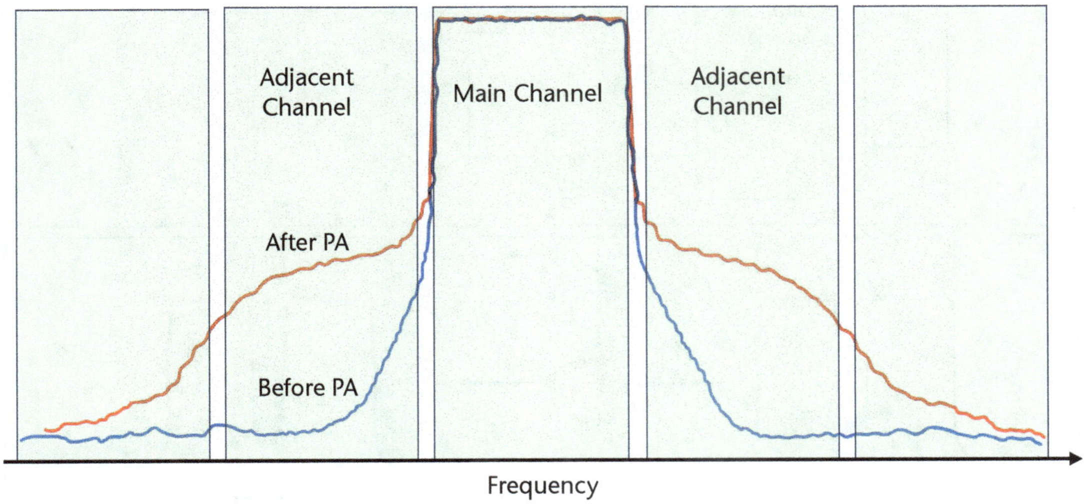

Figure 4.33 Spectral regrowth.

Nonlinearity in Multicarrier RF Systems

Multicarrier systems, especially those using OFDM in 5G and satellite communication systems, pack more data into a given bandwidth by employing multiple orthogonal subcarriers. They are also less susceptible to the effects of multipath fading, as discussed in Chapter 3. Moreover, multicarrier systems can easily support multiple users by assigning different subcarriers to different users in the frequency domain.

Figure 4.34(a) illustrates the high-level structure of a multicarrier transmitter, which forms the basis of technologies such as OFDM. The individually modulated subcarriers are summed to form a composite multicarrier signal, which is then applied to the PA.

If the PA operates in a nonlinear region, the composite signal is distorted, resulting in intermodulation products, ICI, and spectral regrowth that can extend into adjacent channels, causing adjacent-channel interference, degrading signal quality, and reducing overall system efficiency.

These effects are particularly problematic in systems such as OFDM and 5G, where a large number of subcarriers are tightly packed in frequency. For this reason, PA linearity and linearization techniques such as DPD are of critical importance in multicarrier systems.

Noise power ratio (NPR) is a key linearity metric for PAs, particularly suited for multicarrier and wideband systems. The concept is illustrated in Figure 4.34(b). The input test signal used for NPR measurement is a band-limited white-noise waveform containing a narrow spectral notch within the signal band. Nonlinearities in the PA cause spectral energy to leak into this notch.

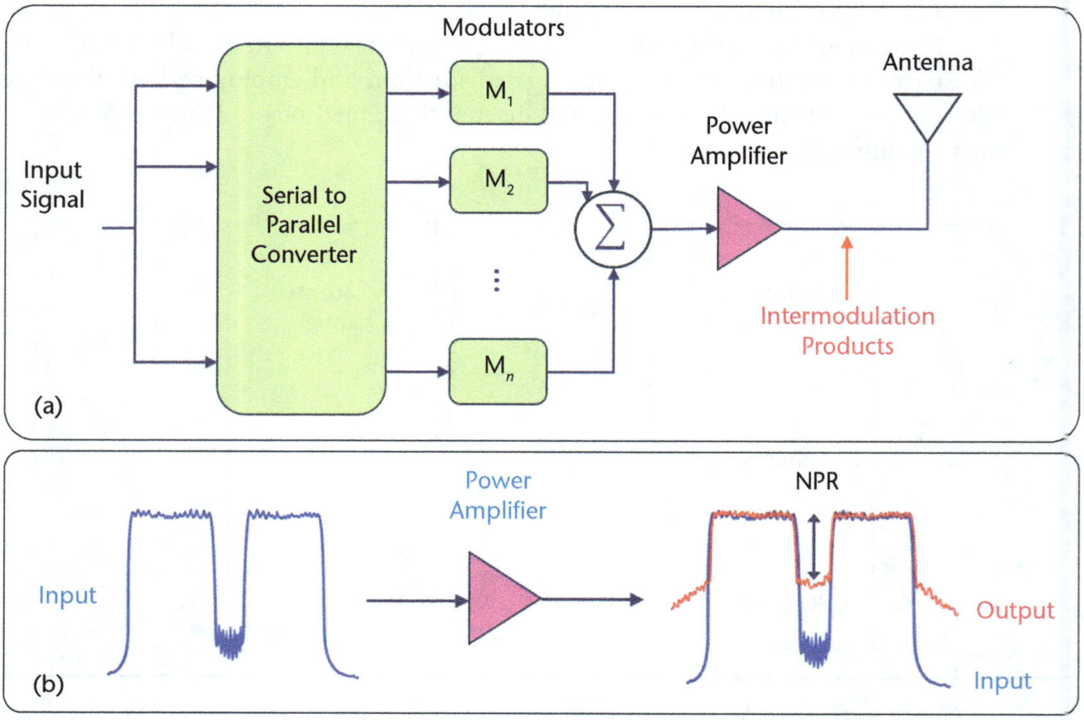

Figure 4.34 (a) Architecture of a multicarrier transmitter. (b) Noise power ratio (NPR) measurement.

As shown in Figure 4.34(b), NPR is defined as the ratio of the power spectral density in the occupied channel to the power spectral density inside the notch. Expressed in decibels, it is the difference between the output channel level and the increased notch level caused by the amplifier's nonlinear distortion.

Passive Intermodulation (PIM)

Passive intermodulation occurs when two or more strong RF signals mix within a passive component, such as connectors, cables, antennas, or waveguides, producing intermodulation products. Unlike active nonlinearities in amplifiers or mixers, passive intermodulation arises from imperfections in passive devices, including microscopic corrosion, poor or contaminated contacts, loose mechanical joints, surface roughness, or the presence of ferromagnetic materials.

These intermodulation products can fall within the operating band of the receiver, raising the noise floor and degrading system sensitivity. Passive intermodulation is especially problematic in cellular base stations, satellite communication systems, and radar applications, where high transmit powers and sensitive receivers, which may not belong to the same system, operate in close proximity.

Mitigating passive intermodulation requires careful material selection (e.g., nonferrous metals), high-quality manufacturing, proper installation, and periodic maintenance of RF connections. If these steps are not followed, interference hunting may become necessary to detect and fix the source of passive intermodulation, which is a process that is often time-consuming, costly, and frustrating.

4.9.6 Cross-Modulation

Cross-modulation occurs when the modulation of a strong signal, usually an interferer, transfers onto a weaker signal as both pass through a nonlinear device. This is shown in Figure 4.35. Unlike intermodulation, which generates new spurious tones, cross-modulation corrupts existing carriers by imposing unwanted modulation. This effect is most common in wideband receivers and multicarrier systems where strong and weak signals share the same front end. The result can be corrupted data, false alarms, or degraded spectral purity. Mitigation relies on highly linear components, effective preselection filtering, and sufficient dynamic range to keep

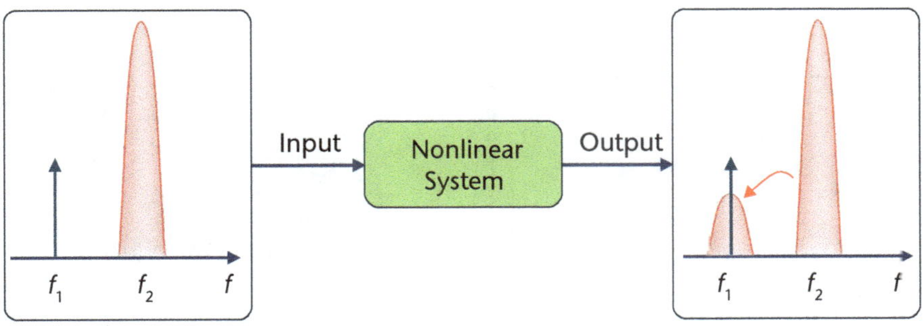

Figure 4.35 Cross-modulation: a strong carrier at f_2 modulates the gain of the nonlinear device, causing amplitude variations that are transferred onto the weaker signal at f_1. As a result, the modulation on the strong signal appears as unwanted modulation (distortion) on the weaker carrier.

strong interferers from distorting weaker signals. A large dynamic range ensures that the receiver front end remains in its linear operating region even in the presence of strong signals, preventing the nonlinear behavior that leads to cross-modulation of weaker desired signals.

4.9.7 AM-AM and AM-PM Conversion

AM-PM conversion occurs when amplitude variations at the input are unintentionally translated into phase variations at the output. This effect can arise from several mechanisms, including fluctuations in the DC supply voltage, instability in the amplifier's bias network, or impedance mismatches that create AM ripple through constructive and destructive interference. These mechanisms introduce AM that is subsequently converted into PM.

AM-PM conversion is particularly critical in systems that rely on accurate phase information, such as QPSK and QAM. Because AM-PM distortion directly rotates constellation points, as illustrated in Figure 4.36(a), it is one of the key contributors to the BER in communication systems. Quantifying AM-PM conversion is therefore essential when evaluating the linearity of RF transmitters and PAs. Similarly, AM-AM conversion refers to the nonlinear relationship between the input and output amplitudes, where changes in input level produce disproportionate or compressed variations at the output. This amplitude distortion typically arises from amplifier nonlinearity or component saturation and leads directly to signal distortion and increased EVM. Instead of preserving a proportional relationship, the amplifier either compresses or expands the signal envelope as the input power varies.

AM-AM distortion affects communication systems in several important ways. For example, because the transmitted constellation points no longer fall at their ideal amplitudes, the constellation becomes "squeezed" or "stretched," which increases EVM and degrades modulation accuracy.

To measure AM-AM and AM-PM characteristics, a power sweep is required, as shown in Figure 4.36(b). Using a vector network analyzer (VNA), the amplifier

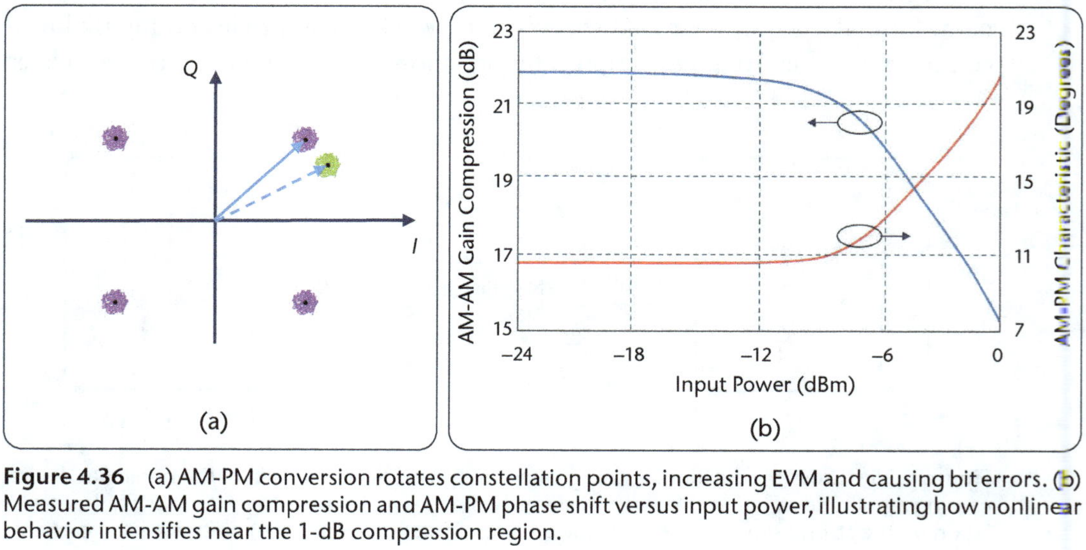

Figure 4.36 (a) AM-PM conversion rotates constellation points, increasing EVM and causing bit errors. (b) Measured AM-AM gain compression and AM-PM phase shift versus input power, illustrating how nonlinear behavior intensifies near the 1-dB compression region.

or the DUT is first connected and properly calibrated. A power sweep is then performed to record both the S_{21} amplitude and S_{21} phase as functions of input power. By examining how a 1-dB change in input power affects the S_{21} amplitude, we obtain the AM-AM conversion, while the corresponding change in the S_{21} phase yields the AM-PM conversion. These measurements are typically performed in the nonlinear region, often around the 1-dB compression point, because, in the linear region, the response is nearly flat, with negligible changes in amplitude or phase, as shown in Figure 4.36(b).

4.9.8 Memory Effects in Nonlinear Systems

Memory effects refer to a class of nonlinear behaviors in RF components, particularly PAs, where the output at a given moment depends not only on the current input but also on past input values. Unlike "memoryless" nonlinearities (such as AM-AM and AM-PM), memory effects are dynamic and time-dependent, often caused by thermal, electrical, or bias-related mechanisms inside the device. Increased spectral regrowth, constellation distortion, and reduced DPD effectiveness are among the key side effects of memory effects in RF amplifiers. Table 4.9 shows some of the memory effects in RF amplifiers.

4.10 Dynamic Range

So far, we have examined how the performance of an RF system is limited at both small and large signal levels. The upper limit corresponds to large signals where nonlinear effects, such as gain compression and intermodulation, begin to dominate. In contrast, the lower limit corresponds to weak signals, where the amount of noise determines the ability to detect or demodulate the signal successfully.

Table 4.9 Memory Effects in RF Amplifiers: Sources, Impact, and Mitigation

Source of Memory Effect	*Description*	*Impact on Signal*	*Mitigation Techniques*
Thermal effects	Heating of transistor junctions or substrate causes time-varying changes in gain and phase	Dynamic AM-AM and AM-PM distortion; slow drift in output power	Improve thermal management and heat sinking to minimize temperature variation
Bias circuit dynamics	Bias networks with large time constants lead to slow changes in the device operating point	Gain and phase drift; increased spectral regrowth	Design bias networks with optimized time constants and decoupling
Trapping effects	Charge trapping in active devices (e.g., GaN HEMTs) causes slow recovery and hysteresis	Distortion dependent on input history; degraded EVM and ACLR	Use materials and device structures with reduced trapping; apply bias conditioning
Impedance mismatch/ frequency-dependent loads	Frequency-selective matching or reflective environments introduce time-varying impedance, especially for wideband signals	Envelope distortion and reduced linearity across frequency; poorer DPD convergence	Optimize matching networks across the full signal bandwidth; use broadband or adaptive matching

The dynamic range of an RF system is defined as the difference between the largest signal that it can process without entering compression or other nonlinear regions and the weakest signal that it can reliably detect, known as its sensitivity. Figure 4.37(a) illustrates this concept: the lower bound is set by the receiver noise floor (i.e., the integrated noise power), and the sensitivity corresponds to the minimum detectable signal discussed in Section 4.11.1.

As an example, a modern cellular receiver can achieve a dynamic range on the order of 100 dB.

Another important definition of dynamic range is the spurious-free dynamic range (SFDR), which accounts for the combined effects of noise and nonlinear distortion. In this case, the lower limit is still set by the receiver sensitivity. However, the upper limit is defined by the maximum input level in a two-tone test for which the IM_3 products remain below the receiver's noise floor.

As illustrated in Figure 4.37(b), SFDR represents the range of input signal levels over which the system operates without spurious components rising above the noise floor, and is given by [6]

$$\text{SFDR} = P_{\text{in,max}} - (-174 \text{ dBm} + \text{NF} + 10\log B + \text{SNR}_{\text{min}}) \quad (4.48)$$

Here, $P_{\text{in,max}}$ is the maximum input signal power before spurious components exceed the noise floor, NF is the noise figure of the system, −174 dBm is the thermal noise power in a 1-Hz bandwidth at room temperature, and SNR_{min} is the minimum SNR required to meet the specified performance.

For a system limited by third-order intermodulation distortion, the SFDR can be expressed as

$$\text{SFDR} = \frac{2}{3}(P_{\text{IIP3}} + 174 \text{ dBm} - \text{NF} - 10\log B) - \text{SNR}_{\text{min}} \quad (4.49)$$

For example, a GSM receiver with NF = 7 dB, $P_{\text{IIP3}} = -15$ dBm, and $\text{SNR}_{\text{min}} =$ 12 dB achieves an SFDR of approximately 55 dB, which is substantially lower than the dynamic range without interference.

Similar to dynamic range, the EVM of a system is also limited by both noise and linearity. When EVM is plotted against operating power, the well-known bathtub curve shown in Figure 4.38 emerges. At low power levels, EVM performance

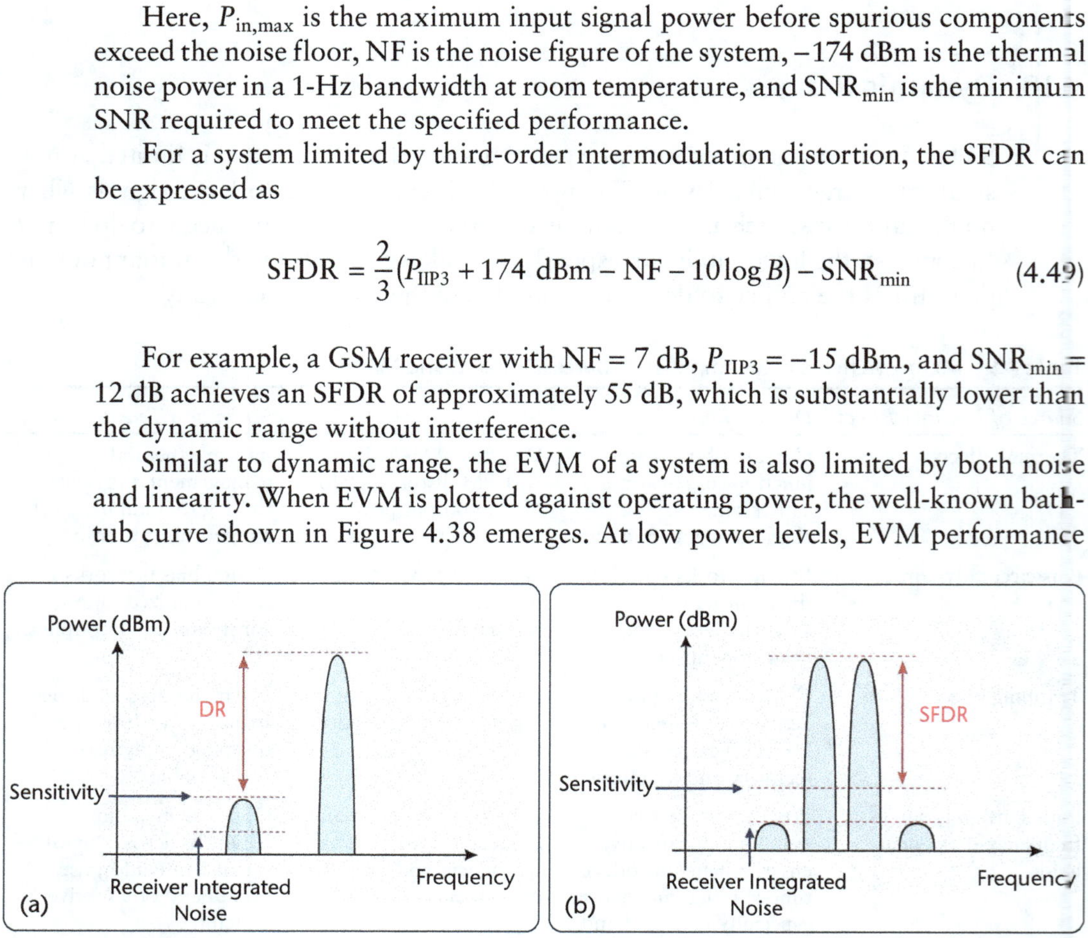

Figure 4.37 (a) Dynamic range. (b) Spurious free dynamic range (SFDR).

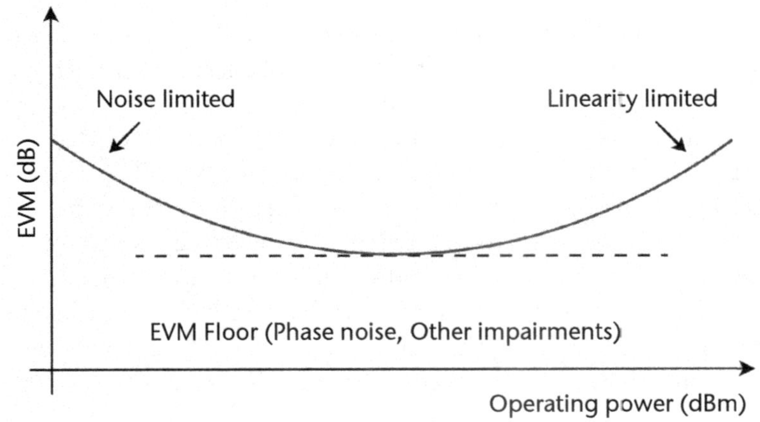

Figure 4.38 Bathtub EVM curve.

is dominated by noise, whereas, at high power levels, it is dominated by nonlinear distortion.

For a given operating point, the total EVM is the combination of contributions from white noise, phase noise, and linearity-related distortion, with their sum determining the overall EVM level of the system.

$$\mathrm{EVM}^2_{\mathrm{total}} = \mathrm{EVM}^2_{\mathrm{White\ noise}} + \mathrm{EVM}^2_{\mathrm{phase\ noise}} + \mathrm{EVM}^2_{\mathrm{linearity}} \tag{4.50}$$

4.11 Detection in Receivers

So far, we have examined several signal degradation mechanisms, attenuation, interference, distortion, and noise, and their impact on overall system performance. In this section, we introduce key detection-related parameters, such as the MDS, signal-to-interference ratio (SINR), signal-to-noise and distortion ratio (SINAD), carrier-to-noise ratio (C/N), and gain-to-noise temperature ratio (G/T). We discuss how these metrics are interrelated and clarify which parameter is most appropriate for various applications. Table 4.10 summarizes these parameters along with their typical use cases.

Table 4.10 Representative System Metrics and Their Primary Roles in Performance Evaluation

System Type	*Key Metrics*	*Primary Purpose*
Satellite systems	C/N, G/T, sometimes SNR	Link budget, receiver sensitivity, system performance over propagation losses
Analog systems	SNR, SINAD	RF front-end quality, receiver sensitivity, and distortion performance
Digital systems	EVM, SNR, SINR, E_b/N_0	Modulation accuracy, BER, throughput, interference resilience

4.11.1 Minimum Detectable Signal

The minimum detectable signal, also known as receiver sensitivity, defines the minimum input signal power required for reliable detection under specified conditions. The MDS depends on the modulation scheme and the acceptable level of signal degradation, typically specified by a target BER or SINAD.

Let's start by quantifying the impact of noise. The receiver's noise floor is determined as follows. At room temperature ($T \approx 290$K), the thermal noise power density (i.e., the noise power per unit bandwidth) is approximately $kT \approx -174$ dBm/Hz. For a system with bandwidth B, the total thermal noise power increases to $-174 + 10\log_{10}(B)$ dBm, since a wider bandwidth admits more noise into the detector. Finally, adding the noise figure NF of the receiver, which represents the noise contributed by the receive chain itself, yields the total system noise floor, given by [1]

$$\text{Noise floor (dBm)} = -174 + 10\log(B) + \text{NF} \tag{4.51}$$

To achieve reliable detection under the specified criterion, the received signal must exceed the receiver noise floor by a margin set by the minimum required SNR for the chosen modulation and the allowable impairment level (e.g., target BER or SINAD). For example, QPSK may require approximately 10-dB SNR to achieve BER < 10^{-5} in an additive white Gaussian noise (AWGN), channel without coding. The required SNR typically varies between 3 dB and 25 dB for different types of receivers. We define the MDS or sensitivity as follows:

$$\text{MDS (dBm)} = \text{Noise floor (dB)} + \text{SNR (dB)} = -174 + 10\log(B) + \text{NF} + \text{SNR} \tag{4.52}$$

For example, consider a receiver with a bandwidth of 100 MHz and a noise figure of 2 dB, resulting in a noise floor of −92 dBm. If the minimum required SNR for reliable detection is 10 dB, then the receiver's sensitivity, or MDS, is −82 dBm. This means that the power arriving at the receiver must exceed −82 dBm to ensure reliable signal detection.

This requirement highlights the importance of link budget analysis, which determines the necessary transmitted power and system gains and losses to achieve the required MDS at the receiver. Table 4.11 shows the MDS values for GSM and Wi-Fi standards. Although GSM exhibits better sensitivity, Wi-Fi supports significantly higher bandwidth and data rates. Therefore, higher sensitivity alone does not imply superior overall performance; both sensitivity and data rate must be considered when comparing different systems [6].

Table 4.11 Comparison of GSM and Wi-Fi Standards

Wireless Standard	*Channel Bandwidth*	*Minimum SNR*	*Receiver NF*	*Sensitivity (MDS)*	*Data Rate*
GSM	200 kHz	10 dB	5 dB	–106 dBm	270 kbps
Wi-Fi	20 MHz	20 dB	5 dB	–76 dBm	55 Mbps

4.11.2 SNR and (E_b/N_0) Review

We discussed SNR in Section 4.7.3 and the energy per bit to noise power spectral density ratio (E_b/N_0) in Section 3.5.4. The ratio E_b/N_0 is a normalized measure of SNR used in digital communications to compare link performance independently of bandwidth or bit rate. If R_b is the bit rate and B is the bandwidth, the SNR and (E_b/N_0) are related as SNR = (E_b/N_0) × (R_b/B).

E_b/N_0 is useful because:

- It removes the effect of symbol rate and bandwidth from SNR comparisons.
- It enables direct comparison of different modulation and coding schemes for the same BER target.
- It serves as a key parameter in link budget calculations when determining required signal levels.

We saw in Section 3.5.4 that, for various modulations, the BER is closely linked to the E_b/N_0.

4.11.3 Signal-to-Interference-Plus-Noise Ratio (SINR)

If we extend the concept of SNR to include the effect of interference, we obtain a new metric: the SINR. It quantifies the ratio of the desired signal power to the combined power of interference and noise:

$$\text{SINR} = \frac{P_{\text{signal}}}{P_{\text{interference}} + P_{\text{noise}}} \tag{4.53}$$

It represents a fundamental performance metric in modern wireless systems, especially in interference-limited environments such as cellular networks, dense IoT deployments, or satellite constellations, where cochannel and adjacent-channel interference often exceed thermal noise.

When interference becomes significant, the MDS expression generalizes by replacing the required SNR with a required SINR, ensuring that the combined effect of thermal noise and interference is accounted for when determining the minimum detectable signal.

A high SINR allows the receiver to use higher-order modulation schemes, improving throughput and spectral efficiency. Conversely, a low SINR forces the system to fall back to more robust modulation and coding formats to maintain reliable communication.

Because of its direct link to achievable data rate and link robustness, SINR is widely used in adaptive modulation and coding (AMC), link-adaptation, and handover algorithms, serving as a dynamic indicator of link quality in real-time channel management.

4.11.4 Signal-to-Noise and Distortion Ratio (SINAD)

Unlike SNR, which considers only noise and is commonly used in systems where linearity is assumed or distortion is negligible, and unlike SINR, which includes noise and interference, the SINAD accounts for both noise and distortion. The SINAD is defined as:

$$\text{SINAD} = \frac{P_{\text{signal}} + P_{\text{noise}} + P_{\text{distortion}}}{P_{\text{noise}} + P_{\text{distortion}}} \tag{4.54}$$

This makes it a more realistic performance metric for analog or mixed-signal receivers, such as FM receivers, ADC-based systems, or software-defined radios, where nonlinearities can significantly affect signal quality. A high SNR may still yield poor detection if distortion is present, but a high SINAD ensures that both noise and distortion are sufficiently low.

In many analog communication systems (e.g., FM voice receivers), a SINAD of 12 dB is widely accepted as the minimum required for intelligible audio. For more demanding systems (e.g., high-quality audio, data transmission), a higher SINAD threshold (20–30 dB or more) may be required.

SINAD is also widely used to characterize receiver sensitivity, dynamic range, and ADC performance. In ADCs, SINAD is directly related to the ENOB and serves as a benchmark for overall signal fidelity. The relationship is given by

$$\text{ENOB} = \frac{\text{SINAD} - 1.76}{6.02} \tag{4.55}$$

4.11.5 Carrier-to-Noise Ratio (CNR)

In terrestrial or laboratory systems, engineers typically refer to the SNR. However, in satellite communications, a different metric, carrier-to-noise ratio (C/R), is used to separate the effects of the link (propagation path, RF front end, and antenna system) from those of detection (demodulation and decoding). As we will see shortly, C/N is defined as the ratio of the carrier power to the noise power.

As shown in Figure 4.39, C/N in a typical receiver chain is measured at the output of the analog front end, just before the ADC stage. At this point, the signal has passed through the antenna, LNA, mixers, and IF filters, so the ratio reflects all propagation losses, antenna gains, and the system noise temperature. It quantifies how "clean" the carrier is before any digital processing.

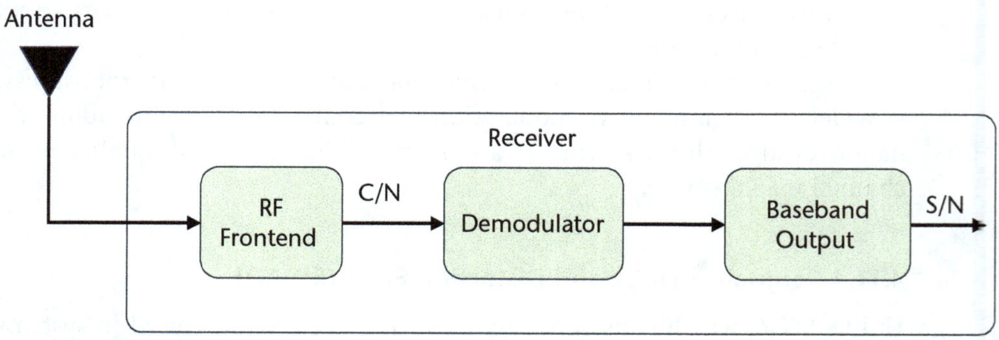

Figure 4.39 Relationship between C/N and S/N measurement points in a receiver chain.

As shown in Figure 4.39, once the signal is digitized, the SNR is evaluated in the baseband domain, after demodulation and decoding. Therefore, SNR includes not only the original C/N but also implementation impairments such as oscillator phase noise, nonlinear distortion, ADC quantization noise, sampling jitter, and other digital imperfections. This makes SNR a more comprehensive and practical metric for assessing end-to-end system performance and user-perceived quality.

Typically, satellite engineers design for specific C/N targets at the receiver input to guarantee a minimum E_b/N_0, which directly determines BER. Because multiple terminals may have different bit rates, bandwidth, modulation, and coding schemes, expressing performance in C/N keeps the metric independent of the particular baseband implementation.

The SNR and C/N are related through the receiver's processing efficiency and implementation quality:

$$\text{SNR} = \frac{C}{N} - L_{\text{impl}} \tag{4.56}$$

where L_{impl} represents implementation losses due to demodulator imperfections, filtering effects, phase noise, ADC quantization, and other receiver nonidealities. The typical values for a well-designed system are L_{impl} = 1–2 dB, while for high-order modulation schemes such as 64-QAM/OFDM, implementation losses are often higher, on the order of L_{impl} = 3–5 dB. This relationship highlights that while C/N characterizes the channel quality, the effective SNR achieved after demodulation depends on the receiver's internal design and signal processing performance.

The CNR is a measure of the strength of the received carrier signal relative to the background noise power within a specified bandwidth, and is defined as:

$$\frac{C}{N} = \frac{P_{\text{carrier,RX}}}{KT_{\text{sys}}B} \tag{4.57}$$

where $P_{\text{carrier, RX}}$ is the received carrier power at the receiver input, and $N = kT_{\text{sys}}B$ is the system noise power. Here k is Boltzmann's constant, T_{sys} is the system noise temperature ($T_{\text{sys}} = T_{\text{ant}} + T_{\text{RX}} + T_{\text{feed}}$), and B is the receiver bandwidth.

It can be easily shown that

$$\left(\frac{C}{N}\right)_{\text{dB}} = \text{EIRP}_{\text{dB}} + \left(\frac{G}{T}\right)_{\text{dB/K}} - L_{p,\text{dB}} - 10\log_{10}(k) - 10\log_{10}(B) \tag{4.58}$$

where $L_{p,\text{dB}}$ is the propagation loss, G/T is the receiver figure of merit (explained in the next section), and EIRP is the equivalent isotropically radiated power expressed in decibels. As we will see shortly, EIRP is simply the sum of the transmit power and the transmit antenna gain, both expressed in decibels: $\text{EIRP}_{\text{dB}} = P_{\text{TX,dB}} + G_{\text{TX,dBi}}$.

Note that measuring C/N with a spectrum analyzer or carrier-monitoring system is relatively straightforward, as it requires only the received carrier power and the noise floor measured over the same bandwidth. In contrast, measuring SNR at the baseband output is more challenging, as it requires demodulation and depends on the internal characteristics and signal-processing performance of the receiver.

4.11.6 Gain-to-Noise Temperature Ratio (G/T)

As shown in the previous section, G/T appears in the C/N expression, which is one of the most critical metrics for assessing the performance of a receiving system in satellite communications, radar, and deep-space links. It represents the ratio of antenna gain to system noise temperature:

$$\frac{G}{T} = \frac{G_{\text{ant}}}{T_{\text{sys}}} \tag{4.59}$$

where G_{ant} is the antenna gain relative to an isotropic radiator (linear scale), and T_{sys} is the system noise temperature (Kelvin), including contributions from the antenna, atmosphere, and receiver noise figure.

In practice, G/T is expressed in decibels per Kelvin (dB/K):

$$\left(\frac{G}{T}\right)_{\text{dB/K}} = G_{\text{dBi}} - 10\log_{10}\left(T_{\text{sys}}\right) \tag{4.60}$$

A higher G/T value indicates a more sensitive receiving system, capable of achieving a higher (C/N) for a given uplink EIRP and path loss. The typical values for G/T are:

- Small satcom ground terminals: 10–20 dB/K;
- Large Earth stations: 25–40 dB/K;
- Deep space networks: >50 dB/K.

Designers can increase G/T by either improving antenna gain or lowering noise temperature. Both paths come with practical constraints (mechanical size, cooling requirements, cost).

1. Antenna gain (G):
   - Larger apertures or higher frequencies improve directivity and thus gain.
   - Mechanical pointing accuracy (due to the very narrow beam) and surface tolerance, the precision with which the antenna's physical surface matches its ideal shape, become limiting factors at high gain.
2. System noise temperature (T_{sys}):
   - Contributions include antenna noise temperature (spillover, sky noise), LNA noise figure, and subsequent receiver stages ($T_{\text{sys}} = T_{\text{ant}} + T_{\text{RX}} + T_{\text{feed}}$).
   - Cooling the RF front end (e.g., cryogenic amplifiers in radio astronomy or deep-space communications) dramatically reduces T_{sys}.

G/T can be measured through the Y-factor method using hot and cold noise sources, where we will discuss this in more detail in Volume 2. Also, the satellite beacon method can be used, where a known EIRP from a satellite is used to back-calculate system G/T.

Figure 4.40 shows a planar phased array located in the XY plane pointing to a satellite. For phased arrays, the gain G of the terminal is calculated as

$$G(\text{dBi}) = \text{Peak Gain}(\text{dBi}) - \text{cosine roll-off} \times 10 \times \log(\cos\theta) \tag{4.61}$$

The peak gain (dBi) represents the maximum antenna gain when the broadside and boresight directions are aligned ($\theta = 0$). The cosine roll-off, also known as scan roll-off, is a coefficient that characterizes the reduction in antenna gain as the scan angle (θ) increases. It is typically provided by the manufacturer and depends on the antenna design and aperture illumination profile. For electronically scanned arrays, a cosine roll-off factor of approximately 1.1 to 1.3 is commonly used.

So far, we have examined how various parameters, such as SNR, C/N, E_b/N_0, G/T, SINR, and SINAD, are used in different contexts to evaluate detection performance. Ultimately, the link budget analysis discussed in Section 4.12.2 will quantify the received signal power, incorporating contributions from the transmitter (antenna gain and radiated power), the propagation channel (losses), and the receiver (antenna gain and internal losses).

4.11.7 Jamming-to-Signal Ratio (J/S)

In EW systems, an additional parameter commonly used to assess the impact of intentional interference is the J/S. The J/S compares the received jammer power to the received desired signal power at the victim receiver. Although it is not discussed in detail here, it is worth briefly introducing this metric; comprehensive treatments can be found in the literature [9].

Consider a representative scenario consisting of a transmitter (e.g., a drone control link), a receiver on board the drone, and a jammer attempting to disrupt the

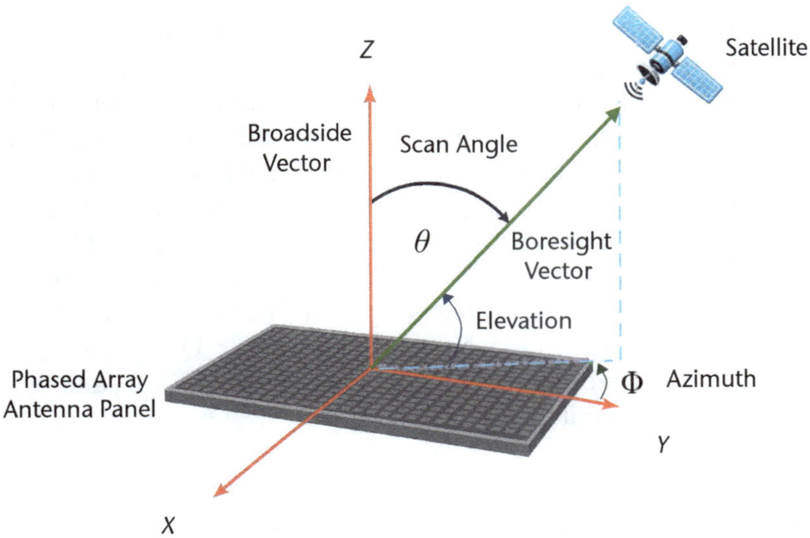

Figure 4.40 A phased array is placed on the XY plane, pointing to a satellite. Theta (θ) is the scan angle. This is the angle between the boresight vector and the broadside vector.

link. To successfully deny or degrade communication between the transmitter and receiver, the jammer must produce a signal at the receiver that exceeds the desired signal by a certain margin. This required margin is quantified by the J/S threshold. Problem 4.11 also walks through the derivation of J/S for a representative drone control-link scenario.

The required J/S depends strongly on the jamming technique employed. Simple noise jamming typically requires a relatively high J/S, as the jammer must raise the interference level across the receiver bandwidth. In contrast, waveform-aware or matched jamming techniques can achieve link disruption at significantly lower J/S values, since the jammer exploits knowledge of the signal structure, modulation, or protocol.

4.12 System Planning: Frequency Management and Link Budget Analysis

Having examined how the key parameters of an RF system shape its performance and influence signal detection at the receiver, we now step back to consider the system from a broader perspective. In particular, we focus on how to select and evaluate fundamental parameters, such as operating frequency, bandwidth, and transmit power, that define the system architecture and ultimately determine overall performance.

Because the ultimate objective is to design a practical RF product, some of the earliest and most consequential decisions involve the choice of modulation scheme, operating frequency, and bandwidth. These parameters fundamentally constrain the system architecture, set performance limits, and even influence the physical realization of the final product.

- *Frequency:* A millimeter-wave design at 28–40 GHz looks nothing like an L-band or S-band RF design, the antennas become tiny and highly directional, losses increase dramatically, packaging shifts to microwave-grade structures, propagation losses become significant, and component cost rises sharply.
- *Bandwidth:* A wideband 2–18-GHz radar forces broadband architecture, distributed matching networks, wideband LNAs and mixers, high-speed digitizers, and challenging calibration, completely different from a narrowband system such as a 25-kHz land-mobile radio or a 200-kHz NB-IoT link, where filters, gain stages, and sampling requirements are far easier and far cheaper.
- *Modulation:* A low-order scheme such as BPSK can tolerate far more noise and nonlinearity, whereas higher-order QAM (64-QAM, 256-QAM) demands excellent SNR, strict linearity, tight EVM, and significant back-off in the transmitter.

These examples illustrate how frequency, bandwidth, and modulation are not abstract choices, they fundamentally dictate the system architecture, component technologies, performance limits, and SWaP-C of the final product.

Ultimately, these parameters establish the foundation for link budget analysis, influencing every aspect of system performance and feasibility.

4.12.1 Spectrum Allocation

Frequency, bandwidth, and modulation are tightly interconnected, and choosing one often constrains the others. Operating frequency dictates propagation loss, antenna size, coverage, component technology, and regulatory limits. Bandwidth sets the ceiling for data rate but also drives requirements for noise performance, filtering, linearity, and ADC/DAC speed. Modulation determines how efficiently the bandwidth is used and directly sets the required SNR and linearity, influencing hardware complexity and link robustness. In practice, selecting the right combination is a multidimensional trade-off, balancing performance targets, environmental conditions, cost, and regulatory constraints to arrive at a feasible product architecture.

Before moving on, it is essential to clarify the meaning of two terms: band and channel. The term "band" refers to the entire frequency spectrum defined by a standard, such as the GSM band, that ranges from 935 MHz to 960 MHz. A "channel" represents a specific portion of a band allocated for the usage of a single user. For example, in the case of GSM, the channel bandwidth is typically 200 kHz.

The choice of frequency and bandwidth for a communication link depends on a combination of technical, regulatory, and practical considerations. In some cases, these parameters are predetermined by the customer or the application, such as an IoT device operating in an Industrial, Scientific, and Medical (ISM) band, which is a license-free frequency range allocated for low-power devices under specific regulatory limits. In other cases, such as a quantum computing control and readout chains, it may fall on the designer to select the operating frequency.

Frequency Selection

It is important to emphasize that acquiring a new frequency allocation for communication applications can be extremely costly and often involves lengthy regulatory approval processes. Whenever possible, it is advisable to use existing, already-allocated bands, unless there is a compelling technical or operational reason to justify the significant expense and effort of securing a new one.

Table 4.12 summarizes various RF bands alongside their typical applications, each imposing distinct constraints on transceiver performance, architecture, and integration strategy.

The key factors influencing the choice of frequency are summarized in Table 4.13.

Table 4.12 An Overview of Various RF Bands and Their Associated Transceiver Applications

Band	*Frequency Range*	*Common Applications*
L-band	1–2 GHz	GNSS, datalinks, identification friend or foe (IFF)
S-band	2–4 GHz	Radar, weather, WiMAX
C-band	4–8 GHz	Satellite communications, radar
X-band	8–12 GHz	Military radar, missile guidance
Ku-band	12–18 GHz	Satellite, tactical communications
K-band	18–27 GHz	Police radar, imaging
Ka-band	27–40 GHz	High-throughput SATCOM, 5G
V/W-band	40–110 GHz	Automotive radar, point-to-point links

Table 4.13 Frequency Selection Factors

Factor	*Description*
Regulatory allocations	Only certain bands are available for specific services (e.g., ITU, FCC rules)
Propagation characteristics	Lower frequencies propagate farther and are more susceptible to Faraday rotation (e.g., SATCOM); higher frequencies suffer greater propagation loss but offer wider bandwidths and higher achievable data rates
Antenna size and gain	Higher frequencies allow smaller antennas for a given gain
Interference environment	Level of congestion and presence of other services in the band
Hardware availability	The cost and maturity of RF components vary significantly across frequency bands; higher operating frequencies require more specialized components and manufacturing processes, increasing hardware and measurement costs; component availability may also be limited, impacting timelines and supply chain reliability
Doppler shift sensitivity	Higher frequencies are more affected by motion-induced frequency changes

Bandwidth Allocation

The total bandwidth required depends heavily on the application. Narrowband systems are generally easier to design and implement, with bandwidths as small as a few megahertz. In contrast, wideband systems, such as ultrawideband (UWB) radars, can span frequencies from 2 GHz to 18 GHz, placing significant demands on the system design. These demands affect nearly every aspect of the RF chain, including filtering, linearity, noise performance, and hardware selection. Wideband designs also require tighter synchronization, higher-performance ADCs and DACs, and more advanced calibration techniques to maintain performance across the entire band.

Table 4.14 summarizes the key factors influencing the choice of bandwidth.

Table 4.14 Bandwidth Selection Factors

Factor	*Description*
Required data rate	Wider bandwidth increases capacity per Shannon's formula
Spectrum availability	Regulatory limits on maximum allowable bandwidth
SNR and link budget	Wider bandwidth reduces SNR because thermal noise power increases linearly with bandwidth, and the signal power is effectively spread over a larger frequency range, unless transmit power is increased to compensate
Adjacent channel interference	Need to meet spectral mask requirements
Modulation and coding scheme	Higher-order modulations may require more bandwidth
Filter and ADC/DAC requirements	Wider bandwidth demands higher-performance ADCs and DACs
Multipath and delay spread	Wideband signals may require countermeasures against frequency-selective fading

Spectral Mask

A spectral mask defines the maximum allowed signal power at various frequency offsets from the carrier, typically specified by regulatory bodies or communication standards. It ensures that the transmitted signal stays within its assigned channel and does not interfere with adjacent channels.

The spectral mask is usually expressed as a plot of relative power (dBc) or power density (dBm/Hz) versus frequency offset (Hz) from the carrier, as shown in Figure 4.41. Compliance is verified through spectrum analyzer measurements with defined resolution bandwidths.

The spectral mask will affect:

- *Filter requirements:* The RF chain must include adequate filtering to meet out-of-band emission limits.
- *PA linearity:* Nonlinear distortion can generate spectral regrowth, pushing emissions outside the mask.
- *LO purity:* Phase noise close to the carrier can cause violations near the mask's edge.
- *Regulatory compliance:* Failure to meet the spectral mask can lead to fines, certification failure, or denied spectrum access.

For example, an LTE 20-MHz channel might require emissions to be at least 45 dB below the carrier at ±10-MHz offset, and 50 dB down at ±15 MHz.

4.12.2 Link Budget Analysis

Before implementing any RF system, whether it is a jammer, a quantum computer, or a cellular communication link, it is essential to perform a preliminary system

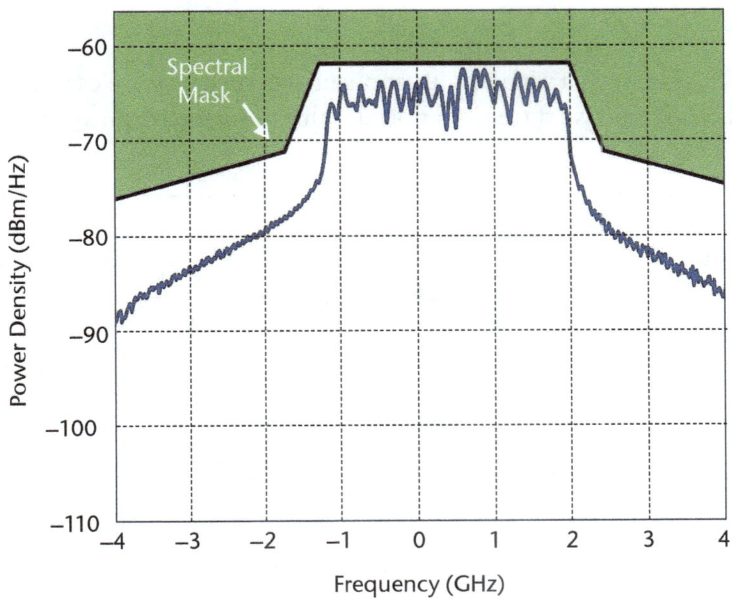

Figure 4.41 Transmitter spectral mask.

evaluation through link budget analysis. The analysis is strongly influenced by requirements such as modulation scheme, throughput, and spectral efficiency from the signal-chain layer, as well as operating frequency and available bandwidth determined by the physical channel. In this section, we will explore the core concepts underlying link budget calculations, including the Friis transmission formula and the EIRP.

Friis Formula

The Friis transmission formula describes the received power at an antenna under free-space, LOS conditions. Figure 4.42(a) illustrates a communication link between a transmitter and a receiver separated by a distance r, with transmit power P_{TX}, receive power P_{RX}, and antenna gains G_{TX} and G_{RX}, operating at the wavelength λ. The received power is given by [4]

$$P_{RX} = P_{TX} G_{TX} G_{RX} \left(\frac{\lambda}{4\pi r} \right)^2 \tag{4.62}$$

In logarithmic (dB) form:

$$P_{RX(dB)} = P_{TX(dB)} + G_{TX(dB)} + G_{RX(dB)} - 20\log_{10}\left(\frac{4\pi r}{\lambda} \right) \tag{4.63}$$

The last term represents FSPL.

Consider a drone-controller link at 2.4 GHz under free-space, LOS conditions. Both ends use antennas with $G_{TX} = G_{RX} = 3$ dBi. The drone transmits $P_{TX} = 1\text{W} = 30$ dBm. Figure 4.42(b) plots the received power P_{RX} versus separation from 100m to 2 km using the Friis transmission formula.

EIRP

In RF systems, particularly in applications such as wireless communication, radar, and electronic warfare, it is not just the transmitted power that matters, but how

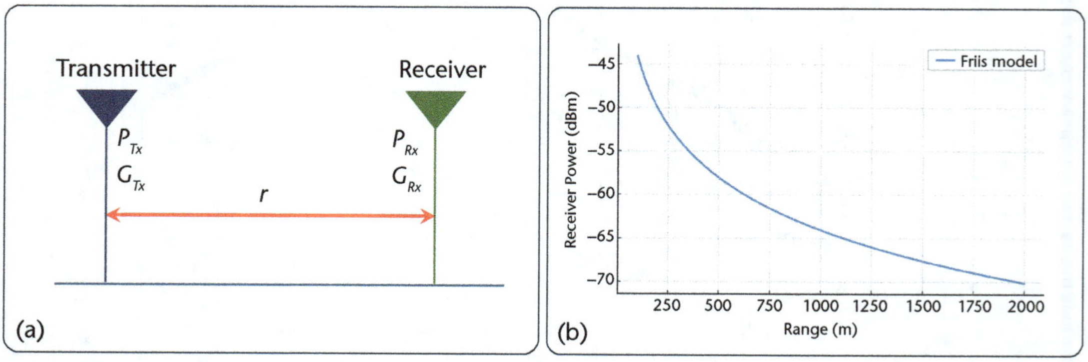

Figure 4.42 (a) Free-space communication link illustrating the Friis transmission model. A transmitter with power P_{TX} and antenna gain G_{TX} sends a signal across a distance r to a receiver with antenna gain G_{RX}, resulting in received power P_{RX}. (b) Received power versus range for a 2.4-GHz drone-controller link.

effectively that power is radiated in the desired direction. Two commonly used metrics for quantifying are effective radiated power (ERP) and EIRP.

ERP is defined as the transmitted power referenced to a half-wave dipole antenna, whereas EIRP is referenced to an ideal isotropic radiator and is more commonly used in RF system design.

Both metrics account for the gain of the transmitting antenna and are calculated as:

$$\text{EIRP}(\text{dBm}) = P_{\text{out}}(\text{dBm}) + G_{\text{ant}}(\text{dBi}) = P_{\text{out}}(\text{dBm}) + G_{\text{ant}}(\text{dBd}) + 2.15 \quad (4.64)$$

$$\text{ERP}(\text{dBm}) = P_{\text{out}}(\text{dBm}) + G_{\text{ant}}(\text{dBd}) \quad (4.65)$$

As can be seen from (4.64) and (4.65), EIRP is always 2.15 dB higher than ERP for the same setup. Let's consider a transmitter with a final-stage PA output of 50 dBm (100W). If the antenna provides a gain of 9 dBi in the desired direction, the resulting EIRP becomes 59 dBm, which corresponds to approximately 794W. This does not mean that additional power has been created; rather, the antenna focuses the available power into a narrower beam. Through this focusing effect, constructive interference in the main lobe increases the power density in the desired direction, while reducing it elsewhere. In other words, antenna gain redistributes power spatially without violating energy conservation.

In applications such as jamming, achieving sufficiently high EIRP is critical. The higher the EIRP, the further the jamming signal can reach, disrupting the target receiver at long distances. In such scenarios, regulatory limits or operational goals are often expressed in terms of EIRP rather than raw transmitter power.

From the EIRP formula, achieving a target EIRP can be done through a combination of:

- Transmitter output power (from the PA);
- Antenna gain (which focuses the energy in a specific direction).

This trade-off is fundamental in practical RF product design. Generating high RF output power is often costly, in terms of:

- DC power consumption;
- Thermal design (requiring larger heat sinks or active cooling);
- Size and cost of the PA (high-power devices are typically larger and more expensive).

However, a higher-gain antenna can achieve the same EIRP with lower transmitter power. For instance, increasing antenna gain by 10 dB enables a tenfold reduction in required output power while maintaining the same EIRP. This reduction can yield significant savings in SWaP-C. However, achieving higher antenna gain is not trivial, as such designs often involve larger apertures, tighter tolerances, or more complex structures, which may introduce their own SWaP-C challenges. Therefore, it is essential to carefully evaluate the trade-offs between transmitter power and antenna design.

In summary, understanding and designing for EIRP rather than just raw transmit power enable more efficient system architectures, especially in applications such as radars and jamming, where long-range performance is essential, but power and thermal budgets are constrained.

Link Margin

Link margin is the amount of excess SNR available in a communication link beyond the minimum required to achieve the target BER or quality of service. It serves as a safety buffer to absorb unpredictable impairments such as fading, interference, weather effects, hardware aging, or alignment errors. The portion of the margin specifically allocated to withstand fading mechanisms (e.g., rain fade, multipath, shadowing) is commonly referred to as the fade margin.

The link margin is calculated as follows

$$\text{Link Margin (dB)} = \text{Available SNR (dB)} - \text{Required SNR (dB)} \tag{4.66}$$

A positive and sufficient link margin indicates that the link should operate reliably under expected conditions, while a small or negative margin suggests potential outages. In high-reliability systems (e.g., satellite links), a link margin of 3–6 dB is common, while critical links in harsh environments may require 10 dB or more.

Typical considerations when determining link margin include:

- *Environmental factors (fade margin):* Rain fade, multipath fading, shadowing, and temperature variations.
- *Hardware degradation:* Component aging, reduced amplifier output power, and connector wear.
- *Operational variability:* Antenna misalignment and interference from other systems.

Link Budget Analysis: An Example

Figure 4.43 illustrates a typical communication link and how the signal level changes as it propagates through the system. As the signal travels along the RF link, it passes through various components, such as cables, amplifiers, filters, and other passive and active elements, that either attenuate or amplify it. To ensure that the signal remains detectable at the receiver, a link budget analysis is performed, which is fundamentally based on the Friis transmission equation. This analysis accounts for all gains and losses in the system and is essential to guarantee a sufficient signal level above the MDS.

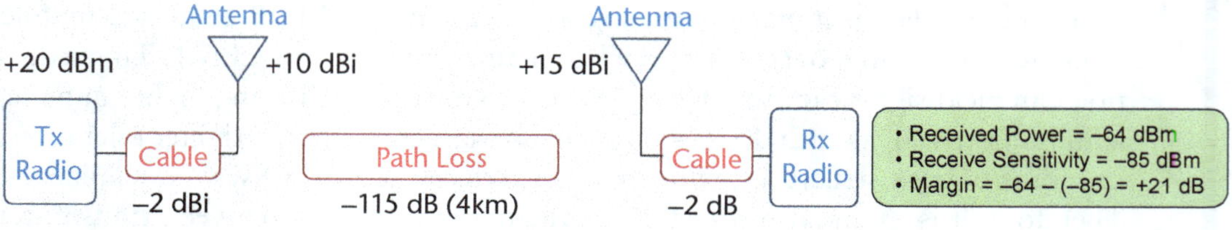

Figure 4.43 Link budget analysis of a communication link.

Let's look at a simple example. The total gain of the link in Figure 4.42 is calculated as

$$G = 10(\text{TX Antenna Gain}) + 15(\text{RX antenna gain}) = 35 \text{ dB} \tag{4.67}$$

The losses are

$$L = 2(\text{TX cable loss}) + 115(\text{Free space loss}) + 2(\text{RX cable loss}) = 119 \text{ dB} \tag{4.68}$$

The received signal level is

$$P_{RX} = P_{TX} + G - L = -64 \text{ dBm} \tag{4.69}$$

The receiver's MDS level is specified to be −85 dBm. Therefore, we have a −64 − (−85) = 21 dB link margin, and the received signal can be detected reliably. This example serves as a simple illustration to demonstrate how a link budget calculation is performed.

Table 4.15 summarizes the key parameters involved in a link budget analysis. Since all quantities are expressed in decibels, the received power can be obtained by summing the gains and subtracting the losses.

Table 4.15 Link Budget Parameters and Their Descriptions

Category	*Parameter*	*Description*
Transmitter	Transmit power (P_{TX}): component gains (drivers, PAs), component losses (cables, passives)	Power delivered to the transmit antenna [dBm or W]; losses from cables, connectors, and passive components between the transmitter and antenna [dB]
	Transmit antenna gain (G_{TX})	Directional gain of the transmit antenna [dBi]
Propagation path	FSPL	Loss due to distance and operating frequency in free space [dB]
	Atmospheric loss	Attenuation due to oxygen, water vapor, rain, fog, or humidity [dB]
	Obstruction/clutter loss	Loss from buildings, vegetation, vehicles, or terrain [dB]
	Multipath fading/shadowing margin	Statistical margin to compensate for multipath fading or shadowing effects [dB]
	Polarization mismatch loss	Loss due to misaligned transmit/receive antenna polarizations [dB]
	Antenna misalignment loss	Loss from pointing errors, especially in directional antennas [dB]
Receiver	Receive antenna gain (G_{RX})	Directional gain of the receive antenna [dBi]
	Component losses (cables, passives)	Losses between the receive antenna and the receiver front end [dB]
	Component gains (LNAs)	Gain provided by LNAs or other front-end components [dB]
Other factors	Diversity or MIMO gain	Gain from antenna diversity, spatial combining, or MIMO techniques [dB]
	Modulation and coding margin	Additional margin required due to modulation scheme and coding performance [dB]

4.12.3 Overall System Performance

To quantify overall system performance, several key aspects must be evaluated. Figure 4.44 summarizes the factors that influence overall link performance. Together, these parameters determine whether an RF system can meet its operational requirements under real-world conditions and provide a framework for system-level design trade-offs:

- Signal degradation factors, such as path loss, noise, interference, and distortion;
- Receiver performance, such as sensitivity, selectivity, and dynamic range;
- Transmitter performance, such as output power, ACLR, and linearity;
- Link performance, such as range, throughput, latency, and link availability.

Therefore, an important step in establishing the RF product blueprint is a thorough evaluation of these metrics to balance competing constraints such as power consumption, size and weight, regulatory limits, and cost. Ultimately, these quantifications establish clear criteria for determining whether the RF system meets its performance objectives.

4.13 What's Next?

We have discussed the fundamental aspects of RF systems, including signal degradation mechanisms, nonlinear effects, and link-budget analysis. In Chapter 5, we shift our focus to practical RF architectures for transmitters and receivers, as well as protection, monitoring, and control systems. These architectures are examined in the context of modern RF products, with particular emphasis on implementations that leverage digital platforms such as FPGAs.

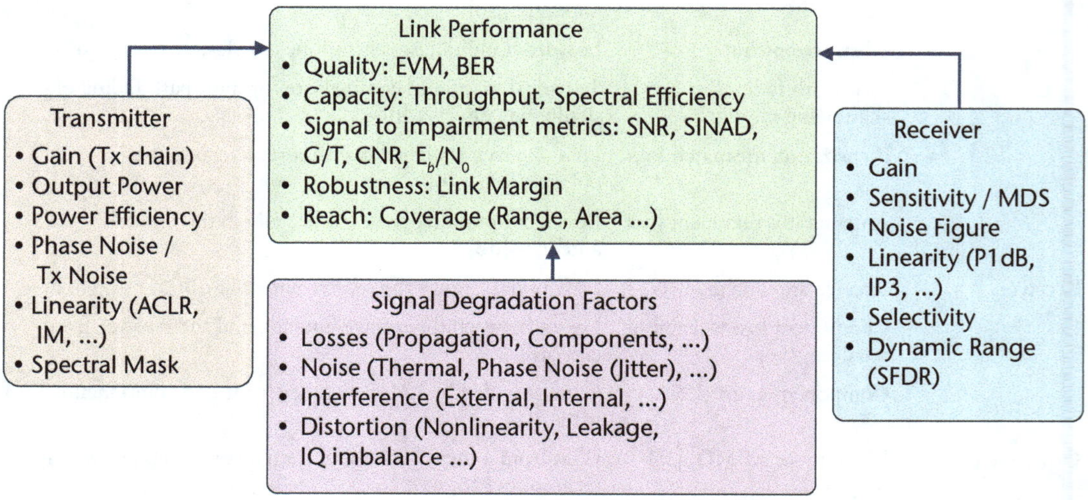

Figure 4.44 End-to-end RF link performance framework highlighting key impairments and metrics.

Problems

4.1 Two point-to-point wireless links operate over the same LOS distance d in free space: Link A: f_A = 900MHz and Link B: f_B = 28 GHz. Assume that both links use: the same transmit power P_T, the same antenna gains G_T and G_R, and the same receiver noise figure and bandwidth.

a. Using the FSPL relationship, determine which link has larger FSPL and compute the difference in FSPL (dB) between Link A and Link B.
b. If both links must achieve the same target BER, explain how the higher FSPL link affects the required: receive SNR margin, modulation order selection (e.g., QPSK versus 16-QAM versus 64-QAM), and coding strength (e.g., higher versus lower code rate).
c. You are allowed to change only one of the following to recover the lost link margin: P_T, G_T, G_R, or bandwidth B. Which would you choose for the 28-GHz link and why? State one practical constraint (hardware/regulatory/implementation) that limits your choice.

4.2 A receiver front end has an initial noise figure of 6 dB. After redesign, the noise figure is reduced to 4 dB, while bandwidth, detection requirements, and all other parameters remain unchanged.

a. Assuming sensitivity is defined at a fixed minimum required SNR, determine the improvement in receiver sensitivity (dB).
b. Interpret the result in two ways (qualitative is fine, quantitative if you want):
   i. *Range impact:* For a free-space link with all else fixed, what does a sensitivity improvement of Δ dB imply about maximum range trend (increase/decrease, and why)?
   ii. *Throughput impact:* If the system uses AMC, explain two different ways you could "spend" the extra link margin to increase throughput (e.g., higher modulation order, higher code rate, wider bandwidth, reduced retransmissions).
c. Give one practical technique that could reduce NF by ~2 dB in a front end (architecture/component choice), and name one trade-off it introduces (linearity, power, cost, stability).

4.3 A receiver front end consists of the stages shown in Table 4.16.

a. Calculate the cascaded gain and noise figure of the receiver.
b. Determine the input-referred P_{1dB} of the full chain.

Table 4.16 Receiver Front-End Stages

Stage	*Gain (dB)*	*NF (dB)*	*IP1dB (dBm)*	*IIP3 (dBm)*
LNA	+16	1.1	−12	−2
RF filter	−2	2	+30	+45
Mixer	−6	8	−5	+5
IF amplifier	+20	4	−15	+5

c. Calculate the cascaded input-referred IP_3.
d. Identify which stage dominates noise performance and which dominates linearity.
e. Propose one modification to improve SFDR without increasing total DC power.

4.4 A wireless link operates at 5.8 GHz over a distance of 2 km.
Given:
- Transmit power: +27 dBm;
- TX antenna gain: +14 dBi;
- RX antenna gain: +12 dBi;
- Receiver noise figure: 5 dB;
- Channel bandwidth: 10 MHz;
- Required SNR: 12 dB.

a. Calculate the FSPL.
b. Compute the received power at the RX input.
c. Determine whether the link meets the required SNR.
d. Identify two practical ways to increase link margin without increasing transmit power.

4.5 A receiver has:
- Noise floor: −140 dBm/Hz;
- IF bandwidth: 1 MHz;
- Input-referred IP_3: −5 dBm.

a. Calculate the integrated noise floor.
b. Determine the SFDR in dB for a two-tone test.
c. Explain how increasing bandwidth impacts SFDR.
d. Discuss why SFDR is often more critical than sensitivity in congested RF environments.

4.6 A receiver operates with:
- Noise figure: 3.5 dB;
- Bandwidth: 200 kHz;
- Required SNR for demodulation: 10 dB.

a. Calculate the MDS.
b. Identify two non-RF factors that can degrade MDS in a real product.
c. If the bandwidth is reduced by a factor of two while all other parameters remain unchanged, calculate the new MDS, and explain one practical consequence of this bandwidth reduction on data rate or latency.

4.7 Two strong interferers at the receiver input are located at f_1 = 2.410 GHz and f_2 = 2.430 GHz.

a. Calculate the frequencies of the second-order and third-order intermodulation products.
b. Identify which products are most likely to fall into the desired signal band at 2.420 GHz.
c. Propose a filtering strategy to mitigate these intermodulation products.
d. Explain why filtering alone may be insufficient in high-power environments.

4.8 A receiver LO has the following phase noise performance: −85 dBc/Hz at 10-kHz offset and −110 dBc/Hz at 1-MHz offset.

a. Explain how phase noise degrades receiver performance in the presence of a strong adjacent signal.
b. Describe the impact of phase noise on EVM in a QAM-based system.
c. Explain the relationship between phase noise and timing jitter.
d. Identify two system-level techniques to reduce phase noise impact without changing the LO source.

4.9 An RF signal travels through the following path: PCB microstrip (10 cm), SMA connector pair, coaxial cable (1m), antenna with VSWR = 2:1, and free-space propagation.
a. Identify all loss mechanisms in this link.
b. Classify each loss as radiation, mismatch, dielectric, ohmic, or propagation.
c. Estimate which loss dominates at microwave frequencies and why.
d. Explain how poor impedance matching can indirectly increase the system noise figure.

4.10 A receiver processes a 100-MHz-wide OFDM signal requiring an SNR of at least 35 dB at the ADC input.
a. Estimate the minimum required ENOB of the ADC.
b. Explain why increasing ADC resolution significantly impacts power consumption.

4.11 A drone is controlled by a ground transmitter (control station) while a hostile jammer attempts to disrupt the link. Assume all quantities are referenced to the drone receiver input and expressed in dB/dBm/dBi, unless stated otherwise. Assume the following scenario:
- Control station transmit power: $P_{S,\text{TX}} = +30$ dBm;
- Control station antenna gain toward the drone: $G_{S,\text{TX}} = +3$ dBi;
- Jammer transmit power: $P_{J,\text{TX}} = +40$ dBm;
- Jammer antenna gain toward the drone: $G_{J,\text{TX}} = +6$ dBi.

Propagation:
- Control link distance: $d_S = 5$ km;
- Jammer distance to drone: $d_J = 2$ km;
- Operating frequency: $f = 2.4$ GHz.

Drone receiver:
- Drone receive antenna gain toward the control station: $G_{\text{RX}}(\theta_S) = -2$ dBi;
- Drone receive antenna gain toward the jammer: $G_{\text{RX}}(\theta_J) = -1$ dBi;
- Front-end insertion loss before the LNA (same for both): $L_{\text{FE}} = 2$ dB;
- Additional polarization mismatch loss: $L_{\text{pol},S} = 0$ dB, $L_{\text{pol},J} = 1$ dB.

a. Compute EIRP of the control station and the jammer.
b. Using the link-budget form, write expressions for the received powers $P_{S,\text{RX}}$ and $P_{J,\text{RX}}$ at the drone receiver input, including gains and losses.
c. Derive the expression for J/S in decibels and show explicitly which terms cancel when the same receiver chain is used.
d. Numerically evaluate $(J/S)_{\text{dB}}$ for the given scenario.
e. If the jammer is a wideband noise jammer with $B_J = 20$ MHz and the receiver bandwidth is $B_R = 1$ MHz, compute the effective $(J/S)_{\text{dB,eff}}$. Explain the relative effectiveness of a 20-MHz wideband noise jammer versus a 1-MHz jammer matched to the receiver bandwidth.

References

[1] Salari, A., *Microwave Techniques in Superconducting Quantum Computers*, Norwood, MA: Artech House, 2024.

[2] Pozar, D. M., *Microwave Engineering*, Fourth Edition, New York: John Wiley & Sons, 2012.

[3] Lee, T. H., *Planar Microwave Engineering: A Practical Guide to Theory, Measurement, and Circuits*. Cambridge, UK: Cambridge University Press, 2004.

[4] Kraus, J. D., and R. J. Marhefka, *Antennas and Wave Propagation*, Fifth Edition, New York: McGraw-Hill, 2002.

[5] Ishimaru, A., *Electromagnetic Wave Propagation, Radiation, and Scattering: From Fundamentals to Applications*, Second Edition, New York: John Wiley & Sons/IEEE Press, 2017.

[6] Razavi, B., *RF Microelectronics*, Second Edition, Upper Saddle River, NJ: Prentice Hall, 2012.

[7] Lizon, B., *Fundamentals of Precision ADC Noise Analysis: Design Tips and Tricks to Reduce Noise with Delta-Sigma ADCs*, Texas Instruments, Application Report, September 2020.

[8] Kaufman, R., et al., "Josephson Parametric Amplifier with Chebyshev Gain Profile and High Saturation," *Physical Review Applied*, Vol. 20, No. 5, November 2023, p. 054053.

[9] Adamy, D., *EW 101: A First Course in Electronic Warfare*, Second Edition, Norwood, MA: Artech House, 2004.

CHAPTER 5

RF Systems: Architectures

I do not think that the wireless waves I have discovered will have any practical application.

—*Heinrich Hertz*

This chapter introduces the fundamental RF architectures that underpin modern RF products. We begin by examining duplexing and multiplexing schemes, followed by a detailed review of core receiver and transmitter architectures, including direct conversion, superheterodyne, and direct RF-sampling, and extend the discussion to multiconversion architectures used in high-performance radar, satellite communications (SATCOM), and quantum computing systems. We then explore transceiver structures for signal routing using circulators, RF multiplexers, and switched-filter banks, and discuss how modern RF transceivers interface with processing platforms such as field programmable gate arrays (FPGAs) and digital signal processing (DSP) chips. Finally, we cover essential supporting mechanisms for monitoring, control, protection, and compensation, including automatic gain control (AGC), digital predistortion (DPD), crest factor reduction (CFR), and equalization.

5.1 Introduction

In RF systems, architecture refers to the high-level organization of how signals, control, and supporting functions are arranged and interact within a product. Broadly, architectures consist of two complementary aspects, as illustrated in Figure 5.1:

- *Core architecture:* This defines the fundamental signal path and how RF energy flows from input to output. Key architectural choices include:
  - Selecting receiver and transmitter architectures (e.g., direct conversion versus heterodyne);
  - Establishing frequency planning and LO distribution strategies;
  - Applying duplexing, multiplexing, and filtering schemes;
  - Interfacing the RF transceiver with processing platforms such as FPGAs and DSP chips.

  These decisions directly influence noise performance, linearity, dynamic range, frequency selectivity, processing capability, and overall integration complexity.
- *Supporting architecture:* This encompasses the mechanisms that preserve performance and stability without directly participating in the primary signal

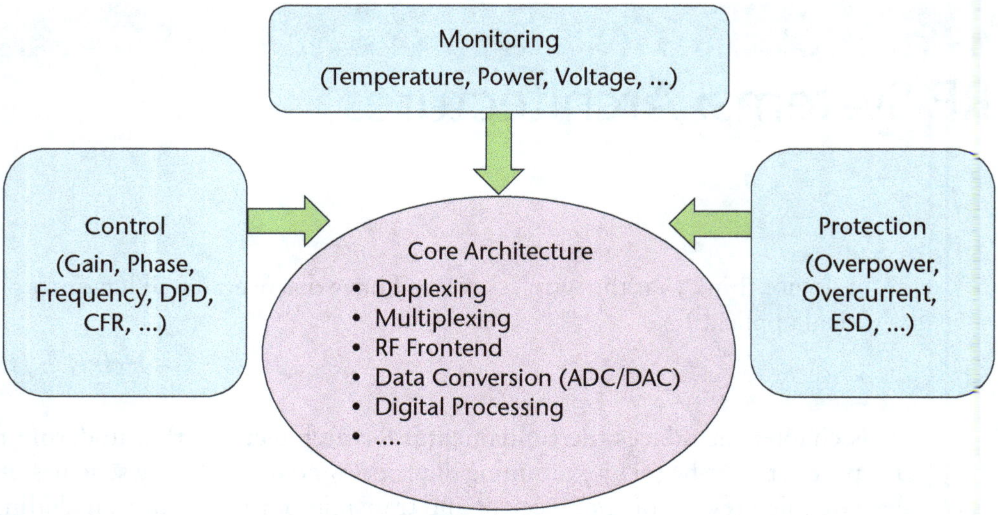

Figure 5.1 The core RF architecture is supported by three essential subsystems, monitoring, control, and protection, which maintain performance, stability, and reliability across operating conditions.

or computational path. It includes monitoring, protection, control, and compensation functions. Typical examples include:

- Power, voltage, current, and temperature monitoring;
- Automatic gain control for dynamic-range management;
- Digital predistortion for amplifier linearization;
- Protection mechanisms for PAs, LNAs, and other sensitive components.

These functions ensure reliable operation across varying conditions, prevent damage, and maintain consistent performance throughout the signal chain. Such functions may seem auxiliary, but they are critical for meeting real-world specifications.

The choice of architecture is one of the most consequential decisions in RF product design. It determines achievable performance as well as SWaP-C tradeoffs. For instance, a double conversion superheterodyne receiver can deliver excellent selectivity but requires multiple oscillators and filters, raising cost and power consumption. A direct-conversion receiver simplifies hardware but introduces challenges such as DC offsets.

Architecture also affects time-to-market. Leveraging a proven architecture with available intellectual property (IP) blocks, components, and test strategies accelerates development, whereas adopting a novel architecture may yield performance gains at the cost of longer validation cycles. Striking the right balance is central to turning a concept into a manufacturable product.

5.2 Duplexing Architectures

One of the first architectural decisions in an RF system is the choice of duplexing mode, as it defines how transmission and reception occur and directly influences requirements for hardware, software, timing, and TX/RX leakage control. The structure of the transmit and receive paths is determined by the selected duplexing method, simplex, half-duplex, or full-duplex, as illustrated in Figure 5.2 [1]. Each mode introduces distinct trade-offs in system complexity, latency, interference management, and the degree of isolation or cancellation required to maintain receiver sensitivity during transmission.

- *Simplex communication:* As shown in Figure 5.2(a), data flows in only one direction. A classic example is radio broadcasting, where a station transmits audio to many receivers, but listeners cannot send information back. Simplex systems are inherently simple, requiring no duplexing hardware or switching, but they offer no interactivity or feedback capability.
- *Half-duplex communication:* As shown in Figure 5.2(b), half-duplex systems allow bidirectional communication, but not simultaneously. Transmission and reception share the same channel, occurring in alternating time slots. A common example is walkie-talkies, where one user speaks while the other listens before roles are reversed. Half-duplex systems are less complex and typically more power-efficient than full-duplex designs but require coordination to prevent transmission collisions.
- *Full-duplex communication:* As shown in Figure 5.2(c), full-duplex systems allow simultaneous transmission and reception. A familiar example is a modern telephone call, where both participants can speak and listen concurrently. Implementing full-duplex communication, however, introduces several challenges:
  - Requires very high isolation between transmit and receive paths to prevent self-interference from desensitizing the receiver.
  - May rely on frequency separation or strong self-interference cancellation, depending on the duplexing architecture.
  - Often necessitates analog, digital, or hybrid self-interference cancellation techniques to suppress transmit leakage over a wide dynamic range.
  - May require separate antennas for transmit and receive, or specialized RF components such as circulators or duplexers, to improve isolation.

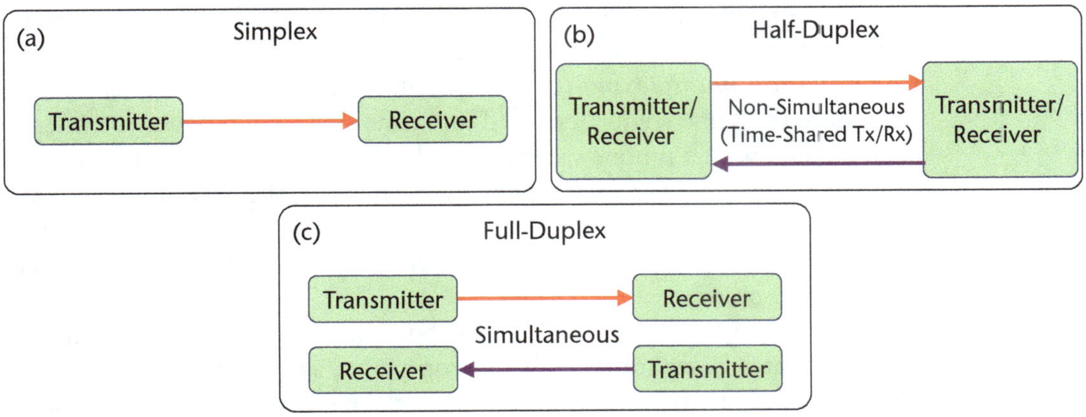

Figure 5.2 Three fundamental duplexing modes: (a) simplex, (b) half-duplex, and (c) full-duplex.

- Demands precise control of linearity, phase, timing, and reflections, as any distortion, mismatch, or group-delay error can limit cancellation depth and degrade receiver sensitivity.

Despite these complexities, full-duplex operation is increasingly adopted in 5G networks, software-defined radios (SDRs), and SATCOM systems. Through active cancellation, adaptive filtering, and antenna isolation methods, these systems achieve reliable performance even under strong self-interference conditions.

5.3 Multiplexing Architectures

The next key architectural choice in RF product design is the multiplexing strategy, as it directly influences RF chain requirements such as filtering, oscillator stability, and timing synchronization. Multiplexing refers to the process of combining multiple signals for transmission over a shared medium, enabling more efficient use of bandwidth and hardware resources. In RF systems, multiplexing can occur in the frequency, time, code, space, or polarization domains, each offering distinct trade-offs in complexity, spectral efficiency, and hardware implementation [1].

- *Frequency-division multiplexing (FDM):* As illustrated in Figure 5.3(a), multiple signals share the same time slot but transmit in distinct frequency bands, shown as channels 1, 2, ... N, within the available spectrum. FDM is common in analog broadcast radio, satellite transponders, and legacy telecommunication systems. It enables simultaneous transmission of multiple signals but requires precise filtering and guard bands to prevent adjacent-channel interference and maintain spectral purity. Orthogonal frequency-division multiplexing (OFDM), discussed in Chapter 3, is a specialized form of FDM in which closely spaced, orthogonal subcarriers transmit data in parallel, enabling high spectral efficiency.
- *Time-division multiplexing (TDM):* As shown in Figure 5.3(b), multiple signals share the same frequency channel but transmit in distinct time slots. TDM is widely used in digital communication systems, including cellular networks and telemetry links, where precise timing and synchronization are essential to prevent overlap and ensure seamless data separation at the receiver.
- *Code-division multiplexing (CDM):* As illustrated in Figure 5.3(c), all signals share the same frequency band and time slot simultaneously, but each is modulated with a unique spreading code. This coding enables the receiver to identify and recover each signal by correlating the received waveform with the code assigned to that user. CDM is common in cellular and spread-spectrum systems, offering strong resilience to interference, multipath fading, and eavesdropping.
- *Space-division multiplexing (SDM):* As shown in Figure 5.3(d), SDM employs multiple spatially separated transmission paths, such as distinct antennas, beams, or propagation modes, to carry independent signals over the same frequency band. In a 2 × 2 system (two transmit and two receive antennas), for example, antenna 1 transmits data stream s_1 while antenna 2 transmits s_2,

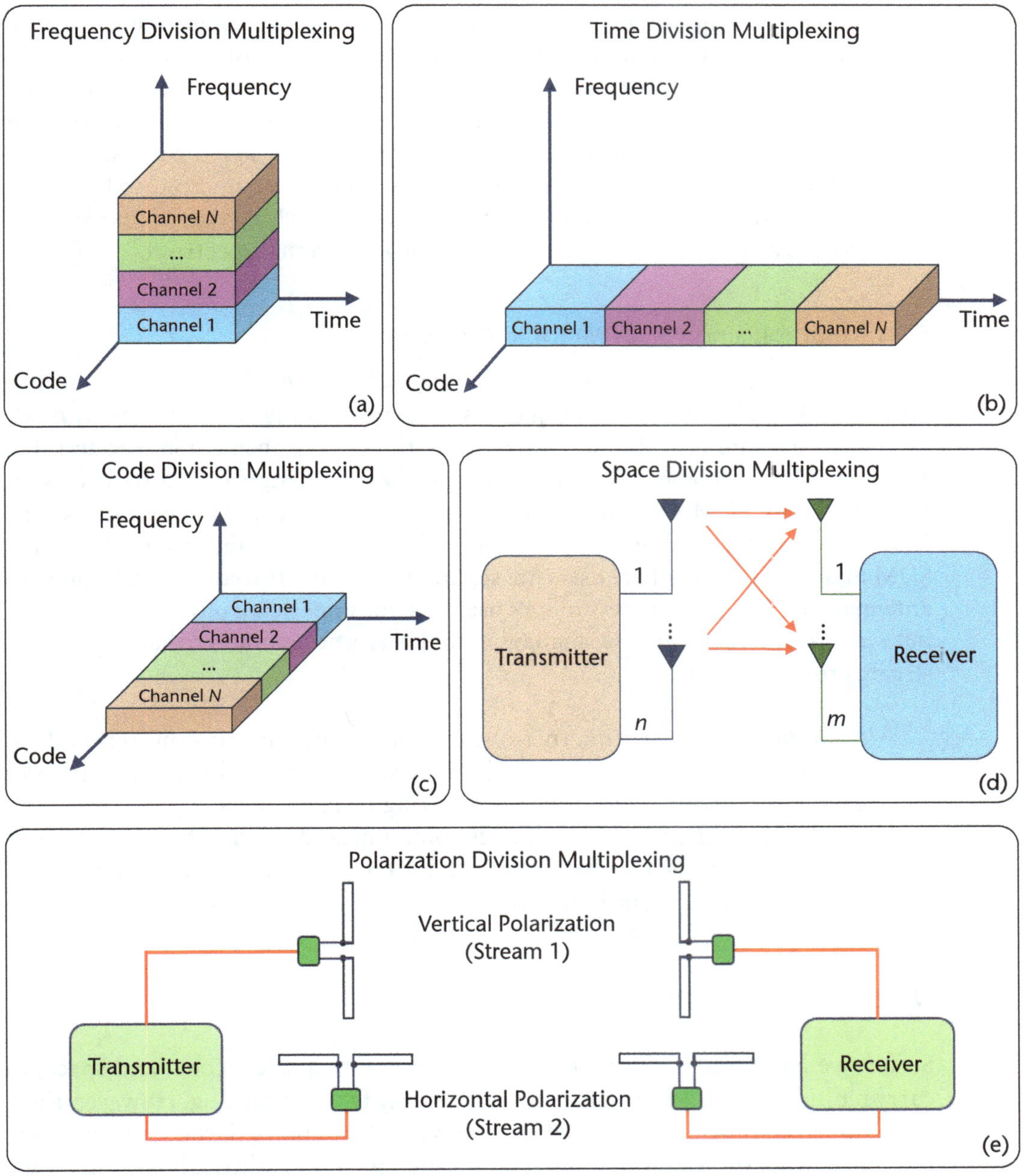

Figure 5.3 (a) FDM, (b) TDM, (c) CDM, (d) SDM, and (e) PDM.

both occupying the same frequency at the same time. Although the streams overlap in the air, each receive antenna observes a different combination of s_1 and s_2 because the channel imposes distinct amplitude and phase characteristics on each path. These differences allow the receiver to mathematically "undo the mixing" and separate the individual data streams.

- *Polarization-division multiplexing (PDM):* As shown in Figure 5.3(e), PDM uses orthogonal polarization states, such as vertical and horizontal or right-hand and left-hand circular, to transmit independent signals over the same frequency band. In a simple example, one antenna transmits data stream s_1 using vertical polarization, while another transmits s_2 using horizontal polarization. Although both signals occupy the same frequency at the same time, their orthogonal polarization states allow the receiver to separate the streams using polarization-selective antennas or appropriate signal-processing techniques. PDM is widely used in satellite communication, microwave backhaul, and advanced multiple-input multiple-output (MIMO) systems to effectively double channel capacity without requiring additional spectrum.

5.3.1 Multiplexing Versus Multiple Access

Although multiplexing focuses on combining multiple signals from a single transmitter, each multiplexing technique has a corresponding multiple-access mode used to coordinate how multiple users share the same transmission medium. For example, TDM becomes TDMA when time slots are assigned to different users, FDM becomes FDMA when frequency channels are allocated across users, and CDM leads to CDMA when unique spreading codes separate users. Likewise, SDM maps to SDMA, where users are separated spatially through beamforming or antenna patterns. These access modes use the same underlying physical dimensions but serve a different purpose, managing multi-user access rather than combining streams from one source.

- *Example of TDM:* An internet-of-things (IoT) gateway transmits sensor data, diagnostics, and control information in separate time slots, with each slot carrying a distinct data stream originating from the same device.
- *Example of TDMA:* Three devices communicating with a base station take turns transmitting. Device A uses time slot 1, Device B uses slot 2, and Device C uses slot 3, ensuring orderly access to the shared channel without interference.

5.4 Diversity Architectures

So far, we have used additional dimensions, time, frequency, code, and space, to carry multiple independent signals simultaneously for multiplexing. However, these same dimensions can also be used to send redundant copies of the same information, improving robustness to fading rather than increasing throughput. This is the basis of diversity.

Diversity mitigates multipath fading by providing the receiver with multiple, independently faded versions of the signal. By exploiting different domains, space, frequency, time, polarization, or angle, diversity ensures that not all copies fade simultaneously.

For example, consider a transmitter sending a single data stream through two antennas: one vertically polarized and one horizontally polarized. Both antennas carry the same information, but multipath reflections rotate polarization differently along each path. At a given moment, the vertical component may arrive in a deep fade (≈–20 dB) while the horizontal component remains strong (≈–3 dB), or vice versa. By receiving both copies and combining them (e.g., using maximal ratio combining

(MRC)) the receiver reconstructs the symbol using the stronger polarization component. Because the two polarizations fade independently, the probability that both experience a deep fade simultaneously is low, providing a robust form of diversity.

To combine these versions effectively, receivers use techniques such as selection combining, equal gain combining (EGC), and maximal ratio combining (MRC), each improving signal quality or maximizing the effective SNR.

5.5 MIMO Architectures

Figure 5.3(d) shows a MIMO architecture. MIMO extends the idea of diversity by using multiple transmit and receive antennas to exploit the spatial dimension. Depending on channel conditions, MIMO can provide spatial diversity (multiple independently faded copies of the same stream) or spatial multiplexing (multiple independent streams to increase throughput). In this way, MIMO unifies both concepts, diversity and multiplexing, and has become a cornerstone of modern RF communication systems.

Complementing both diversity and MIMO is beamforming, where antenna arrays steer energy in desired directions while suppressing interference. Beamforming may be implemented in analog, digital, or hybrid architectures, each offering different trade-offs in flexibility, complexity, and performance.

A single MIMO hardware platform can deliver all three benefits, and modern systems dynamically decide which combination to use based on channel conditions.

- At high SNR → spatial multiplexing increases data rate;
- At low or moderate SNR → diversity improves reliability;
- In directional or interference-limited scenarios → beamforming provides gain and interference suppression.

MIMO is foundational in modern wireless standards such as LTE and 5G, enabling high data rates, wide coverage, and robust performance in multipath-rich environments.

5.6 RF Signal-Chain Architectures

After examining duplexing and multiplexing architectures, we now turn our attention to RF architecture, beginning with the most fundamental element: the signal chain. It is useful to distinguish among the three dominant signal-chain architectures found in RF products:

1. *Fully analog signal chain:* In this architecture, shown in Figure 5.4(a), the signal remains entirely in the analog domain from input to output. These systems were common in legacy RF platforms and in applications where digital conversion was impractical. They are still used where simple modulation schemes, fixed-function filtering, or ultralow-latency analog processing are required.

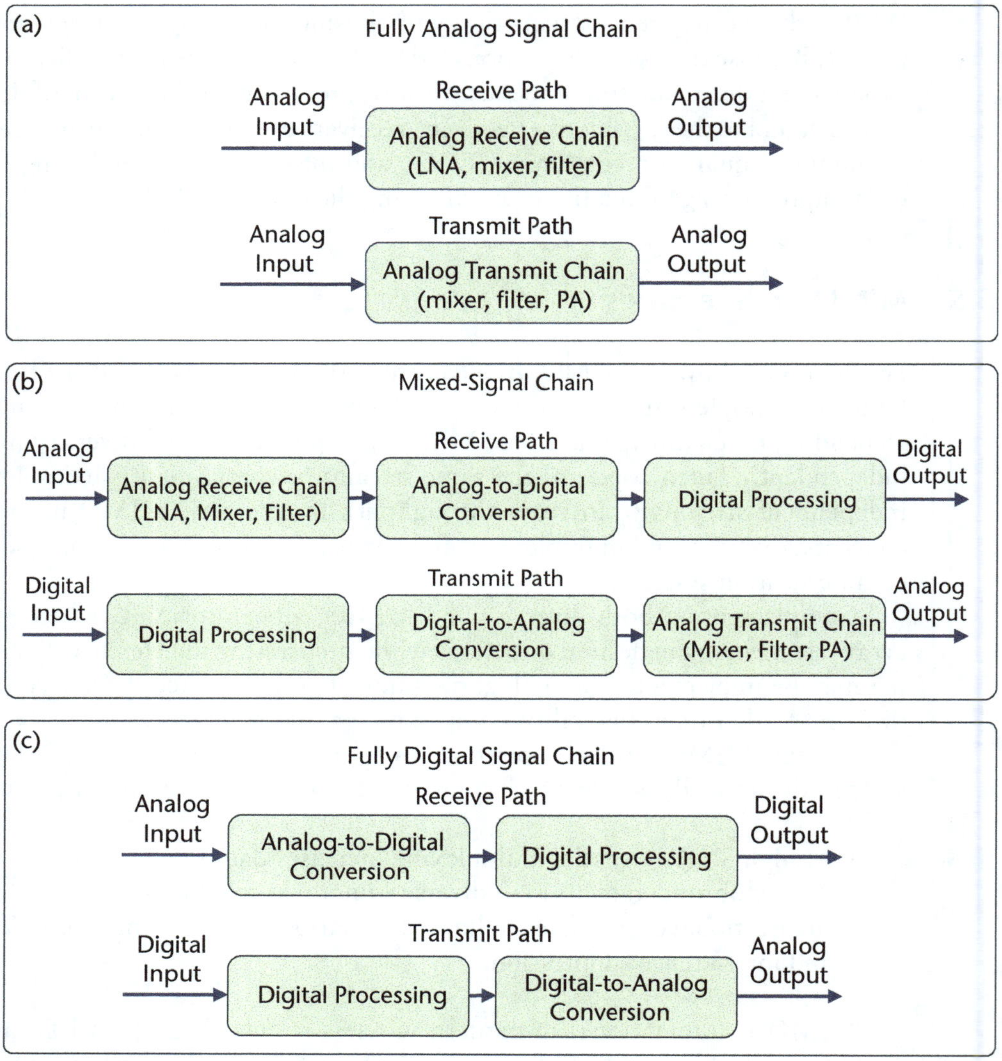

Figure 5.4 (a) Fully analog signal chain, (b) mixed signal chain, and (c) fully digital signal chain.

2. *Mixed-signal chain:* In this architecture, shown in Figure 5.4(b), part of the signal chain operates in the analog domain and part of it operates in the digital domain. After the analog front-end performs operations such as amplification, filtering, and frequency conversion, the signal is digitized by an analog-to-digital converter (ADC) and is handed off to a digital baseband processor, typically an FPGA, DSP, or application-specific integrated circuit (ASIC). On the transmit side, digitally generated or processed baseband signals are converted back to analog using a digital-to-analog converter (DAC).

   The benefit of this digitized approach is that many signal-processing functions, such as filtering, modulation and demodulation, channelization, decimation, and equalization, can be shifted into the digital domain, where they can leverage the flexibility and computational power of modern digital back ends. This allows for a streamlined and highly integrated RF front end and enables reconfigurable architectures such as wideband SDRs, provided the required frequency range, bandwidth, and dynamic range remain within the capabilities of the ADC, DAC, and digital processing chain.

Most modern RF products adopt mixed-signal architectures because they offer an optimal balance of performance, cost, power consumption, and design flexibility.

3. *Fully digital signal chain:* In this architecture, shown in Figure 5.4(c), the analog-to-digital boundary is pushed as close as possible to where the physical signal is generated or measured. High-speed ADCs and DACs digitize or synthesize signals at RF or near RF, greatly reducing the amount of analog circuitry required. With most processing performed digitally, these architectures offer maximum flexibility for waveform synthesis, signal shaping, calibration, beamforming, and wideband spectral analysis. Although constrained by converter performance, clocking, and power consumption, fully digital approaches are increasingly used in SDRs, direct-sampling receivers, instrumentation front ends, radar systems, and quantum control hardware. In these designs, the analog front end is minimal, often limited to a filter, an amplifier, or both.

5.6.1 The Three Pillars: RF Front End, Data Conversion, and Digital Baseband

In the previous section, we examined how mixed-signal and fully digital signal chains leverage increasing amounts of digital processing. It is useful to partition the signal chain into three major domains:

- *RF front end:* Encompasses all analog processing between the antenna and the ADC input in the receive path, or between the DAC output and the antenna in the transmit path. This includes impedance matching, switching or duplexing, amplification, filtering, and frequency conversion.
- *Data conversion:* Converts analog RF or IF waveforms into digital samples (ADC) and digital samples back into analog signals (DAC). This stage defines the analog-digital boundary of the system and places fundamental limits on dynamic range, noise performance, and instantaneous bandwidth.
- *Digital baseband:* Handles all DSP performed after the ADC (in the receive path) or before the DAC (in the transmit path). This includes synchronization, equalization, modulation/demodulation, channel estimation, and error correction, as well as interfacing with processing platforms such as FPGAs and dedicated DSP chips.

In the following sections, we will examine these three pillars in detail. Although this chapter focuses on modern architectures with digital back ends, it is important to note that legacy systems and some specialized receivers still rely on analog demodulation and detection, operating entirely without digitizing the signal.

5.7 RF Front End: Receiver Architectures

Over the decades, a wide range of receiver architectures has been developed, each offering its own balance of performance, complexity, and cost. This section introduces the principal approaches, traditional, modern, and hybrid architectures, and highlights their defining features and trade-offs. Factors such as sensitivity, linearity, selectivity, dynamic range, size, power consumption, and ease of integration all influence the choice of architecture. These considerations, along with practical implementation constraints, will be discussed in detail in the sections that follow.

5.7.1 General Overview of a Receiver Chain

We begin with a general overview of the signal detection process common to most receiver architectures, as illustrated in Figure 5.5. The chain typically starts with amplification to boost weak incoming signals, followed by filtering to suppress spurious responses, harmonics, and out-of-band interference. The filtered signal is then downconverted to an IF, which may be 0 in a direct-conversion receiver or involve multiple stages in a dual-conversion architecture, before being conditioned for detection or digitization by an ADC. This conditioning stage may include additional filtering and gain control to achieve the desired signal level while minimizing spurious content introduced during frequency translation [2].

Each receiver architecture adapts these stages to meet specific objectives, such as mitigating image frequencies, improving selectivity, enhancing dynamic range, or enabling higher integration and flexibility.

Let's take a closer look at each block next.

Amplification

In Chapter 4, we introduced the concept of the minimum detectable signal (MDS), emphasizing that detection fails if the input falls below this threshold. Amplifiers

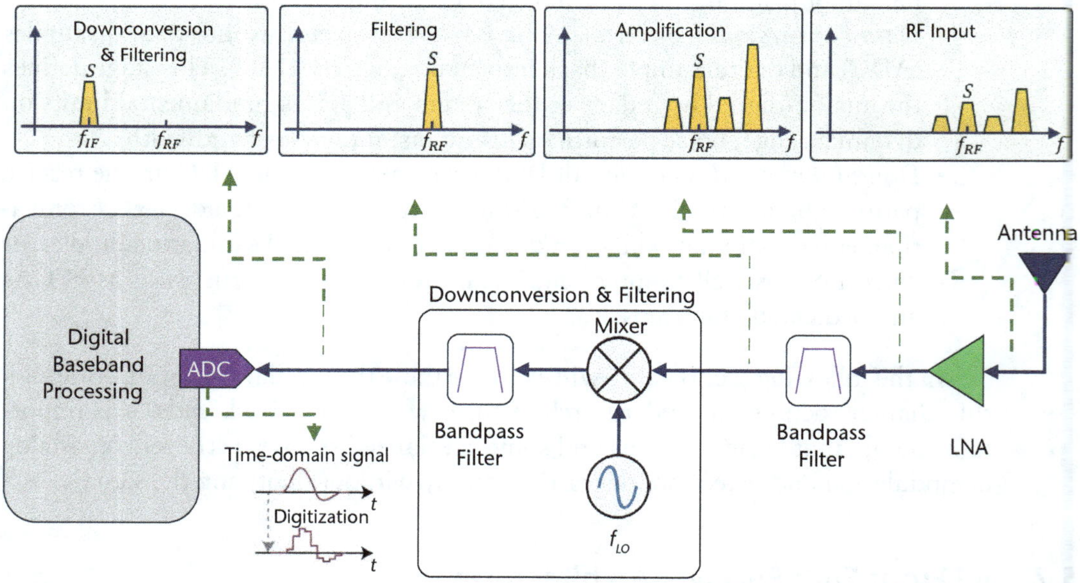

Figure 5.5 The desired signal is denoted by *S*. Signal flow in a general receiver architecture. All blocks are shown with idealized behavior: the amplifier provides flat gain, the filter exhibits a brick-wall response, and the mixer and subsequent filtering produce perfectly clean outputs. This figure is intended only to illustrate the conceptual signal flow.

are therefore essential for raising signal levels above MDS. At the receiver front end, low-noise amplifiers (LNAs) deliver the necessary gain while contributing minimal noise, thereby preserving sensitivity and keeping the overall system MDS as low as possible.

Yet amplification is not without trade-offs. As discussed in Chapter 4, nonlinearities in amplifiers or mixers generate intermodulation products that can fall directly into the desired channel and corrupt the received signal. Beyond nonlinearities and noise figure, other amplifier characteristics, such as gain flatness, group delay, and impedance matching, also influence detection performance. Gain variations can distort the signal or complicate AGC operation; impedance mismatches introduce reflections and passband ripple; and group-delay variations can distort broadband or phase-sensitive waveforms. Together, these factors determine how faithfully weak or complex signals are amplified prior to detection.

In crowded spectral environments or systems with tightly spaced channels, designers must carefully balance gain and noise figure against linearity, ensuring that the receiver remains sensitive enough to detect weak signals while simultaneously preventing intermodulation distortion from stronger adjacent signals.

Filtering

Filtering plays a pivotal role in the RF detection chain, with two primary objectives: selecting the desired channel and suppressing out-of-band signals, including interference, image frequencies, and spurious tones generated by nonlinear devices such as amplifiers and mixers.

In a heterodyne architecture, a bandpass filter is typically employed to isolate the channel of interest at the IF. By contrast, a direct-conversion (homodyne) receiver relies on a single lowpass filter to remove out-of-band components after downconverting directly to baseband.

Maintaining a flat group delay and amplitude response across the filter passband is essential to preserve waveform integrity. In wideband and multicarrier systems such as OFDM, even small variations in group delay or passband ripple can introduce intersymbol interference (ISI) and constellation distortion. Another critical property is out-of-band rejection and roll-off, which determine the filter's selectivity and its ability to suppress adjacent-channel interference (see Section 6.8.1).

The placement of the filter relative to the LNA introduces important design trade-offs. Two possibilities are shown in Figure 5.6. When the LNA precedes the

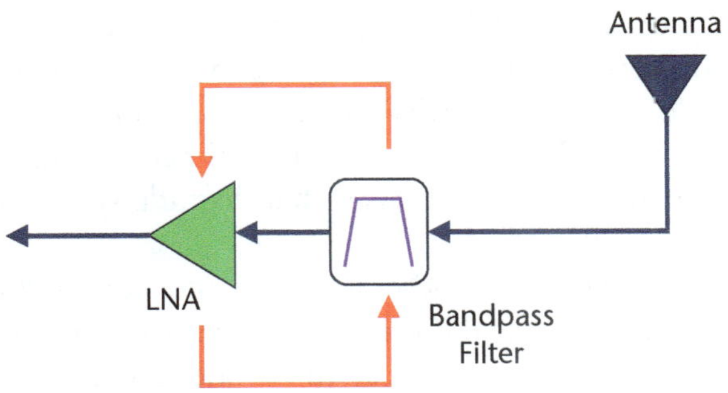

Figure 5.6 Two possible filter-LNA configurations.

filter, the noise figure of the overall receiver chain is dominated by this first stage, as discussed in Chapter 4 and captured by Friis's formula. For this reason, a low noise figure, high-gain LNA is often placed directly at the input to minimize the effective RF chain noise figure. The drawback is that, without prior filtering, the LNA may be driven into compression by strong out-of-band signals, and image responses or other unwanted components may pass into later stages and corrupt detection.

If the order is reversed, with the filter preceding the LNA, the system becomes more robust against overload and image issues. However, because a passive filter introduces loss, its noise figure equals its insertion loss. This degrades the chain noise figure and, consequently, the overall sensitivity, especially in low-interference conditions.

In practice, designers must choose the arrangement based on the interference environment. In clean-spectrum conditions, an LNA-first approach maximizes sensitivity. In high-interference or image-sensitive applications, a filter-first design offers better protection against overload and spurious responses, albeit at the expense of a higher noise figure. As we will see later in this chapter, more advanced front ends often employ filter banks or tunable filters to partition a wideband input into multiple simultaneous narrowband channels. Tracking filters may also be used to follow the LO, providing tunable selectivity across the operating band.

Downconversion

This is a fundamental step in RF receivers, shifting the incoming signal to a lower-frequency range suitable for digitization or demodulation. This is typically achieved using a mixer and an LO, as illustrated in Figure 5.5. Depending on the architecture, downconversion may occur in a single stage or multiple stages. In direct-conversion receivers, the signal is directly converted to baseband, while superheterodyne receivers often employ two or more conversion stages to achieve better selectivity and image rejection.

- *LO phase noise:* The LO plays a critical role by tuning the receiver into different channels. As discussed in Chapter 4, its spectral purity is paramount: excessive phase noise broadens the LO spectrum, leading to reciprocal mixing and energy leakage into adjacent channels. In addition, LO harmonics may interact with spurious input signals, producing unwanted IF components.
- *LO leakage:* Another concern is LO leakage, as shown in Figure 5.7(a) [3]. If LO energy couples into the mixer input and reflects back, it can mix with itself, creating a DC offset, commonly referred to as self-mixing. Leakage can also propagate toward the antenna, radiating into free space and potentially desensitizing nearby receivers operating in the same band, as shown in Figure 5.7(b). Likewise, interferer leakage into the mixer path can generate additional spurious responses that degrade system sensitivity. This is shown in Figure 5.7(c).
- *Quadrature error:* Another important concern in downconversion is quadrature error. In an ideal IQ receiver or transmitter, the in-phase (I) and quadrature (Q) paths have identical amplitudes and a perfect 90° phase separation. In practice, mismatches in mixers, phase shifters, and gain stages introduce

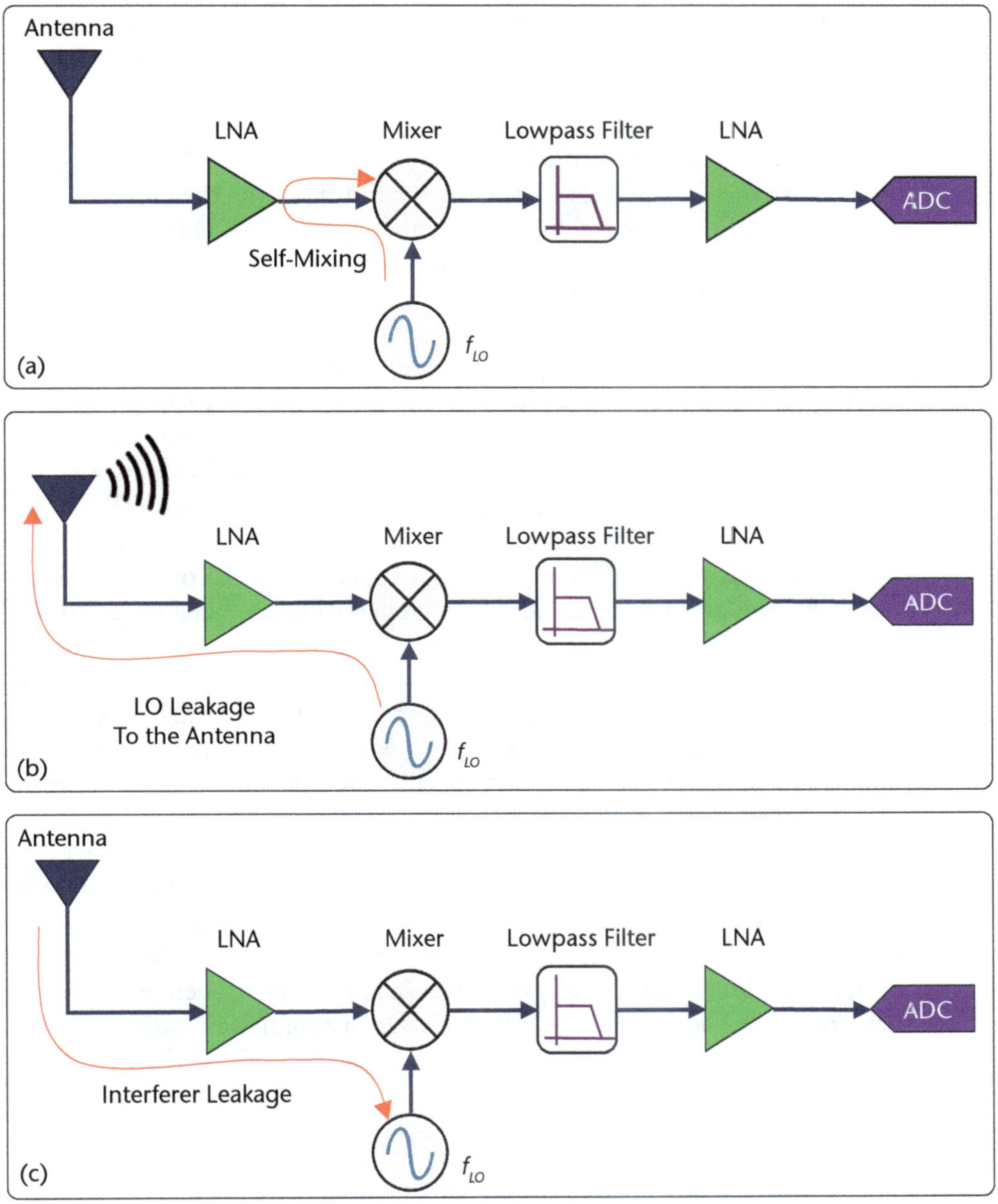

Figure 5.7 (a) Self-mixing effect due to LO leakage, (b) LO leakage to the antenna, and (c) interferer leakage.

amplitude and phase imbalance. These quadrature errors distort the signal constellation, degrade image rejection, and increase the error vector magnitude (EVM). As we will see in Chapter 6, modern systems mitigate these impairments using quadrature error correction, implemented in hardware, software, or a combination of both.

As we will see in Section 5.9.2, careful frequency planning can also mitigate these concerns by selecting LO frequencies and IF placements that minimize image interference, reduce sensitivity to quadrature errors, and avoid spurious mixing products during downconversion.

- *Image problem:* Another important concern is the image problem. In an ideal frequency conversion process, only the desired signal band should be translated during upconversion or downconversion. In practice, during downconversion, any signal located at a frequency symmetric to the desired signal with respect to the LO, known as the image frequency, is also translated to the same IF. As illustrated in Figure 5.8, the desired band at f_{RF} and the unwanted image band at f_{IM} are symmetrically positioned about the LO. During downconversion, both signals produce energy at the same IF f_{IF}, making them indistinguishable without additional filtering or image-rejection techniques.

This undesired frequency translation introduces interference and degrades receiver performance. Beyond classical communication systems, the image problem is also relevant in qubit control architectures, where unwanted sidebands generated during upconversion can be coupled to neighboring qubits or excite higher-order transitions, leading to reduced gate fidelity and increased crosstalk.

A dual problem occurs during upconversion. In a conventional mixer, both the desired signal at f_{RF1} and its image at f_{RF2} appear at the RF output. This configuration is known as double-sideband (DSB) upconversion. The unwanted sideband at f_{RF2} can interfere with adjacent channels or generate spurious emissions.

To mitigate this, single-sideband (SSB) upconversion introduces phase shifting and quadrature mixing techniques to suppress one of the sidebands, allowing only the desired frequency component to pass. We will discuss this technique in Section 6.8.2. The downconversion counterpart of an SSB mixer is known as an image-reject mixer.

Another way to address the image problem is to use an image-reject filter, essentially a bandpass filter designed to suppress signals at the image frequency before they reach the mixer. In many systems, the image lies far outside the desired signal bandwidth and can be removed with appropriately designed RF or IF filters. However, this approach is not always feasible. In applications where the image frequency lies close to the desired band, or where wide instantaneous bandwidth is required,

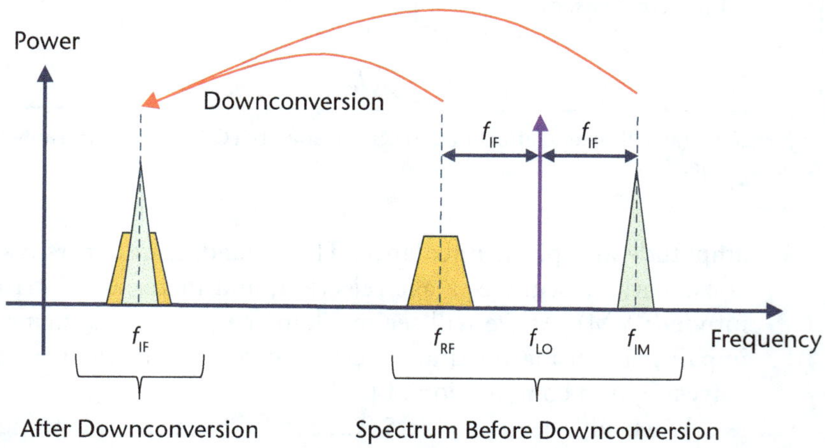

Figure 5.8 The image frequency band is downconverted to the same IF as the desired band, causing both to overlap after mixing.

filtering alone may be insufficient. In such cases, image-reject mixers or other techniques are employed to suppress the unwanted sideband at the mixing stage.

Digitization/Demodulation

As illustrated in Figure 5.5, once downconverted, the signal is typically sampled by an ADC, producing a stream of digital samples. These samples form the basis for digital demodulation, where the modulation format is recovered, as well as additional DSP operations such as filtering, equalization, and error correction. Shifting these functions into the digital domain provides modern receivers with greater flexibility, precision, and the ability to implement advanced algorithms that would be impractical in analog circuitry.

The performance of the ADC has a profound impact on signal detection because it determines how faithfully the analog waveform is captured for digital processing. The ADC's resolution sets the quantization noise floor and therefore the effective sensitivity of the digital system (see Section 4.7.6). Its sampling rate defines the maximum signal bandwidth that can be digitized without aliasing, while real-world imperfections, summarized by the effective number of bits (ENOB), govern how cleanly weak or closely spaced tones can be resolved [2].

Equally important are the ADC's dynamic and temporal characteristics. Limited spurious-free dynamic range (SFDR) or poor linearity can introduce spurious tones and distortions that obscure genuine signals, while aperture jitter introduces timing uncertainty that becomes amplitude or phase error at high frequencies. In essence, the ADC's precision, speed, and stability together define the clarity of the digital representation, and therefore how reliably a receiver can detect and discriminate signals in a noisy or crowded spectrum.

5.7.2 Direct-Conversion/Homodyne Receiver

The term homodyne comes from "homo" (same) and "dyne" (mixing), reflecting that the LO operates at the same frequency as the received signal. As shown in Figure 5.9, this architecture downconverts the input directly to baseband, centering the desired signal at DC. In IQ detection, two parallel branches, the in-phase (I) and quadrature (Q) paths, are generated using an IQ mixer rather than a single scalar mixer. A lowpass filter then suppresses harmonics and spurious components introduced in earlier stages. With relatively few components, the homodyne receiver is attractive for its simplicity, low cost, and high level of integration, though it also presents several well-known drawbacks.

One of the primary limitations of the homodyne architecture is the presence of DC offsets, primarily caused by LO self-mixing and LO leakage into the RF path, as discussed earlier (see Figure 5.7(a)). Although analog circuit techniques can reduce this effect, it remains a persistent challenge. Modern receivers typically mitigate such impairments by combining quadrature downconversion with digital calibration and compensation algorithms. Moreover, at very low frequencies (near DC), amplifiers and mixers exhibit significant flicker noise. Homodyne architecture places the signal exactly where the flicker noise is strongest.

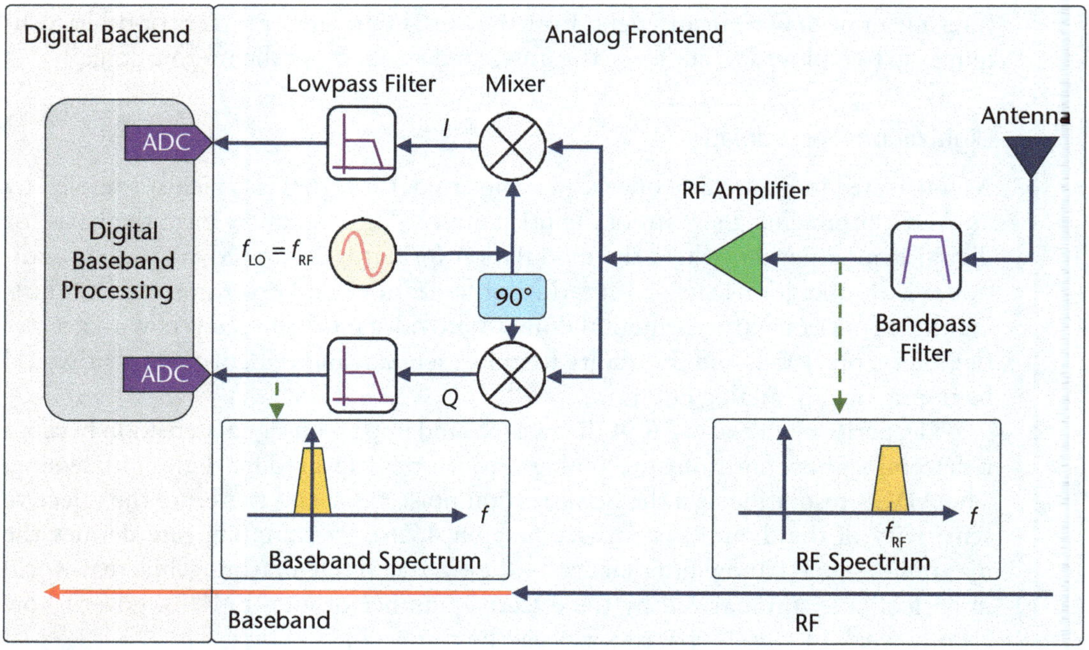

Figure 5.9 Homodyne receiver with quadrature downconversion.

Another limitation of homodyne receivers is that even-order nonlinearities in the mixer or preceding analog stages generate distortion products that fall near DC, where the desired signal resides. These second-order components manifest as DC offsets, low-frequency spurs, and baseline shifts, ultimately degrading receiver sensitivity. As discussed in Chapter 4, the second-order intercept point (IP_2) is therefore a critical metric for homodyne architectures, as it quantifies the receiver's susceptibility to these even-order nonlinear effects.

Another drawback is the architecture's inherent susceptibility to image signals. In a homodyne (zero-IF) receiver, the LO is tuned exactly to the carrier frequency ($f_{LO} = f_{RF}$), so the desired signal at $+\Delta f$ (upper sideband) and its image at $-\Delta f$ (lower sideband) are symmetrically located around the LO. When the RF signal is mixed with the LO, both frequency components translate directly to baseband, causing the positive and negative sidebands to overlap. Unlike a heterodyne receiver, where a bandpass IF filter can isolate one sideband and suppress its image, a homodyne receiver uses only lowpass filtering, which cannot distinguish between the desired and image components. As a result, any interference or noise in the image band folds into the desired signal, degrading sensitivity and increasing EVM.

To address this issue, Campbell introduced a modified direct-conversion architecture that achieves SSB downconversion. By adding controlled phase and amplitude relationships in the I/Q paths, the receiver can cancel the unwanted sideband while preserving the desired one. This approach improves image rejection without requiring a high-Q analog IF filter, making it compatible with integrated implementations.

As discussed in Section 5.7.4, much of the analog front end can be replaced by digital processing, leaving a minimal analog front end consisting primarily of an amplifier and a filter. Modern RF systems are increasingly moving in this direction.

5.7.3 Superheterodyne Receiver

We will discuss variations of the superheterodyne receiver in this section.

Single-Conversion Superheterodyne Receiver

A typical single-conversion superheterodyne receiver architecture is shown in Figure 5.10. Unlike a homodyne receiver, it uses a nonzero IF, which naturally prevents DC offsets, minimizes the impact of flicker noise, and enables the use of high-Q filters that provide significantly better selectivity.

The front-end RF filter typically has a wider bandwidth and lower selectivity and provides initial suppression of out-of-band and image signals to protect the mixer from strong interferers. After downconversion, a high-Q channel-selection IF filter, such as a surface acoustic wave (SAW), bulk acoustic wave (BAW), ceramic, or mechanical filter, precisely isolates the desired channel. These IF filter technologies achieve much higher Q than RF filters, enabling very narrow bandwidths and excellent selectivity.

IF filters achieve higher Q in part because frequency-dependent losses, such as skin effect, dielectric loss, and radiation loss, are much smaller at lower frequencies, and because technologies such as SAW and ceramic resonators offer extremely high Q values that are impractical to realize at RF. As a result, IF filters provide far sharper selectivity than typical RF filters. Moreover, because the IF is significantly lower than the RF carrier, even if the RF and IF filters had the same Q-factor, the

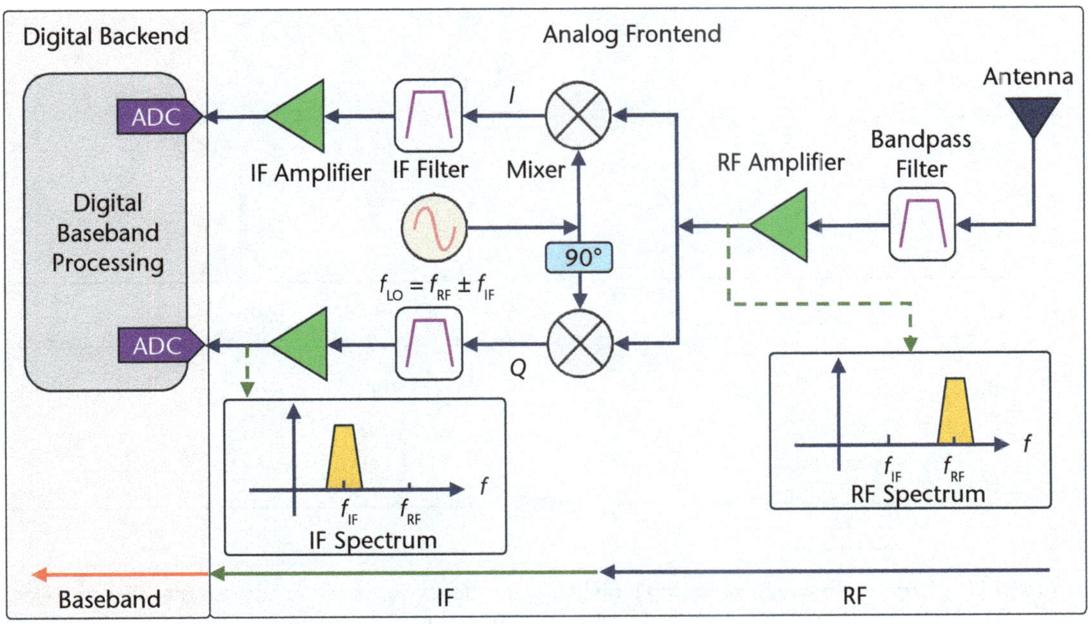

Figure 5.10 Single-conversion superheterodyne receiver. The LO frequency is set to $f_{LO} = f_{RF} + f_{IF}$ for high-side injection, and $f_{LO} = f_{RF} - f_{IF}$ for low side injection.

absolute bandwidth of the IF filter would still be much smaller (BW = f_0/Q), enabling a tighter passband and steeper roll-off than an RF filter with equivalent Q.

In a superheterodyne receiver, the first LO can be placed either above or below the incoming RF signal, leading to high-side or low-side injection, respectively. With high-side injection, the LO frequency is set to $f_{LO} = f_{RF} + f_{IF}$, so the IF is formed by the difference $f_{LO} - f_{RF}$. In low-side injection, the LO is instead placed below the RF, $f_{LO} = f_{RF} - f_{IF}$, and the mixer converts upward to the same IF. Both approaches create the desired IF, but they produce different image frequencies and different filtering requirements. High-side injection places the image at $f_{RF} + 2f_{IF}$, whereas low-side injection places it at $f_{RF} - 2f_{IF}$. The choice between them is therefore driven by practical considerations, such as available filter bandwidth, proximity of interfering signals, LO tuning range, and spurious-response management, rather than by performance differences in the mixing process itself.

The choice of IF involves a trade-off between image rejection and channel selectivity, as illustrated in Figure 5.11, where Figure 5.11(a) shows the RF front end. The frequency separation between the desired IF signal f_{IF} and the image frequency f_{IM} is equal to twice the IF, $2f_{IF}$. A higher IF increases this separation, enabling more effective image rejection by the RF image-reject filter (BPF1). However, it also results in a lower quality factor for the channel-selectivity filter (BPF2), making it less effective at suppressing in-band interferers, as shown in Figure 5.11(b). Conversely, as illustrated in Figure 5.11(c), a lower IF reduces the separation from the image frequency, making image suppression more challenging. However, it allows the use of a higher-Q channel-selectivity filter, improving suppression of in-band

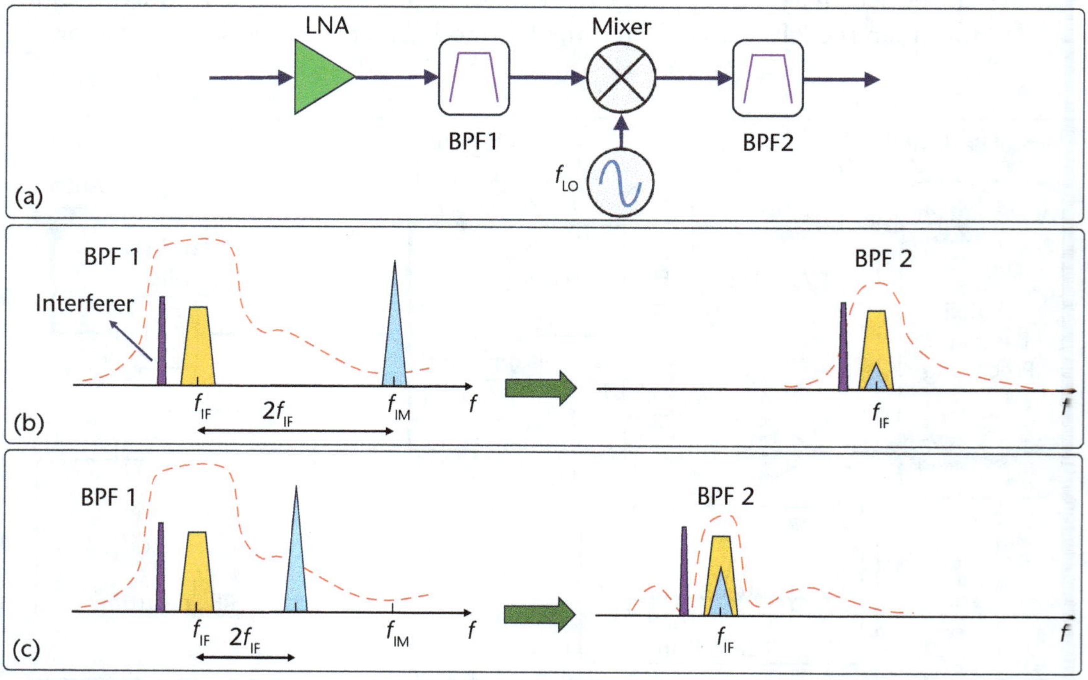

Figure 5.11 Trade-off between achieving high image frequency suppression and optimal channel selectivity. (a) RF front end. (b) A higher IF provides significant rejection of the image frequency but suffers from a lower Q of the channel selectivity filter. (c) A lower IF is more effective in suppressing in-band interferers and benefits from a higher Q of the channel selectivity filter.

interferers. Consequently, IF selection requires balancing image-frequency suppression against achievable channel selectivity.

Low IF Superheterodyne Receiver

In a classic superheterodyne receiver, the IF is relatively high (e.g., 10.7 MHz, 21.4 MHz, 70 MHz, 140 MHz). A widely used variation of this approach is the low IF architecture, in which the signal is shifted to a small but nonzero IF, typically from a few hundred kilohertz to several tens of megahertz. Low IF designs help to avoid severe DC offsets and flicker-noise issues that commonly affect zero-IF (homodyne) receivers, while still simplifying analog filtering compared to high IF superheterodyne architectures.

However, low IF receivers introduce their own trade-offs. Because the IF is nonzero, they continue to suffer from image frequency interference, and the image lies very close to the desired signal, making analog filtering far less effective than in high IF designs. Therefore, achieving high image rejection requires precise I/Q demodulation and careful amplitude and phase matching between the I and Q paths. Any I/Q imbalance results in incomplete image cancellation, leading to image leakage and degraded signal quality. Although DC offset is reduced relative to zero-IF, near-DC spurs can still impair performance, particularly when the IF is very low. These imperfections often necessitate digital compensation, increasing DSP complexity. LO leakage and mixing spurs may also fall near the signal band, further complicating receiver design.

Despite these challenges, low IF receivers are widely used in compact, low-power, and cost-sensitive wireless systems because they offer a practical compromise between the simplicity of zero-IF architectures and the strong image rejection of superheterodyne designs. They are predominantly applied in narrowband and low to moderate bandwidth systems such as Bluetooth, ZigBee/802.15.4, NB-IoT transceivers, FM broadcast receivers, GPS/GNSS chipsets, wireless meter reading devices, and other short-range ISM-band radios.

Double-Conversion Superheterodyne Receiver

A double-conversion architecture can achieve both strong suppression of the image frequency and high channel selectivity. However, these benefits come at the cost of additional components, higher power consumption, greater design effort, and increased system complexity. As the saying goes, there is no free lunch; improving performance in one area often requires trade-offs elsewhere in the design.

Figure 5.12(a) illustrates a two-stage IF architecture. In this design, the first frequency conversion translates the incoming RF signal to a high IF, pushing the image frequency far away and greatly reducing image-related interference. A second conversion then brings the signal down to a low IF, where a high-Q filter can be applied to achieve excellent channel selectivity and narrowband performance.

Figure 5.12(b) shows a double-conversion receiver commonly used in qubit readout chains, where staged frequency translation is required to place the readout tone within the optimal bandwidth of the analog and digital signal-processing components [2]. A quadrature (I/Q) variant of the double-conversion architecture is also widely used when complex IF processing or single-sideband operation is desired.

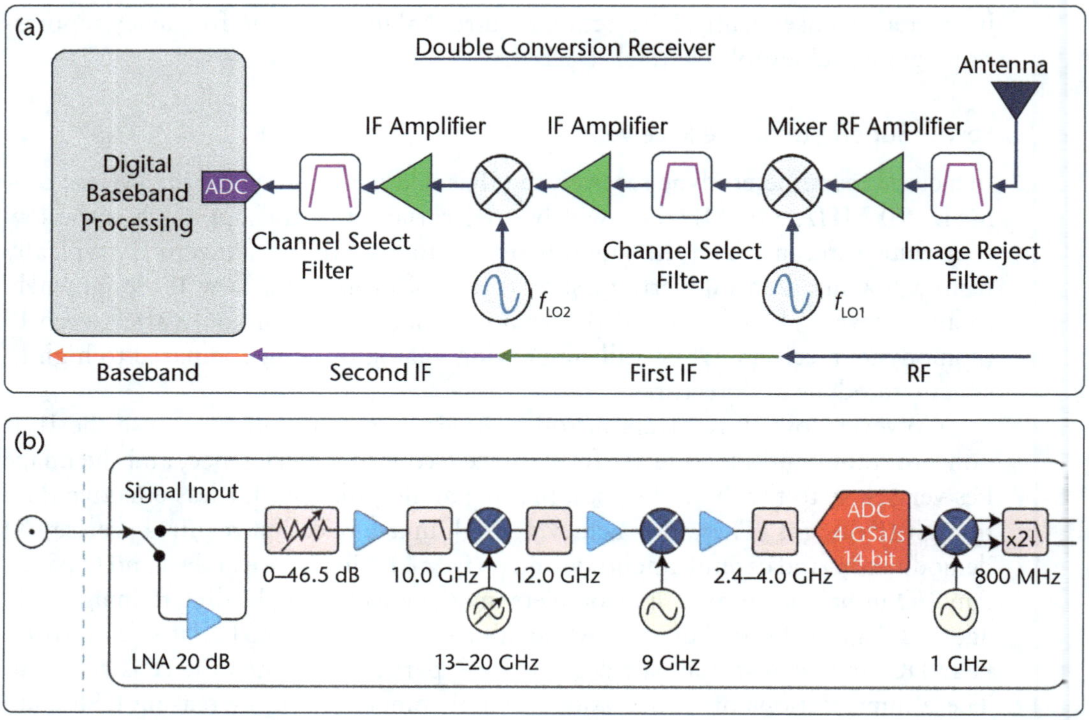

Figure 5.12 (a) Architecture of a double-conversion receiver. (b) An example of a double-conversion receiver used for detecting qubit readout signals.

5.7.4 Direct RF Sampling Receiver

We saw in the previous architectures that downconversion is performed using analog mixers and LOs. This approach introduces traditional impairments such as image responses, mixer spurs, LO leakage, and quadrature mismatches. Modern receivers address these limitations by employing high-speed ADCs operating at gigasample-per-second rates, enabling direct RF sampling. In this approach, the intermediate downconversion stages are bypassed, and the incoming RF signal is digitized directly. As a result, IQ mixer calibration and multistage synchronization are no longer necessary.

The concept of direct RF sampling is shown in Figure 5.13. The analog front end consists of a band-select filter and an RF amplifier that condition the incoming signal before digitization by a high-speed ADC. Once in the digital domain, digital downconversion (DDC) is used to translate the sampled RF signal to baseband. The DDC employs a direct digital synthesizer (DDS) or numerically controlled oscillator (NCO) to generate the quadrature LO signals for digital mixing, which is implemented simply as a point-by-point digital multiplication. The resulting I/Q signals then pass through a lowpass decimation filter that removes unwanted spectral components such as spurs, harmonics, and out-of-band noise, after which a downsampler reduces the sample rate to match the bandwidth requirements of the baseband processing chain.

Downsampling is required because, after translating the signal to baseband, the bandwidth of interest becomes much smaller than the original ADC sampling

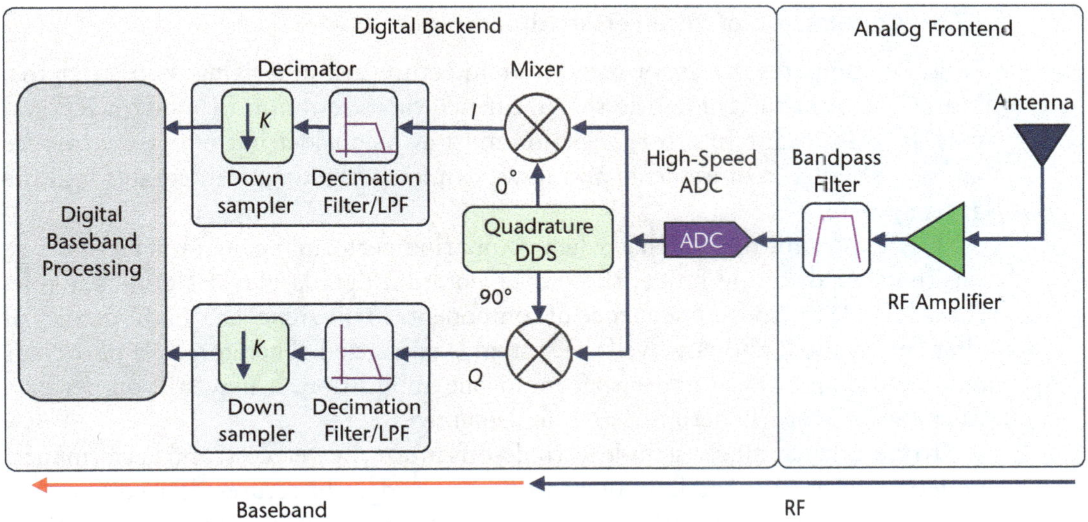

Figure 5.13 Direct RF sampling receiver architecture.

rate. Reducing the sampling rate after appropriate filtering lowers computational load, prevents aliasing, reduces data throughput, and aligns the sample rate with downstream baseband processing requirements. The combination of the decimation filter and the downsampler is collectively referred to as a decimator.

Direct sampling is practical only when both the carrier frequency and the signal bandwidth lie within the ADC's performance envelope. At higher frequencies, particularly in millimeter-wave SDR applications, the limitations of current ADC technology, sampling rate, ENOB, jitter, and power consumption, make direct RF sampling impractical. In such cases, the heterodyne architecture remains the preferred solution: mixers first translate the RF signal to a lower IF, which is then digitized by the ADC. This mixed-signal approach continues to be indispensable for wideband, high-frequency systems that operate beyond the reach of today's ADC technology.

While direct RF sampling offers compelling advantages, several drawbacks must be considered. High-speed ADCs capable of sampling RF and microwave frequencies remain expensive, and achieving the required dynamic range is technically challenging. Wideband RF environments often contain signals spanning many tens of decibels, and accommodating this requires ADCs with very high ENOB, an increasingly difficult and costly requirement at multigigahertz sampling rates.

Direct RF sampling systems also rely heavily on advanced digital signal processing and calibration techniques, increasing implementation complexity. In addition, they are more susceptible to phase noise and jitter in the ADC sampling clock and other system components, which can significantly degrade performance, especially at high input frequencies.

Finally, direct RF sampling architectures typically consume more power than traditional heterodyne systems because of the high sampling rates and the substantial digital processing workloads required. These power demands can be prohibitive in battery-powered, size-constrained, or thermally limited applications.

5.7.5 Comparison of Receiver Architectures

Table 5.1 compares the major receiver architectures, summarizing their strengths, limitations, and the trade-offs that influence their adoption in modern RF systems. In practice, architecture selection must also consider not only performance but also SWaP-C requirements and how seamlessly the design integrates into the larger system.

It is important to note that when comparing performance parameters such as sensitivity or dynamic range, the implementation details play a significant role. These include the specific sequence of components in the signal chain, the quality of components used, and how well the system is optimized. Therefore, if a particular architecture appears to underperform in one application, it may still outperform others when properly designed and implemented.

To meet demanding system-level trade-offs in size, power, cost, and performance, modern receivers often adopt alternative or hybrid architectures derived from the classical types discussed earlier. These architectures are tailored to specific application requirements, leveraging a mix of analog and digital techniques.

5.8 RF Front End: Transmitter Architectures

Transmitter design centers on two complementary challenges:

- Signal generation and frequency conversion, encompassing modulation, waveform generation, and upconversion architectures;

Table 5.1 Receiver Architecture Comparison

Architecture	*Sensitivity*	*Selectivity*	*Image Rejection*	*Dynamic Range*	*Pros*	*Cons*
Direct conversion (homodyne)	Good	Moderate	Limited (IQ balance dependent)	Moderate-high	Simple, compact, low cost, CMOS-friendly, good for SDR	DC offset, IQ imbalance, 1/f noise, calibration needed
Superheterodyne (single conversion)	Excellent	Excellent	Good (with IF filters)	High	Proven, high-performance, excellent filtering and linearity	More components, LO leakage, spurs, bulkier design
Superheterodyne (low IF)	Good	Moderate-good	Better than zero IF	Moderate	Avoids DC offset; reduced analog filtering burden	Residual image issues; less selectivity than superheterodyne
Superheterodyne (double/triple conversion)	Excellent	Excellent	Excellent	Very high	Extreme performance; suitable for radar, EW, satellite comms	Complex, expensive, higher power consumption, large size
Direct sampling	Excellent (ADC-limited)	Flexible (digital filtering)	Excellent (digital domain)	High (if ADC supports)	Wideband, reconfigurable in software, highly flexible SDR design	Requires cutting-edge ADCs/DSP; high power/cost; clock jitter sensitive

- Efficient and linear amplification, addressing PA efficiency, linearity, and the associated techniques required to meet spectral, modulation, and power requirements.

Over time, a variety of transmitter architectures have been developed to address these challenges, each with distinct trade-offs in performance, efficiency, and implementation complexity. Modern systems often integrate multiple approaches, combining high-efficiency amplification with advanced linearization strategies, to achieve optimal overall performance.

5.8.1 Signal Generation and Frequency Conversion Architectures

Signal-generation and frequency-conversion architectures define how a baseband signal is translated to its final RF carrier. These architectures are the transmit-side counterparts of the receiver architectures discussed earlier. Although the underlying principles are similar, transmitter design introduces unique constraints. A well-engineered transmitter must:

- Generate spectrally clean signals with minimal spurious emissions to avoid adjacent-channel leakage and comply with regulatory spectral masks.
- Maintain high PA efficiency and linearity to deliver the required output power while controlling distortion such as EVM, spectral regrowth, and adjacent-channel leakage ratio (ACLR).

The main signal-generation and upconversion architectures are described next.

Direct Upconversion (Homodyne Transmitter)

The homodyne transmitter is the dual of a homodyne receiver. Baseband I/Q signals are mixed directly to the RF carrier. This architecture reduces component count and simplifies calibration but places strict requirements on LO isolation, quadrature balance, and DC offset control, as any imbalance directly leads to carrier leakage and image distortion.

Superheterodyne Upconversion

As shown in Figure 5.14, a two-step conversion chain is: Baseband $\rightarrow$ IF $\rightarrow$ RF. Upconverting through an IF allows the designer to leverage high-Q IF filters, providing excellent image rejection, spurious suppression, and spectral purity. The trade-off is increased complexity, more components, and frequency-planning constraints.

In Figure 5.13, the reconstruction filter after the DAC output is used to suppress spectral images and smooth the discrete-time output into a continuous-time waveform.

DDS + Upconversion

A DDS acts as a digital function generator capable of producing finely tunable sinusoidal or modulated waveforms with subhertz resolution, deterministic phase, and extremely fast frequency hopping (see Section 6.4.5). DDS-based architectures come in two variants:

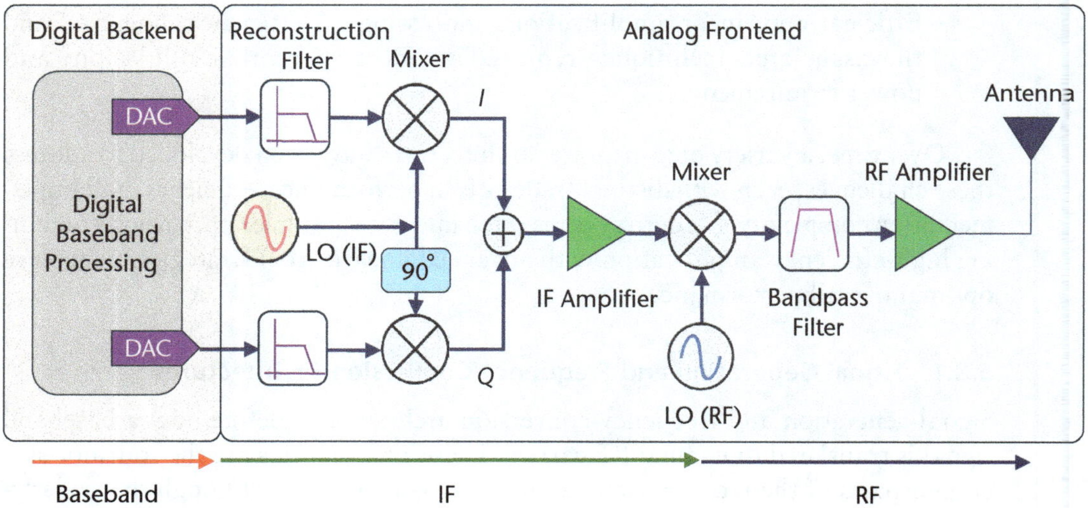

Figure 5.14 Architecture of superheterodyne upconversion.

- *DDS with analog upconversion:* The DDS drives a DAC to produce a low-frequency or mid-frequency analog waveform. This signal is then mixed with an analog LO and is filtered to generate the final RF output. Therefore, the RF front end consists of analog mixers, LO generation, filtering, and gain stages.

  This approach is flexible and high-performance but inherits the usual analog-mixing challenges: image rejection, LO feedthrough, and spur management.
- *DDS + digital upconversion (DUC):* Figure 5.15 illustrates this architecture. In the digital domain, the counterpart of the decimator used in DDC is the interpolator, used in digital upconversion to raise the sample rate before digital mixing.

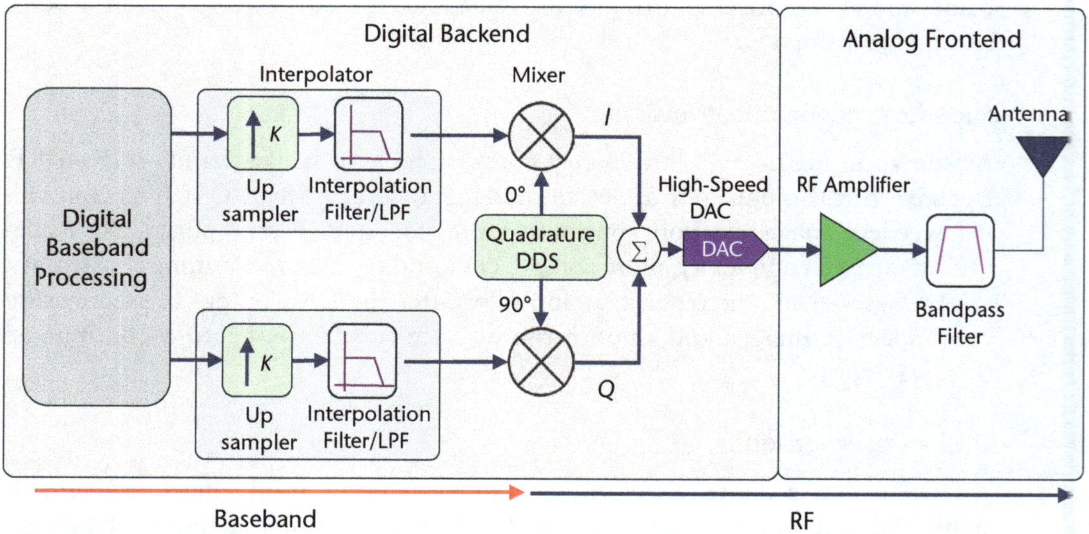

Figure 5.15 Direct RF synthesis transmitter architecture.

The interpolation stage consists of two steps:

1. *Upsampling:* Insert $K - 1$ zeros between consecutive samples to increase the sampling rate by a factor of K. This produces spectral images at multiples of the original bandwidth.
2. *Interpolation (reconstruction) filtering:* A digital lowpass filter removes the spectral images created during upsampling and shapes the waveform so that it meets bandwidth, spectral-purity, and dynamic-range requirements.

After interpolation, a digital mixer (with an NCO-generated digital LO) shifts the waveform to the desired digital carrier frequency. A high-speed DAC then converts this digitally upconverted signal to the analog domain, producing the RF waveform directly. Only analog filtering and amplification are required afterward.

This architecture offers very fine frequency resolution, deterministic phase alignment, rapid frequency hopping, and minimal analog components. Its limitations arise from DDS spurious tones and the finite output bandwidth of the DAC, which may necessitate an additional analog upconversion stage for very high-frequency (e.g., millimeter-wave) transmitters.

Direct DAC RF Synthesis

Modern ultrahigh-speed DACs can generate RF carriers directly, operating in higher Nyquist zones to synthesize multigigahertz signals without analog mixers (see Section 6.4.7). This is the transmit-side analog of a direct-RF sampling receiver architecture.

As long as the DAC supports the required carrier frequency and modulation bandwidth, the RF waveform is produced directly at its final frequency.

For very high bands (e.g., Ka or millimeter-wave), an external upconversion stage may still be necessary.

Direct-RF DAC architectures provide fine frequency and phase resolution, extremely fast hopping, excellent channelization flexibility, and simplified analog front end (filters + PA only). This architecture is increasingly used in SDRs, test instrumentation, phased-array systems, and electronic-warfare transmitters.

5.8.2 Comparison of Transmitter Architectures

Table 5.2 compares the transmitter architectures used for frequency conversion.

5.8.3 PA Efficiency and Linearity Architectures

When designing RF transmitters, the PA is the dominant factor governing efficiency, linearity, output power capability, and overall SWaP-C. Modern communication standards such as OFDM, LTE, and 5G introduce high peak-to-average power ratio (PAPR) signals that demand highly linear amplification across a wide dynamic range. Although PAs achieve their maximum efficiency near saturation, operating in this region causes severe nonlinear distortion, including spectral regrowth, EVM degradation, and violation of spectral masks.

Table 5.2 Comparison of Frequency Conversion Transmitter Architectures

Architecture	*Principle*	*Advantages*	*Drawbacks*	*Common Applications*
Direct upconversion	Baseband I/Q signals are mixed directly to the RF carrier	Simple design, low parts count, compact implementation	Sensitive to LO leakage and DC offsets; limited filtering options	Mobile handsets, SDRs, compact radios
Superheterodyne upconversion	Baseband is first upconverted to an IF, then to RF	Excellent filtering flexibility and image suppression	Higher complexity, more components, larger footprint	High-performance radios, satellite communications
DDS + upconversion	A DDS generates a low-frequency to mid-frequency signal (such as a digital AWG), then shifted to RF using analog or digital upconversion	Ultrafine frequency resolution, agile phase control, rapid hopping	Spurious outputs from DDS; limited by DAC output bandwidth; often requires an extra analog upconversion stage for higher frequencies (e.g., millimeter-wave)	Agile radios, test equipment, EW systems
Direct DAC RF synthesis	A high-speed DAC generates the RF carrier waveform directly, bypassing analog mixers	Simplified architecture; no mixers required; precise frequency and phase agility	Limited by DAC speed, linearity, and ENOB; millimeter-wave still needs external upconversion	SDRs, advanced instrumentation, EW systems

To preserve linearity, PAs are typically backed off from saturation, often by 6 to 12 dB, depending on the modulation. However, back-off drastically reduces efficiency, creating a fundamental and unavoidable trade-off in transmitter design: high linearity requires back-off, but high efficiency requires saturation.

To mitigate this trade-off, several advanced PA architectures and linearization techniques have been developed. These methods aim to maintain linearity under high PAPR signals while preserving efficiency, especially at the power levels where modern transmitters spend most of their operating time. The most widely adopted approaches include [4]:

- *Envelope tracking:* Envelope tracking continuously modulates the PA supply voltage to follow the instantaneous envelope of the transmitted signal, as illustrated in Figure 5.16(a). While the PA bias point remains fixed, the available voltage headroom dynamically adjusts with the envelope. This improves efficiency by reducing power dissipation associated with excess voltage headroom at low signal levels and enhances linearity by providing sufficient supply voltage during signal peaks to prevent distortion.
- *Envelope elimination and restoration:* Decomposes the signal into separate amplitude and phase components. A highly efficient nonlinear PA amplifies only the constant-envelope phase signal, while the amplitude component is reapplied through high-bandwidth supply modulation. This technique, shown in Figure 5.16(b), enables both high efficiency and linearity when properly synchronized.
- *Doherty amplification:* Uses two amplifiers working together, shown in Figure 5.16(c): a carrier (main) amplifier, typically biased in Class A/B, that handles

low and medium power levels, and a peaking amplifier, usually Class C, that turns on only when the signal becomes strong. As the output power increases, the peaking amplifier activates and helps by dynamically adjusting the load seen by the main amplifier. This load modulation keeps the main amplifier operating in a more efficient region, even though most communication signals spend the majority of their time at lower power levels.

- *Outphasing amplification:* This technique represents the original amplitude-modulated and phase-modulated waveform as two constant-envelope signals [5]. Each branch is amplified using a highly efficient switching-mode PA, and the outputs are recombined through a Chireix (outphasing) combiner to reconstruct a linear RF signal, as shown in Figure 5.16(d). Rather than amplifying amplitude directly, outphasing converts amplitude information into a phase difference between the two constant-envelope signals. A large output amplitude corresponds to a small phase separation, with the signals nearly in phase, while a small output amplitude corresponds to a large phase separation, with the signals nearly out of phase. When recombined, constructive and destructive interference between the two signals reproduces the desired output amplitude, enabling high efficiency while maintaining linearity.

Table 5.3 compares the advantages, drawbacks, and common applications of the architectures discussed above.

5.8.4 Baseband Techniques for PA Efficiency and Linearity

In addition to architectural solutions such as Doherty, envelope elimination and restoration, envelope tracking, and outphasing, modern transmitters rely heavily on DSP methods at baseband to further improve PA performance. These techniques do

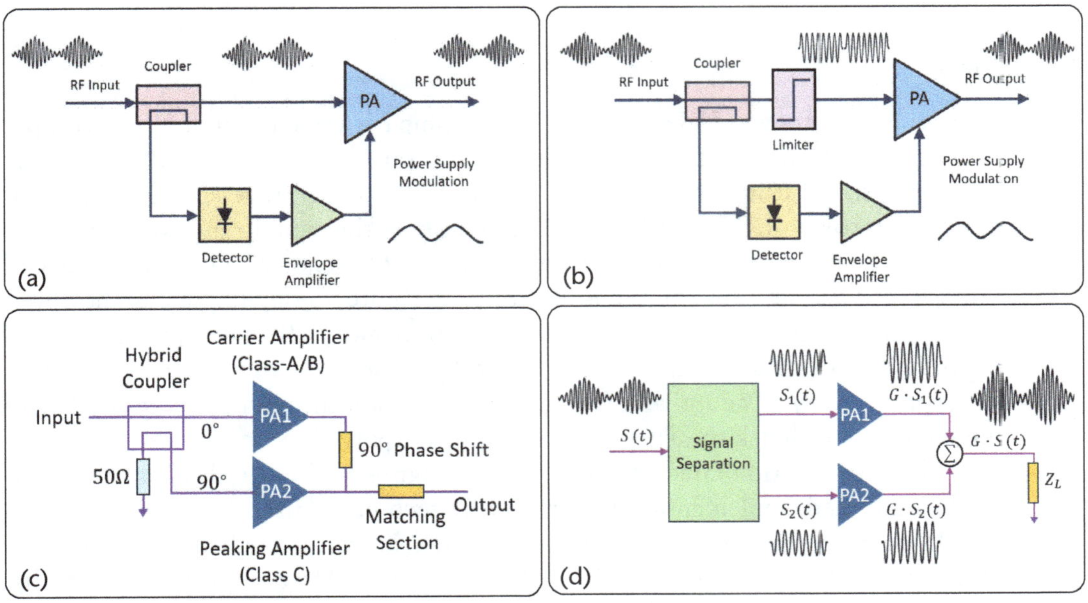

Figure 5.16 Simplified block schematics of transmitters based on: (a) envelope tracking (ET) and (b) envelope elimination and restoration (EER) technique, (c) Doherty amplifier, and (d) outphasing amplifier.

Table 5.3 Comparison of PA Efficiency and Linearity Transmitter Architectures

Architecture	*Principle*	*Advantages*	*Drawbacks*	*Common Applications*
Envelope tracking	PA supply voltage tracks envelope in real time	Improves efficiency over wide power range	Requires very fast, wideband supply modulators; design complexity	4G/5G smartphones, Wi-Fi, LTE base stations
Envelope elimination and restoration	Splits signal into amplitude and phase; PA handles only constant-envelope phase, envelope restored via supply modulation	Allows highly efficient nonlinear PAs (Class E/F); good efficiency for high PAPR signals	Sensitive to delay/phase mismatch; limited bandwidth due to supply modulation speed	Legacy mobile (2G), research prototypes
Doherty	Carrier and peaking amplifiers use dynamic load modulation	High efficiency at power back-off	Complex design; matching networks critical; limited bandwidth	Cellular base stations, 5G, broadcast transmitters
Outphasing	Signal decomposed into two constant-envelope signals, amplified separately, recombined at RF	Each PA operates at high efficiency; theoretically linear	Very sensitive to phase imbalance; recombiner losses; large hardware overhead	Research, specialized high-linearity systems

not alter the PA's physical design but instead optimize the signal before or during amplification. CFR lowers the peak-to-average power ratio, reducing the required back-off, while DPD precompensates for the PA's nonlinearities to preserve signal fidelity. Together, these baseband methods are essential companions to advanced PA architectures, enabling transmitters to simultaneously meet efficiency, linearity, and spectral mask requirements.

DPD

If the transfer function H describes the relationship between the input $x(t)$ and output $y(t)$ of a system such that $y(t) = H(x(t))$, then it is possible to intentionally predistort the input signal using the inverse function H^{-1}, provided that H is invertible (or can be approximated by an invertible model) over the signal range of interest. In this way, the cascaded response becomes $H(H^{-1}(x(t))) \approx x(t)$, which ideally yields an undistorted output. This concept is illustrated in Figure 5.17(a). In practice, however, the system is rarely perfectly invertible due to memory effects, clipping, thermal drift, and model mismatch. Therefore, digital predistortion algorithms aim to learn or approximate H^{-1} adaptively, minimizing distortion through iterative coefficient updates and feedback measurements from the output.

The predistortion concept, illustrated in Figure 5.17(b), finds a practical and widely adopted implementation in the DPD technique, used in modern PAs. In DPD, a correction function, known as the predistorter, is applied at baseband to precompensate for the nonlinearities of the amplifier. After amplification, the distortions introduced by the PA are effectively canceled by the predistortion, yielding an output that closely replicates the desired input signal.

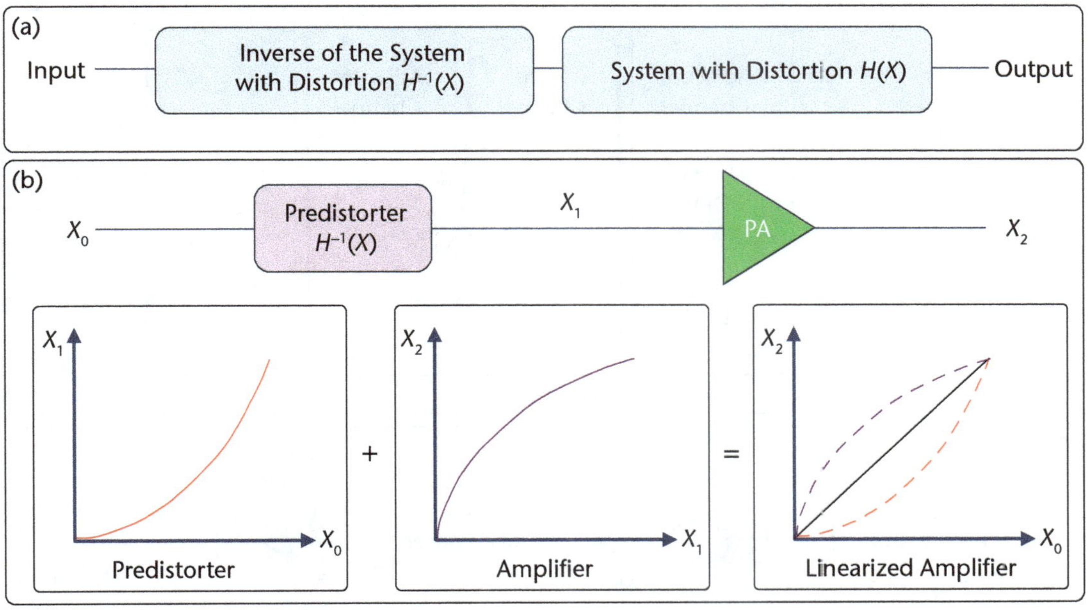

Figure 5.17 Distortion compensation using predistortion techniques by applying the inverse function of the system with distortion.

Without DPD, the spectral regrowth of a PA is significantly higher, as illustrated in Figure 5.18(a), causing unwanted leakage into adjacent channels. When DPD is applied, the nonlinear distortion is greatly reduced, ensuring compliance with linearity, ACLR, and spectral mask requirements, key metrics for transmitter performance and regulatory approval.

The architecture of a PA employing DPD is shown in Figure 5.18(b). The predistorted baseband signal is upconverted and amplified before transmission, while a directional coupler samples the PA output and routes it through a downconverter and ADC to the parameter-extraction block. This feedback path enables continuous adaptation of the predistorter model under varying signal and operating conditions.

The practical implementation of DPD will be examined in greater detail in Volume 2.

CFR

As discussed in Chapter 3, communication waveforms such as OFDM exhibit inherently high PAPR, meaning that the instantaneous signal peaks can be much larger than the average power. To amplify these peaks without distortion, a PA must operate with substantial output back-off from saturation, an operating point that significantly reduces efficiency. CFR alleviates this challenge by intentionally reducing the amplitude of these peaks before the signal reaches the PA.

As shown in Figure 5.19(a), CFR is applied entirely in the baseband domain. By lowering the peak excursions, the PA can operate closer to saturation for a greater portion of the time, markedly improving efficiency. The resulting waveform, illustrated in Figure 5.19(b), exhibits a reduced PAPR with minimal degradation in key transmitter metrics such as EVM and ACLR.

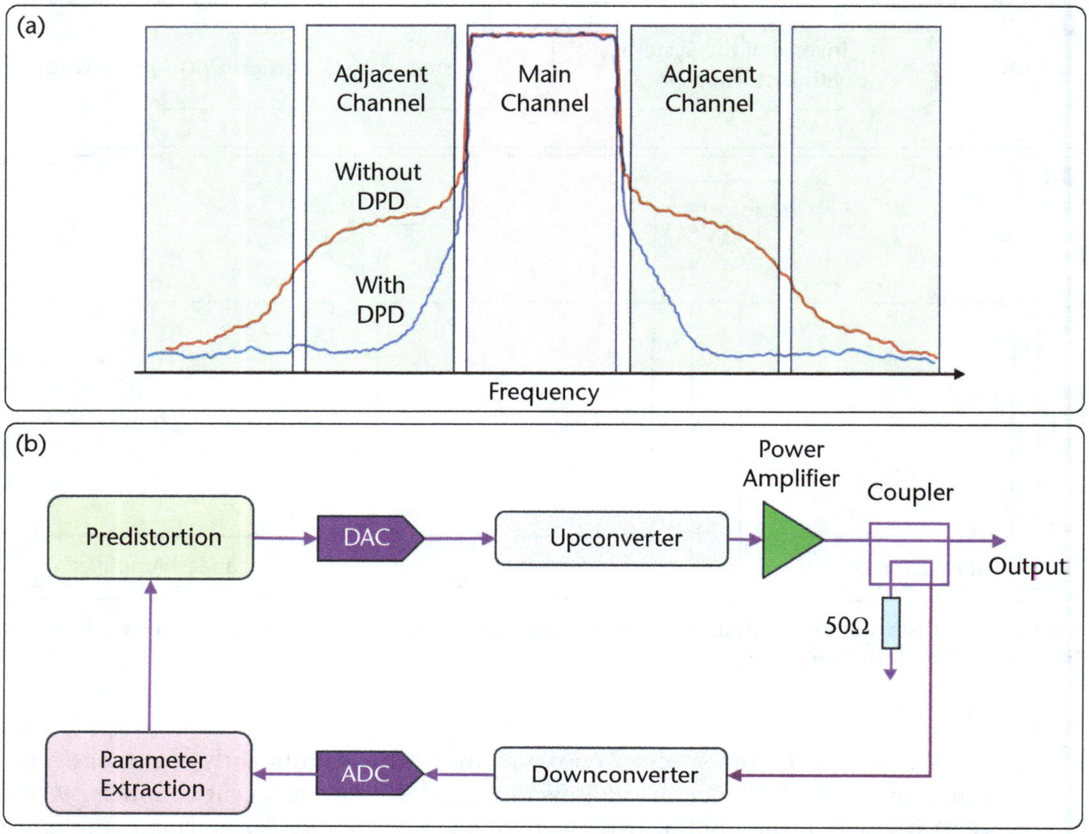

Figure 5.18 (a) The output spectrum with and without DPD. (b) Block diagram of a PA with DPD.

CFR is seldom used in isolation. It is typically combined with techniques such as envelope tracking, Doherty architectures, and DPD. CFR reduces the PAPR and relaxes the PA's efficiency-linearity trade-off, while DPD subsequently compensates for the nonlinear distortion introduced by the PA. Together, these complementary methods enable the deployment of highly efficient yet linear PAs in modern wireless communication systems.

Several CFR methods exist, including peak clipping, peak windowing, and peak cancellation using auxiliary signals. The goal is to reduce PAPR while keeping in-band distortion and out-of-band spectral regrowth within acceptable limits.

The clipping mathematical equation is as follows [6]:

$$x_{\text{clip}}(n) = c(n)x(n)$$
$$c(n) = \begin{cases} \dfrac{A}{|x(n)|} & |x(n)| \geq A \\ 1 & |x(n)| < A \end{cases} \tag{5.1}$$

where $x(n)$ is the input signal and $c(n)$ is the clipping algorithm. As shown in Figure 5.18(b), clipping introduces sharp discontinuities in the waveform, which manifest

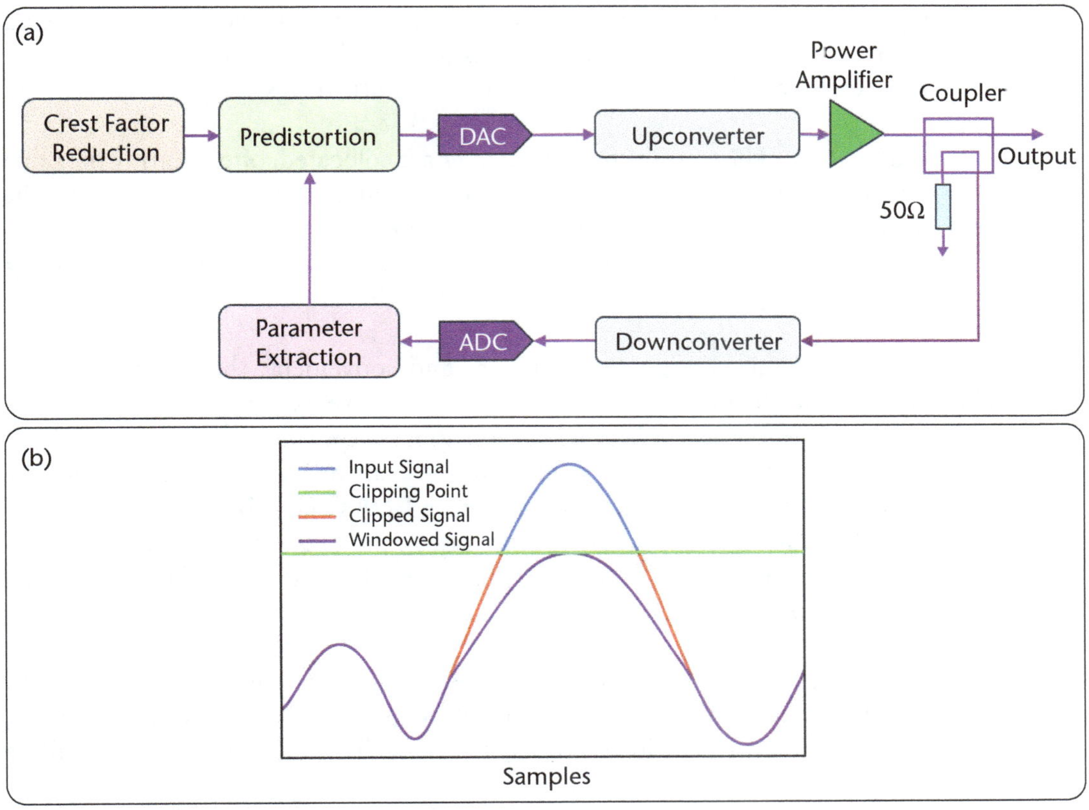

Figure 5.19 (a) Block diagram of digital enhancement techniques, including CFR and DPD. (b) CFR reduces the PAPR by clipping or reshaping signal peaks. The unclipped signal exhibits high peaks (red), while the clipped signal $|x_{\text{clipping}}[n]|$ (blue) is constrained below the clipping threshold A (green), reducing PAPR before amplification.

as high-frequency components in the spectrum. These high-frequency components lead to unwanted out-of-band emissions and an increase in adjacent-channel power.

To mitigate this effect, the clipped signal is passed through a lowpass filter, which attenuates the high-frequency components generated by the sharp corners. Thus, in the clipping-and-filtering method, the key design parameters include the target PAPR and the lowpass filter characteristics, maximum filter order, passband frequency, stopband frequency, passband ripple, and stopband attenuation.

Peak windowing provides an alternative approach by smoothing the sharp transitions introduced by hard clipping. Instead of abruptly limiting the peaks, the algorithm applies a windowing function to the regions surrounding the peaks, producing a softer, more controlled reduction in amplitude and reducing spectral regrowth.

5.9 RF Front End: Transceiver Architectures

A transceiver integrates both a transmitter and a receiver into a single system, enabling bidirectional communication over a shared medium. This section examines signal path partitioning, duplexing methods, and transmit/receive isolation, followed by system-level challenges such as interference management and antenna sharing strategies using components such as RF switches and duplexers.

Transceivers can be classified as either monostatic or bistatic:

- *Monostatic:* The transmitter and receiver are colocated, often sharing the same antenna or enclosure. A common example is an automotive radar, where the transmitter and receiver share the same antenna.
- *Bistatic:* The transmitter and receiver are spatially separated, ranging from a few wavelengths to several kilometers, as in bistatic radar systems.

While monostatic designs are compact and convenient, they present unique challenges due to the proximity of transmit and receive paths:

- *Self-interference and leakage:* High-power transmit signals may leak into the receiver chain, desensitizing or saturating LNAs and mixers.
- *Front-end protection:* Sensitive receiver components require protection from strong transmit signals, often using fast-acting limiters, transmit/receive (T/R) switches, or isolators.

Addressing these challenges requires careful attention to isolation, timing, and front-end architecture. Practical mitigation strategies include the use of circulators, directional couplers, T/R switches, and active cancellation methods, which will be discussed in the following sections.

Although transmitters and receivers are often studied independently, in compact or monostatic systems they must be codesigned as part of a shared platform. Integration on a single PCB with common components, timing references, and control loops demands a transceiver codesign approach, where the performance and isolation of both paths are considered together. Table 5.4 summarizes key TX/RX codesign principles.

5.9.1 Signal Routing for Various Transceiver Architectures

Depending on system requirements, RF designs may use either a single shared antenna for both transmit (TX) and receive (RX) paths or separate antennas for each. The choice between these configurations depends not only on the duplexing mode (half or full) but also on whether TX and RX occur in the same or different frequency bands. Accordingly, various signal routing techniques are employed, including T/R switches, filter-based diplexers, circulators, and even directional couplers combined with cancellation strategies.

Table 5.4 Transceiver Codesign Principles

Design Aspect	*TX-RX Shared Advantage*	*Codesign Challenge*
LO sharing	Reduced complexity	Requires isolation and careful routing
Antenna sharing	Smaller form factor	Increased isolation and switching demands
Frequency plan	Simplified filtering	Risk of interference and LO/RF overlap
PCB layout	Compact design	Increases the risk of mutual coupling

Transmit/Receive Switch

In half-duplex transceiver systems, a transmit/receive (T/R) switch is commonly used to alternate the connection of the RF front end between the transmitter and receiver paths. As shown in Figure 5.20(a), when the transmit and receive functions operate at different frequencies, separate antennas are often employed. An example is a Ku-band satellite communication system, where reception occurs in the 10.7–12.7-GHz band and transmission occurs in the 14–14.5-GHz band using two distinct antennas or apertures. Additional internal switches, not shown in Figure 5.20(a), are typically used to route the transmit signal to the TX chain and the received signal to the RX chain. In satellite communication systems, this routing is commonly performed within the frequency converter.

In contrast, when the TX and RX share the same frequency band, a single antenna can be used, as illustrated in Figure 5.20(b). In this case, the T/R switch toggles between the transmit and receive chains, enabling antenna sharing while ensuring proper signal routing and isolation.

The switching between the transmitter and receiver chains is based on timing control signals, often derived from the system's baseband processor or control logic.

T/R switching can be implemented using various technologies depending on frequency, power handling, timing requirements, isolation, and integration requirements. Common options include electromechanical, PIN diode switches, GaAs FETs, and emerging MEMS-based switches for improved linearity and low loss. Table 5.5 compares the characteristics of these technologies.

While electromechanical switches offer advantages such as high power-handling capability, low insertion loss, and excellent isolation, they are relatively slow, often operating on the order of milliseconds, and have a limited life cycle due to mechanical wear. In contrast, semiconductor switches, including PIN diodes and GaAs FETs, provide much faster switching speeds and significantly longer lifespans, as they have no moving parts. However, they typically exhibit higher insertion loss, lower isolation, and reduced power handling compared to their electromechanical

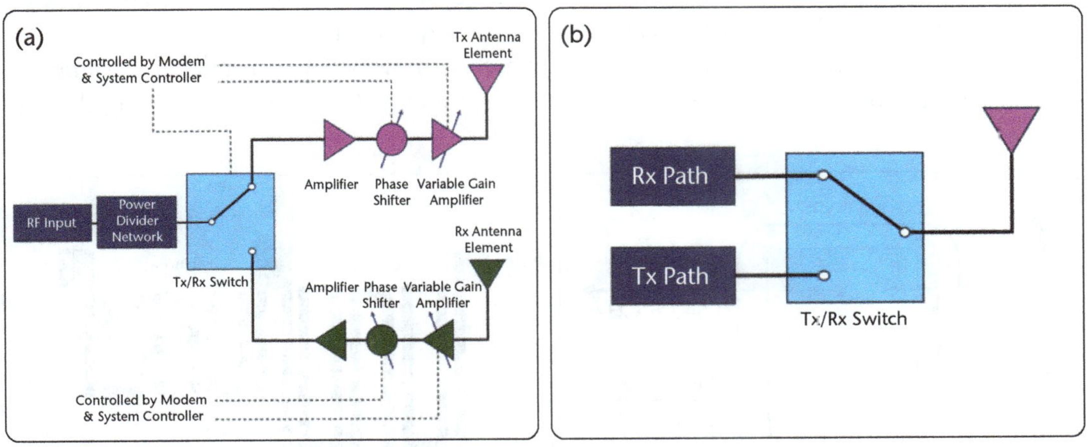

Figure 5.20 (a) Half-duplex phased array system using a T/R switch and two separate antennas, each dedicated to a different frequency band for transmit and receive operations. (b) Half-duplex system with a T/R switch sharing a single antenna for both transmission and reception.

Table 5.5 Comparison of Common T/R Switch Technologies

Switch Type	*Switching Speed*	*Power Handling*	*Insertion Loss*	*Isolation*	*Linearity*	*Complexity*	*Lifetime*
Electromechanical	Slow (ms range)	Very high	Very low	Excellent	Excellent	Simple control, but bulky	Limited (mechanical wear)
PIN diode	Fast (μs range)	Moderate to high	Moderate	Moderate to good	Good	Complex biasing and thermal management	Very long
GaAs FET	Very fast (ns–μs range)	Low to moderate	Moderate to high	Moderate	Moderate to good	Compact, simple control	Very long
MEMS	Moderate (μs–ms range)	Moderate	Very low	Very good	Excellent	Emerging technology, integration challenges	Moderate (improving)

counterparts. Additionally, high-power PIN diode switches can be challenging to implement, requiring complex biasing and thermal management.

Multiplexing in RF Front Ends

Multiplexers are frequency-selective components that allow multiple RF signals, typically from different frequency bands, to share a common signal path, such as a single antenna or transceiver chain. Multiplexers incorporate band-specific filters at each input or output port. These filters allow only a designated frequency range to pass while rejecting others, providing both channel isolation and frequency discrimination. This makes multiplexers especially valuable in multiband transceivers, tunable radios, and spectrum monitoring systems.

In RF front-end architectures, multiplexers play a crucial role in band selection and signal routing, reducing system complexity by eliminating the need for dedicated signal chains for each frequency band. Unlike switched signal routing, a multiplexer provides simultaneous access to all desired frequency bands.

Figure 5.21(a) shows a diplexer, which is a two-way multiplexer designed to separate or combine exactly two distinct frequency bands while allowing them to

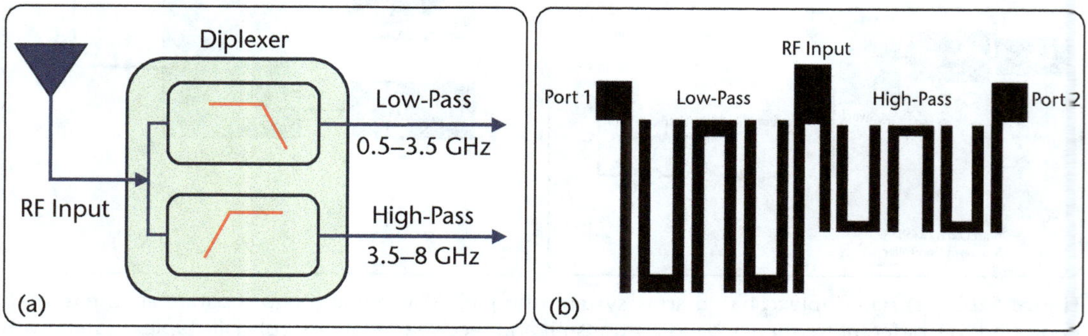

Figure 5.21 (a) A diplexer, and (b) implementation of a hairpin diplexer.

share a common port, typically an antenna. Figure 5.21(b) shows a hairpin diplexer: the left side, with larger resonators, is tuned for the lower-frequency band, while the right side, with smaller resonant structures, covers the higher-frequency band.

Another example is that a sub-6-GHz receiver may need to operate across multiple bands, such as 915 MHz, 2.4 GHz, and 5.8 GHz. Instead of using three separate antennas, a multiplexer can route each band through a shared wideband antenna. This approach minimizes component count, conserves space, and simplifies system integration.

All multiplexers introduce some signal loss, primarily due to the filter structures. In sensitive receiver chains, minimizing insertion loss is essential to preserve noise figure and maintain receiver sensitivity. So the loss-noise figure trade-off needs to be incorporated in system design.

Additionally, high port-to-port isolation is critical to prevent signal leakage, which can lead to intermodulation distortion, receiver desensitization, or cross-band interference. Achieving this isolation requires careful filter design, shielding, and layout practices. Parameters such as group delay are also important to ensure low distortion and maintain signal integrity.

In transmit applications, such as base stations, military communication systems, and EW platforms, the multiplexer must also withstand high RF power levels without degradation or failure.

Filter Banks

In systems where multiple frequency bands must be monitored, processed, or selected dynamically, filter banks offer a practical and scalable solution. As shown in Figure 5.22, a filter bank consists of multiple bandpass filters arranged in parallel, with each filter covering a distinct sub-band. These filters may be implemented using discrete components, integrated RFICs, or switched modules, depending on requirements such as operating frequency, insertion loss, and switching speed. A multipole switch at the input and output selects which filter, and thus which band,

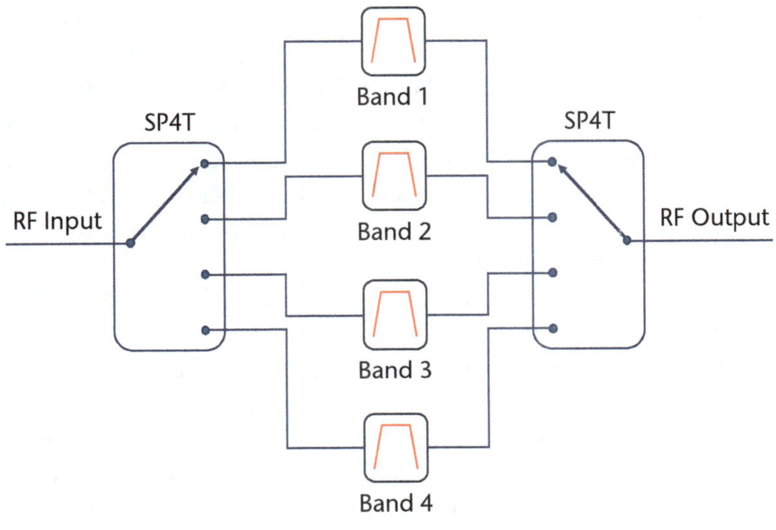

Figure 5.22 A switched filter bank covering five bands.

is active at any given time. Unlike a multiplexer, a switched filter bank does not provide simultaneous access to all bands; only one path is available at a time.

Filter banks are used in several key applications:

- *Multiband receivers:* Enable reception or selection across multiple communication bands (e.g., cellular, satellite, or broadband systems).
- *Spectrum monitoring:* Isolate and analyze different sub-bands in parallel, supporting EW, interference detection, and wideband spectrum sensing.
- *Transmitter preselection:* Limit out-of-band emissions and suppress harmonics by selecting only the desired transmit band before amplification.

Here are some considerations for the filter banks:

- *Isolation:* Crosstalk between adjacent channels must be minimized, especially in high dynamic range systems.
- *Insertion loss:* Varies with frequency and the type of switching used. MEMS and PIN-diode switches offer trade-offs in speed and loss.
- *Group delay and phase matching:* Critical in systems requiring coherent processing or beamforming.
- *Size and integration:* Miniaturization via LTCC, MMIC, or SiP packaging is often necessary in SWaP-constrained systems.
- *Reconfigurability:* Modern systems may require tunable filter banks, implemented via varactors, MEMS, or RF-silicon-on-insulator (RF-SOI) switches. In more advanced architectures, especially for frequency-agile systems such as cognitive radios, reconfigurable filter banks are used. These may employ tunable resonators, digitally controlled capacitors, or even software-defined digital filter banks post-ADC.

Circulator

As we will see in Chapter 6, a circulator is a passive, nonreciprocal three-port device commonly used in full-duplex systems. As illustrated in Figure 5.23, it routes RF power in a fixed, unidirectional sequence, typically from Port 1 to Port 2, Port 2 to Port 3, and Port 3 back to Port 1. This directional behavior enables simultaneous transmission and reception through a shared antenna while maintaining high isolation between the TX and RX paths to reduce self-interference. Circulators are widely used in radar systems, high-power RF transmitters, and reflectometry applications, where simultaneous transmit-receive operation and strong port-to-port isolation are essential.

Circulators support full-duplex operation by allowing the transmitter and receiver to operate concurrently, with high isolation between them. As shown in Figure 5.23, in a typical configuration, port 1 is connected to the transmitter, port 2 to the shared antenna, and port 3 to the receiver. The circulator directs energy from the transmitter to the antenna (Port 1 → Port 2) and routes the received signal from the antenna to the receiver (Port 2 → Port 3). However, it is important to note that any reflected power from the antenna, such as due to impedance mismatch,

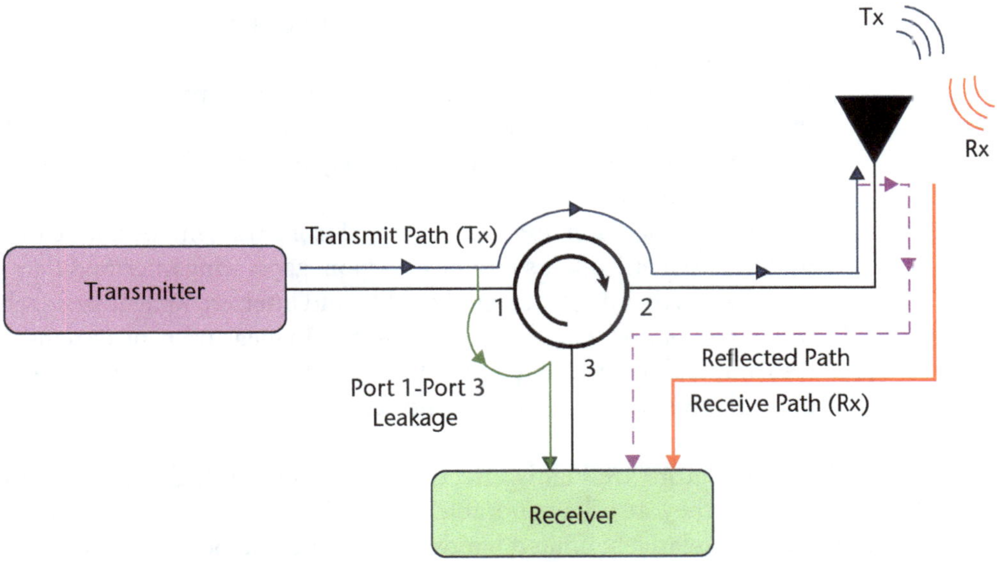

Figure 5.23 A circulator enables a transmitter and receiver to share a single antenna. The signal flows from the transmitter to the antenna, and incoming signals from the antenna are routed to the receiver. Any reflected transmit power is diverted toward the receive port. The leakage between Port 1 and Port 3 is also coupled to the receiver.

can also be coupled into the receive path. Moreover, the isolation between Port 1 and Port 3 is finite, typically on the order of 20 to 50 dB.

This becomes especially critical in high-power applications. For example, consider a 100-W (+50-dBm) PA connected to an antenna through a circulator where the antenna presents a return loss of 10 dB. A 10-dB return loss corresponds to roughly 10% of the incident power being reflected, meaning approximately 10W (+40 dBm) of RF energy could be coupled into the receiver path. This is a significant amount of power that can damage sensitive components in the receiver front end or drive them into saturation, severely degrading system performance.

Therefore, it is essential to account for this scenario during the design phase, particularly in high-power transmit systems. Appropriate protection mechanisms, such as limiters, blanking circuits, or additional filtering, should be implemented to safeguard the receiver from potential damage and to maintain overall system performance.

Operating the transmit and receive paths at different frequencies can further aid in improving isolation, as it allows for frequency-selective filtering on the receive path. However, this approach requires that the antenna and the circulator support both the transmit and receive frequency bands to ensure efficient operation across the entire system.

Minimizing leakage from the transmit port to the receive port is essential for system reliability and performance, for several key reasons:

1. *Receiver front-end protection:* In high-power systems, even moderate levels of reflected power from the antenna can couple into the receive path. This

can exceed the power-handling capability of LNAs or mixers, potentially leading to permanent damage.
2. *Preservation of receiver sensitivity:* Leakage from the transmitter can raise the noise floor or introduce spurious signals at the receiver input. This can mask weak incoming signals and severely limit the dynamic range and sensitivity of the receive chain.
3. *Intermodulation and nonlinear effects:* Strong transmit leakage can drive the receiver front end into its nonlinear region, generating intermodulation products that further degrade signal quality and interfere with adjacent channels.
4. *Impact on calibration and measurements:* In measurement systems or SDRs, transmit leakage can corrupt calibration routines or create false readings, undermining system accuracy and reliability.

While circulators are highly effective for enabling full-duplex operation with good isolation, they also have notable limitations. These components tend to be bulky and offer relatively limited bandwidth. Achieving both high power handling and wide bandwidth in a single design is challenging, making circulators less suitable for compact, lightweight, or broadband systems.

Directional Coupler with Cancellation

A directional coupler used in self-interference cancellation (SIC) configuration is an advanced approach for enabling full-duplex operation on the same frequency band. Unlike traditional methods that rely on frequency or time separation, this technique allows simultaneous transmission and reception by isolating the strong transmit signal from the weak received signal using a combination of hardware and signal processing techniques.

As shown in Figure 5.24, a directional coupler, or a similar structure such as a hybrid coupler or balun, is used to tap off a portion of the transmit signal. This reference signal is then used to cancel self-interference at the receiver input. Cancellation can be performed in the analog domain (e.g., with tunable delay lines or vector modulators) or in the digital domain using adaptive algorithms.

SIC typically works across three domains, shown in Table 5.6.

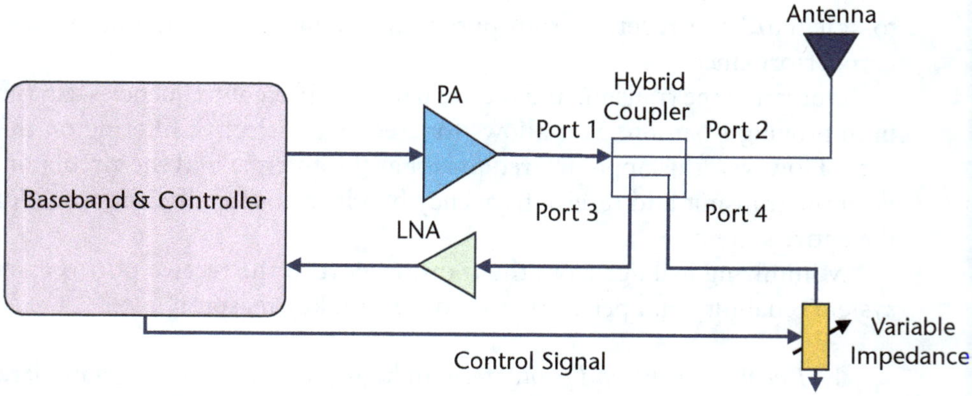

Figure 5.24 Single antenna, hybrid coupler full-duplex system.

Table 5.6 SIC Techniques

Domain	*Goal*	*Methods/ Implementation*	*Advantages*	*Limitations*
Passive suppression	Reduce SI before it reaches the receiver	Antenna isolation (cross-polarization, spatial separation), circulators/directional couplers, differential signaling and shielding	Simple, always-on baseline suppression (~20–40 dB)	Limited cancellation; insufficient alone for high-SNR systems
Analog cancellation	Subtract TX replica before LNA	Sample TX via coupler, tunable delay, phase, amplitude path, subtraction via coupler/summing junction	Protects LNA from saturation; works in RF domain (~20–40 dB)	Requires precise matching; performance drifts with temperature/ frequency
Digital cancellation	Remove residual SI after ADC	Reconstruct TX signal digitally, model leakage channel including nonlinearities/ multipath, subtract synthesized interference	High flexibility; adaptive; enables nonlinear cancellation (~20–50 dB)	Dependent on ADC resolution; computationally intensive; residual noise floor

By using a combination of techniques, many practical systems aim for ~100-dB total cancellation to suppress a strong TX signal (e.g., +20 dBm) down to the receiver noise floor (e.g., −90 dBm).

Here are some design considerations:

- Linearity of the analog path is crucial; nonlinearities can introduce harmonics that are harder to cancel.
- Channel modeling accuracy directly impacts digital cancellation performance.
- Latency and synchronization between TX and RX chains are critical for effective subtraction.
- Calibration loops may be needed for adapting to time-varying environments.

Note that this technique requires highly precise calibration to maintain cancellation performance. It is sensitive to amplitude and phase mismatches, temperature drift, and nonlinearities in the transmit chain. Maintaining consistent performance across wide bandwidths and dynamic conditions remains a significant challenge.

5.9.2 TX/RX Integration Considerations

In transceiver systems where the TX and RX chains share the same enclosure, such as in a cell phone, they cannot be designed independently. Their performance is tightly coupled through shared components, frequency plans, linearity requirements, noise budgets, and filtering strategies. A codesign approach ensures that trade-offs made in one path do not degrade overall system performance but instead contribute to a balanced and optimized RF architecture.

Shared Resource Optimization

In integrated transceiver systems, certain hardware blocks are often shared between the transmit and receive paths to reduce cost, footprint, and power consumption. While this shared architecture improves integration, it also imposes stringent requirements on isolation, timing alignment, and interference management to prevent cross-coupling and performance degradation.

1. *Shared antennas:* A common shared resource is the antenna port. As discussed in earlier sections, a single antenna may serve both transmission and reception through RF switches, duplexers, or circulators. This configuration reduces hardware complexity and board area, but requires:
   - High isolation to protect the sensitive RX chain from high TX power.
   - Robust protection against leakage, transient artifacts, or reflected power that could damage PAs or LNAs or degrade system performance.
   - Low insertion loss to preserve link budget and sensitivity.
   - Fast switching speed (when switches are used) to minimize dead time in time-division duplex (TDD) systems.
2. *Shared LO and clock sources:* Another critical shared resource is the LO. Figure 5.25 illustrates the shared LO synthesizer, where both upconversion and downconversion stages rely on a common LO. This approach simplifies the architecture and ensures coherent frequency planning, but it also imposes key requirements on noise, isolation, and calibration.

   The benefits of LO and clock sharing include reduced component count, lower power and cost, and more compact layouts. A shared LO also provides

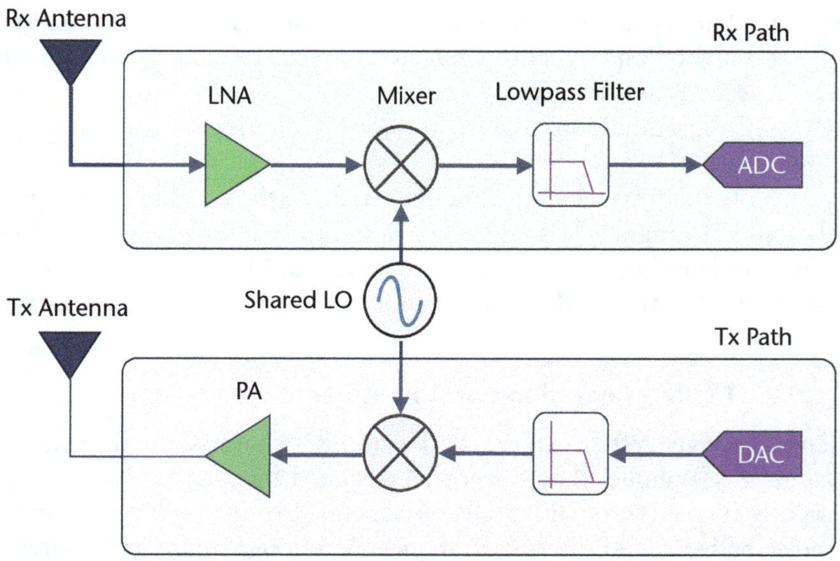

Figure 5.25 The LO is shared between the RX and TX mixers.

tight phase and frequency alignment between the TX and RX paths, greatly improving synchronization, an essential requirement for coherent architectures such as full-duplex radios, phased arrays, and MIMO systems.

Here are some design considerations and challenges for shared LO architectures:

- *Isolation:* LO leakage from TX into RX can degrade receiver sensitivity. Careful shielding, filtering, and buffer stages are often required.
- *Spurious performance:* Shared LOs can create intermodulation products or spurs that affect both chains. PLL design and filtering must be optimized.
- *Clock jitter:* Shared clocks must have low jitter to support high-speed data converters (ADCs/DACs), especially in wideband systems.
- *Switching artifacts:* In TDD systems, rapid LO path switching between TX and RX can introduce transients or phase discontinuities if not properly managed.

In some designs, spread-spectrum clock generators (SSCGs) are employed to further mitigate interference challenges. An SSCG intentionally modulates the frequency of a clock signal slightly (often within ±0.5% to ±2%), spreading its spectral energy over a wider band rather than concentrating it at discrete harmonics to reduce EMI peaks. As we will see in Volume 2, spread-spectrum techniques are also used in voltage regulators to reduce EMI peaks.

3. *Other shared blocks:* Beyond antennas and LOs, advanced transceivers often share additional resources:
   - *Filters:* Reconfigurable or tunable filters serve both TX and RX paths, although they require linear tuning elements and careful biasing.
   - *ADCs/DACs:* In time-multiplexed systems, high-speed data converters may be shared, demanding precise timing control and low-noise switching.
   - *Power management:* Regulators and bias networks are commonly shared, where noise coupling at these levels can ripple across both chains.

When resources are shared, shielding becomes a critical EMI-mitigation technique. Sensitive RF sections, or noisy digital circuits, are enclosed with conductive materials to block unwanted electromagnetic fields and prevent cross-coupling. With proper isolation, grounding, and shielding practices, shared-resource architectures can still achieve high performance.

Gain and Dynamic Range Coordination

Another key aspect of transceiver integration is coordinating the gain and dynamic range across both the TX and RX paths. The receiver must be robust enough to tolerate leakage from the transmitter output, particularly in high-power systems such as radars, while still maintaining high sensitivity for weak-signal reception. At the same time, the transmitter must exhibit sufficient linearity to prevent intermodulation products from falling into the RX band, where they can degrade sensitivity or desensitize the front end. Achieving this balance requires careful frequency planning, comprehensive testing, and calibration to ensure stable performance under real operating conditions.

Timing Coordination

In systems relying on accurate RX and TX timing, such as TDD, precise timing coordination between transmit and receive intervals is essential. Even small mismatches can lead to incomplete transitions between TX and RX modes, causing data loss, corrupted symbols, or elevated error rates. Critical timing factors include PA turn-on/turn-off behavior, switching latencies, receiver recovery time, software-induced delays, and hardware propagation delays through cables, PCB traces, and logic gates. These elements must be carefully measured, estimated, and synchronized to ensure smooth and reliable mode transitions across time slots.

We have discussed several key integration considerations; however, implementing a robust integrated TX/RX system requires attention to many additional details that cannot be fully covered here. Multiple layers, including hardware, firmware, and software, must be precisely coordinated for the system to operate reliably. Successful transceiver integration is ultimately a multidimensional effort, where small oversights in any layer can compromise overall performance.

Calibration and Tuning in Transceivers

Accurate calibration and tuning are essential to maintain transceiver performance, ensuring linearity, minimizing distortion, and meeting metrics such as EVM, ACLR, and spectral mask compliance. Modern transceivers employ both analog and digital calibration loops to correct imperfections introduced by component tolerances, temperature drift, and aging effects.

- *IQ imbalance correction:* Compensates for amplitude and phase mismatches between the I and Q paths, performed in both analog and digital domains
- *DC offset calibration:* Removes unwanted DC components, particularly critical in direct-conversion receivers where self-mixing can create large offsets.
- *LO leakage suppression:* Minimizes carrier leakage from the LO into the RF output or input, improving spectral purity.
- *Phase and gain calibration across channels:* Ensures coherent signal transmission and reception by aligning the phase and gain across multiple transceiver paths.

In modern SDR and RFIC-based transceivers, these calibration routines are often executed automatically at startup or periodically during operation, allowing the system to maintain optimal performance across varying temperature, frequency, and operating conditions.

Frequency Planning

Frequency planning involves the strategic selection of IFs, LO frequencies, sampling clock rates, and filter cutoff points to ensure that spurious signals, harmonics, and intermodulation products fall outside the critical passbands, far enough to be effectively filtered. This process becomes particularly important in direct-conversion and wideband architectures, where the close spectral spacing of unwanted products can complicate filtering and degrade overall performance.

In practice, even seemingly simple frequency plans can become highly complex in modern multiband communication systems. The designer must carefully balance numerous trade-offs, such as image rejection, spur placement, spectral efficiency, and filter design complexity, to arrive at an optimized solution that meets both performance and implementation requirements.

Figure 5.26(a) illustrates a frequency-planning scenario in which the desired signal band is shown in blue, while the harmonic distortion bands, such as the second (HD2) and third (HD3) harmonics, and clock mixing (CLK Mix) spurs are shown in purple. These unwanted components arise from nonlinearities in active devices and can overlap with the desired signal band, causing interference, distortion, and overall degradation in system performance.

When a signal is mixed down to IF or baseband, the mixer generates sum and difference frequency components. The foundation of frequency planning lies in the well-known equation describing mixer action:

$$f_{\text{spur}} = m \times f_{\text{RF}} \pm n \times f_{\text{LO}}$$

where f_{spur} is the spurious output frequency, f_{RF} is the RF input frequency, f_{LO} is the LO frequency, and m and n are integers representing the mixing order. A spur table is typically constructed (see Section 6.8.2) to identify specific nonlinear mixing products and their frequency locations, with spur levels expressed in dBc relative to the desired signal.

The integers m and n represent the harmonic orders of the RF and LO signals, respectively. Certain combinations of m and n can produce spurious responses that "fold back" into the desired band, making them indistinguishable from legitimate signals. Effective frequency planning seeks to identify and avoid these combinations so that spurious products fall outside the passband and can be filtered.

A related concept arises in digital systems: aliasing, the discrete-time equivalent of fold-back, occurs when frequency components above half the sampling rate (the Nyquist frequency) reflect back into the baseband, distorting the sampled signal spectrum.

The goals of frequency planning differ between transmitters and receivers. In receivers, the primary objective is to preserve sensitivity by minimizing the effects of interference from both in-band and out-of-band signals. Out-of-band interferers,

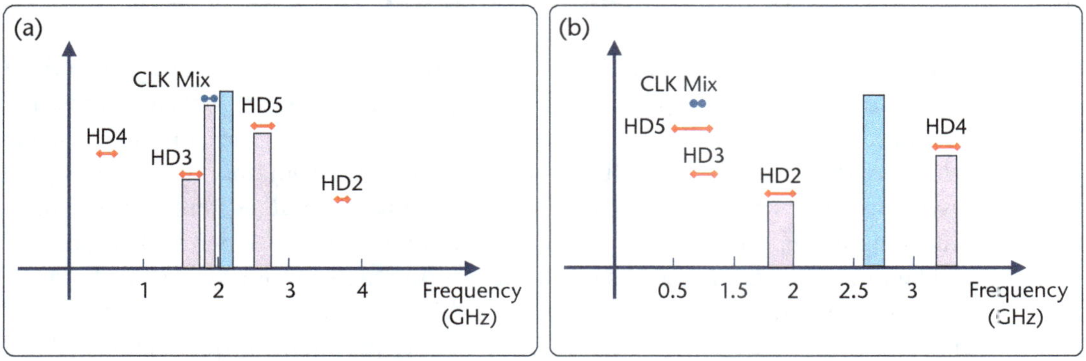

Figure 5.26 Spectrum chart with: (a) an 8-GHz clock and (b) a 5.7-GHz clock.

such as signals from adjacent users or transmitter leakage, can generally be attenuated through proper band-limiting filters at the RF input. In contrast, in-band interferers cannot be filtered out, making precise frequency planning critical. An effective frequency plan ensures that harmonic content from in-band interferers does not fold back into the desired band during frequency conversion. In digital receivers, for example, this may involve carefully selecting the ADC clock rate to ensure that spurious or aliased components do not fold back into the desired signal band.

To illustrate this concept, consider a transmitter operating in the 2.14-GHz band with a 60-MHz-wide signal. Figure 5.26(a) shows the first Nyquist zone spectrum for a high sampling clock rate of 8 GHz. In this case, the desired band (highlighted in blue) is well defined, but third-order and fifth-order harmonics, as well as clock-mixing spurs, appear close to the band of interest. These unwanted products would be difficult to filter.

In Figure 5.26(b), the same band is shown with a reduced clock rate of 5.7 GHz. With this lower sampling rate, neither high-order harmonics nor clock-mixing spurs fall near the desired band. In this example, the lower sampling rate provides a more favorable frequency plan because the high-order spurs are easier to filter, leading to improved spectral purity and overall system performance.

With traditional architectures, modifying the frequency plan required changes to synthesizers and adjustments to IF or baseband filter stages. In contrast, RF sampling architectures enable straightforward frequency-planning modifications simply by adjusting the sampling rate. Because operation in different frequency bands only requires a clock-frequency change, RF sampling architectures offer exceptional flexibility and are easily adapted to various applications and standards.

Modern design tools simplify the frequency planning process by allowing designers to model both single-tone and modulated carrier inputs, visualizing how signals fold across multiple Nyquist zones. Such visualization helps to identify potential spectral collisions early in the design phase, well before hardware implementation.

Some tools provide tabulated results, which, while precise, can be tedious to interpret. Others generate graphical outputs that are visually appealing for high-level presentations but may obscure fine details. The best approach is to use both types of tools in combination, tabular outputs for accuracy and graphical displays for intuition, to achieve a comprehensive understanding of the frequency plan.

5.10 Data Conversion: Mixed Signal Subsystem

So far, we have examined the RF front end and various receiver and transmitter architectures. The next step is understanding how analog RF signals are converted into digital form so they can be processed by FPGAs, DSP engines, or CPUs. Sitting at the boundary between two fundamentally different domains, the mixed signal subsystem is one of the most critical contributors to overall system performance, influencing noise, linearity, dynamic range, bandwidth, and synchronization.

The mixed-signal domain is often the limiting factor in RF systems. The ADC noise floor constrains receiver sensitivity; DAC spurs degrade transmitter spectral purity; and sampling-clock jitter directly affects modulation quality and EVM. Properly designing this subsystem determines the system's ability to detect weak signals, maintain linearity, meet spectral masks, and support high-order modulation formats.

As shown in Figure 5.27(a, b), a typical mixed-signal chain includes data converters and the required analog conditioning:

- *ADC:* Digitizes incoming analog RF, IF, or baseband signals.
- *DAC:* Reconstructs analog waveforms from digital samples.
- *IF/baseband analog filters:* Provide anti-alias filtering before the ADC and reconstruction filtering after the DAC.
- *Baluns and differential drivers:* Interface single-ended RF signals with the differential inputs and outputs of data converters.
- *Sampling-clock PLL/jitter cleaner:* Generates and conditions the sampling clock; because converter performance is extremely sensitive to jitter, clock generation and distribution must be handled with great care.
- *Variable-gain amplifiers (VGAs) or line drivers:* Although not shown in Figure 5.26, VGAs or line drivers typically appear just before the ADC to optimize input swing, or just after the DAC to drive filters and loads. They may also appear in the IF/baseband chain as part of AGC and dynamic-range management.

In Volume 2, we will examine the implementation details, performance considerations, and practical design techniques needed to build a high-performance mixed-signal subsystem.

5.11 Digital Baseband Stage: Processing Architectures for RF Systems

The next piece of the puzzle sits behind the ADCs and DACs, where the core digital signal processing occurs. This stage is responsible for modulation, demodulation,

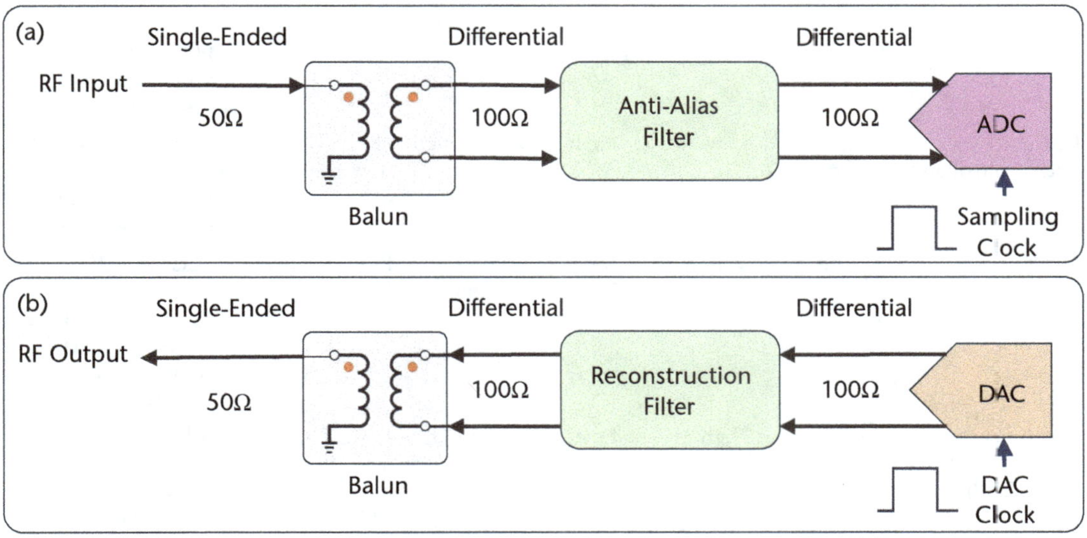

Figure 5.27 (a) Mixed-signal interface for digitizing an RF/IF signal using an ADC. (b) Mixed signal interface for reconstructing an analog waveform using a DAC.

decoding, equalization, digital upconversion and downconversion, SIC, and many other essential techniques used in modern RF systems. Different digital backend platforms that interface with data converters are summarized in Table 5.7 and fall into three major families:

1. Fixed-function processing chips (DSPs, ASICs, embedded RF chipsets):
   - Designed for low latency, high power efficiency, and low cost;
   - Provide limited flexibility, mostly fixed DSP functions with only configurable parameters;
   - Common in embedded radios, cellular modems, IoT modules, and high-volume consumer devices.

Table 5.7 Digital Backend Architectures in RF Systems

Digital Backend	*Flexibility*	*Throughput*	*Latency*	*Power Efficiency*	*Typical ADC/ DAC Interface*	*Typical RF Use Cases*
DSP chips	Medium	Medium	Low	High	LVDS/CMOS parallel (moderate-speed converters)	Channel filtering, equalization, channel estimation, low-rate modem functions
Embedded RF chipsets	Low-medium	Medium	Low	Excellent	Internal proprietary interface (converters on-chip)	Wi-Fi, Bluetooth, GNSS, LTE/5G IoT modules
ASICs (application-specific integrated circuit)	None (fixed)	Very high	Very low	Excellent	Integrated converters/ custom internal bus	Cellular modems, phased-array beamformers, EW systems
FPGAs	Very high	Very high	Very low	Medium-high	JESD204B/C, sometimes LVDS	DUC/DDC, MIMO, beamforming, digital predistortion, radar/SATCOM processing
CPUs/ microcontrollers	High (software)	Low	High	Very high	SPI/LVDS (low-speed converters)	Control loops, calibration, protocol handling, supervisory DSP
GPUs	High	Very high	Medium-high	Low-medium	PCIe (fed from FPGA)	Massive-MIMO baseband, ML-based receivers, large-scale radar/ communications processing
Heterogeneous platforms (CPU + FPGA/DSP)	High	High	Low	Medium	JESD204B/C into FPGA, internal bus to CPU/DSP	Advanced SDRs, 5G baseband units, digital beamforming arrays, radar

2. FPGAs:
   - Offer the highest signal-processing capability and deterministic low latency;
   - Ideal for DUC/DDC, MIMO, beamforming, DPD, radar, and SATCOM applications;
   - Provide extreme parallelism and reconfigurability, but are power-hungry, expensive, and require specialized development.
3. General-purpose processors (CPUs) and accelerators (GPUs):
   - CPUs excel at control-plane processing, scheduling, calibration, and supervisory DSP.
   - GPUs deliver massive parallel throughput for algorithms such as massive-MIMO, machine learning-based receivers, and radar imaging.
   - Not suitable for direct ADC/DAC interfacing, data must first be streamed through an FPGA or a network interface card (NIC) using high-speed transport such as Ethernet, optical links, SFP+, or QSFP, and the resulting latency is nondeterministic.

5.11.1 Data-Plane Interfaces for Signal Processing

When designing an RF product, it is essential to understand the interfaces that connect key subsystems, including data converters, FPGAs, CPUs/GPUs, and networking infrastructure.

Dedicated Converter Interfaces (ADC/DAC ↔ FPGA/SoC)

1. *JESD204:* Modern high-speed data converters used in RF systems generate and consume data at rates reaching tens of gigabits per second. Transferring this data between the ADC and the FPGA, or from the FPGA to the DAC, therefore requires specialized physical interfaces and protocols capable of sustaining extremely high throughput with tight timing constraints.

   A straightforward approach is to use wide parallel data buses. However, parallel interfaces require a large number of signal lines, significantly increasing pin count and PCB routing complexity. More critically, they suffer from signal integrity challenges and skew, defined as the difference in arrival time between signals that are intended to be sampled simultaneously. As data rates increase, even small mismatches in trace length or loading can reduce timing margins and limit achievable performance.

   To overcome these limitations, modern converter interfaces rely on high-speed serial links rather than wide parallel buses. Using serializer/deserializer (SERDES) technology, parallel data is transmitted over one or a small number of high-speed differential lanes. This approach dramatically reduces the number of required traces, minimizes skew and pin count, improves signal integrity, and enables much higher data rates.

   Figure 5.28(a) illustrates how SERDES is used to transfer data between the FPGA and a high-speed data converter using the JESD204 protocol. JESD204B supports data rates of up to approximately 12.5 Gbps per lane, while JESD204C extends this capability to about 32 Gbps per lane, enabling significantly higher aggregate throughput and improved scalability.

The device clock provides the continuous timing required for sampling, serialization, and data transport within the JESD204 link. SYSREF is a timing signal that tells all devices when to align their internal timing, ensuring that data latency through the system is fixed and repeatable (deterministic). Deterministic latency is critical in applications such as beamforming and MIMO, where precise phase alignment between channels is required, as well as in quantum computing systems, where repeatable timing relationships directly affect qubit control and readout fidelity.

The physical connection between the data converter and the FPGA is implemented using high-speed differential pairs routed on the PCB, or through high-speed connectors such as an FMC interface in board-level implementations, as illustrated in Figure 5.28(b).

2. *Low-voltage differential signaling (LVDS):* In legacy systems or lower-data-rate applications, such as earlier high-speed data-converter interfaces, and FPGA parallel links, LVDS is commonly used. Typical LVDS interfaces support lane data rates on the order of ~1 Gbps per lane. However, LVDS does not provide deterministic latency and lacks built-in mechanisms for multidevice synchronization, which limits its suitability for modern multichannel and beamforming systems.
3. *CMOS:* This interface (e.g., 1.8V, 3.3V) is single-ended and primarily used for low-speed control and configuration rather than high-speed data transfer. They are commonly employed to program data converters and related devices, such as setting clock rates, reference levels, operating modes, and calibration parameters, using low-speed serial protocols such as SPI and I^2C.

On-Chip/Board-Level Digital Fabric Interfaces (FPGA ↔ CPU/GPU/Memory)

1. *Peripheral Component Interconnect Express (PCIe):* PCIe is a board-level interface used to transfer large volumes of digitized or processed data between

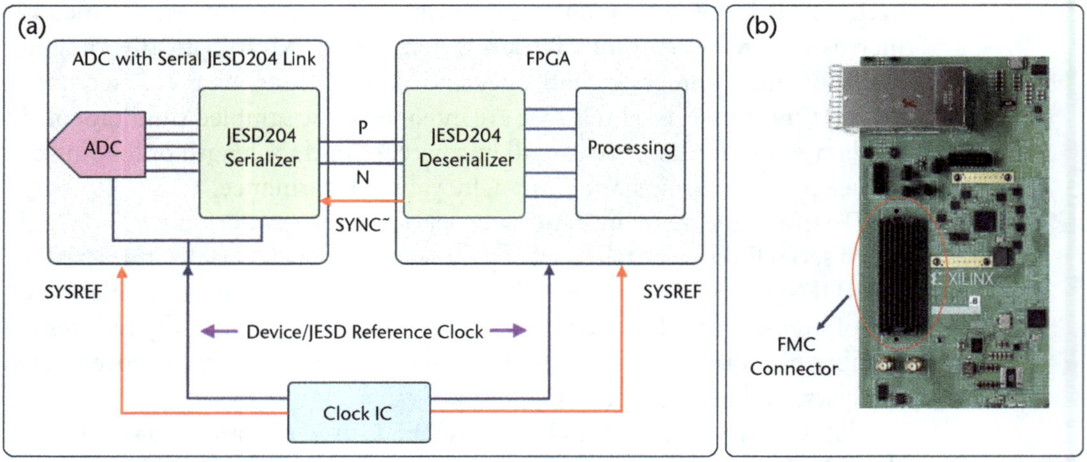

Figure 5.28 (a) A JESD204-based SERDES interface used to transfer high-speed data between an ADC and an FPGA, including differential data lanes (P stands for positive and N for negative), synchronization, and clocking required for deterministic operation. (b) FMC connector on a Xilinx ZCU216 board.

an FPGA and host CPUs, GPUs, or accelerators, providing high bandwidth and low latency for data-intensive RF applications.

2. *Advanced eXtensible Interface (AXI):* AXI is an on-chip interconnect used within FPGAs and system-on-chips (SoCs) to move data between IP blocks, DSP pipelines, processors, and memory. AXI-Stream supports high-throughput IQ data streaming, while AXI and AXI-Lite are used for memory-mapped data transfer and control.

High-Speed Serial Transport Interfaces (FPGA ↔ Module/System)

SFP, SFP+, and QSFP modules support data rates of approximately 1 Gbps, 10 Gbps, and 40 Gbps or higher, respectively, enabling high-bandwidth, packetized data transport over optical fiber or copper media. These modules provide the physical-layer electrical-optical (or electrical-electrical) conversion, while interfacing to the host system through high-speed differential serial lanes. In most systems, the high-speed SERDES and protocol logic reside in the host FPGA, ASIC, or network PHY, with the SFP/SFP+ module acting as a pluggable transceiver that handles modulation, detection, and media adaptation.

Figure 5.29(a) illustrates a typical SFP optical connection using an LC duplex fiber connector inserted into an SFP cage (jack) on the host PCB. Figure 5.29(b) shows an SFP connector on the Xilinx ZCU216 evaluation board.

These interfaces are packet-based, exhibit nondeterministic latency, and are not designed for sample-accurate or synchronous data streaming, as they introduce protocol layers that are not understood by ADC or DAC devices. As a result, SFP/SFP+ links are used to transport already-digitized data in applications such as SDRs streaming I/Q data over 10 GbE, wideband real-time spectrum analyzers, distributed radar systems using fiber links, and data-center accelerators. Although many SDR platforms rely on SFP+ for high-speed packetized data transfer and remote monitoring or data acquisition, the converter-to-FPGA interface within the system still uses JESD204, not SFP.

5.11.2 Integrated RF Platforms

In many modern RF systems, the processing platform is embedded within highly integrated architectures that combine RF front ends, data converters, digital baseband

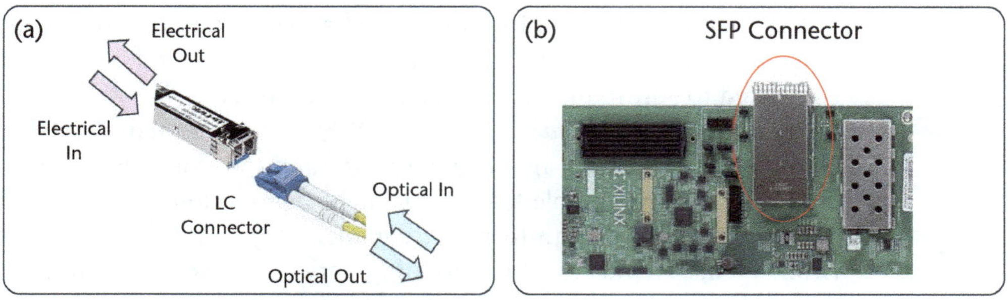

Figure 5.29 (a) Conceptual view of an SFP module showing the electrical SERDES interface to the host system and the optical transmit/receive paths through an LC duplex fiber connector. (b) Physical SFP cage and connector as implemented on the Xilinx ZCU216 RFSoC evaluation board.

processing, and embedded processors such as ARM Cortex cores. Two widely used examples of this level of integration are RF system-on-chip devices (RFSoCs) and highly integrated RF transceivers, such as the ADRV900x family from Analog Devices.

These integrated RF platforms significantly reduce board area, accelerate time to market, and lower bill of materials, while minimizing RF layout complexity and tuning effort. Their digital programmability enables flexible multiband and multimode operation, and standardized interfaces such as SPI, LVDS, and JESD204 allow seamless integration with FPGAs, DSPs, and processors. As a result, much of the traditional analog signal chain becomes digitally configurable and reconfigurable, enabling scalable, multistandard RF systems.

For most RF engineers, this shifts the primary design effort toward developing external front-end modules, such as PAs, LNAs, switches, and filters, that complement these integrated devices. Only in highly customized, high-performance applications, such as radar, EW, or satellite communications, do designers typically move beyond integrated ICs to more specialized or discrete architectures. Consequently, modern RF engineers must not only understand RF hardware, but also be proficient in software, firmware, and the programming and control of integrated transceiver platforms.

RFSoC Platforms

RFSoC platforms represent the highest level of integration among modern RF platforms [7]. As shown in Figure 5.30, an RFSoC combines multiple traditionally discrete subsystems into a single device, including:

- *Data converters:* High-speed ADCs and DACs capable of multigigabit-per-second sampling rates, enabling direct interfacing between the RF domain and the digital signal-processing domain.
- *Fixed-function DSP (digital front end):* Fixed-function DSP refers to hardware-accelerated signal-processing blocks with predefined functionality and programmable parameters, in contrast to custom DSP, where the signal-processing algorithms are fully user-defined and implemented in programmable logic. In Zynq RFSoC devices, the digital front end integrates critical RF signal-conditioning functions as hardened IP blocks. By moving high-throughput, latency-critical, and power-sensitive DSP functions, such as DDC/DUC, interpolation and decimation, filtering, and channelization, into fixed silicon, the platform achieves significantly higher performance and energy efficiency compared to soft implementations in FPGA fabric.
- *Programmable FPGA fabric (custom DSP):* Overall system flexibility is preserved by implementing application-specific and waveform-dependent processing in the programmable logic. This includes functions such as modulation and demodulation, beamforming, equalization, and proprietary DSP algorithms. This architectural partitioning between fixed-function digital front end and custom DSP makes RFSoCs particularly well-suited for wideband, multichannel systems, such as 5G New Radio, radars, SATCOM, and quantum computing.

- *Processing system:* The processing system consists of ARM Cortex-A53 application processors and ARM Cortex-R5F real-time processors, along with memory controllers and I/O peripherals such as SPI, I^2C, UART, and Ethernet. These processors handle system configuration, control, calibration orchestration, networking, and higher-level software functions, but typically do not operate in the real-time signal path.

Integrated RF Transceivers

Integrated RF transceivers come in different levels of integration. We begin with the ADRV900x family, which represents one of the highest levels of integration, and then contrast it with devices such as the AD9361, which offer a lower degree of on-chip digital processing.

Analog Devices ADRV900x Family

The ADRV900x family, shown in Figure 5.31, integrates complete receive and transmit signal chains within a single device. As an example, the ADRV9009 supports an instantaneous bandwidth of up to 200 MHz on the receive path and up to 450 MHz on the transmit path, covering the sub-6-GHz frequency range [8]. These bandwidths make the device well-suited for wideband SDR platforms, cellular and private-network infrastructure, EW systems, phased-array and beamforming architectures, and other high-performance coherent RF applications.

The ADRV900x integrates analog frequency conversion, high-speed ADCs and DACs, and fixed-function digital front-end processing blocks, including digital filtering, gain control, IQ and DC offset correction, and LO leakage correction. While the device performs extensive RF front-end and pre-baseband digital processing, it

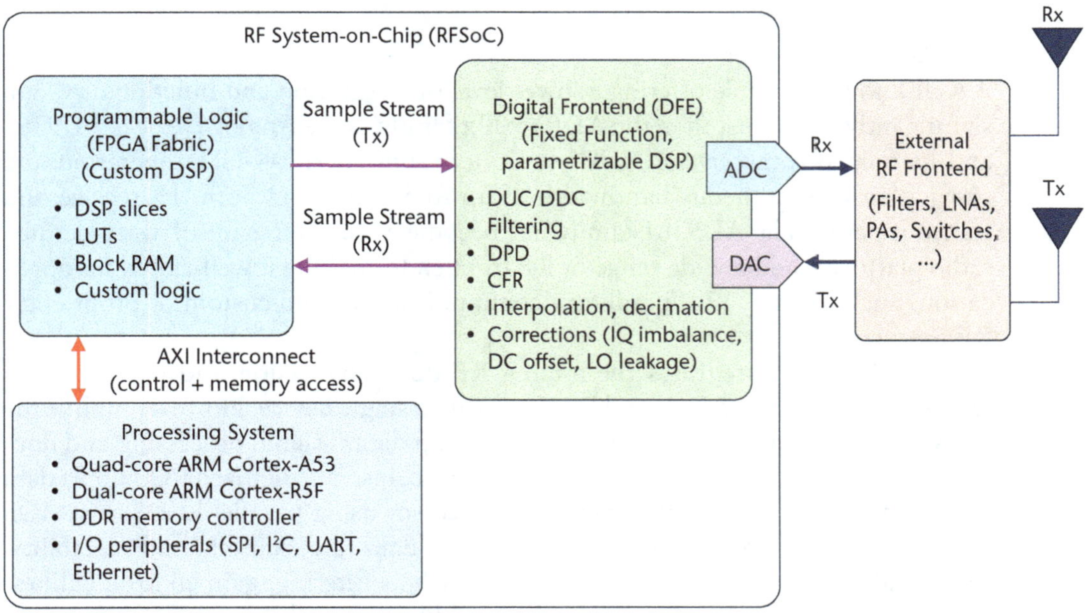

Figure 5.30 The digital front end implements fixed-function, hardware-accelerated DSP tightly coupled to the data converters, while application-specific signal processing is implemented as custom DSP in programmable logic.

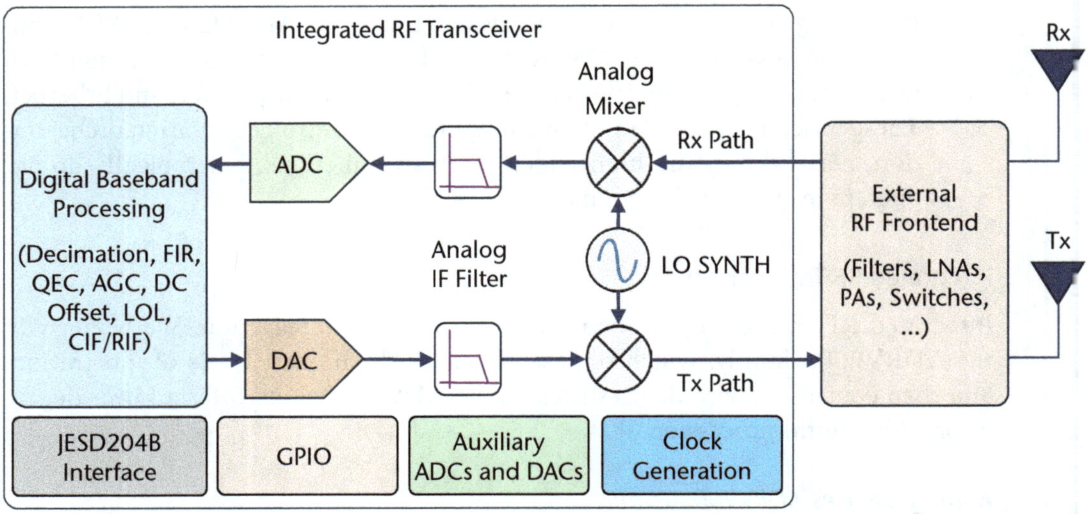

Figure 5.31 Architecture of an integrated RF transceiver showing analog frequency conversion, IF filtering, data converters, digital baseband processing, and the interface to an external FPGA and RF front end.

does not implement higher-level baseband DSP functions such as symbol processing, equalization, framing, or modulation and demodulation. Instead, digitized I/Q sample data is transferred to an external FPGA or SoC via a high-speed JESD204B interface. Deterministic latency and phase alignment are achieved using reference clocks and SYSREF systems. Device configuration and real-time control are handled through SPI and GPIO interfaces. This high level of integration significantly reduces system complexity, improves channel-to-channel coherence, and enables compact, power-efficient RF platforms.

Analog Devices AD9361

A well-known example offering a lower level of integration and functionality, at a significantly lower cost than the ADRV900x family, is the Analog Devices AD9361 and its related transceiver ICs. These devices support sub-6-GHz operation and provide an instantaneous bandwidth of up to 56 MHz on both the receive and transmit paths. The AD9361 family has become a workhorse in software-defined radio platforms and a wide range of RF front ends, making it well-suited for applications such as LTE, Wi-Fi, military communications, and custom or proprietary RF protocols.

The AD9361 integrates the analog RF downconversion and upconversion stages, along with the ADCs and DACs, within a single device. However, unlike the ADRV900x family, it provides minimal on-chip digital signal processing and does not support a JESD204 high-speed serial interface. Instead, digitized I/Q sample data is exchanged with an external FPGA or processor using parallel LVDS or CMOS interfaces, which places stricter constraints on data rate, routing, and scalability. As a result, higher-level digital processing, such as filtering, gain control, calibration, and demodulation, must be implemented largely in the external digital logic.

Tables 5.8 and 5.9 summarize the key receive and transmit performance parameters and specifications of the AD9361 family [9].

Table 5.8 AD9361 Receiver Performance Parameters and Specifications

Parameter	*Typical Value*	*Unit*	*Condition/Notes*
Receivers, general			
Center frequency	70–6,000	MHz	
Gain			
Minimum	0	dB	At 800 MHz
Maximum	74.5	dB	At 800 MHz
Gain step	1	dB	
Received signal strength indicator (RSSI)			
Range	100	dB	
Accuracy	±2	dB	
Receivers, 800 MHz			
Noise figure	2	dB	Maximum RX gain
Third-order input intermodulation intercept point (IIP_3)	−18	dBm	Maximum RX gain
Second-order input intermodulation intercept point (IIP_2)	40	dBm	Maximum RX gain
LO leakage	−122	dBm	At RX front-end input
Quadrature gain error	0.2	%	
Quadrature phase error	0.2	Degrees	
Modulation accuracy (EVM)	−42	dB	19.2-MHz reference clock
Input S_{11}	−10	dB	
RX1 to RX2 isolation	70	dB	

Table 5.9 AD9361 Transmitter Performance Parameters and Specifications

Parameter	*Typical Value*	*Unit*	*Condition/Notes*
Transmitters, general			
Center frequency	46.875–6,000	MHz	
Power control range	90	dB	
Power control resolution	0.25	dB	
Transmitters, 800 MHz			
Output S_{22}	−10	dB	
Maximum output power	8	dBm	1-MHz tone into 50Ω load
Modulation accuracy (EVM)	−40	dB	19.2-MHz reference clock
Third-order output intermodulation intercept point (OIP_3)	23	dBm	
Carrier leakage	−50	dBc	0-dB attenuation
	−32	dBc	40-dB attenuation
Noise floor	−157	dBm/Hz	90-MHz offset
TX1 to TX2 isolation	50	dB	

5.11.3 SDRs Enabled by Integrated RF Platforms

Integrated RF platforms not only simplify hardware design but also form the technological foundation for SDR and cognitive radio systems by tightly coupling RF front ends, data converters, and programmable digital processing.

In SDRs, most signal-processing functions are implemented in software rather than fixed hardware. By digitizing signals as close to the antenna as practical, SDR architectures provide a high degree of flexibility, allowing the operating frequency, bandwidth, waveforms, modulation and demodulation schemes, signal-processing functions, and communication protocols to be modified with little or no hardware change. This reconfigurability makes SDRs central to modern agile communication platforms, used in both commercial and military applications, including multistandard communication systems.

Building on the SDR concept, cognitive radios introduce awareness and autonomy through mechanisms such as spectrum sensing, dynamic frequency selection, and adaptive power control. These systems monitor the RF environment in real time and automatically reconfigure their operating parameters to reduce interference and improve spectral efficiency. Such capabilities are increasingly important in dense wireless environments and are closely aligned with emerging spectrum-sharing and dynamic spectrum access policies.

5.12 Monitoring, Control, and Protection in RF Systems

Effective monitoring, control, and protection mechanisms are essential in RF products. Monitoring provides the observability required to enable both control and protection functions, and it can be used in either an open-loop or closed-loop fashion. Figure 5.32 illustrates representative open-loop and closed-loop control and protection architectures.

1. *Open-loop architecture:* In this architecture, shown in Figure 5.32(a), the element for control or protection, such as an RF switch, digital step attenuator (DSA), or VGA, is driven by an operator or by predefined logic. For example, an RF switch may be programmed to toggle every 1 millisecond according to a fixed timing sequence. In this architecture, no real-time decision-making based on the instantaneous state of the system is performed, and no automatic corrective action is taken in response to measured conditions.

   Monitoring elements such as power detectors, temperature sensors, and voltage or current monitors are optional and, when present, are used primarily for observability, logging, and post-event analysis. Their outputs are not fed back into the control path and therefore do not influence system behavior in real time.

   For protection, open-loop systems typically rely on nontunable, passive components such as RF limiters, ESD diodes, or fixed clamping circuits. Because these elements operate without sensing, feedback, or programmability, the protection function is inherently passive and always activated, independent of the system's operating state.

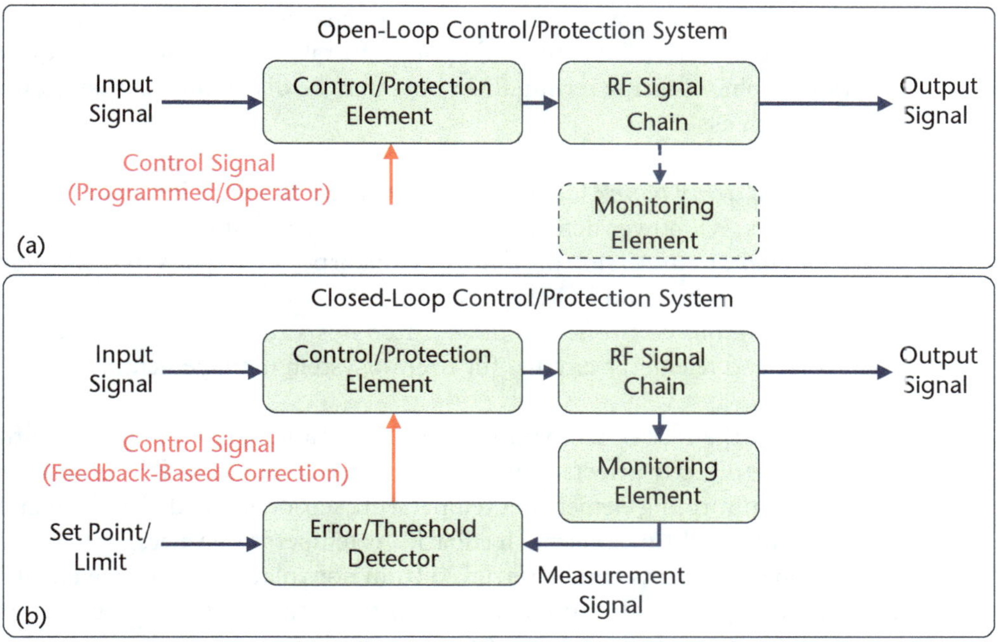

Figure 5.32 Comparison of (a) open-loop and (b) closed-loop control and protection architectures in RF systems.

Open-loop architectures are simple, deterministic, and easy to implement and validate. However, because they lack feedback, they cannot compensate for dynamic variations such as temperature drift, load mismatch, component aging, or fast fault conditions.

2. *Closed-loop architecture:* In this architecture, shown in Figure 5.32(b), monitored signals are actively fed back into a control or protection loop. The measured parameter is continuously compared against a reference or threshold using an error or decision mechanism, and corrective action is applied automatically. For example, if the measured reflected power exceeds a predefined threshold, the RF switch acting as the control element can be opened to protect the PA from damage.

The same closed-loop principle is widely used to optimize RF signal-chain performance through functions such as AGC, DPD, CFR, and phase-locked loops (PLLs), where feedback enables real-time correction of gain, linearity, peak power, or phase and frequency errors.

Closed-loop architectures significantly improve system robustness, performance stability, and long-term reliability. However, they also introduce additional complexity, loop latency, and potential stability concerns, all of which must be carefully analyzed and managed during system design, implementation, and validation.

5.12.1 Monitoring in RF Systems

Monitoring is implemented using various types of sensors, including temperature, RF power, voltage, and current sensors. A practical example can be found in phased-array systems, which are inherently sensor-rich architectures due to their large number of active RF paths and tight performance and reliability requirements. In a typical phased-array terminal, sensors may be distributed across multiple subsystems, such as:

- *Phased array panel:* Contains beamforming ICs with integrated temperature sensors, RF power detectors, and bias current/voltage monitors.
- *Frequency conversion module:* May incorporate sensors to track supply voltages, currents, and LO chain health.
- *Modem and controller module:* Often hosts thermal sensors, voltage monitors, and telemetry circuits for overall system management.

Consider the case of temperature sensing, a fundamental monitoring function. Temperature directly affects the gain and phase response of RF components, particularly beamforming elements. A temperature sensor embedded in the phased array panel provides real-time thermal feedback. As temperature varies, pre-characterized calibration coefficients can be retrieved from nonvolatile memory (e.g., EEPROM) and applied dynamically to correct beam-steering errors. This ensures accurate beam pointing across diverse operating conditions.

Another example is found in beamforming ICs, which integrate current monitors and RF power detectors. These readings serve multiple purposes:

- *Current monitoring:* Enables monitoring of bias currents to optimize noise and linearity, detect early signs of device degradation or thermal stress, and provide insight into the health of each beamforming path.
- *Power detection:* Supports per-channel output power estimation, assisting in calibration and ensuring compliance with regulatory requirements (e.g., effective isotropic radiated power limits).

By strategically monitoring these parameters, phased array systems can maintain calibration, optimize performance, and protect themselves from conditions such as overheating, power mismatch, or device failure.

5.12.2 Control in RF Systems

Control functions in RF systems serve two primary purposes: protecting hardware and optimizing performance. As mentioned earlier, control can be implemented using open-loop or closed-loop architectures.

Performance-oriented control mechanisms focus on maintaining optimal operation rather than preventing damage. These techniques continuously fine-tune key parameters such as amplitude, phase, frequency, and linearity to ensure stable, efficient, and specification-compliant operation under varying temperature, load, and channel conditions. Common examples include AGC, DPD, and CFR.

In contrast, control for protection focuses on preventing damaging operating conditions such as overdrive, overheating, or excessive reflected power. These mechanisms typically respond by adjusting system gain, modifying attenuation using VGAs or DSAs, or toggling RF switches to isolate or shut down vulnerable hardware.

Table 5.10 summarizes key open-loop and closed-loop control functions commonly employed in RF systems.

We discussed CFR and DPD earlier. We now consider AGC, which dynamically adjusts the system gain to maintain a relatively constant output amplitude, even when the input signal strength varies significantly. AGC is commonly used in both the receive and transmit signal paths.

A familiar example of AGC is listening to the radio while driving. As the vehicle moves, the received signal strength changes due to variations in distance from the broadcast station, multipath fading, shadowing, and blockage. AGC compensates for these effects, allowing the listener to perceive a nearly constant audio volume.

From a system perspective, AGC prevents the signal chain from being overdriven, which could lead to compression or distortion, while also avoiding underutilization of the available dynamic range. By continuously adapting the gain, AGC ensures reliable and consistent demodulation across a wide range of signal conditions.

Figure 5.33 illustrates a basic AGC loop. The input signal passes through a VGA, whose gain is controlled by a feedback loop. The output signal is monitored by a power detector that measures the signal amplitude. This measurement is smoothed by a lowpass filter to remove fast fluctuations and is then compared against a predefined target level (set point) using an error amplifier. The resulting error signal drives the VGA control input, adjusting the gain to maintain a stable output level. In some implementations, the VGA is replaced by a fixed-gain amplifier combined with a DSA, which regulates the overall gain in discrete steps while preserving the same closed-loop control principle.

Table 5.10 Open-Loop and Closed-Loop Implementations of Common RF Control Objectives

Control Objective	*Open-Loop Implementation*	*Closed-Loop Implementation*	*Purpose/Typical Use*
Amplitude control	DSA, VGA, bias control	AGC	Set or maintain signal level at IF/baseband or ADC input
Frequency control	LO/PLL programming	Automatic frequency control (AFC)	Set carrier frequency or correct frequency error
Phase control	Tunable phase shifter	Automatic phase control (APC)	Phase alignment for beamforming or coherent reception
Linearity control	Fixed power back-off, bias tuning	DPD	Improve PA linearity
Peak power control	Fixed power back-off	CFR	Reduce PAPR and improve PA efficiency
Path/mode control	RF switches, T/R switches	Fault-triggered path control	Route or disconnect RF paths under normal or fault conditions
Impedance/filtering control	Tunable matching network, tunable filters	Adaptive impedance/filter tuning	Minimize reflections and optimize power transfer
Calibration	Factory or manual calibration	Continuous/adaptive calibration loops	Compensate for drift, temperature, and aging
Link adaptation	Fixed modulation and coding	Adaptive modulation and coding (AMC)	Optimize throughput versus link quality

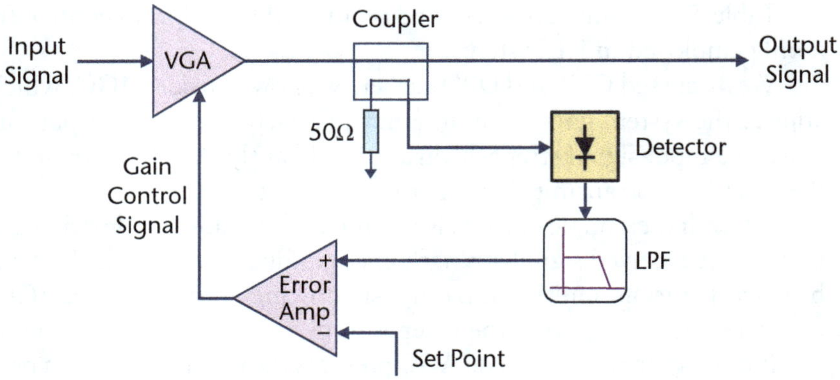

Figure 5.33 Automatic gain control feedback loop.

The AGC loop must be carefully tuned: it should respond quickly enough to track genuine changes in signal power while avoiding reaction to the modulation envelope itself. If the loop bandwidth is set too wide, the AGC may respond to symbol-level amplitude variations, leading to gain pumping, a condition in which the control loop unintentionally suppresses the modulation. This behavior degrades signal fidelity and increases EVM, an effect that becomes particularly pronounced in high-order QAM systems with large amplitude dynamics.

5.12.3 Protection in RF Systems

Protecting RF systems can be implemented using both open-loop and closed-loop approaches.

Open-Loop Protection

It relies on passive or fixed-response mechanisms, such as RF limiters placed at the front end to clip excessive signal amplitudes before they reach sensitive components (see Figure 5.34(a)).

Unlike closed-loop methods, open-loop protection mechanisms are either always active or triggered automatically by an external stimulus such as overvoltage, overcurrent, or excessive RF power. These protections are simple, fast-acting, and highly

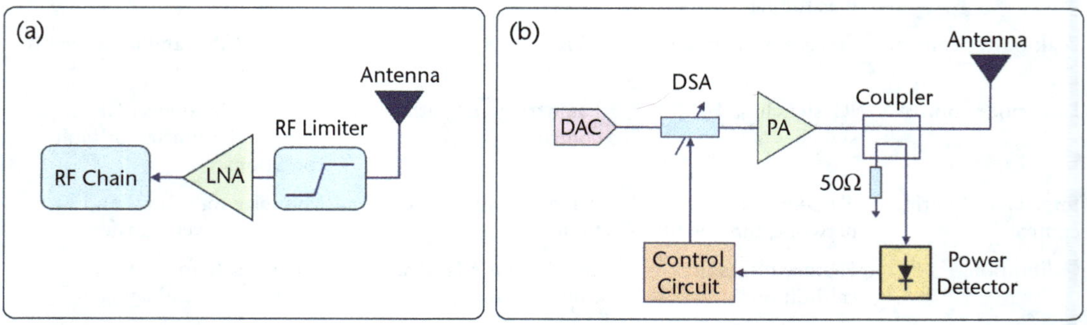

Figure 5.34 (a) Open-loop protection of an LNA using an RF limiter and (b) closed-loop PA protection system.

reliable, making them well-suited for safeguarding sensitive RF hardware. However, they lack adaptability, as they cannot dynamically adjust their behavior in response to changing operating conditions.

Table 5.11 summarizes key open-loop protection methods commonly used in RF systems.

Table 5.12 categorizes common closed-loop protection functions used in RF systems.

Closed-Loop Protection

While open-loop protection mechanisms are often the first line of defense, they are most effective when combined with closed-loop protection strategies. Closed-loop protection plays a critical role in safeguarding RF components and systems by continuously monitoring key operating parameters and applying active control based on feedback. Table 5.12 shows some of the closed-loop control functions for protection.

Let us consider a representative example. PAs are particularly susceptible to overheating, which may result either from insufficient thermal management or from load mismatch that reflects RF power back into the device, elevating the junction

Table 5.11 Open-Loop Protection Techniques

Protection Type	*Mechanism Action*	*Typical Applications*
RF limiters/PIN diodes	Automatically attenuate high-power RF surges to protect sensitive stages	LNAs, receivers, test instruments
Fuses/resettable limiters	Interrupt or limit the supply current during fault conditions	DC bias networks, PA supply rails
ESD protection devices	Passive clamps that shunt static discharge energy without active monitoring	RFIC inputs, LNAs, test equipment connectors
Surge/transient suppression	Absorbs or clamps lightning, electrical fast transient (EFT), or surge events	Outdoor radios, satellite ground stations, radars

Table 5.12 Closed-Loop Control for Protection

Protection Type	*Mechanism/Action*	*Typical Applications*
VSWR protection	Monitors reflected power or mismatch and reduces output power, detunes, or disables transmission to protect the PA	PAs, transmitters, antenna systems, phased arrays
RF overdrive/ overpower protection	Detects excessive RF power and dynamically reduces gain, increases attenuation, or disables the RF path to prevent damage	PAs, LNAs, and receivers exposed to strong interferers (overdrive)
Over-voltage protection (OVP)	Monitors supply or node voltage and limits, clamps, or shuts down the circuit when voltage exceeds safe thresholds	PAs, LNAs, converters, bias networks
Over-current protection (OCP)	Monitors current consumption and limits or shuts down operation under fault or overload conditions	PAs, power supplies, DC-DC converters, RF front ends
Thermal management	Monitors temperature sensors and actively regulates operation through cooling control, power derating, or automatic shutdown	PAs, LNAs, phased arrays, SDRs

temperature. In such cases, reaction time is critical, as permanent damage can occur within just a few microseconds.

For this reason, protection schemes often rely on fast analog circuitry, such as comparators, rather than slower digital or microcontroller-based feedback paths. While these protection mechanisms are discussed in more detail in Volume 2, a simplified example is presented here.

The example is based on the closed-loop protection architecture shown in Figure 5.34(b). A directional coupler samples the PA output power, which is measured by a power detector and fed back to a control circuit. Based on the detected power level, the control circuit adjusts a DSA ahead of the PA to regulate the drive level, thereby preventing overdrive and mitigating excessive reflected power at the antenna interface. It is important to note that the protection loop shown in Figure 5.33(b) is not intended to provide the fastest possible response. Because the attenuation is adjusted using a DSA, the control action typically involves a microcontroller or FPGA, introducing latency associated with detection, decision-making, and digital control. As a result, this approach is slower than purely analog protection circuits based on comparators or bias shutdown mechanisms. Instead, it is well-suited for mitigating sustained overdrive or mismatch conditions and is often used in combination with faster analog protection layers that provide immediate fault response.

Additional strategies are required when addressing protection, as recovery is as important as fault prevention. These include software-based safeguards and fault-handling procedures that define how the system responds once a fault has occurred. Together, these approaches provide both immediate protection against damage and a structured recovery process, ensuring safe operation and long-term system reliability.

Multilayer Protection Mechanisms

While open-loop and closed-loop mechanisms address individual fault conditions, robust RF systems rely on multiple coordinated protection layers to handle faults, recovery, and safe operation across all operating states. The key pillars of RF protection are shown in Table 5.13.

Let's look at each layer in more depth. Multilayer protection is most effective when the layers are intentionally partitioned by response time, failure mode coverage, and independence (i.e., ensuring a software fault cannot disable the protection path).

Interlocks

Interlocks are essential safety and sequencing mechanisms used in RF systems to ensure that components operate only under predefined safe conditions. They act as hardware or software gates that "lock" or "unlock" power paths, high-voltage supplies, RF signal chains, or hazardous operations based on real-time status monitoring. Table 5.14 shows various types of interlocks.

Emergency Stop (E-Stop)

The emergency stop (E-stop) provides an immediate, human-initiated shutdown path that is independent of firmware, software, and supervisory control. In RF systems,

Table 5.13 Multilayer Protection Mechanisms

Protection Element	*Role in the System*	*Example*
Interlocks	Prevent system operation unless all safety conditions are satisfied	Disable RF transmission unless the antenna is connected and the enclosure is closed
Emergency stop (E-Stop)	Provide immediate, human-initiated shutdown independent of software or control logic	Mechanical stop button that removes PA bias and disables RF power; essential in high-power systems like radars
Watchdog circuits	Monitor software or control logic responsiveness and initiate a reset or shutdown upon failure	Reset the control FPGA if no heartbeat signal is detected
Fail-safe defaults	Force the system into a safe state under power loss or undefined conditions	Default RF path to off and disable PA bias if an interlock is not asserted
Redundancy/ backup	Eliminate single-point failures by duplicating critical protection paths	Redundant over-temperature sensors in satellite payloads

Table 5.14 Types of Interlocks

Type	*Description*	*Example*
Hardware interlock	Physical signals (e.g., contact closures, voltage levels) that must be present before a subsystem can power up or operate; common in high-power RF amplifiers and test fixtures	RF amplifier cannot enable unless the enclosure door switch is closed
Software interlock	Logical conditions enforced in firmware or software to ensure proper sequencing, often with state-machine enforcement	Signal generator firmware prevents RF-on until bias rails are stable
Logical interlock	Combinations of sensor readings or flags (e.g., "temperature <80°C AND door closed") required to enable operation	PA bias only enabled when cooling fan is running and temperature is safe

an E-stop typically removes PA bias power and disables RF output through a hardwired, safety-rated circuit, ensuring a deterministic transition to a safe state under uncertain or escalating conditions. Unlike software-based shutdown mechanisms, which may be delayed by processing latency or impaired by control-plane faults, the E-stop acts directly at the hardware level and remains effective even in the presence of software or communication failures.

E-stops are a mandatory safety element in high-power RF systems such as radars, EW systems, and high-power phased-array terminals. In these systems, abnormal operating conditions can escalate rapidly, and failure to react within a very short time window can result in catastrophic damage, including thermal runaway, arcing, or fire. Given that transmitted power levels may reach kilowatts or even megawatts, immediate hardware-level shutdown is a fundamental safety requirement rather than an optional design feature.

From an implementation perspective, E-stop circuits are typically designed as normally closed loops, such that any cable break, connector failure, or loss of control power automatically forces the system into a safe, nontransmitting state. This fail-safe behavior aligns with common industrial safety standards and ensures that loss of integrity in the shutdown path cannot inadvertently enable RF transmission.

Watchdogs

In the world of embedded control for RF systems, a watchdog timer is a dedicated hardware or software-based timer that is used to detect system malfunctions, particularly software hangs or infinite loops.

The concept is simple: the software is expected to periodically reset (often called "kicking" or "petting") the watchdog timer during normal operation. If this reset does not occur within a specified time window, the watchdog assumes that the system has become unresponsive. In response, it takes corrective action, typically forcing a system reset or placing the system in a known safe state.

In RF systems, real-time responsiveness is often mission-critical. A failed control loop, stuck state machine, or corrupted firmware could result in:

- Continuous RF transmission beyond allowed limits (violating regulatory compliance);
- Thermal runaway of PAs;
- Unresponsive tuning loops in frequency synthesizers;
- Loss of communication in satellite or remote RF links.

These failures not only degrade system performance but can also cause hardware damage or safety violations. A properly configured watchdog provides a last-resort safety mechanism to bring the system back online or prevent further damage.

Example applications are as follows:

- In a base station transmitter, if the microcontroller controlling the PA biasing gets stuck, a watchdog can reset it to prevent damage from uncontrolled power levels.
- In a remote RF sensor node, a watchdog can reboot the system in case of software failure due to radiation or power glitches.
- In FPGA-based phased arrays, watchdog logic can monitor heartbeat signals from each beamforming core and reset specific blocks that go silent.

Here are some best practices for implementing watchdogs:

- The watchdog timeout should be long enough to allow legitimate operations to complete, but short enough to catch meaningful failures quickly.
- Use hardware watchdogs where possible for higher reliability (many MCUs offer built-in watchdog peripherals).
- Integrate watchdog handling with system diagnostics to record fault information before reset.

Fail-Safe Defaults

Fail-safe defaults ensure that an RF system enters a safe operating state whenever conditions are uncertain, such as during power loss, initialization, or unexpected faults. The guiding principle is "if in doubt, stay safe." Instead of leaving circuits in an undefined state, fail-safe logic guarantees that sensitive components are protected and regulatory compliance is maintained.

Typical strategies include:

- *Default to off:* RF outputs, bias supplies, or transmit enables remain off unless explicitly commanded on. *Example: A PA bias circuit that remains disabled until interlocks confirm antenna connection and enclosure closure.*
- *Disable high-risk functions:* Functions that could damage hardware or violate safety rules are locked out in ambiguous states. *Example: A synthesizer that holds its output muted if the reference clock is missing.*
- *Controlled power-up sequences:* Ensure systems only energize in the correct order, avoiding conditions such as bias-before-cooling or RF-before-load. *Example: RF-on is inhibited until all supply rails and cooling fans report healthy states.*
- *Safe state on reset or loss of communication:* If a controller crashes or communication is lost, hardware defaults prevent runaway behavior. *Example: Defaulting an RF switch to "terminate" instead of "through" when control signals are absent.*

Fail-safe defaults are often implemented in hardware logic (pull-downs, relays, default switch states) as well as in firmware/FPGA state machines. Unlike redundancy, which covers failures, fail-safe design anticipates ambiguity and uncertainty and prevents the system from ever turning unsafe by default.

Redundancy/Backup

Redundancy is a cornerstone of protection in mission-critical RF systems where single-point failures are unacceptable. The principle is simple: provide a backup path or duplicate component so that if one protection mechanism fails, another takes over. This approach increases system reliability at the cost of additional complexity, weight, and expense.

There are several levels of redundancy used in RF protection:

- *Component-level redundancy:* Duplicate sensors, fuses, or limiters ensure that failure of a single device does not disable protection. *Example: Two independent over-temperature sensors on a satellite payload, each capable of triggering a shutdown.*
- *Path-level redundancy:* Entire RF chains are duplicated so that one path can be taken offline if it malfunctions. *Example: Redundant transmitter chains in satellite communication payloads, where a "cold spare" is switched in when the primary fails.*

- *Control/decision redundancy:* Multiple controllers or voting logic ensure that no single stuck processor or faulty sensor drives the wrong action. *Example: Triple-modular redundancy (TMR) in FPGA control logic, where majority voting mitigates single-bit errors due to radiation.*

Redundancy is especially critical in spaceborne, defense, and high-reliability terrestrial systems, where repair or reset may be impossible once deployed. However, designers must carefully weigh the benefits against increased SWaP-C.

5.13 What's Next?

We have examined several RF system architectures in this chapter. Implementing these architectures requires both the appropriate components and a clear understanding of how to select them and which parameters must be considered. In Chapter 6, we focus on the RF components used for the generation, transmission, processing, and detection of RF signals.

Problems

5.1 You are designing a sub-6-GHz wideband SDR intended for field deployment and remote updates. Assume channel bandwidths up to 20 MHz and frequency coverage of 0.7–6.0 GHz (TDD operation).

a. Compare direct-conversion, low IF, and direct RF-sampling receiver architectures for this SDR.
b. Select one architecture and justify your choice in terms of performance, calibration effort, SWaP-C, and time to market.
c. Draw a block diagram including RF front end, clocks, converters/transceiver, FPGA/DSP, processing system, and power.
d. Specify key interfaces and what they carry (e.g., JESD204 to FPGA, SPI control, I^2C sensors, UART/Ethernet management).
e. Propose a minimal set for monitoring, control, and protection (e.g., OVP/OCP/thermal, VSWR/overpower, telemetry, fault handling, safe shutdown).
f. Identify the top two risks and how you would verify them during EVT and DVT.
g. Describe how you would update the SDR in the field (software/firmware/FPGA), including security, rollback/fail-safe, and how you would verify the update remotely.

5.2 A compact product must support simultaneous transmit and receive using a single antenna. The TX and RX operate at different frequencies (FDD). The antenna return loss is ≈10 dB over the operating band. Assume a transmitter output power of +30 dBm and a receiver sensitivity of –100 dBm and ignore the AGC for the initial isolation estimate.

a. Compare the following approaches for sharing one antenna: TDD (RF switch-based), FDD with circulator/duplexer, and full-duplex with active SIC.

b. Estimate the required TX-RX isolation such that transmitter leakage at the receiver input is at or below the receiver sensitivity. State the value in decibels.
c. For each architecture (TDD, FDD, full-duplex), propose a practical isolation strategy and give numerical examples using typical values, such as RF switch isolation (e.g., 20–50 dB) and circulator isolation (e.g., 20–50 dB).
d. With a circulator, estimate how much TX power leaks to the RX port during transmit. Assume that the circulator provides 40-dB isolation from the TX port to the RX port.
e. Propose a protection approach to ensure the leaked power does not damage the receiver front end (e.g., limiter, attenuator, blanking, front-end switch, LNA bypass) and briefly justify your choice.

5.3 Given a receiver chain consisting of an LNA → mixer → IF amplifier → ADC → digital filtering:
a. Identify the dominant contributors to noise floor and linearity limits (noise figure cascade, compression, IP_3, blocker effects).
b. Explain how AGC placement (RF/IF/analog versus digital) affects ADC headroom, quantization noise, and intermodulation under blockers.
c. Describe how you would verify dynamic range during validation (e.g., two-tone IMD, blocker desense, SFDR, noise floor/MDS, gain-step tests, EVM versus input power, swept-level and swept-frequency tests).

5.4 A wireless link must operate reliably in both high SNR LOS and low SNR multipath environments.
a. Explain how a MIMO system can dynamically switch between spatial multiplexing, diversity, and beamforming.
b. Identify which real-time metrics should drive the mode selection (e.g., SNR/SINR, channel rank/condition number, EVM, PER/block error rate (BLER), delay spread, Doppler, channel quality indicator (QCI)).
c. Propose a simple decision policy (rules or thresholds) and describe how often it should update.

5.5 You are free to choose where the ADC/DAC boundary lies in a new RF product.
a. Compare analog IF, low IF, and direct RF-sampling approaches in terms of clocking/jitter sensitivity, analog filtering requirements, and digital processing load (DDC/DUC, channelization, sample rates).
b. For a battery-powered product, explain how converter choice impacts power budget (converter power + FPGA/DSP load) and thermal design (hot spots, heat spreading, derating).

5.6 An OFDM transmitter must meet strict ACLR and EVM requirements while maximizing PA efficiency.
a. Explain the trade-offs between power backoff, CFR, and DPD.
b. Describe how CFR changes the PAPR distribution, impacts spectral regrowth, and shifts the PA operating point.
c. Identify where CFR and DPD are best implemented (FPGA, DSP, or RFSoC) and justify your choice in terms of latency, throughput, update rate/adaptation, and integration effort.

5.7 Compare the following transmitter architectures for a frequency-agile RF test instrument (e.g., a signal generator or vector signal generator (VSG)). Assume a 0.1–6-GHz tuning range, 100-MHz instantaneous bandwidth, and a typical spur target of ≤–60 dBc.

a. Direct IQ upconversion;
b. Superheterodyne upconversion;
c. Direct-DAC RF synthesis.

For each, discuss spectral purity (phase noise, images), tuning agility (hop speed, retune time), spur management (LO feedthrough, harmonics, mixing spurs, DAC images), and calibration complexity (IQ imbalance, DC offset, LO leakage). Recommend one architecture and justify your choice for a test instrument.

5.8 A development team must choose between an RFSoC-based platform and a discrete transceiver + FPGA solution.

a. Compare the two approaches in terms of integration risk, flexibility, debugging/test access, and scalability (channels, bandwidth, power, cost).
b. Identify which product stages (PoC, EVT, DVT, PVT) benefit most from each approach and explain why (speed of iteration versus optimization and manufacturability).

5.9 You are extending a proven sub-6-GHz architecture to 28 GHz.

a. Identify the top five architectural challenges introduced at millimeter-wave frequencies (e.g., path loss and link margin, packaging and interconnect loss, antenna integration, phase-noise and jitter sensitivity, thermal density, calibration and array coherence, and testability via over-the-air (OTA) verification).
b. Explain how antenna integration approaches, such as Antenna-in-Package (AiP) and Antenna-on-Package (AoP), along with packaging and beamforming, change system-level design decisions (partitioning, interfaces, calibration strategy, and test methodology).
c. For each challenge, propose one mitigation strategy and state its associated cost or trade-off (e.g., added noise figure, power consumption, size, calibration time, or bill of materials).

5.10 An RF system is deployed in a remote, nonserviceable environment.

a. Propose a monitoring, protection, and redundancy strategy covering PA protection (VSWR/overpower/over-temp), temperature drift (calibration/compensation), and component failure (degraded modes, redundancy, watchdogs).
b. Identify what you would monitor (telemetry signals) and how each signal is reported (e.g., SPI/I^2C sensors, internal ADC monitors, GPIO fault pins, Ethernet/serial). For each interface, comment on reliability (noise susceptibility, fault detection, and behavior during power or clock faults).
c. Explain how these supporting architectures influence cost, power, SWaP, and complexity, even though they do not carry user data.

References

[1] Carlson, B. A., *Communication Systems: An Introduction to Signals and Noise in Electrical Communication*, Fifth Edition, New York: McGraw-Hill, 2009.

[2] Salari, A., *Microwave Techniques in Superconducting Quantum Computers*, Norwood, MA: Artech House, 2023.

[3] Razavi, B., *Microelectronics*, Second Edition, New York: Wiley, 2014.

[4] Cripps, S. C., *RF Power Amplifiers for Wireless Communications*, Second Edition, Norwood, MA: Artech House, 2006.

[5] Raab, F. H., *RF and Microwave Power Amplifiers*, Norwood, MA: Artech House, 2025.

[6] Keysight Technologies, "Crest Factor Reduction (CFR) Concept," *Keysight Help Files*. https://helpfiles.keysight.com/csg/n7614/Content/Main/Crest_Factor_Reduction_Concept.htm.

[7] AMD, *Zynq™ UltraScale+™ RFSoC ZCU216 Evaluation Board User Guide*, UG1390. https://docs.amd.com/r/en-US/ug1390-zcu216-eval-bd.

[8] Analog Devices, *ADRV9009 Integrated Dual RF Transceiver Datasheet*. https://www.analog.com/en/products/adrv9009.html.

[9] Analog Devices, *AD9361: RF Agile Transceiver Datasheet*, Analog Devices, Norwood, MA, https://www.analog.com/media/en/technical-documentation/data-sheets/AD9361.pdf.

References

CHAPTER 6

RF Components

Details make perfection, and perfection is not a detail.

—Leonardo da Vinci

This chapter reviews RF components with an emphasis on the four fundamental functions of RF systems: signal generation, transmission, processing, and detection. For each category, the most commonly used components are introduced, along with their practical applications and key performance parameters. Particular attention is given to how these components integrate into the system architectures discussed in Chapter 5, bridging theoretical concepts with real-world design considerations.

6.1 Introduction

After establishing the signal chain layer, core RF concepts, and RF architectures, the next step toward a complete system blueprint is the selection of appropriate RF components. Up to this point, RF systems have been treated largely as black boxes, with emphasis placed on system-level behavior and performance rather than on the internal building blocks themselves. In this chapter, we open that black box and examine the RF components that enable real-world implementation.

To provide a clear structure, RF components are organized into four functional categories: signal generation, transmission, processing, and detection. For each category, the most widely used components are introduced, along with practical examples illustrating how they are deployed in real systems. Typical performance parameters and design considerations are highlighted to support informed component selection.

Although antennas are often treated as a standalone topic, a brief introduction and practical considerations are included at the end of this chapter. Antennas are tightly coupled to RF system performance, influencing critical aspects such as link budget, bandwidth, efficiency, and SWaP-C. A basic understanding is therefore essential at this stage. Antenna design and advanced topics are covered in greater depth in Volume 2.

6.2 S-Parameter Terminology

Before discussing individual RF components, it is useful to briefly review scattering parameters (S-parameters), which are the primary metrics used to characterize RF and microwave networks. The key points to recall are:

1. *Linear versus logarithmic interpretation:* In linear form, S-parameters represent voltage ratios. When expressed in logarithmic units (decibels), they correspond to power ratios, which is why S-parameters in decibels can be directly used in link-budget calculations alongside powers, gains, and losses expressed in dB, dBm, or dBW.
2. *Transmission and reflection terminology:* The reflection coefficient Γ and transmission coefficient T are represented by specific S-parameters. For a two-port network, transmission is typically described by S_{21} (forward transmission) or S_{12} (reverse transmission).

In passive components, transmission loss is commonly referred to as insertion loss (IL), defined as:

$$\mathrm{IL} = -20\log_{10}|T| = -20\log_{10}|S_{21}|\,(\mathrm{dB}) \tag{6.1}$$

Insertion loss is conventionally expressed as a positive value in decibels. In active components such as amplifiers, the same transmission parameters represent gain rather than loss.

Depending on the component type and application, transmission-related S-parameters are described using different terminology, as illustrated in Figure 6.1 [1]:

- *Insertion loss:* Used primarily for passive components such as filters, cables, and power dividers. It quantifies signal attenuation through the device.
- *Gain:* Used for active components such as amplifiers, describing the increase in signal power.
- *Coupling:* Used for multiport devices such as directional couplers. It describes the intentional transfer of a fraction of the signal to a coupled port.
- *Isolation:* Describes the suppression of unwanted signal transfer between ports. High isolation is critical in devices such as power dividers, isolators, and circulators to prevent signal leakage and interaction.

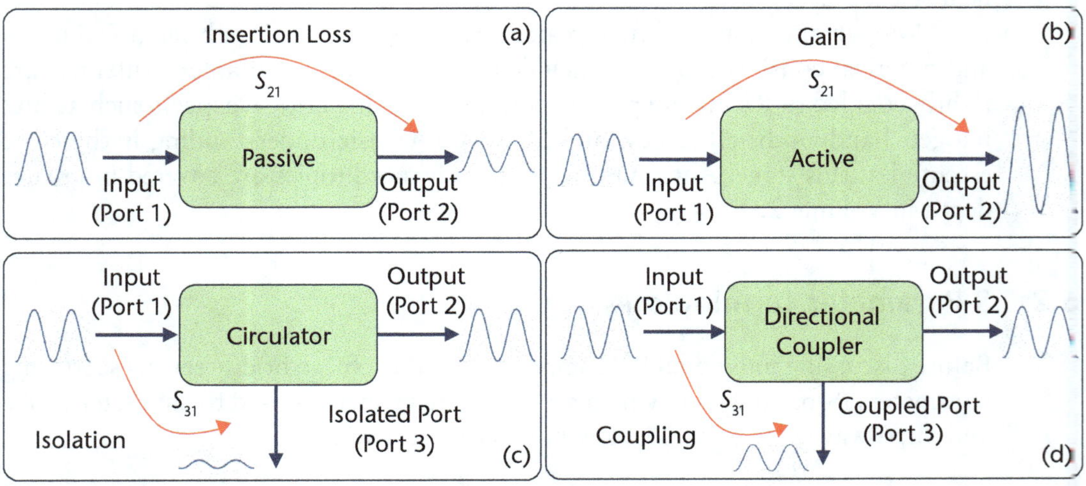

Figure 6.1 Terminologies used for the transmission coefficient in various microwave components.

The return loss (RL) is related to the reflection coefficients (S_{11}, S_{22}) and, for port 1, is defined as follows:

$$RL = -20\log|\Gamma|\,dB = -20\log|S_{11}|\,dB \tag{6.2}$$

For a passive network, return loss is always a non-negative quantity. A return loss of 0 dB corresponds to total reflection $|S_{11}| = 1$, whereas larger return loss values indicate better impedance matching and lower reflected power. In practical RF designs, a return loss of 10 dB or higher is commonly considered acceptable. Table 6.1 summarizes the fraction of reflected power corresponding to different return loss values.

Return loss is a critical performance parameter in RF components and systems, as it reflects the quality of impedance matching between interconnected stages. While low return loss (good matching) is generally desirable, there are cases where high reflection is intentional, such as in the stopband of reflective filters.

It is important to recognize that a low return loss alone does not guarantee optimal system performance. Low reflection can also result from internal losses, where incident energy is dissipated within the component rather than transmitted to the output. For this reason, return loss must always be evaluated together with insertion loss when assessing RF component performance.

From an energy conservation perspective, a two-port passive network must satisfy the unitary property of its S-parameters, meaning that the incident power is partitioned between reflected and transmitted components. This relationship provides a consistency check between return loss and insertion loss and is fundamental to the interpretation of measured S-parameters.

$$|S_{11}|^2 + |S_{21}|^2 = 1 \tag{6.3}$$

and

$$|S_{22}|^2 + |S_{12}|^2 = 1 \tag{6.4}$$

For a reciprocal network $|S_{21}|^2 = |S_{12}|^2$ and therefore $|S_{11}|^2 = |S_{22}|^2$ [2].

In general, low insertion loss (minimal signal attenuation) and high return loss (good impedance matching) are desirable across the intended frequency range of operation.

Table 6.1 Amount of Reflected Power for Various Values of Return Loss

Return Loss	*Means*
0 dB	100% reflection, no power flows into the device
3 dB	50% reflection, 50% power flows into the device
10 dB	10% reflection, 90% power flows into the device
15 dB	3% reflection, 97% power flows into the device
20 dB	1% reflection, 99% power flows into the device

In addition to S-parameters, several extended parameter sets are used in RF system analysis. X-parameters model nonlinear behavior under large-signal excitation, mixed-mode S-parameters describe differential and common-mode behavior, and active (hot) S-parameters characterize small-signal performance under bias and steady-state operating conditions, such as power amplifiers, oscillators, or biased microwave sources.

6.3 RF Link Elements

The upcoming sections will delve into the constituents of an RF link, as shown in Figure 6.2, encompassing signal generation, signal transmission, RF signal processing, and signal detection [1]. These four functional domains will be introduced along with their fundamental building blocks and the key performance indicators used to characterize each block. In addition, their roles and interactions within a complete RF system will be illustrated, including the system architectures discussed in Chapter 5.

6.4 Signal Generation

Most RF systems rely on three primary forms of signal generation:

- Tone generation, such as continuous-wave (CW) carriers and single-frequency test tones;
- Waveform generation, including modulated signals and arbitrary test patterns;
- Clock and timing generation, such as reference clocks and synchronization signals.

There are numerous techniques for generating RF tones and waveforms, each suited to different performance, flexibility, and implementation requirements. Table 6.2 summarizes the primary signal-generation techniques and highlights their key advantages and limitations.

We will discuss the techniques listed in Table 6.3; however, multivibrators, being limited to low-frequency operation, are generally not relevant for RF applications.

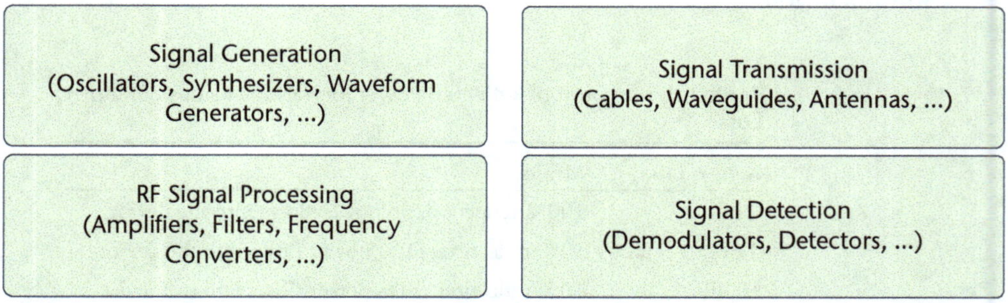

Figure 6.2 Core functional domains of an RF system, illustrating the four primary building blocks, signal generation, RF signal processing, signal transmission, and signal detection, that collectively define the end-to-end RF signal chain.

Table 6.2 Common Methods of Tone and Waveform Generation

Method	*Description/Typical Waveforms*	*Typical Frequency Range*	*Advantages*	*Limitations*
Multivibrators/ relaxation oscillators	Simple analog circuits producing square, triangle, sawtooth, or pulse waveforms	Hertz to a few megahertz	Very low cost, easy to implement	Poor frequency accuracy and stability; limited spectral purity
Oscillators (crystal, YIG, DRO)	Using gain and feedback with a frequency-selective resonator to generate RF/microwave tones with high stability and accuracy	Kilohertz to tens of gigahertz (extendable into millimeter-wave)	Excellent spectral purity, wide tuning range, fine resolution	Higher complexity and cost; tuning speed depends on technology
Harmonic/nonlinear generation	Frequency multipliers, mixers, subharmonic oscillators	Extension of lower-frequency source	Enables higher frequencies from low-frequency references	Adds spurious tones; requires filtering
Optoelectronic/ photonic sources	Optical heterodyning or mode-locked lasers generating RF/terahertz signals	10 GHz to terahertz	Ultralow phase noise, very high frequency range	Expensive, complex, specialized applications
Digital generation (DDS, RFSoC/FPGA DACs, AWGs)	Digitally defined waveforms (stored or real-time computed) converted to analog via DAC; supports arbitrary signals and modulation formats	Baseband to hundreds of megahertz (DDS); several gigahertz (RFSoC/FPGA); tens of gigahertz (high-end direct-RF AWGs)	Extremely flexible, highly programmable, and well-suited for advanced modulation schemes	Constrained by DAC resolution and sampling rate, clock jitter, spectral images, and output reconstruction filtering

Optoelectronic sources are employed only in specialized cases, such as photonic generation of millimeter-wave carriers, radio-over-fiber links, and ultralow phase-noise reference distribution.

6.4.1 Key Performance Parameters of RF Signal Generators

Signal generators, synthesizers, and timing sources are specified by a common set of performance parameters that directly impact system accuracy, robustness, and suitability for a given RF application. While the relative importance of each parameter depends on whether the generator is used for tone generation, waveform synthesis, or clocking, the following characteristics are broadly applicable across most RF systems.

Frequency Range

The frequency range specifies the span of frequencies that a signal generator can produce. Many cellular, PCS, and SATCOM systems operate over relatively narrow bands, typically on the order of 3%–10% fractional bandwidth. In contrast, wideband applications, such as general-purpose RF test systems, EW platforms, or research instrumentation, may require tuning ranges that extend to tens of gigahertz and, in specialized cases, hundreds of gigahertz for millimeter-wave and subterahertz signal generators.

Frequency Resolution

Frequency resolution refers to the smallest frequency increment that a signal generator can produce. Fine resolution is particularly important in applications requiring precise channel spacing, small frequency offsets, or Doppler compensation, where even minor frequency errors can degrade system performance.

As discussed in the following sections, in integer-*N* phase-locked loop (PLL) architectures, the frequency step size is fundamentally set by the reference frequency, often limiting resolution to tens of kilohertz or larger. Fractional-*N* PLLs overcome this constraint by allowing fractional division ratios, enabling frequency resolution on the order of a few hertz to a few kilohertz, depending on loop design, reference quality, and modulation strategy.

Direct digital synthesis (DDS) provides even finer resolution by generating frequencies numerically, with resolution determined by the phase accumulator and clock frequency. Subhertz frequency steps are readily achievable, and tuning is deterministic since no analog loop settling is required. In many modern systems, DDS or numerically controlled oscillators (NCOs) are combined with PLLs to achieve fine resolution, fast tuning, and low phase noise.

Frequency Stability

Frequency stability describes how much an oscillator's output frequency deviates from its nominal value over temperature, time, and environmental conditions [3]. It is typically specified in parts per million (ppm) or parts per billion (ppb). For example, a 10-MHz oscillator with ±1-ppm stability may drift by up to ±10 Hz from its nominal frequency.

Oscillator performance is commonly evaluated in terms of short-term and long-term stability, reflecting frequency behavior over different time scales.

- *Short-term stability* (milliseconds to seconds) is dominated by phase noise and timing jitter. It directly affects spectral purity, modulation accuracy, and error performance in communication systems.
- *Long-term stability* (minutes to days or longer) is determined by aging and environmental influences such as temperature, vibration, and pressure. It is usually specified as frequency drift in ppm or ppb per day, month, or year. In navigation and timing systems, such as GPS, 5G base stations, and phased arrays, long-term stability is critical for maintaining phase coherence and timing alignment across distributed nodes.

A widely used metric that captures both effects is the Allan deviation, which characterizes frequency fluctuations as a function of averaging time and helps identify whether noise or drift dominates for a given application. A phase noise analyzer can be used to measure the Allan deviation.

Here are some practical guidelines:

- Communication systems are primarily limited by short-term phase noise.
- Navigation, timing, and precision test instruments require strong long-term stability in addition to low phase noise.
- The choice of frequency reference (crystal oscillator (XO), temperature-compensated crystal oscillator (TCXO), oven-controlled crystal oscillator (OCXO), or atomic (Rubidium or Cesium clocks)) reflects the required trade-off between short-term spectral purity and long-term drift. Table 6.3 compares the most commonly used reference types.

Atomic references offer unmatched long-term stability and accuracy but typically exhibit worse close-in phase noise than high-quality OCXOs, which remain the preferred choice when short-term spectral purity is critical. OCXO absolute accuracy can be significantly improved when disciplined to an external reference (e.g., GPS).

Phase Noise

Phase noise, typically specified in dBc/Hz, characterizes the spectral spreading of a signal around the carrier. As discussed in detail in Chapter 4, both close-in and far-out phase noise directly impact spectral purity, adjacent-channel leakage, and EVM. Consequently, phase noise is often one of the dominant performance metrics when selecting or designing a signal generator for communication, radar, or sensing applications.

Output Power and Buffering

Typical synthesizer output power levels range from approximately 0 to +5 dBm. Driving downstream components such as mixers, switches, power splitters, or multiple loads often requires buffer amplification. Careful buffer design is essential, as poorly designed output stages can introduce additional spurious tones, degraded phase noise, or excessive power consumption, offsetting the performance of the core signal source.

Flatness of Frequency Response

Amplitude flatness refers to how consistently the output power is maintained across the tuning range. Good flatness is critical for linear system operation and distortion-free performance, particularly in wideband applications. Many modern synthesizers employ automatic level control (ALC) circuitry to compensate for frequency-dependent gain variations and maintain a nearly constant output amplitude.

Table 6.3 Comparison of Frequency Reference Types

Reference Type	*Short-Term Stability*	*Long-Term Stability*	*Absolute Accuracy*
XO	Poor	Poor	Poor
TCXO	Moderate	Moderate	Moderate
OCXO	Excellent	Good	Good
Atomic (Rb/Cs)	Moderate (worse than OCXO)	Excellent	Excellent

Spurious Signal Level

Spurious signals are deterministic, discrete spectral components that appear in addition to the desired carrier. They are distinct from broadband noise and are typically categorized as harmonics, spurs, or subharmonics.

- Harmonics are integer multiples of the fundamental frequency (e.g., $2f_0$, $3f_0$) and typically result from nonlinearities in oscillators, buffers, or output stages.
- Spurious tones (spurs) arise from mixing products within PLLs, fractional-N modulation, digital clocks, nonlinearities in output stages, or power-supply coupling.
- Subharmonics are tones occurring at fractional frequencies of the carrier (e.g., $f_0/2$, $f_0/3$). These commonly appear in divider-based architectures, injection-locked systems, or due to nonlinear behavior in frequency synthesis chains.

Typical performance levels are:

- *Harmonics:* −30 to −50 dBc (often improving at higher frequencies due to filtering);
- *Spurs:* −80 to −50 dBc, depending on architecture and loop design;
- *Subharmonics:* −90 to −50 dBc, with levels often increasing at higher output frequencies.

Effective suppression of these unwanted components usually requires a combination of careful architecture selection, loop optimization, filtering, and output-stage linearity.

Switching Speed

Switching speed defines the time required for a frequency synthesizer or PLL to retune and settle to a new frequency. Because PLL dynamics converge asymptotically, switching speed is typically specified as the settling time to within a defined frequency or phase error tolerance.

DDS-based architectures can achieve submicrosecond switching, since frequency changes are implemented digitally without the need for analog loop settling. In contrast, voltage-controlled oscillator (VCO) or yttrium-iron-garnet (YIG)-tuned oscillators typically achieve submillisecond switching, depending on loop filter design, charge-pump current, tuning sensitivity, and mechanical or magnetic dynamics.

For agile radios, radar systems, and frequency-hopping applications, lock time is often as critical as frequency accuracy. Excessive settling times can disrupt timing budgets, reduce throughput, or limit system responsiveness. Despite being slower than DDS solutions, YIG-tuned oscillators (YTOs) remain indispensable in high-frequency, low-phase-noise synthesizers, with modern driver circuits enabling sub-millisecond switching even at microwave and millimeter-wave frequencies.

Startup Time and Power

Startup time refers to the interval required for a signal source to reach its specified operating conditions after power is applied. This parameter is especially critical for temperature-stabilized references, such as OCXOs, which require a warm-up period to reach thermal equilibrium. Startup time directly impacts system availability, duty cycling, and power management, particularly in battery-powered or intermittently operated systems.

Frequency Pulling and Frequency Pushing

Even with frequency stabilization loops such as PLLs in place, oscillators are not immune to external disturbances. Two important mechanisms that affect frequency stability are frequency pulling and frequency pushing.

Frequency pulling occurs when changes in load impedance shift the oscillator's frequency. Because oscillators are sensitive to their load, impedance mismatches can reflect energy back into the circuit and alter the effective resonant conditions. For example, a VCO driving a poorly matched mixer may experience a frequency shift away from its nominal value. A typical specification is on the order of less than 1 MHz peak-to-peak for a load reflection level of 12 dBr, where dBr denotes a level measured relative to the oscillator's nominal output power.

Frequency pushing is caused by variations in supply or tuning voltage. Changes in bias conditions modify device operating points and effective capacitances, resulting in a frequency shift. As an example, a 2.4-GHz VCO may exhibit several megahertz of frequency change per volt of supply variation, with a typical pushing sensitivity of a few tenths of MHz/V.

Both effects underscore the importance of good impedance matching and well-regulated bias and supply networks to maintain stable frequency generation in RF systems.

Other Practical Considerations

Beyond core RF performance metrics, practical system design must also account for factors such as module size, power consumption, supply voltage, control and interface protocols, operating temperature range, and long-term reliability. While these considerations lie outside the scope of this section, they are often decisive during system-level integration and product selection.

6.4.2 Tone Generation

In this section, we discuss the methods for generating tones or CWs at different frequencies. Tone generation represents the most fundamental form of signal generation, providing single-frequency outputs that serve as carriers, LOs, or reference signals in RF systems. The quality of these tones, particularly their phase noise, frequency accuracy, and spectral purity, directly impacts the performance of communication links, radar systems, and measurement setups.

6.4.3 Tone Generation: Oscillators

Sustained oscillation requires three elements: an active device to provide gain (e.g., a transistor), a positive feedback path that returns a portion of the output signal in phase, and a frequency-selective element, typically a resonator, that defines the oscillation frequency. According to the Barkhausen criterion, oscillation occurs when the loop-gain magnitude equals unity and the total loop phase shift is an integer multiple of 2π. In practical RF oscillators, the resonator is passive and cannot generate a signal on its own; instead, it enforces the phase condition and strongly influences frequency stability and phase noise. Well-known resonator-based oscillator topologies include the Colpitts, Clapp, Hartley, and Pierce oscillators. In these topologies, the resonator may be implemented using a quartz crystal, a SAW resonator, or a dielectric resonator.

Different types of oscillators offer varying levels of stability, cost, tunability, and performance trade-offs. Table 6.4 compares the features of widely used oscillator types.

These characteristics make them ideal for fixed-frequency applications, such as reference clocks, local oscillators in narrowband radios, and stable frequency sources in communication systems. However, their limited tunability makes them unsuitable for applications that require wide frequency coverage or frequency sweeping, such as signal generators, spectrum analyzers, or wideband radar systems.

Table 6.4 Comparison of Common Oscillator Types

Oscillator Type	*Frequency and Phase Stability*	*Oscillation Mechanism*	*Frequency Tuning Range*	*Key Advantage*	*Limitation*	*Typical Use Cases*
XO	Good (±10–50 ppm), moderate phase noise	Quartz crystal resonance (piezoelectric effect)	Fixed (single frequency)	Low cost, simple	Sensitive to temperature	General electronics, clocks
TCXO	Better (±0.5–2 ppm), moderate phase noise	Quartz crystal with temperature compensation	Fixed (single frequency)	Improved stability over XO	More complex circuitry	GPS/GNSS, telecom
OCXO (oven-controlled XO)	Excellent (±0.01–0.1 ppm), very good phase noise	Quartz crystal stabilized by temperature-controlled oven	Fixed (single frequency)	High stability, excellent long-term performance	Higher power, warm-up time	Test equipment, satellite systems
DOCXO (double-oven XO)	Ultra-stable (sub-ppb), excellent phase noise	Dual or multiple quartz crystals with oven compensation	Fixed (single frequency)	Metrology-grade stability	Expensive, bulky	Frequency standards, precision timing
MEMS oscillator	Good (±0.5–5 ppm), phase noise varies (generally worse than XO)	Micro-electromechanical resonator	Fixed (single frequency)	Very small, low power, shock resistant	Higher phase noise than quartz	Portable electronics, IoT, consumer devices
LC oscillator	Poor stability, poor phase noise	Inductor-capacitor tank resonance	Wide (kilohertz to hundreds of megahertz), but drifts	Tunable, simple	Susceptible to drift (temperature, aging)	Low-cost RF, VCO cores

Table 6.4 *(Continued)*

Oscillator Type	*Frequency and Phase Stability*	*Oscillation Mechanism*	*Frequency Tuning Range*	*Key Advantage*	*Limitation*	*Typical Use Cases*
Gunn diode oscillator	Moderate stability, relatively high phase noise	Negative differential resistance in GaAs (transit-time domains)	Limited (~5%–10% tunability)	Simple, compact, millimeter-wave-capable	Limited output power, high noise	Police radar, motion sensors, microwave links
IMPATT diode oscillator	Poor stability, very high phase noise	Avalanche breakdown + transit-time effect	Limited, narrowband	Higher RF power (up to several watts at millimeter-wave)	High bias voltage, noisy, generates heat	High-power millimeter-wave transmitters, jammers
Dielectric resonator oscillator (DRO)	Very good stability, excellent (low-noise) phase noise	High-Q dielectric resonator	Narrow (few megahertz around resonance)	Very low phase noise	Bulky resonator	Radar, microwave links
YTO	Very good (ppm to ppb when PLL-locked), excellent phase noise	Ferromagnetic resonance of YIG sphere tuned by magnetic field	Very wide (gigahertz to 40+ GHz)	Wide tuning, superb phase noise	Bulky, power-hungry, magnetic sensitivity	Signal generators, EW systems, test equipment
SAW oscillator	Good stability, good phase noise	SAW resonator	Fixed or very narrow (approximately tens of kilohertz)	Compact, stable	Limited frequency range	Portable RF systems
Atomic reference (Rb, Cs, H-maser)	Exceptional (10^{-11} to 10^{-13}), excellent phase noise	Atomic transition frequency locked via spectroscopy	Fixed (atomic transition frequency)	Ultimate long-term stability	Expensive, size/power constraints	GNSS master clocks, telecom timing, scientific instrumentation

For applications requiring large tuning ranges, YTOs are widely used. YTOs offer a unique combination of:

- Very wide frequency tuning range;
- Exceptional phase noise performance;
- High spectral purity across the tuning band.

In a YTO, a DC current applied to a tuning coil generates a magnetic field that alters the resonance of a YIG sphere. By varying this magnetic field, the oscillator's frequency can be smoothly tuned over several octaves, making YTOs especially valuable in swept-frequency instruments and agile RF systems.

Oscillator performance is defined not only by instantaneous frequency accuracy, but also by how well that accuracy is maintained over time and under varying environmental conditions such as temperature, aging, and supply variation.

In crystal oscillators, frequency varies with temperature, and the exact behavior depends strongly on the crystal cut and orientation. Different quartz cuts exhibit distinct temperature-frequency characteristics, making crystal selection a critical design consideration for precision timing and frequency control [4].

Despite their excellent phase noise performance, YTOs and tunable oscillators in general exhibit limited absolute frequency accuracy and long-term stability when operated in open-loop mode. Several mechanisms contribute to frequency drift, including tuning coil drift, temperature variation, aging effects, and control electronics instability. As a result, the output frequency can deviate over time, even if short-term noise performance remains excellent.

To mitigate temperature-induced and aging-induced frequency drift, several compensation techniques are commonly employed in crystal-based oscillators. TCXOs apply electronic correction to reduce thermal drift across temperature. OCXOs maintain the crystal at a constant elevated temperature to improve short-term and long-term stability, while double-oven crystal oscillators (DOCXOs) add an additional layer of thermal isolation to further enhance long-term stability and aging performance.

Even with advanced compensation methods, open-loop oscillator operation is typically insufficient for applications requiring high absolute frequency accuracy, long-term stability, or repeatability across environmental conditions. To overcome these limitations, oscillators are commonly stabilized using a PLL. A PLL locks the oscillator output to a highly accurate reference source, ensuring long-term frequency accuracy while preserving the oscillator's low phase noise characteristics. The principles and architectures of PLL-based frequency stabilization are discussed in detail next.

6.4.4 Tone Generation: Nonlinear Methods

As discussed in Chapter 4, nonlinear mechanisms generate new frequency components, such as harmonics and mixing products, that are not present at the input. This property can be exploited to extend lower-frequency sources into higher-frequency bands. However, nonlinear generation inherently produces unwanted spurious tones that must be carefully filtered. In addition, harmonic signal generation is associated with high conversion loss, resulting in much lower output power levels and necessitating subsequent amplification to achieve usable output power.

In the context of nonlinear tone generation, two methods dominate modern RF systems: frequency multiplication and frequency division. Both rely on nonlinear circuit behavior to generate new frequency components that are not present at the input, either as integer multiples or integer fractions of a reference frequency.

Frequency Multiplication

Nonlinear devices are used to generate integer multiples of a fundamental frequency ($2f$, $3f$, ...), followed by filtering and amplification. Frequency multipliers are extensively employed in RF, microwave, and millimeter-wave systems to reach frequency bands that are difficult or impractical to generate directly with oscillators. Depending on the circuit topology and the degree of nonlinearity, the multiplication factor may be 2 (doubler), 3 (tripler), or higher.

The performance of a frequency multiplier is characterized by several important factors. Conversion loss quantifies how much of the input power is transferred to the desired harmonic, as shown in Figure 6.3(a). The input drive level is also critical; as illustrated in Figure 6.3(a), conversion loss is a strong function of input power. Frequency multipliers typically require relatively high drive levels to achieve acceptable conversion efficiency. In general, conversion loss increases as the multiplication factor increases.

Spectral purity, shown in Figure 6.3(b), characterizes the relative levels of the desired harmonic and unwanted spectral components, including the fundamental and higher-order harmonics. Poor spectral purity can introduce spurious tones that degrade system performance, necessitating effective filtering at the multiplier output.

Several implementation approaches exist. Diode-based multipliers are widely used at high frequencies, often employing Schottky diodes because of their fast-switching characteristics and low junction capacitance. Transistor-based multipliers can provide both frequency multiplication and gain, making them attractive at lower frequencies or in applications where additional amplification is required. Integrated multiplier chains combine multiple stages, for example, a ×2 stage followed by a ×3 stage to realize an overall ×6 multiplier, enabling operation at very high frequencies within a compact MMIC implementation.

Frequency Division and Subharmonic Generation

A frequency divider is a circuit that reduces the frequency of an input signal by an integer or fractional ratio. In RF systems, frequency dividers are most commonly used in PLLs, frequency synthesizers, and clock generation circuits to derive stable, lower-frequency signals from a high-frequency source.

In millimeter-wave systems, subharmonic techniques are widely used to ease oscillator and mixer design by operating key building blocks at a fraction of the RF and exploiting harmonic generation to reach the desired band. For example, in a 60-GHz millimeter-wave synthesizer, a PLL may lock a 30-GHz VCO to a

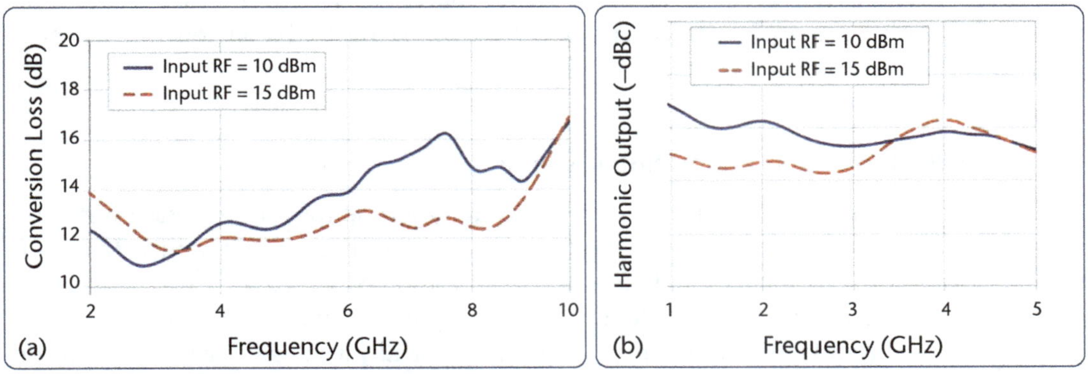

Figure 6.3 (a) Conversion loss of a frequency doubler. The input range is 1 to 5 GHz, and the output range is 2 to 10 GHz. (b) Harmonic output of a frequency multiplier.

100-MHz reference using a ÷300 frequency divider, and the locked signal is then frequency-doubled to generate the final 60-GHz output.

Key performance parameters include the division ratio N, input frequency range, phase noise contribution, and lock range. In integer-N frequency synthesizers, the divider ratios determine both the output frequency and the achievable frequency resolution according to

$$f_{out} = \frac{f_{ref} \times N}{R} \tag{6.5}$$

where f_{ref} is the reference frequency, N is the feedback divider, and R is the reference divider (see Figure 6.6). Because in-band PLL phase noise scales as $+20\log_{10}N$, large divider ratios severely amplify reference noise, forcing a trade-off between frequency resolution, loop bandwidth, and achievable phase-noise performance. This phase-noise scaling is a primary reason that millimeter-wave systems avoid very large divider ratios and instead rely on subharmonic architectures and frequency multiplication.

Two main classes of frequency dividers are commonly used: digital and analog (regenerative) dividers.

Analog or regenerative dividers employ mixers, amplifiers, and feedback paths to achieve frequency division at microwave and millimeter-wave frequencies, where conventional digital logic becomes impractical due to speed and power limitations. These dividers rely on nonlinear mixing products and selective feedback to lock onto a subharmonic of the input signal.

Digital dividers, implemented using counters or flip-flop chains, are widely used in PLLs and digital clocking systems. They provide precise integer division, exhibit excellent repeatability, and are easily integrated into CMOS ASICs, RFICs, or field-programmable gate arrays (FPGA)-based designs.

Modern frequency dividers may also support fractional-N division, enabling fine frequency steps and fast frequency agility. Such architectures are essential in contemporary communication, radar, and instrumentation systems that require wide tuning ranges combined with high spectral purity.

Together, these nonlinear frequency generation and division techniques extend the practical operating range of RF signal sources and are often used in combination with PLLs or DDS to cover wide frequency spans with controlled phase noise and resolution.

6.4.5 Tone Generation: Digital Methods

Digital techniques have become the dominant method for tone generation in modern RF systems due to their flexibility, precision, and programmability. Instead of relying solely on analog oscillators, an NCO combined with a high-speed digital-to-analog converter (DAC) enables DDS. In this approach, the NCO generates a digital representation of the desired periodic signal, typically a sinusoid, which is then converted to the analog domain by the DAC. This concept is illustrated in Figure 6.4.

The output frequency of a digitally generated tone is determined by incrementing a phase accumulator with a programmable step size, commonly referred to as

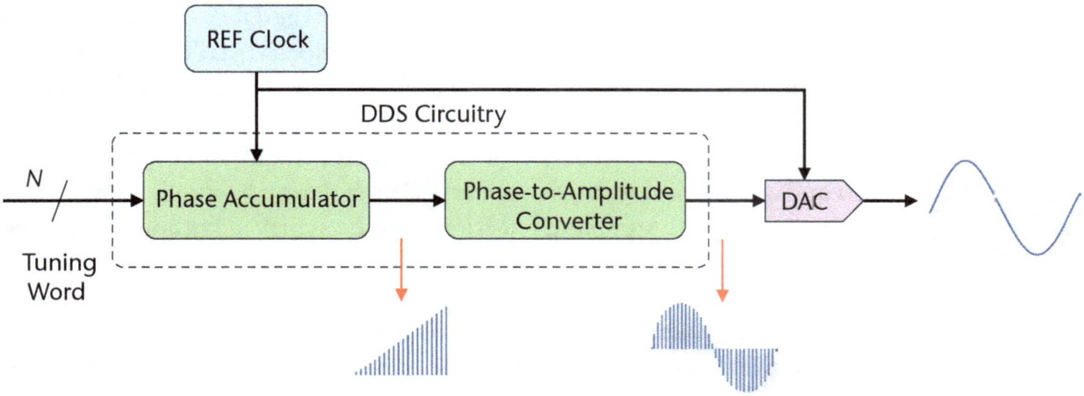

Figure 6.4 Functional modules of a DDS.

the frequency tuning word (FTW). Each accumulated phase value is mapped to a corresponding waveform amplitude using a phase-to-amplitude conversion method, such as a sine lookup table (LUT), CORDIC algorithm, or polynomial approximation. These digital amplitude samples are then converted to the analog domain by a high-speed DAC. If required, the resulting signal may be further translated to the desired RF using the digital and analog upconversion techniques discussed in Chapter 5.

By adjusting the FTW, the generated frequency can be tuned with extremely fine resolution, often reaching the subhertz range, while maintaining phase coherence.

The FTW specifies the output frequency as a fraction of the reference clock frequency and is given by

$$f_{\text{out}} = \frac{\text{FTW}}{2^N} f_{\text{clk}} \tag{6.6}$$

where N is the number of bits in the phase accumulator and f_{clk} is the reference clock frequency driving the DDS.

DDS offers several compelling advantages in modern RF and microwave systems:

- *High-frequency resolution:* Extremely fine-tuning steps are possible, often reaching subhertz resolution due to the large phase accumulator word length.
- *Fast switching speed:* Frequency changes occur instantaneously through digital updates, without the settling time associated with PLLs.
- *Arbitrary phase control:* Output phase can be set and adjusted digitally, enabling phase-coherent operation across multiple channels.
- *Seamless integration with modulation:* IQ samples for digital modulation schemes can be generated directly in the digital domain, simplifying transmitter architectures.

The primary limitations of DDS-based systems stem from DAC performance and sampling effects:

- *Spurious signals:* Quantization, phase truncation, and nonlinearities introduce spurs that limit spurious-free dynamic range (SFDR).

- *Clock-related phase noise:* The spectral purity of the DDS output is fundamentally constrained by the reference clock and clock distribution network.
- *Bandwidth limitations:* The maximum usable output frequency is limited by the DAC sampling rate and its analog bandwidth.
- *Output filtering:* A reconstruction (lowpass) filter is required to suppress DAC images and harmonics. At high clock rates, filter design becomes increasingly challenging, as strong stopband rejection must be achieved without degrading passband flatness or phase linearity.

As discussed next, higher Nyquist zones of the DAC may also be exploited to generate high-frequency signals, extending the usable output range of DDS-based sources beyond the fundamental band.

In practice, DDS is frequently combined with the PLL-based frequency synthesizers discussed in the next section in hybrid architectures. This approach leverages the fine frequency resolution and fast tuning capability of DDS at lower frequencies, while relying on PLLs to provide superior phase noise performance and frequency translation at higher RF and microwave bands.

6.4.6 PLLs for Frequency and Phase Stabilization

Before discussing waveform generation techniques, it is essential to examine how instabilities in tone generation are mitigated through closed-loop control using PLLs.

Integer-*N* PLL

In various applications, it is essential to construct a tunable oscillator that can adjust its frequency across a specific range. For example, when selecting different channels in a receiver, such as an FM radio, the LO frequency must be tuned accordingly for each channel. A VCO serves as an electronically tunable oscillator. The frequency of a VCO is adjusted by modifying its resonant frequency, for example, by varying the capacitance of a tank circuit with a varactor diode, or in a YIG-based oscillator, by changing the current through the tuning coil.

A VCO block diagram is shown in Figure 6.5(a). A DC supply provides the required biasing for the circuit, while a control voltage determines the output frequency of the generated sinusoidal signal.

Figure 6.5(b) illustrates the transfer function of an ideal VCO. In the ideal case, this relationship is linear, and the output frequency f_{out} varies from f_1 to f_2 (defining the tuning range) as the control voltage V_c changes from V_1 to V_2. The slope of this characteristic, denoted K_{VCO}, is referred to as the VCO gain and is typically expressed in units of Hz/V. The relationship can be written as

$$f_{\text{out}} = K_{\text{VCO}} V_c + f_0 \tag{6.7}$$

where f_0 represents the free-running (or offset) frequency at zero control voltage [5].

However, in practice, the tuning characteristic is not perfectly linear, and deviations from an ideal straight-line response are commonly observed due to device nonlinearities, temperature variations, and process tolerances.

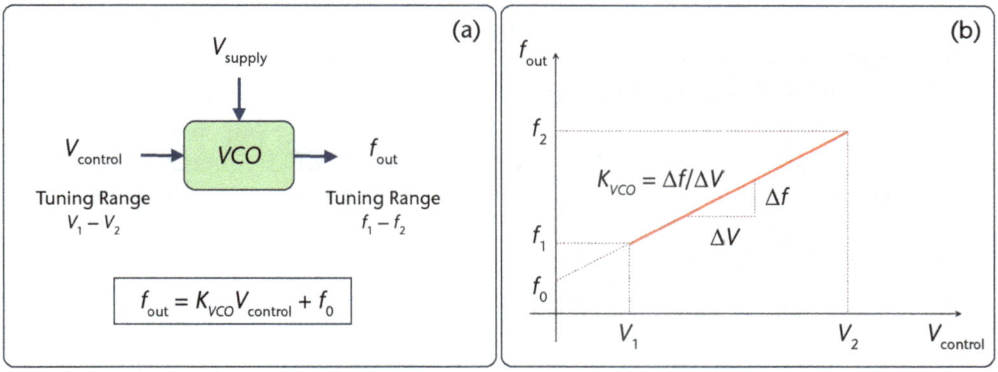

Figure 6.5 (a) High-level VCO block diagram with control voltage and frequency tuning ranges, and (b) ideal linear tuning curve defining the VCO gain K_{VCO}.

The frequency of a VCO can drift due to variations in temperature, supply voltage, and bias conditions. To achieve a stable and accurate output frequency, a PLL is commonly employed. A PLL is a feedback control system that locks the VCO output to a precise reference source, such as a TCXO. Figure 6.6 illustrates the basic block diagram of a PLL.

To understand how a PLL operates, it is helpful to trace the signal flow starting from the output. Because the VCO typically operates at much higher frequencies, often in the gigahertz range, than the reference oscillator, its output is first divided by N using a programmable integer divider. This division scales the VCO frequency down to a level suitable for accurate phase comparison with the reference.

The phase detector compares the phase of the divided VCO signal with that of the reference and generates an error signal proportional to their phase difference. This error signal is passed through a loop filter, which suppresses high-frequency components and sets the dynamic behavior of the loop. The filtered control signal drives the VCO, adjusting its frequency in a direction that reduces the phase error.

The R-divider at the input, referred to as the reference divider, scales down the reference frequency before it is compared with the divided VCO output in the phase detector. This block is typically included when:

- The reference frequency is too high for direct comparison.
- Specific output frequency grids are required.
- Legacy or multistandard compatibility must be supported.

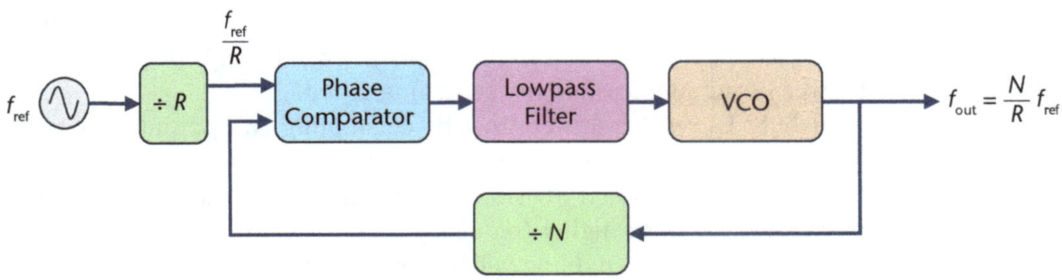

Figure 6.6 Block diagram of a PLL.

By scaling the reference applied to the phase detector, the reference divider R enables flexible frequency synthesis and provides control over the phase detector frequency. This, in turn, directly impacts loop bandwidth, lock time, and in-band phase noise.

In a steady state, the PLL forces the divided VCO frequency to match the divided reference frequency. As a result, the VCO output frequency is given by

$$f_{out} = \frac{N}{R} \cdot f_{ref} \tag{6.8}$$

where N is the feedback divider ratio and R is the reference divider ratio.

In an integer-N PLL, the achievable output frequency resolution is equal to the phase detector frequency $f_{PFD} = f_{ref}/R$. Each increment of the divider ratio N increases the output frequency by one phase-detector-frequency step. For example, using a 10-MHz reference with $R = 1$ results in frequency steps of 10 MHz, whereas choosing $R = 10$ reduces the step size to 1 MHz.

Because in-band PLL phase noise scales as $+20\log_{10}N$, large divider ratios degrade phase-noise performance, motivating the use of fractional-N PLLs to reduce N while maintaining frequency resolution.

Another important parameter is the loop bandwidth, which governs the dynamic response of a PLL and defines how rapidly the loop can track reference changes or suppress disturbances. The choice of loop bandwidth directly affects key trade-offs between phase noise, spurious suppression, and frequency switching speed. A wider loop bandwidth improves settling time and enables faster reference tracking, but allows more reference noise to pass through, whereas a narrower bandwidth improves close-in phase noise at the expense of slower tuning.

Fractional-*N* PLL

The phase detector in a PLL is a major contributor to close-in phase noise, since its digital circuitry generates significant transient noise at the reference frequency f_{ref}. Because frequency multiplication by N raises the phase noise by $+20\log_{10}N$ dB, large divider values can severely degrade performance; for instance, N = 30,000 adds about 90 dB to the noise floor, a typical case in integer-N PLLs for cellular transceivers [6]. The problem is that N is tied to channel spacing, so reducing N to improve noise is not straightforward. In addition, the phase detector noise superimposed on the control voltage to the VCO and modulates the VCO RF output accordingly. This appears as spurious signals at offsets of $+/-f_{ref}$ (and its harmonics) around f_{VCO}, requiring a narrow loop filter to suppress them. However, narrowing the loop bandwidth slows the synthesizer's ability to switch frequencies, creating a trade-off between spur suppression and switching speed.

Fractional-N PLLs resolve this conflict by decoupling N, f_{ref}, and the channel spacing. By dynamically alternating the divider between N and $N + 1$ in the correct proportion, an average fractional division ratio is achieved. This allows the phase detector to operate at a much higher frequency than the channel spacing, improving phase noise performance, enabling wider loop bandwidths, and allowing faster switching without sacrificing spur suppression. As a result, fractional-N synthesis provides finer frequency resolution and improved spectral purity compared to traditional integer-N designs.

The overall architecture of integer-N and fractional-N PLLs is largely the same. The key difference lies in the divider block. In an integer-N PLL, the divider is a simple counter that divides the VCO frequency by an integer N. In a fractional-N PLL, the divider is enhanced with additional circuitry that rapidly switches between N and $N + 1$ (or other values) in a controlled sequence, creating an effective fractional division ratio. To suppress the spurious tones that result from this switching, fractional-N architectures typically incorporate modulators such as sigma-delta, which spread the spur energy and improve spectral purity.

Let's look at an example. In an integer-N PLL, if $f_{ref} = 10$ MHz and $N = 240$, the output frequency is $f_{out} = N \cdot f_{ref} = 240 \times 10$ MHz $= 2.40$ GHz The smallest achievable step size is 10 MHz, since incrementing N by 1 changes the output frequency by one reference-frequency step. In a fractional-N synthesizer, the divider ratio can assume fractional values, commonly expressed as $N + K/M_{frac}$, where K and M_{frac} are integers. The output frequency is then given by

$$f_{out} = \left(N + \frac{K}{M_{frac}}\right) f_{ref} \tag{6.9}$$

The effective channel spacing is now a fraction of the reference frequency, $\Delta f = f_{ref}/M_{frac}$ Using the same reference frequency $f_{ref} = 1$ MHz, let $N = 240$, $M_{frac} = 1{,}000$, and $K = 123$. The output frequency becomes $f_{out} = (240 + 123/1{,}000) \times 10$ MHz $= 240.123 \times 10$ MHz $= 2.40123$ GHz. In this case, the frequency step size is only 10 MHz/1,000 = 10 kHz, compared to 10 MHz in the integer-N case. Depending on the architecture and modulation scheme, the resolution of a fractional-N PLL can reach the hertz or even subhertz range. Modern fractional-N devices routinely achieve step sizes below 1 Hz, even when using reference frequencies as high as 100 MHz.

PLLs not only stabilize the frequency and phase of oscillators, but also serve as the foundation of frequency synthesizers capable of generating a wide range of output frequencies from a single reference source. Frequency synthesizers are indispensable in RF transceivers for channel selection and frequency hopping and are a key building block in signal generators and other RF test equipment.

6.4.7 Waveform Generation

Now, let's go from tone generation to a more complex task of generating waveforms. As we have seen on many occasions earlier, waveforms are essential for modulated signals in telecommunications, quantum control systems, and radar.

Signal Generation Architectures: Scalar and Vector Approaches

Waveform generation can be broadly classified into two categories:

1. ***Scalar signal generators:*** Scalar signal generators provide control over frequency and amplitude (or output power), but not over phase. They are capable of producing CW signals as well as conventional analog-modulated

waveforms such as amplitude modulation (AM) and frequency modulation (FM).

Because phase is neither explicitly controlled nor preserved, scalar generators are best suited for:

- Basic RF testing and troubleshooting;
- Alignment and calibration tasks;
- Gain, compression, and frequency response measurements.

They are simple to use, robust, and widely available, but inherently limited for modern digitally modulated systems where phase accuracy is critical.

2. ***Vector signal generators (VSGs):*** VSGs extend waveform generation by providing full control over amplitude, phase, and frequency through in-phase (I) and quadrature (Q) modulation. By independently manipulating the I and Q baseband components, VSGs can synthesize complex modulation formats such as QPSK, QAM, and other multicarrier or spectrally efficient waveforms.

   Vector generators are essential for:

- Modern wireless communication systems (e.g., LTE, 5G, satellite links);
- Radar pulse and chirp waveform generation;
- Quantum control and qubit drive signals;
- Any application requiring precise phase coherence and modulation accuracy.

This capability makes VSGs a cornerstone of contemporary RF and microwave test setups.

Figure 6.7(a) shows a digital VSG based on digital upconversion. Baseband I and Q signals are interpolated, digitally mixed with a quadrature DDS, summed, and converted directly to RF using a high-speed DAC. The analog frontend selects the desired band (Nyquist zone), suppresses DAC images and spurious content, and provides amplification if required.

Figure 6.7(b) shows an analog vector signal generator, where digitally generated baseband I and Q signals are first converted to analog, mixed with quadrature LO signals, and summed to produce a modulated RF output. A bandpass filter is used to suppress image components and LO leakage, and amplification may be applied to reach the desired output signal level.

In practice, IQ signals for vector modulation can be generated in three main ways:

1. *Internal IQ generation:* The signal generator digitally synthesizes the I and Q waveforms internally using built-in modulation engines. This method is commonly used for standard modulation formats (e.g., QPSK, QAM, LTE, 5G) and offers tight integration, ease of use, and good repeatability.
2. *Uploaded (offline generated) IQ waveforms:* The I and Q data are generated offline using software tools such as MATLAB or Python and then uploaded to the signal generator for playback. This approach provides maximum flexibility for custom waveforms, algorithm development, and research applications.
3. *External IQ sources:* An external instrument, such as an arbitrary waveform generator (AWG), supplies the analog I and Q signals directly to the signal generator. The generator then uses these external baseband signals to modulate its internally generated RF carrier. This method is typically used

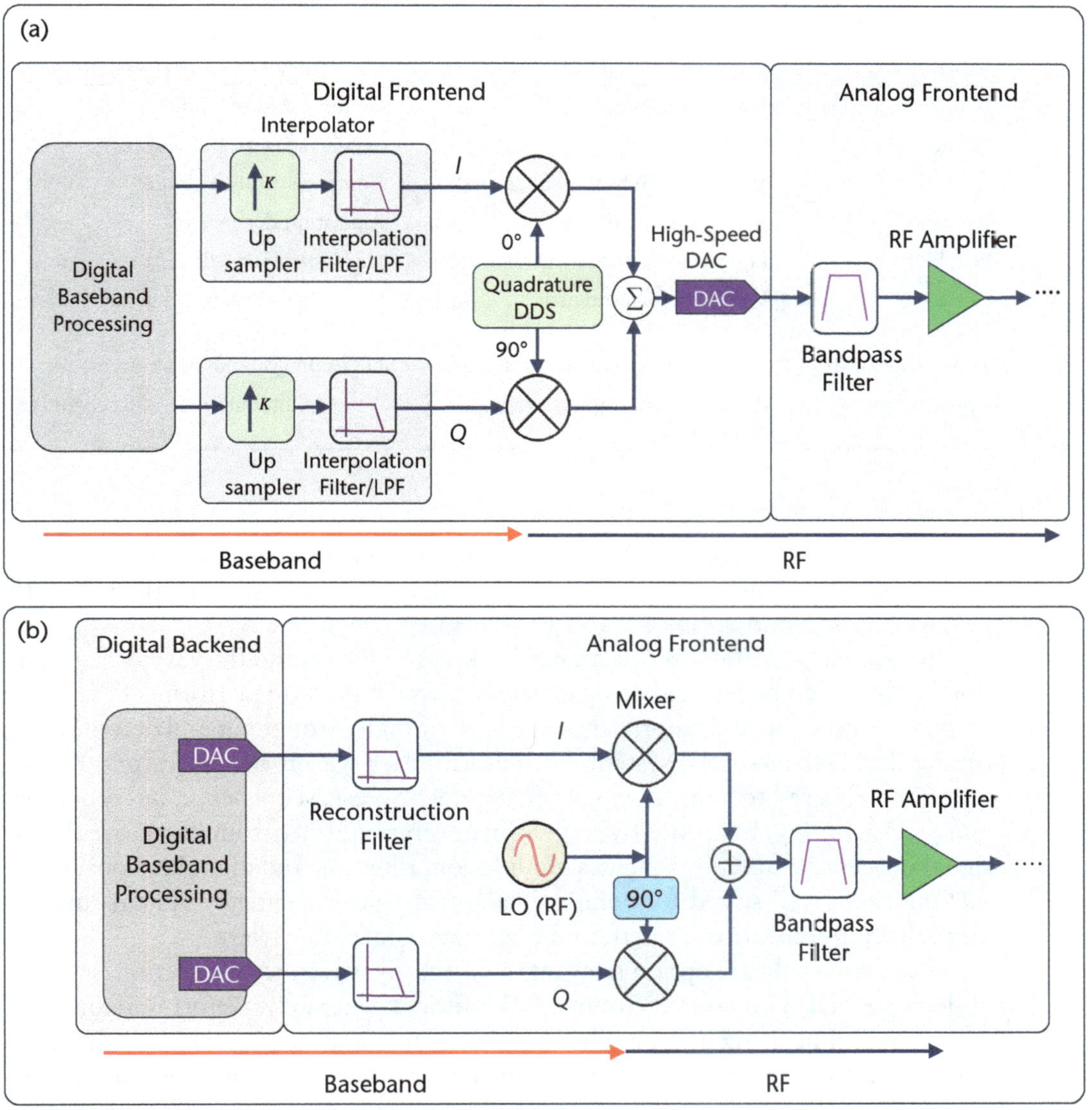

Figure 6.7 Block diagram of (a) digital VSG and (b) analog VSG.

for ultrawideband signals, high sample-rate applications, or when the IQ generation must be tightly synchronized with other hardware.

Waveform Generation Engine

Having examined scalar and vector waveform generators, along with their analog and digital implementations, the discussion now shifts from system-level architectures to the waveform generation engine itself. Regardless of whether a waveform is ultimately produced using a direct-RF, analog IQ, or superheterodyne architecture, the signal must first be represented and manipulated in the digital domain.

First, let us examine two distinct waveform generation methods based on real-time synthesis and sample playback, as summarized in Table 6.5. AWGs store waveform samples explicitly in memory and reproduce them through deterministic

Table 6.5 AWG Versus On-the-Fly (Engine-Level View)

Aspect	*Playback (AWG)*	*On-the-Fly Waveform Generation*
Waveform samples	Stored in memory	Computed in real time
Memory size	Large (MB-GB)	Small (registers, buffers)
Computation	Minimal (playback, sequencing)	Continuous (algorithmic)
Latency	Very low and deterministic	Algorithm-dependent
Flexibility	Limited to stored waveforms	High (parameter-driven)
Fidelity	Very high (deterministic reproduction)	Dependent on processing and algorithms
Typical hardware	FPGA with deep waveform memory	FPGA and/or DSP processors
Applications	Predefined test signals, standards compliance	Adaptive modulation, SDR, cognitive radio

playback. In contrast, on-the-fly waveform generation computes the waveform samples in real time from a set of parameters or algorithms and streams them directly to the DAC without storing the full waveform.

In the context of Table 6.6, both playback-based and on-the-fly waveform generation methods can be realized using different digital hardware platforms. FPGAs are commonly used for both approaches due to their high throughput and deterministic timing. In playback-based systems such as arbitrary waveform generators, FPGAs are primarily used to control waveform memory access, sequencing, interpolation, and precise timing. In on-the-fly waveform generation, FPGAs implement real-time signal processing functions such as modulation, filtering, and digital upconversion, making them well-suited for complex, adaptive, and low-latency systems used in modern communications, radar, and quantum control hardware.

A second hardware option relevant to on-the-fly waveform generation is the use of dedicated DDS integrated circuits. DDS devices generate waveforms algorithmically in real time using internal phase accumulators and lookup tables, rather than stored waveform samples. They are optimized for producing phase-continuous periodic and parametrically waveforms with fine frequency resolution, fast tuning, and excellent phase coherence. DDS ICs are typically programmed via SPI or parallel interfaces and offer a compact, low-complexity solution for stable tone and simple waveform generation. However, when operated at high clock rates, the associated DAC power consumption can become a practical concern in SWaP-limited systems.

AWG and Its Key Parameters

To understand how an AWG produces signals, it is helpful to examine its internal signal flow. Figure 6.8(a) shows a simplified block diagram of an AWG. In operation, an AWG reads precomputed waveform samples from memory at a precisely controlled clock rate, sequentially loads them into a shift register, converts the digital samples into an analog waveform using a DAC, and finally conditions the signal through an analog output stage before presenting it at the output.

The output of an AWG is not limited to the desired signal at the target frequency f_0. Owing to the discrete-time sampling process and the zero-order hold behavior of

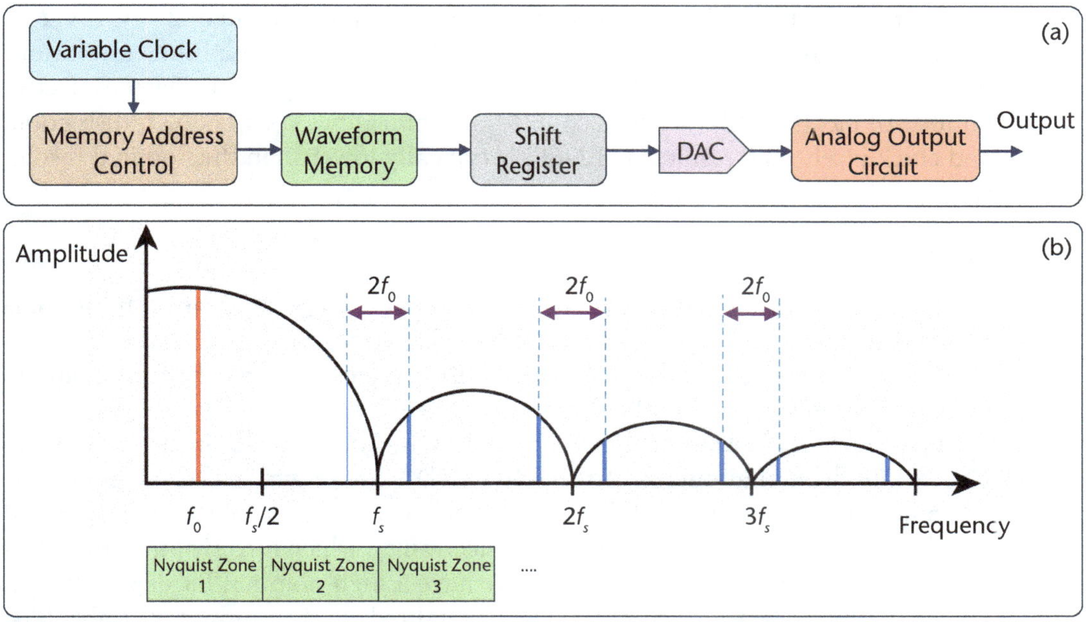

Figure 6.8 (a) Block diagram of an AWG. (b) Frequency-domain representation of an AWG output showing the sinc-shaped DAC roll-off and periodic spectral images of the desired tone f_0 across successive Nyquist zones.

the DAC, the output spectrum is shaped by a characteristic sinc envelope, as illustrated in Figure 6.8(b). This inherent frequency roll-off leads to amplitude droop at higher frequencies and the presence of unwanted spectral components, making analog reconstruction filtering essential for both image suppression and amplitude equalization [1].

This behavior can be deliberately exploited to extend the effective output frequency range of a DAC by selecting and filtering a desired image in a higher Nyquist zone. As shown in Figure 6.8(b), each Nyquist zone spans a bandwidth of $f_s/2$. The desired tone at f_0 is periodically replicated around integer multiples of the sampling frequency f_s. These replicas, commonly referred to as spectral images, occur at frequencies $|kf_s \pm f_0|$, where k is an integer. The images are organized into distinct Nyquist zones, with alternating spectral inversion between adjacent zones. This framework is fundamental to understanding direct digital synthesis, intentional image selection, and undersampling operation. In undersampling, aliasing is deliberately used to translate a band-limited RF signal to a lower frequency during sampling, provided the signal bandwidth is preserved, and spectral overlap is avoided. This topic is discussed in more detail in Volume 2.

Understanding the key specifications of an AWG is essential for selecting the most suitable instrument for a given application. In this section, we review the most important figures of merit that directly impact waveform fidelity, bandwidth, and timing performance.

Vertical Resolution

The vertical resolution of an AWG determines how accurately the amplitude of a waveform can be represented. For example, a 10-bit AWG with a 2-V peak-to-peak output range can resolve voltage steps as small as approximately 2 mV ($2V/2^{10}$). Each additional bit of vertical resolution ideally doubles the amplitude resolution.

In practice, however, the effective amplitude accuracy is governed by the effective number of bits (ENOB), which accounts for nonidealities such as noise, distortion, and clock jitter. As a result, the ENOB is typically lower than the nominal resolution specified in the datasheet.

ENOB

In practical AWGs, impairments such as thermal noise, quantization noise, harmonic distortion, and spurious tones reduce the usable resolution of the DAC. As a result, the effective number of bits (ENOB) is typically lower than the nominal DAC resolution specified in the datasheet.

Figure 6.9(a) illustrates the concept of ENOB. The ideal waveform is shown in blue, while the actual output of the AWG or DAC is shown in red, including the combined effects of noise and distortion. The vertical spread of the signal around the ideal waveform represents amplitude uncertainty, which directly limits the effective resolution. Although a converter may have a high nominal bit depth, instantaneous output deviations from the ideal signal reduce the number of usable bits. ENOB quantifies how many bits of resolution remain after these impairments are taken into account.

ENOB can be calculated from the measured signal-to-noise-and-distortion ratio (SINAD) using

$$\text{ENOB} = \frac{\text{SINAD} - 1.76}{6.02} \tag{6.10}$$

where SINAD is expressed in decibels. We discussed SINAD in Chapter 4.

Moreover, ENOB generally degrades with increasing output frequency, as illustrated in Figure 6.9(b), where a DAC with a nominal resolution of 14 bits may exhibit an ENOB of only 8 bits at 2.5 GHz.

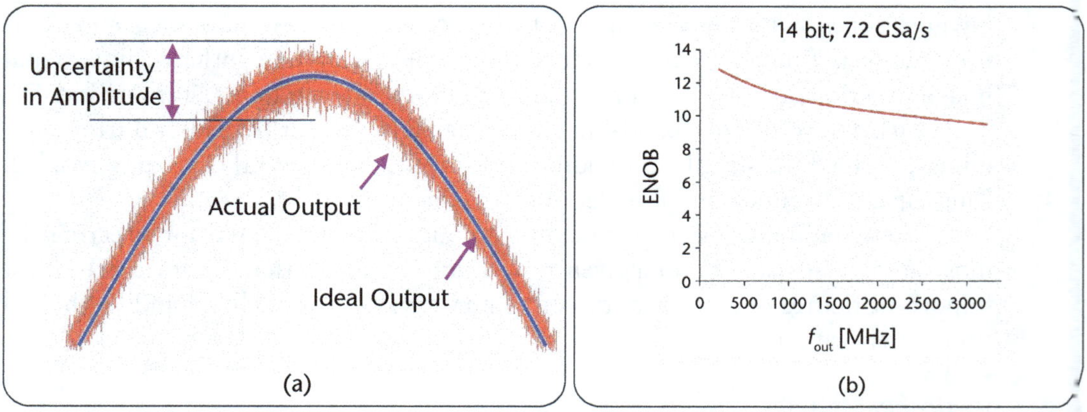

Figure 6.9 (a) Illustration of ENOB in the time domain. (b) ENOB as a function of output frequency for a 14-bit, 7.2-GSa/s system, showing the degradation of effective resolution at higher frequencies due to noise, distortion, and clock jitter.

ENOB provides a more realistic measure of amplitude resolution than the nominal DAC bit depth, as it captures the combined effects of noise and distortion. In addition, sampling clock jitter plays a critical role in limiting ENOB at high frequencies and should be carefully considered to ensure accurate waveform reproduction.

Memory Size

Memory size determines how many waveform samples can be stored and replayed by the AWG. Sufficient memory depth is especially important when operating at high sample rates or when long, nonrepeating waveforms are required. Memory size is typically specified in gigasamples (GSa).

Larger memory enables longer waveform records, higher time resolution, and greater flexibility in waveform design, particularly for complex pulse sequences and long modulation patterns.

Sample Rate

The sample rate of an AWG, specified in gigasamples per second (GSa/s), defines how frequently samples are delivered to the DAC. A higher sample rate improves time resolution and allows higher-frequency content to be generated, but it also increases memory consumption.

The play time (or waveform duration) of an AWG is given by:

$$\text{Play time} = \frac{\text{Memory size}}{\text{Sample rate}} \tag{6.11}$$

For example, a memory size of 1 GSa and a sample rate of 8 GSa/s results in a play time of 0.125 second. Therefore, increasing the sample rate requires a proportional increase in memory size to maintain the same waveform duration.

Bandwidth

The usable bandwidth of an AWG is limited by both the sampling rate and the analog reconstruction filter following the DAC. According to the Nyquist criterion, frequency components up to one-half of the sample rate can, in principle, be generated; for example, a sample rate of 4 GSa/s corresponds to a Nyquist frequency of 2 GHz.

In practice, the usable bandwidth is further constrained by the intrinsic frequency roll-off of the DAC, primarily due to zero-order hold behavior and by the analog reconstruction filter used to smooth the output waveform. These effects typically reduce the achievable bandwidth by approximately 10% relative to the theoretical Nyquist limit. As a practical rule of thumb, the maximum usable output frequency of an AWG is usually around 40%–50% of the sampling rate. For instance, an AWG operating at 10 GSa/s generally provides a usable bandwidth of approximately 4–5 GHz, depending on DAC performance and output filtering.

SFDR

We defined the SFDR for RF systems in Chapter 4. For AWGs, it is the distance in decibels from the desired frequency to the highest visible spur or harmonic within the stated bandwidth.

With this, we conclude the discussion of the first critical RF operation: signal generation. In the next section, we turn our attention to signal transmission, examining how generated signals propagate through transmission lines, interconnects, and free space.

6.5 Signal Transmission

In this section, we focus on signal transmission in guided structures such as cables and waveguides. Transmission in free space was discussed in Chapter 4, where topics such as free-space path loss, propagation loss, and related mechanisms were introduced.

To understand RF circuit behavior, it is essential to distinguish between lumped-element and distributed-element transmission line models. This distinction forms the foundation of RF circuit theory and defines the boundary at which classical low-frequency circuit assumptions cease to be valid.

Table 6.6 compares the key characteristics of lumped-element and distributed-element representations, including physical length relative to wavelength, phase variation along the structure, and the appropriate modeling approach. Figure 6.10 provides a conceptual illustration of these differences, emphasizing how voltage and current are treated as spatially uniform quantities in lumped-element models, but as position-dependent quantities in distributed-element transmission lines.

This distinction becomes increasingly important as operating frequency rises, where circuit dimensions approach a significant fraction of the wavelength and distributed effects can no longer be neglected.

In transmission lines, a fundamental concept is that of propagation modes. A propagation mode describes how the electric field E, magnetic field H, and the direction of wave propagation are oriented with respect to one another as electromagnetic energy travels along a structure. Different geometries and boundary conditions support different field configurations, leading to distinct modes of propagation.

Broadly, transmission lines can be categorized into two main groups: transverse electromagnetic (TEM) transmission lines and non-TEM transmission lines. In TEM structures, both the electric and magnetic fields are entirely transverse to

Table 6.6 Comparison of Lumped-Element and Distributed-Element Transmission Lines

Lumped-Element Transmission Line	*Distributed-Element Transmission Line*
Electrically short: $E = l/\lambda << 1$	Electrically long: $E = l/\lambda \sim 1$ (physical length comparable to wavelength)
Voltage and current are assumed constant along the structure	Voltage and current vary with position along the line
Wave propagation effects are negligible	Wave nature of voltage and current must be considered
Phase variation along the structure is ignored	Phase varies continuously along the line
Impedance matching is generally not required for efficient power transfer	Matching to the characteristic impedance Z_0 of the line is critical to minimize reflections and maximize power transfer

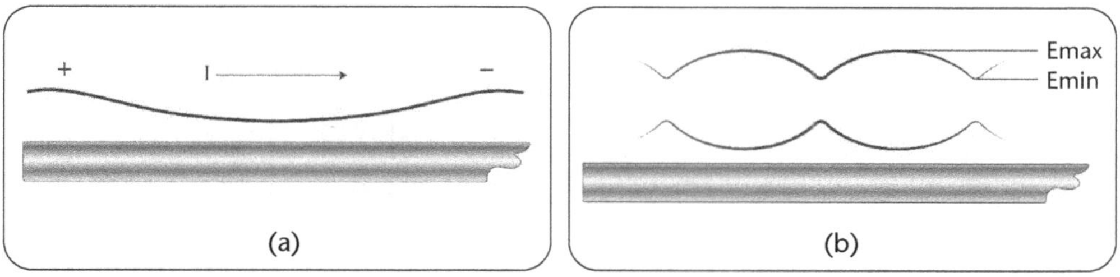

Figure 6.10 Current distribution on: (a) a lumped element transmission line and (b) a distributed element transmission line.

the direction of propagation. In contrast, non-TEM structures support modes in which at least one field component has a longitudinal component along the propagation direction.

In the following sections, we examine these two categories in more detail, discuss where each type arises in practical RF and microwave hardware, and highlight the implications for bandwidth, dispersion, and circuit design.

6.5.1 TEM Transmission Lines

This section focuses on a class of transmission lines for which classical circuit analysis techniques remain applicable, namely, TEM transmission lines. TEM lines are characterized by the presence of two conductors and include structures such as coaxial lines and parallel-plate waveguides. For these transmission lines, circuit-based modeling provides a powerful and intuitive framework for analysis and design. In contrast, other guided structures, such as rectangular waveguides, do not support TEM modes and therefore require full electromagnetic field analysis based on Maxwell's equations.

For TEM propagation along the z-direction, the electric and magnetic fields have no longitudinal components (i.e., $E_z = H_z = 0$). As a result, the governing wave equation reduces to the Laplace equation in the transverse cross-section of the transmission line. This reduction effectively transforms the problem into an electrostatic one, in which the electric field can be expressed as the gradient of a scalar potential $\Phi(x, y)$. The potential satisfies the transverse Laplace equation $\nabla_t^2\Phi(x, y) = 0$, which allows a unique voltage to be defined between the two conductors at every point along the line. This property is fundamental, as it enables the use of circuit quantities such as voltage and current and eliminates the need to directly solve Maxwell's equations for many practical analyses.

A TEM transmission line can therefore be modeled as an infinite cascade of infinitesimally short transmission line segments of length Δz, as illustrated in Figure 6.11. Each segment has a sufficiently small electrical length to be treated as a lumped circuit element, allowing the application of Kirchhoff's Voltage Law (KVL) and Kirchhoff's Current Law (KCL). This distributed-circuit representation forms the basis for transmission line equations and underpins much of RF circuit theory.

It can be shown that the traveling-wave solutions to the one-dimensional transmission-line wave equations for voltage and current are given by [2]

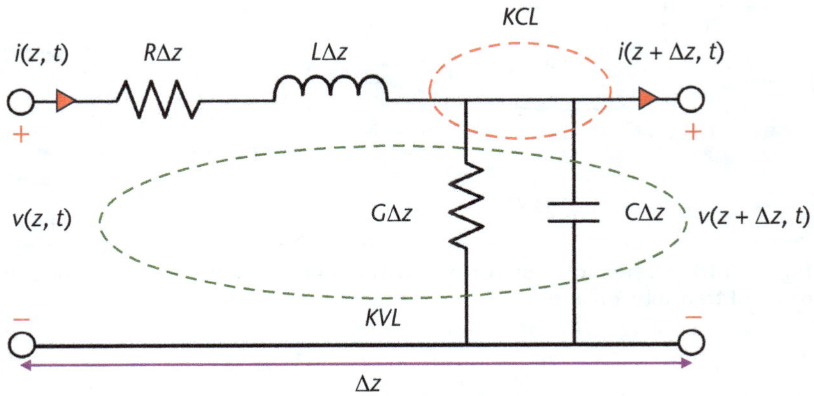

Figure 6.11 Distributed-element unit cell of a transmission line of length Δz, consisting of series per-unit-length resistance R and inductance L, and shunt per-unit-length conductance G and capacitance C. The indicated KVL loop and KCL node illustrate how Kirchhoff's laws are applied to derive the transmission line equations.

$$\begin{aligned} V(z) &= V_0^+ e^{-\gamma z} + V_0^- e^{\gamma z} \\ I(z) &= I_0^+ e^{-\gamma z} + I_0^- e^{\gamma z} \end{aligned} \tag{6.12}$$

where the term $e^{-\gamma z}$ represents a wave propagating in the $+z$ direction, while $e^{\gamma z}$ represents a wave propagating in the $-z$ direction.

The complex propagation constant γ is defined as

$$\gamma = \alpha + j\beta = \sqrt{ZY} = \sqrt{(R + j\omega L)(G + j\omega C)} \tag{6.13}$$

where α is the attenuation constant and β is the phase constant. The attenuation constant quantifies the exponential decay of voltage amplitude along the transmission line, whereas the phase constant describes the rate at which the signal phase progresses with distance.

The characteristic impedance of the transmission line is defined as

$$Z_0 = \frac{R + j\omega L}{\gamma} = \sqrt{\frac{R + j\omega L}{G + j\omega C}} \tag{6.14}$$

and relates the voltage and current amplitudes of the forward- and backward-traveling waves as

$$\frac{V_0^+}{I_0^+} = Z_0 \quad \text{and} \quad \frac{V_0^-}{I_0^-} = -Z_0 \tag{6.15}$$

Table 6.7 summarizes the attenuation constant, phase constant, phase velocity, and characteristic impedance for lossless, low-loss, and distortionless transmission lines.

We now highlight the three defining characteristics of the TEM mode.

Table 6.7 Attenuation and Phase Constants for Different Transmission-Line Conditions

Condition	*Circuit Conditions*	*α (Attenuation Constant)*	*β (Phase Constant)*	*v_p (Phase Velocity)*	Z_C
Lossless	$R = 0$ $G = 0$	0	$\omega\sqrt{LC}$	$1/\sqrt{LC}$	$\sqrt{L/C}$
Low-loss	$R << \omega L$, $G << \omega C$	$\frac{1}{2}\left(\frac{R}{Z_0}+\frac{G}{Y_0}\right)$ where $Y_0 = 1/Z_0$	$\approx \omega\sqrt{LC}$	$\approx 1/\sqrt{LC}$	$\sqrt{L/C}$
Distortionless	$RC = LG$	$\sqrt{RG} = R\sqrt{C/L}$	$\omega\sqrt{LC}$	$1/\sqrt{LC}$	$\sqrt{L/C}$

1. *No cutoff frequency:* TEM waves can propagate from DC to arbitrarily high frequencies.
2. *Frequency-independent phase velocity:* Phase velocity does not vary with frequency, resulting in no dispersion and, consequently, no waveform distortion.
3. *Well-defined voltage and current:* A unique voltage and current can be defined at every point along the line, enabling the use of classical circuit analysis techniques.

In non-TEM transmission lines, the voltage between two points becomes path-dependent, preventing the definition of a unique voltage between conductors. As a result, classical circuit concepts lose their strict physical meaning. Nevertheless, circuit-based analysis techniques can still be applied by introducing appropriate assumptions and approximations.

6.5.2 Non-TEM Transmission Lines

Propagation modes are classified based on the orientation of the electric and magnetic field vectors relative to the direction of wave propagation. The three most commonly encountered propagation modes in guided RF structures are TEM, transverse electric (TE), and transverse magnetic (TM) modes. Table 6.8 summarizes the key characteristics of these modes.

Table 6.8 Propagation Modes in Transmission Structures

Propagation Mode	*Description*	*Transmission Structure*	*Characteristics*
TEM	Electric and magnetic fields are entirely transverse to the direction of propagation	Two-conductor transmission lines (e.g., coaxial line, stripline, parallel-plate line)	No cutoff frequency, phase velocity is constant (nondispersive)
TE	No electric field component in the direction of propagation ($E_z = 0$)	Waveguides and higher-order modes in two-conductor lines	Has a cutoff frequency, phase velocity is frequency-dependent
TM	No magnetic field component in the direction of propagation ($H_z = 0$)	Waveguides and higher-order modes in two-conductor lines	Has a cutoff frequency, phase velocity is frequency-dependent

In addition to these pure modes, hybrid modes, in which both longitudinal electric and magnetic field components are present, can propagate in certain transmission lines such as microstrip lines. While hybrid modes are prevalent in practical planar structures, they are generally less intuitive and are often approximated using quasi-TEM assumptions for circuit-based analysis.

TE and TM modes can propagate in both two-conductor and single-conductor structures, most notably in hollow waveguides. The letter following the "T" specifies which field component lacks a longitudinal component along the direction of propagation. A TE mode has no electric field component in the propagation direction, whereas a TM mode has no magnetic field component in that direction.

Figure 6.12(a) illustrates the propagation of the TEM mode in a parallel-plate waveguide, as well as TE and TM modes in a rectangular waveguide. In the TE mode, the electromagnetic fields undergo reflections from the side walls of the waveguide while propagating longitudinally along the structure. In contrast, for the TM mode, the fields reflect primarily from the upper and lower walls as the wave propagates in the direction of travel.

The TE and TM modes exhibit a cutoff frequency f_c, meaning that only modes with an operating frequency $f > f_c$ can propagate. For frequencies below the cutoff ($f < f_c$), fields do not propagate but instead decay exponentially away from the source, forming evanescent modes. The cutoff frequency is formally obtained by solving Maxwell's equations in the cross-section of the transmission line or waveguide, subject to the appropriate boundary conditions.

An intuitive understanding of cutoff behavior can be gained by considering the relationship between wavelength and the physical dimensions of the guiding structure. For a mode to propagate, the electromagnetic field distribution must "fit" within the waveguide cross-section. More precisely, the transverse dimensions of the structure must be able to support an integer number of half-wavelength variations of the field. In rectangular waveguides, this condition is evaluated independently along the x- and y-dimensions of the cross-section, as illustrated in Figure 6.12(b).

The subscripts in the mode designation indicate the number of half-wavelength field variations at cutoff. For example, the $TE_{m,n}$ mode corresponds to m half-wavelength variations of the electric field across the x-dimension and n half-wavelength

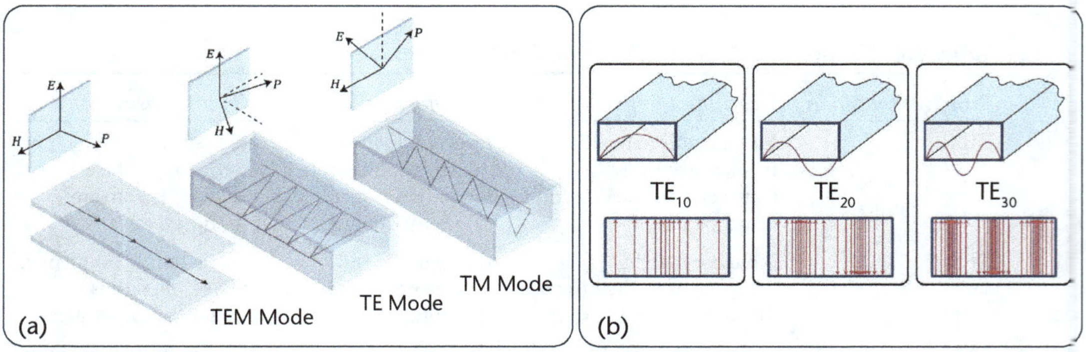

Figure 6.12 (a) Propagation modes in guided structures: TEM mode with transverse electric and magnetic fields; TE mode with fields reflecting from the side walls; TM mode with fields reflecting from the top and bottom walls; and (b) electric field distributions of higher-order modes in a rectangular waveguide, with subscripts denoting half-wavelength variations along the x- and y-dimensions.

variations across the y-dimension of a rectangular waveguide. Figure 6.12(b) illustrates such modes along with their corresponding electric field distributions. As the operating frequency increases (and the wavelength decreases), the same number of half-wavelength variations can be accommodated within a smaller cross-section.

As a specific example, the dominant TE_{10} mode of a rectangular waveguide exhibits one half-wavelength variation in the x-direction and no field variation in the y-direction at cutoff, as shown in Figure 6.12(b). This mode has the lowest cutoff frequency and is therefore the first mode to propagate as frequency increases.

For circular waveguides, the field behavior is described in terms of radial and circumferential variations. At the cutoff, the dominant TE_{11} mode exhibits a circumferential electric field distribution with no angular variation and a single field maximum at the center of the waveguide, with the field magnitude going to 0 at the conducting boundary.

Next we will give the cutoff frequency expressions for TE and TM modes and examine the frequency-dependent phase velocity associated with these modes. Unlike TEM waves, whose phase velocity is frequency-independent, TE and TM waves exhibit dispersion. In such cases, group velocity can be defined to describe the propagation speed of a modulated signal, provided the signal bandwidth is sufficiently narrow.

Useful Bandwidth of a Transmission Line

For most RF applications, single-mode propagation is preferred, since the simultaneous propagation of multiple modes can lead to undesirable effects such as signal distortion, increased reflections, and unpredictable impedance behavior. However, multimode propagation is intentionally exploited in certain applications, such as specific waveguide filters. In such cases, the waveguide is referred to as overmoded. The single-mode bandwidth of a transmission line is defined as the frequency range between the cutoff frequencies of two consecutive modes.

As discussed earlier, two-conductor transmission lines, such as coaxial cables, can support TE and TM modes in addition to the TEM mode. For a coaxial cable, the single-mode bandwidth extends from DC up to the cutoff frequency of the first higher-order mode, which is the TE_{11} mode. The TE_{11} mode begins to propagate when the cutoff wavelength λ_c in an air-filled coaxial cable is equal to the average circumference of the structure, given by

$$\lambda_c = 2\pi\left(\frac{a+b}{2}\right) = \pi(a+b) \tag{6.16}$$

where a and b represent the inner and outer conductor radii, respectively. If the coaxial cable is filled with a dielectric, the cutoff wavelength is modified by the relative permittivity ε_r:

$$\lambda_c = \pi(a+b)\sqrt{\varepsilon_r} \tag{6.17}$$

The cutoff frequency of the first higher-order mode is obtained using $\lambda_c = c/f_c$, yielding

$$f_c = \frac{c}{\pi(a+b)\sqrt{\varepsilon_r}} \tag{6.18}$$

The single-mode bandwidth of a coaxial cable is therefore limited by this cutoff frequency f_c, which depends strongly on the cable geometry and dielectric material. In practice, this cutoff frequency can range from approximately 1 GHz to over 100 GHz. Since frequency and wavelength are inversely proportional, increasing the operating frequency requires progressively smaller cable diameters and connector dimensions to avoid higher-order mode excitation. However, reducing the physical size of the coaxial structure leads to higher losses and reduced power handling capability, introducing important design trade-offs.

For rectangular waveguides, the useful single-mode bandwidth is typically slightly less than one octave, corresponding to a frequency ratio of approximately 2:1. This limitation arises because the TE_{20} mode begins to propagate at twice the cutoff frequency of the dominant TE_{10} mode. To extend the operational bandwidth, ridged waveguides are often employed, as illustrated in Figure 6.13(a).

The propagation of higher-order modes introduces several challenges. Their presence causes variations in the transmission line's characteristic impedance, leading to increased reflections due to impedance mismatch. Since impedance consists of both resistive and reactive components, higher-order modes perturb both. As illustrated in Figure 6.13(b), the onset of these modes introduces a resistive perturbation to the impedance. In a coaxial cable, for example, when the TE_{11} mode begins to propagate, the reactive energy stored in the higher-order field components alters the reactive part of the impedance, resulting in significant mismatch and elevated reflection levels.

For optimal RF performance, cables and connectors should maintain a constant characteristic impedance along their entire length, ensuring a flat frequency response over the widest possible bandwidth and minimizing reflections.

6.5.3 Types of Transmission Lines

Selecting an appropriate transmission line for a given application requires consideration of several factors, including bandwidth, physical size, ease of fabrication and

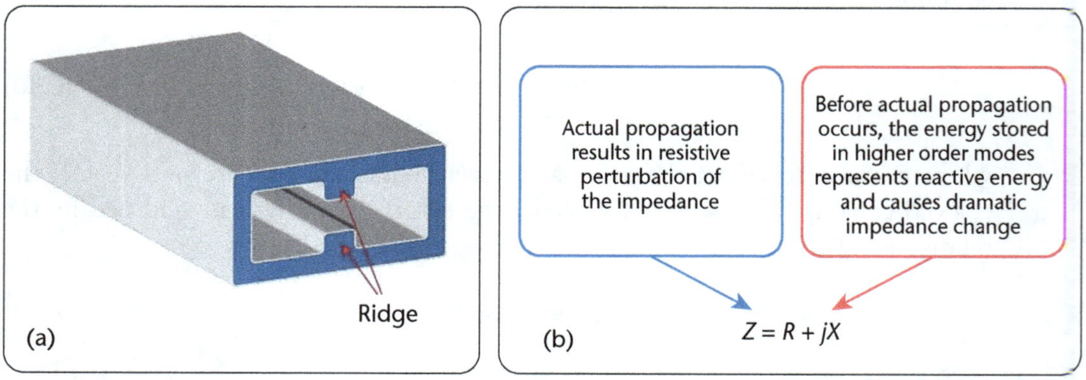

Figure 6.13 (a) Ridged waveguide. (b) Effect of propagating and cutoff (evanescent) modes on the waveguide impedance.

integration, and power handling capability. This section reviews the most commonly used transmission lines and highlights the key parameters that guide their selection.

The following factors are particularly important when choosing a cable or transmission line:

- *Frequency range of operation:* The frequency range of operation defines the range over which signal propagation remains free of higher-order (non-TEM) modes. When selecting a transmission line, the upper frequency limit should exceed the maximum operating frequency of the system to ensure single-mode propagation.
- *Return loss:* Return loss characterizes how well the cable impedance matches the nominal characteristic impedance of 50Ω. Frequency-dependent impedance variations lead to reflections that degrade return loss and can impact system performance. Most high-quality RF cables exhibit return loss values of 30 dB or better across their specified operating bandwidth. An example of a cable return loss measurement is shown in Figure 6.14(a).
- *Insertion loss:* Insertion loss quantifies the attenuation experienced by a signal as it propagates along the cable. As discussed previously, insertion loss generally increases with frequency and arises from a combination of conductor, dielectric, and radiation losses. For example, Figure 6.14(b) shows the insertion loss of a representative cable, which is approximately 0.2 dB at low frequencies and increases to about 1.8 dB at 18 GHz. In practice, insertion loss is often specified in dB per meter, allowing straightforward estimation for a given cable length.
- *Power handling capability:* Power-handling capability specifies the maximum RF power that a transmission line can safely carry without excessive heating, dielectric breakdown, or degradation. This parameter is critical in high-power applications such as LTE base stations, satellite communications, and radar systems.

Coaxial Cable

Unshielded cables, such as twisted pairs, become increasingly disadvantageous at higher frequencies, as they begin to radiate and suffer from electromagnetic

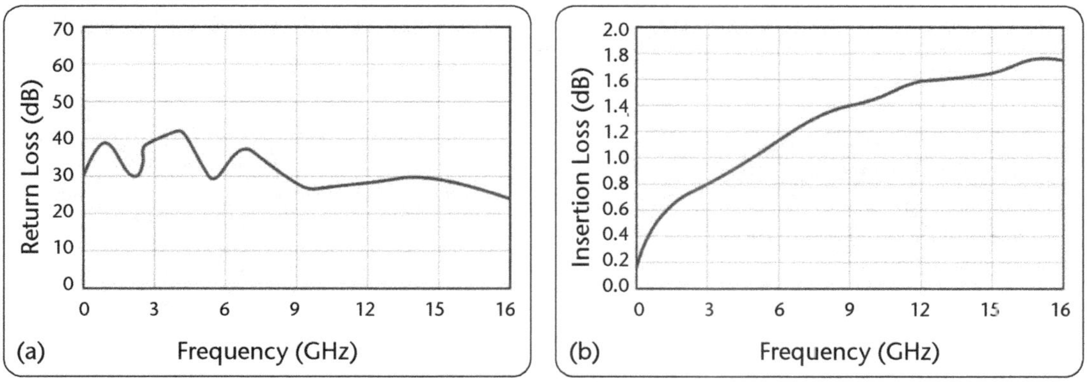

Figure 6.14 RF cable: (a) return loss, and (b) insertion loss.

interference. In contrast, the coaxial transmission line does not experience radiation loss under ideal conditions and inherently provides self-shielding, preventing both radiation of the signal and susceptibility to external interference. The physical structure of a coaxial cable is illustrated in Figure 6.15(a), while the corresponding electromagnetic field configuration is shown in Figure 6.15(b).

As a result, coaxial cables are widely used in RF systems where low radiation loss, high noise immunity, and robust interference rejection are required. This is particularly important in test and measurement setups and in the interconnection of RF subsystems, where signal integrity and repeatability are critical.

As discussed in Chapter 4, resistance and conductance, both of which contribute to signal loss, exhibit frequency dependence. Coaxial cables commonly employ dielectric materials such as PTFE (Teflon), while air dielectric provides the lowest possible attenuation. Although the use of air dielectric and larger physical dimensions reduces loss, these design choices must be balanced against higher-order mode limitations, as discussed previously. Consequently, a fundamental trade-off exists between minimizing attenuation and maintaining single-mode propagation.

In many RF systems, standard coaxial cables introduce unwanted phase drift when subjected to temperature variations or mechanical bending. This effect is particularly problematic in applications where phase coherence and repeatability are critical, such as phased arrays, vector network analyzers (VNAs), calibration setups, and precision measurement systems. In these cases, phase-stable cables are employed to mitigate phase variations.

The primary benefits of phase-stable cables include:

- Maintaining stable phase with temperature changes (low ppm/°C);
- Minimizing phase shift under flexing or rerouting;
- Improving measurement repeatability and reducing the need for frequent recalibration.

When selecting phase-stable cables, the following parameters are especially critical:

- Phase-versus-temperature coefficient (lower is better);
- Phase stability under flexing, typically specified in degrees per defined bend radius or length;

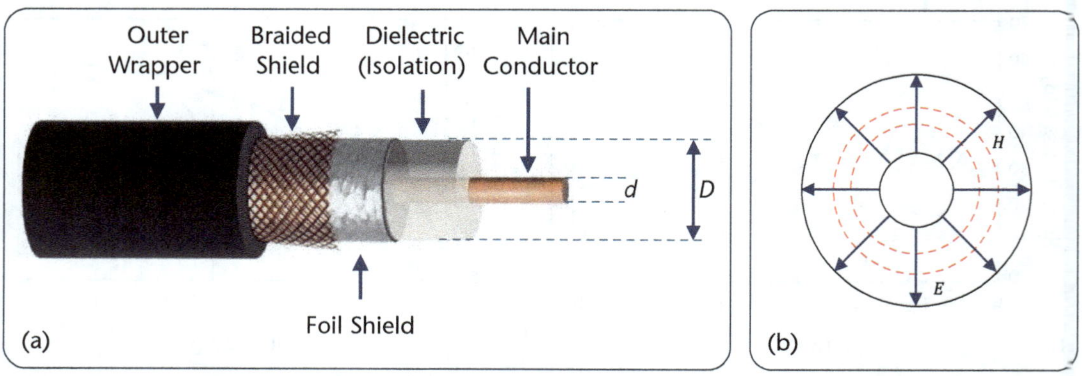

Figure 6.15 (a) Structure of a coaxial cable. (b) Field configuration in the cross-section of a coaxial cable.

- Return loss and insertion loss stability across the operating frequency band;
- Mechanical ruggedness, particularly for production environments or field deployment.

Microstrip Line

A microstrip line is a planar transmission line consisting of a ground plane, a dielectric substrate, and a signal conductor located on top of the dielectric, as illustrated in Figure 6.16(a). The widespread use of microstrip lines arises from their ease of fabrication, low cost, and high level of integration, making them well-suited for implementing integrated microwave structures that combine both active and passive components on a single substrate.

Despite these advantages, microstrip lines exhibit higher dielectric and radiation losses than enclosed structures such as coaxial cables and hollow waveguides. As the operating frequency increases, radiation losses become increasingly significant, which can limit performance in high-frequency and low-loss applications. For this reason, stripline and waveguide structures are often preferred at higher frequencies when minimizing radiation loss and preserving signal integrity are critical.

Microstrip lines have a portion of their electric field lines in air and the majority within the dielectric region between the strip conductor and the ground plane, as illustrated in Figure 6.16(a). In contrast, a stripline structure, shown in Figure 6.16(b), confines the electromagnetic fields entirely within a homogeneous dielectric medium.

Because of this inhomogeneous field distribution, a pure TEM wave cannot be supported by a microstrip line. The reason is that the phase velocity of TEM fields in the dielectric region, $c/\sqrt{\varepsilon_r}$, differs from that in the air region, c, making phase matching at the dielectric-air interface impossible. Consequently, the fields supported by a microstrip line constitute a hybrid TE-TM mode, which, in general, requires more advanced analysis techniques beyond the scope of this discussion.

In most practical applications, however, the dielectric substrate thickness d is electrically thin ($d << \lambda$), resulting in quasi-TEM propagation. Under these conditions, the field distribution closely resembles that of the static (DC) case. As a result, accurate approximations for the phase velocity, propagation constant, and characteristic impedance can be obtained using static or quasi-static analysis.

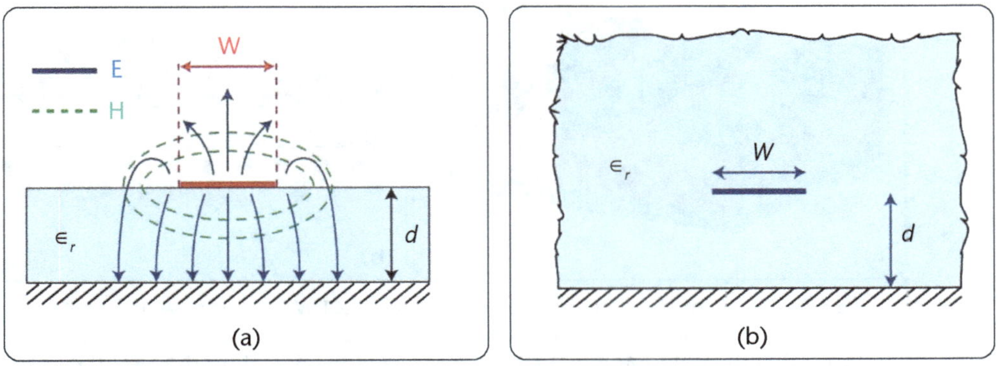

Figure 6.16 (a) Geometry of a microstrip line. (b) Equivalent geometry with a homogeneous medium of effective relative permittivity ε_e.

To account for the mixed air-dielectric field distribution, an effective dielectric constant ε_{eff} is introduced. This parameter represents the dielectric constant of an equivalent homogeneous medium that replaces the combined air and dielectric regions of the microstrip line. Since some field lines reside in air while others are confined to the dielectric, the effective dielectric constant satisfies $1 < \varepsilon_{eff} < \varepsilon_r$ and depends on several factors, including the substrate dielectric constant, substrate thickness, conductor width, and operating frequency [2].

Coplanar Waveguide (CPW)

A CPW is a planar transmission line consisting of a dielectric substrate with a signal trace on its top surface and two ground planes located on either side of the signal trace. A related variant, known as a grounded CPW (GCPW), includes an additional ground plane on the underside of the substrate, as illustrated in Figure 6.17(a). A CPW structure incorporating ground vias is shown in Figure 6.17(b), where the vias connect the top ground conductors to the bottom ground plane. Similar to microstrip lines, CPW structures support a low-loss quasi-TEM mode, since pure TEM propagation is not possible due to the inhomogeneous field distribution.

The CPW is widely used as a transmission line in microwave integrated circuits (MICs), monolithic microwave integrated circuits (MMICs), and superconducting qubit circuits. Compared to microstrip lines, a 50Ω CPW typically requires smaller physical dimensions, making it well-suited for compact and high-density layouts. In addition, the uniplanar geometry of CPW enables straightforward integration of both series and shunt components on the same metal layer, eliminating the need for via holes, which would otherwise introduce parasitic inductance and potentially lead to impedance mismatch.

Unlike microstrip lines, where a significant portion of the electromagnetic field is confined within the substrate, CPW structures confine a larger fraction of the field near the surface. This reduces substrate coupling and improves isolation and overall performance, particularly in multilayer and cryogenic designs.

For symmetric CPW structures, two propagation modes may exist: the even mode, commonly referred to as the CPW mode, and the odd mode, also known

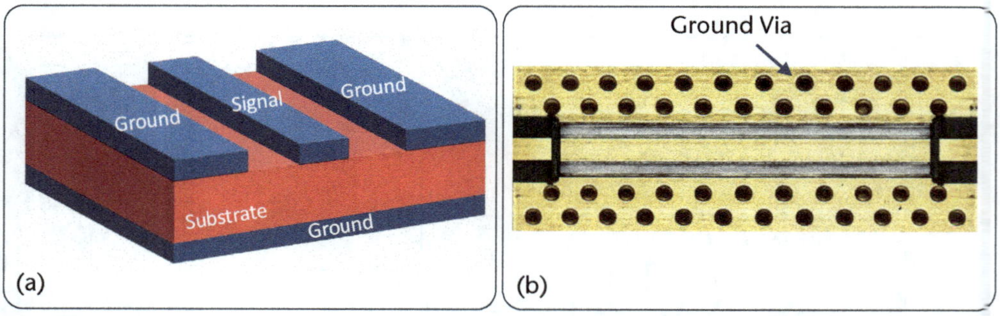

Figure 6.17 (a) Schematic 3D view illustrating the GCPW geometry with a central signal conductor flanked by ground conductors on the same layer and a continuous ground plane beneath the substrate. (b) Photograph of a fabricated CPW implementation showing the via-fence used to connect the top ground conductors to the bottom ground plane, providing improved ground continuity and electromagnetic confinement.

as the slotline mode, as illustrated in Figures 6.18(a, b) [1]. While these modes can propagate independently when selectively excited, structural asymmetries and discontinuities can cause mode coupling, leading to unwanted signal degradation.

In the presence of asymmetry or discontinuities, parasitic slotline modes can be excited in CPW structures. Asymmetrical CPW discontinuities, such as bends in microwave resonators or T-junctions, can couple energy into the odd (slotline) mode, leading to unwanted propagation. To suppress the slotline mode and ensure that the ground planes on either side of a CPW line remain at the same electrical potential, it is essential to connect them using an air bridge, as shown in Figure 6.18(c). In superconducting quantum circuits, air bridges are typically fabricated using a wire bonding machine.

The calculation of characteristic impedance and effective permittivity for CPW structures is nontrivial and generally involves elliptic integrals and hyperbolic functions [2].

Hollow Waveguides

Hollow waveguides offer low loss and high power-handling capability. However, they are bulky, particularly at lower frequencies, and expensive. As a result, these transmission lines are most commonly employed in radar and satellite communication systems, where low attenuation and high-power handling are critical to system performance. Figure 6.19 illustrates the probe-coupling method used to excite a rectangular waveguide.

Waveguides support only TE and TM modes. Each mode exhibits a cutoff frequency $f_{c_{mn}}$, which for a rectangular waveguide is given by [3]

$$f_{c_{mn}} = \frac{1}{2\pi\sqrt{\mu\varepsilon}}\sqrt{\left(\frac{m\pi}{a}\right)^2 + \left(\frac{n\pi}{b}\right)^2} \tag{6.19}$$

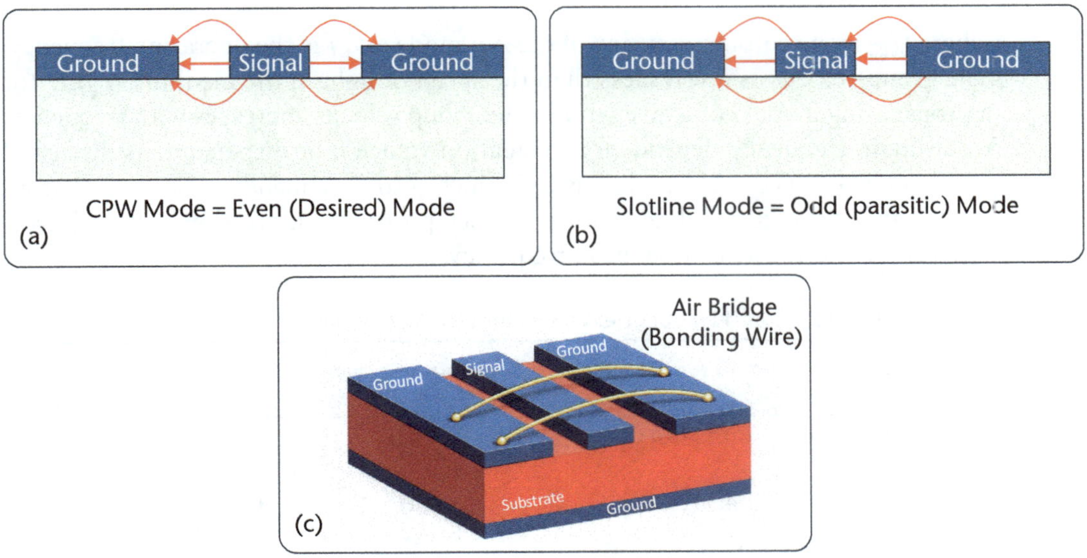

Figure 6.18 (a) CPW mode, (b) coupled slot line mode, and (c) air bridges used to connect the grounds on the sides of the signal line to suppress slotline modes.

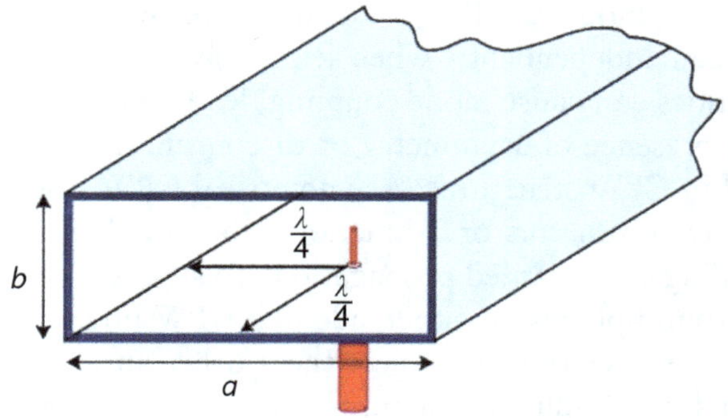

Figure 6.19 Excitation of a waveguide by using a probe.

where m and n denote the number of half-wavelength field variations in the x- (width) and y- (height) directions, respectively.

Rectangular waveguide sizes and performance characteristics are standardized using the WR designation system, established by the American National Standards Institute (ANSI) and the Institute of Electrical and Electronics Engineers (IEEE). In this system, the letters *WR* are followed by a number indicating the approximate inside width of the waveguide in hundredths of an inch. For example, a WR-90 waveguide has an internal width of approximately 0.9 inches. Table 6.9 lists several commonly used rectangular waveguides along with their dimensions and cutoff frequencies.

In a rectangular waveguide, the group velocity represents the speed at which energy, and therefore the envelope of a modulated signal, propagates along the structure. It is given by [3]

$$v_g = \frac{1}{\sqrt{\mu\varepsilon}}\sqrt{1-\left(\frac{f_{c_{mn}}}{f}\right)^2} \tag{6.20}$$

where $f_{c_{mn}}$ is the cutoff frequency of the mn mode and f is the operating frequency. The group velocity is always less than the speed of light in the medium $1/\sqrt{\mu\varepsilon}$ for a propagating mode. For a given mode, the group velocity increases with frequency, resulting in frequency-dependent propagation velocity, or dispersion, as discussed in Chapter 4. Since each mode has a distinct cutoff frequency, assuming that all modes are above cutoff, higher-order modes propagate more slowly than lower-order modes at the same operating frequency.

Table 6.9 Dimensions and Cutoff Frequencies of Rectangular Waveguides

Waveguide	*Recommended Frequency*	*Cutoff Frequency (Lowest Mode)*	*Cutoff Frequency (Next Mode)*	*A (in [mm])*	*B (in [mm])*
WR2300	0.32–0.45 GHz	0.257 GHz	0.513 GHz	23.0 [584.2]	11.5 [292.1]
WR137	5.85–8.20 GHz	4.301 GHz	8.603 GHz	1.372 [34.85]	0.622 [15.80]
WR90	8.20–12.40 GHz	6.557 GHz	13.114 GHz	0.900 [22.86]	0.400 [10.16]

As an example, consider a narrowband signal at 10 GHz propagating in a rectangular waveguide with cutoff frequencies f_{10} = 6.55 GHz and f_{20} = 13.11 GHz. Under these conditions, only the TE_{10} mode propagates. Using (6.20), the group velocity is approximately $v_g \approx 2.26 \times 10^8$ m/s, which corresponds to about 75.5% of the speed of light in free space.

Historically, SF_6 gas was used as a dielectric filling in high-power waveguides due to its high dielectric strength, approximately 2.5 times that of air, and excellent arc-quenching properties. These characteristics enabled higher RF power handling without electrical breakdown, making SF_6 common in radar transmitters, satellite earth stations, and other high-voltage microwave systems. However, SF_6 is an extremely potent greenhouse gas, with a global warming potential roughly 23,500 times that of CO_2, and can form toxic by-products during arcing. As a result, its use is now heavily restricted. Modern waveguide systems instead rely on pressurized dry air or vacuum environments to achieve comparable breakdown performance with reduced environmental impact.

6.5.4 Comparison of Transmission Lines

Several factors must be considered when selecting a transmission line, including bandwidth, loss, power-handling capability, physical size (volume and weight), ease of fabrication, and the potential for integration with active and passive devices. In addition to these technical considerations, nontechnical factors such as cost, availability, and manufacturing constraints should also be taken into account. Table 6.10 compares various transmission line technologies to facilitate the evaluation process.

6.5.5 RF Connectors

Connectors play a vital role as the interface between different transmission lines, enabling reliable interconnection and proper signal transmission. They are available in a wide variety of form factors, performance grades (such as metrology, instrument, and production), frequency ranges, power-handling capabilities, and characteristic impedances (commonly 50Ω and 75Ω), among other attributes. While a

Table 6.10 Comparison of Transmission Lines [2]

Characteristic	*Coaxial Line*	*Waveguide*	*Stripline*	*Microstrip*
Preferred mode	TEM	TE_{10}	TEM	Quasi-TEM
Other modes	TM, TE	TM, TE	TM, TE	Hybrid TM, TE
Dispersion	None	Medium	None	Low
Bandwidth	High	Low	High	High
Loss	Medium	Low	High	High
Power capacity	Medium	High	Low	Low
Physical size	Large	Large	Medium	Small
Ease of fabrication	Medium	Medium	Easy	Easy
Integration with PCBs/ ICs	Hard	Hard	Fair	Easy

comprehensive treatment of microwave connector design and operation is beyond the scope of this discussion, this section highlights practical considerations for commonly used connector types.

When selecting a connector, it is essential to ensure that it meets the specific requirements of the intended application. Table 6.11 summarizes several connector types along with their typical frequency ranges and applications.

Snap-on connectors, such as SMP and MMPX, enable quick and repeatable connections without the need for threaded coupling. This makes them well-suited for applications requiring frequent mating cycles or operating in space-constrained environments, such as sample holders in quantum chip setups. In contrast, threaded connectors, including SMA, N-type, and TNC, are generally bulkier but provide a more robust and mechanically secure connection, making them preferable in environments subject to vibration or mechanical stress.

In many practical situations, it is necessary to interconnect connectors of different types, identical types, or even connectors with incompatible genders. In such cases, RF adapters are employed. Figure 6.20 illustrates several common adapters, including N-to-SMA, BNC-to-SMA, SMA female-to-female (often referred to as a bullet adapter), and SMA male-to-male (commonly called a barrel adapter).

Adapters not only facilitate mechanical and electrical compatibility between connectors but are also frequently used to protect metrology-grade connectors, such as 3.5-mm and 2.92-mm interfaces, from wear or damage. For example, an N-to-SMA

Table 6.11 Connectors with Broad Applications in RF Systems

Connector Type	*Frequency Range*	*Typical Applications*
BNC	DC–4 GHz	Low-frequency and DC connections at room temperature, such as 10-MHz reference clocks, triggers, and control signals routed to breakout boxes
TNC	DC–11 GHz	Threaded variant of BNC offering improved RF performance and vibration resistance; commonly used for control and IF connections
U.FL	DC–6 GHz	Ultraminiature coaxial connector widely used on compact PCBs where space is limited and low mating-cycle operation is acceptable
MCX	DC–6 GHz	Compact snap-on connector used for control lines, compact instrumentation, and dense setups requiring quick mating and demating
N-type	DC–18 GHz	Rugged connector commonly used on RF test and measurement equipment, including spectrum analyzers and network analyzers
SMA/3.5 mm/2.92 mm/2.4 mm/1.85 mm	DC–18 GHz (SMA) up to DC–67 GHz (1.85 mm)	Widely used precision connector family for RF and microwave systems, suitable for most room-temperature and cryogenic interconnections
SMP	DC–40 GHz	Compact snap-on connector enabling fast, threadless connections; well-suited for space-constrained systems with frequent cable changes
MMPX	DC–65 GHz	Miniature snap-on connector supporting very high frequencies; preferred for dense assemblies where compactness and performance are critical

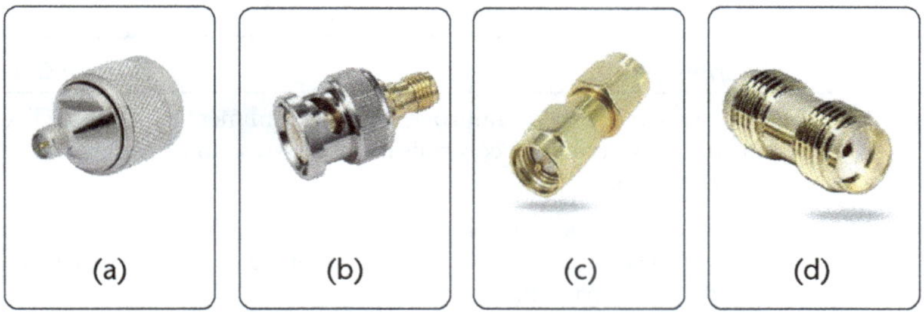

Figure 6.20 Common RF connectors and adapters: (a) N-to-SMA adapter; (b) BNC-to-SMA adapter; (c) SMA male-to-male adapter (barrel); and (d) SMA female-to-female adapter (bullet).

adapter is often installed at the input ports of network or spectrum analyzers to safeguard the instrument's precision connectors from repeated mating cycles.

As discussed earlier, the maximum usable bandwidth of a coaxial cable is fundamentally determined by its cross-sectional dimensions. As the physical size of the connector and cable decreases, the maximum operating frequency increases. For this reason, specialized connectors, such as 3.5 mm, 2.92 mm, 2.4 mm, and 1.85 mm, are used in high-frequency applications, with 1.85-mm connectors supporting operation up to 110 GHz. Metrology-grade connectors, such as the 3.5-mm interfaces commonly used on network and spectrum analyzer ports, offer superior mechanical tolerances, long-term stability, temperature performance, standing wave ratio (SWR), repeatability, insertion loss, and wear characteristics.

When handling RF connectors, it is essential to use a calibrated torque wrench during mating and demating to ensure proper torque application and prevent mechanical damage caused by overtightening. In addition, connectors should be visually inspected before use and cleaned when necessary, as part of routine connector maintenance to preserve measurement accuracy and long-term reliability.

6.6 RF Signal Processing

Up to this point, we have focused on the generation and transmission of RF signals. In practical systems, however, RF signals often require additional RF-domain processing before they can be effectively transmitted or detected. This processing includes intentional manipulation of the signal's amplitude, frequency, and phase, and should not be confused with digital signal processing (DSP) or baseband techniques.

RF signal processing techniques can be broadly classified into three categories: amplitude, phase, and frequency processing. The following sections examine these techniques in more detail and discuss their role and significance within an RF link.

6.7 Amplitude-Domain Techniques

Table 6.12 summarizes the primary amplitude-domain techniques used in RF systems. These operations modify signal amplitude for purposes such as control, protection, routing, biasing, and mode conversion (differential to single-ended). Each technique is elaborated upon in subsequent sections.

Table 6.12 Amplitude Manipulation Techniques

Operation	*Description*	*Relevant RF Components*
Amplification	Increases signal level to improve the SNR and meet detection requirements, compensates for losses along the signal chain	LNA, PA, driver amplifier
Attenuation	Tuning: adjusts signal level to the desired operating range, level setting: provides controlled, linear reduction of signal amplitude	Fixed attenuator, variable attenuator, step attenuator
Limiting	Protection: clamps signal amplitude above a defined threshold, survivability: protects sensitive components from high-power or transient signals	RF limiter, PIN-diode limiter, gas discharge limiter
Power combining/ dividing	Combines multiple signals into a single path or divides one signal into multiple paths	Wilkinson power divider/ combiner, resistive splitter
Power coupling	Samples a defined portion of an RF signal for monitoring, control, or feedback with minimal disturbance to the main signal path	Directional coupler, hybrid coupler, Magic-T
Power circulation/ isolation	Circulation: enables shared-antenna operation by routing signals directionally, isolation: suppresses reflections and reverse power flow	Circulator, isolator
Switching	Selects, routes, or disconnects RF paths dynamically	RF switch (PIN diode, FET, MEMS)
Bias injection/ extraction	Injects or extracts DC bias without disturbing the RF signal path	Bias-tee
DC blocking	Blocks DC while allowing RF signals to pass	DC block (series capacitor)
Mode conversion	Converts between differential and single-ended signal representations	Balun, RF transformer

6.7.1 Amplifier

Amplifiers play a vital role in increasing signal amplitude. Although amplification inherently introduces noise, which degrades the overall SNR, and may introduce nonlinear effects such as intermodulation products, as discussed in Chapter 4, amplification remains essential to ensure that the received signal exceeds the minimum detectable signal threshold and can be reliably detected. In certain applications, such as RF heating, plasma generation, or particle acceleration, amplifiers are used primarily to deliver specified RF power levels rather than to improve signal detectability.

Amplifiers are typically categorized based on their function within the RF signal chain:

- *LNA:* On the receiver side, an LNA is used to boost very weak signals while adding minimal noise. LNAs are characterized by a low noise figure, moderate gain, and low to medium output power.
- *PA:* On the transmitter side, a PA provides high output power and moderate to high gain to ensure that the transmitted signal maintains sufficient strength at the receiver.
- *Driver amplifier:* In some architectures, a driver amplifier is placed ahead of the PA to raise the signal level to a point where the PA can operate efficiently and deliver the required output power. Driver amplifiers typically provide

moderate gain and output power, higher than that of LNAs and lower than that of PAs.

A wide range of semiconductor technologies, such as GaAs and GaN, as well as amplifier topologies, including distributed amplifiers, are used in modern RF systems. Selected technologies and architectures, as well as a detailed treatment of amplifier implementation aspects, including efficiency, thermal management, cooling strategies, and spur optimization, are provided in Volume 2.

Amplifiers differ from passive devices in that they consume DC power and actively deliver RF energy into the circuit. This is achieved by converting DC bias power into RF power using active devices such as transistors. The primary parameter used to characterize an amplifier is its gain, defined as the ratio of output power to input power:

$$G = \frac{P_{\text{out}}}{P_{\text{in}}} \tag{6.21}$$

In practical RF systems, the system gain, often realized through a chain of amplifiers, typically ranges from approximately 20 dB to 50 dB, depending on the design and intended application. Narrowband amplifiers generally achieve higher gain than wideband amplifiers, as increasing bandwidth typically reduces the achievable gain. In many cases, the gain-bandwidth product can be considered approximately constant.

Key performance indicators for amplifiers include gain, noise figure, particularly critical for LNAs, third-order intercept point (IP_3), and the 1-dB compression point (P_{1dB}). Together, these parameters define the amplifier's linearity, noise, and dynamic range. Additional important characteristics include reverse isolation, which quantifies how effectively signal leakage from output to input is suppressed, and the stability factor, which indicates the amplifier's tendency to oscillate under varying source and load conditions.

Finally, a low return loss within the intended frequency band of operation is essential to minimize signal reflections and ensure efficient power transfer.

Figure 6.21(a) shows a typical biasing arrangement for a commercial off-the-shelf (COTS) LNA. A DC block at the LNA input prevents external DC from disturbing the amplifier's biasing or causing damage, while ensuring stable, RF-only signal coupling and optimal noise performance. At the output, a bias tee routes the externally supplied DC bias to the LNA while maintaining isolation from the RF signal path. Figure 6.21(b) shows the gain response of the LNA as a function of frequency.

In the following section, we introduce the DC block and the bias-tee and discuss their role in RF signal chains.

6.7.2 DC Block

DC blocks are passive RF components used to prevent DC voltages or currents from propagating along an RF signal path while allowing the RF signal to pass with minimal degradation.

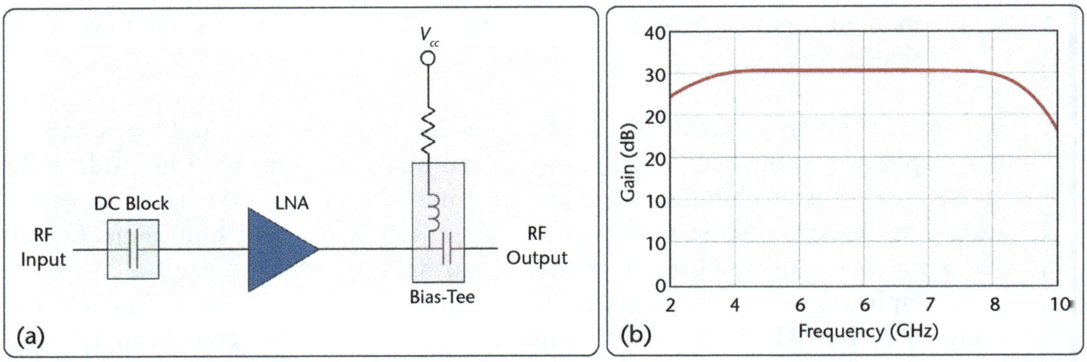

Figure 6.21 (a) Simplified LNA front-end architecture showing a DC block at the RF input and a bias-tee at the output for DC bias injection while preserving the RF signal path. (b) Representative frequency response of the LNA.

Standard DC blocks typically isolate only the center conductor, providing DC isolation between interconnected devices while maintaining a continuous ground reference through the outer conductor, as illustrated in Figure 6.22. This configuration is sufficient for many bench-top measurements and system integrations where ground potential differences are negligible.

However, in certain applications, such as rack-mounted test systems, shielded enclosures, high-power RF chains, or remote measurement setups, DC currents flowing through the outer conductor (cable shield) can create unintended ground loops. These ground current paths may introduce noise, measurement errors, or even damage sensitive equipment.

In such cases, dual DC blocks, which provide DC isolation on both the inner and outer conductors, are preferred. By fully interrupting DC paths while preserving RF continuity, dual DC blocks ensure complete ground isolation and improved RF signal integrity. A practical example is the Inmet 8039, which blocks DC on both conductors and is commonly used in precision RF test environments and isolated system architectures.

6.7.3 Bias-Tee

A diplexer is a passive component used to separate or combine signals occupying different frequency ranges. One important and widely used form of a diplexer in RF systems is the bias-tee. A bias-tee consists of an RF choke (inductor) and a DC block (capacitor), as illustrated in Figure 6.23(a).

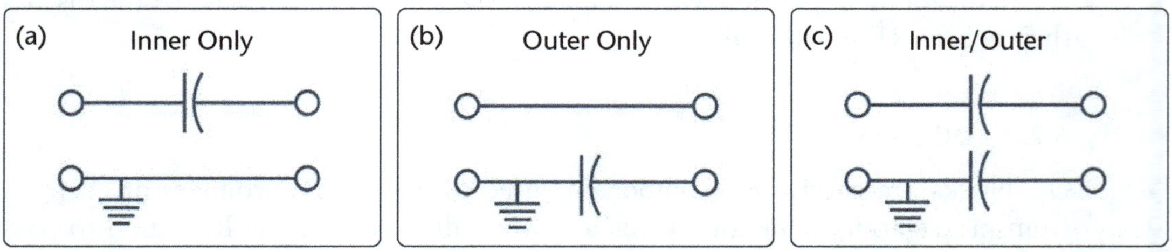

Figure 6.22 (a) DC block on the outer conductor only. (b) DC block on the inner conductor only. (c) DC block on the inner and outer conductors.

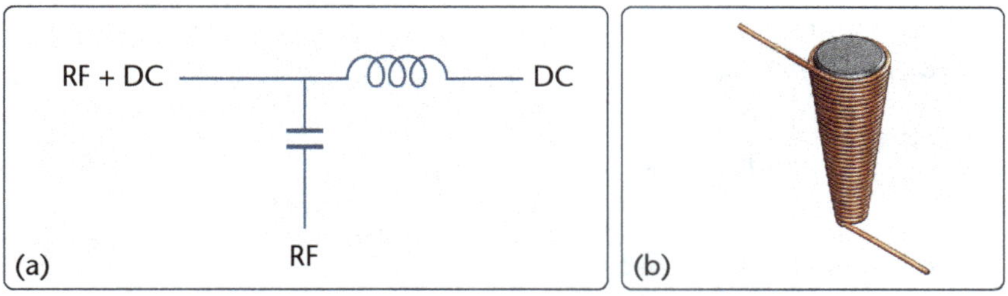

Figure 6.23 (a) Schematic of a bias-tee, and (b) conical inductor used in wideband bias-tees.

When operating as a combiner, the AC (RF) signal passes through the capacitive path, while the inductive path allows the DC signal to pass. At the common port of the bias-tee, the DC and RF signals are combined and delivered to the load. Conversely, when used as a separator, the bias-tee extracts the DC and RF components from a combined signal without significant interaction between the two paths.

Bias-tees are widely used in RF circuits to apply DC bias to active components, such as transistors, amplifiers, mixers, and PIN diodes, while allowing the RF signal to propagate with minimal attenuation or distortion.

Bias-tees can be implemented as passive or active devices. Passive bias-tees do not require an external power supply and rely solely on reactive components to separate or combine the DC and RF paths. Active bias-tees incorporate powered circuitry and may provide additional functionality such as adjustable DC bias voltage, current limiting, or enhanced low-frequency performance.

For wideband bias-tee implementations, conical inductors are often used, as shown in Figure 6.23(b). The conical geometry reduces the effects of stray capacitance by gradually varying the inductance along the structure. This effectively creates a series of narrowband inductive sections, resulting in a high RF impedance over a very wide bandwidth. As a result, conical inductors are well-suited for broadband RF applications where maintaining isolation between DC and RF over multiple octaves is critical.

Key operational parameters of a bias-tee include DC port isolation, RF insertion loss, usable frequency range, maximum allowable DC voltage, and maximum DC current. Proper selection of these parameters is essential to avoid unwanted coupling, signal degradation, or damage to downstream RF components.

6.7.4 Attenuators

In many RF systems, it is necessary to intentionally reduce the signal amplitude. Figure 6.24(a) illustrates the schematic of a π-attenuator. There are four primary reasons for introducing attenuation into an RF signal path [1]:

1. *Protection of sensitive devices:* Attenuators are often used to protect sensitive components or measurement equipment. For example, an attenuator may be placed at the input of a measurement instrument to prevent damage to the front end caused by excessive output power from the device under test (DUT).
2. *Impedance matching improvement:* Attenuators can be used to improve the impedance match seen by an upstream circuit when driving a mismatched

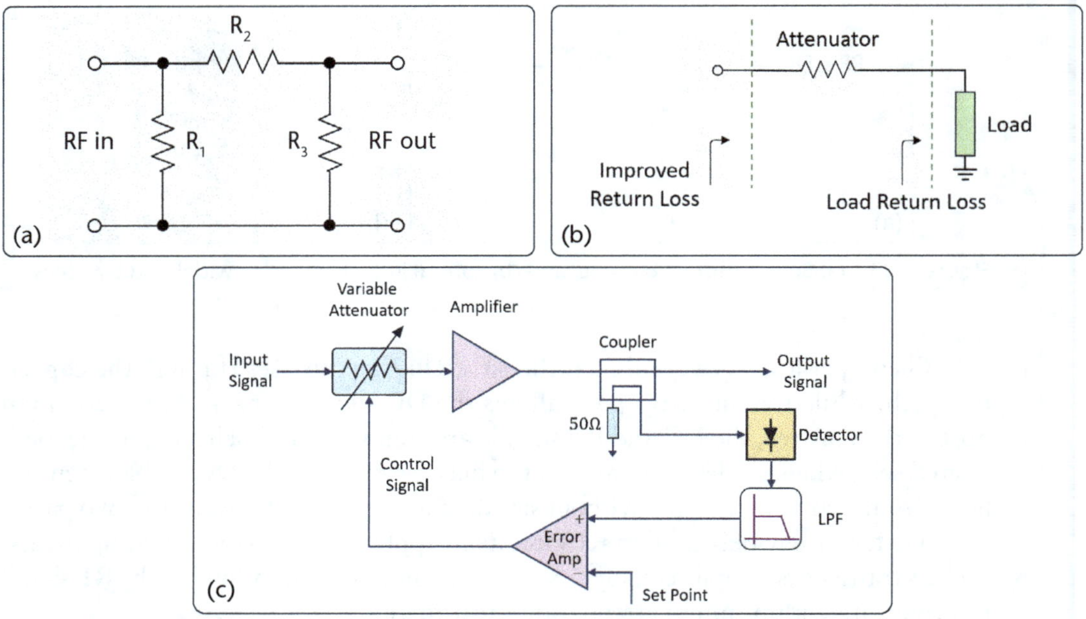

Figure 6.24 (a) A π-attenuator. The symmetrical configuration enables straightforward matching of both the input and output to 50Ω. (b) Use of an attenuator to improve impedance matching and reduce the effect of load return loss. (c) Use of an attenuator to control the output signal level within an AGC feedback loop.

load. As shown in Figure 6.24(b), inserting a matched attenuator ahead of a mismatched load reduces the magnitude of the reflected signal returning to the source. Because the signal traverses the attenuator twice (forward and reflected), the return loss improvement at the input is equal to twice the attenuation value. For example, a 3-dB attenuator improves the input return loss by 6 dB.

This technique does not improve the impedance match of the load itself and introduces additional insertion loss, which must be traded against benefits such as improved source match, reduced reflections, and enhanced stability of the preceding stage.

3. *Signal level adjustment and tuning:* Attenuation is frequently required for signal-level control. Digital step attenuators (DSAs) are tunable attenuators commonly used in RF signal generators or automatic gain control (AGC) loops to precisely adjust signal amplitude, as illustrated in Figure 6.24(c).
4. *Noise suppression in cryogenic environments:* In cryogenic and quantum systems, attenuators are deliberately introduced to reduce the amount of thermal noise coupled into sensitive devices, such as qubits in a dilution refrigerator. This noise reduction is effective only when the attenuators themselves are properly thermalized at cryogenic temperatures.

A 0-dB attenuator is used in specific RF circuits and typically exhibits an insertion loss of less than 0.5 dB. These attenuators function similarly to wideband 0Ω resistors and are commonly employed in cryogenic environments to thermalize the center conductor of RF cables. By providing a direct thermal path between the center conductor and the outer body, they offer significantly better thermal conduction than cables relying solely on dielectric materials. In addition, 0-dB attenuators can enhance mechanical stability by allowing the assembly to be rigidly mounted to a thermal plate.

6.7.5 Power Combiner

A power combiner is a passive RF component capable of combining or dividing RF signals. The Wilkinson combiner shown in Figure 6.25(a) is the most widely used implementation. When operated as a power divider, port 1 serves as the input, while ports 2 and 3 act as the outputs, with the input signal equally split between them. When used for signal combining, ports 2 and 3 function as the inputs, and port 1 serves as the combined output.

A quarter-wavelength transmission-line section is employed in the Wilkinson power combiner, as shown in Figure 6.25(a). This inherently narrowband behavior is evident from the S-parameters in Figure 6.25(b), where the ideal −3-dB transmission (corresponding to equal power division) is achieved only over a limited frequency range. As a result, single-section Wilkinson combiners offer high performance near their design frequency but exhibit reduced effectiveness outside this band. The operational bandwidth can be extended through the use of multisection power combiners, which trade increased complexity for improved frequency coverage. The analysis of power combiners is commonly carried out using even-mode and odd-mode analysis techniques [3].

For efficient power combining, the input signals must be equal in both amplitude and phase. Any deviation from this condition results in combining loss, which reduces the overall efficiency of the network. For an N-way power combiner in which only M input ports are actively driven (with equal amplitude and phase), the combining loss is given by

$$L = 10\log_{10}\left(\frac{N}{M}\right) \tag{6.22}$$

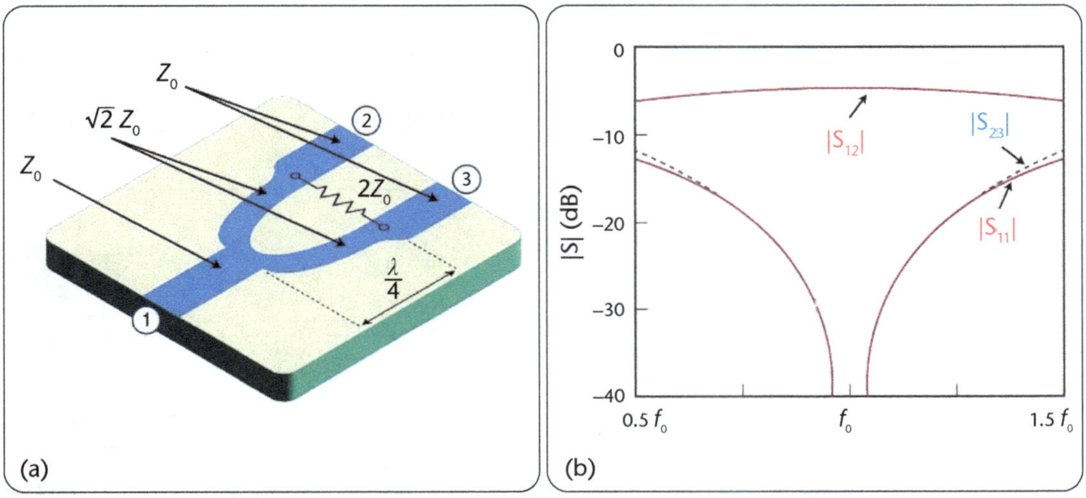

Figure 6.25 (a) Wilkinson power combiner. (b) S-parameters of a Wilkinson power combiner.

This loss represents the insertion loss associated with power being distributed across all N ports, even though only M ports are utilized. As an example, if three inputs are driven in a four-way power combiner, the combining loss is $L = 10\log_{10}(4/3) \approx 1.25$ dB.

Power combiners are essential components in power-amplifier arrays, where the outputs of multiple amplifiers are combined to achieve higher total output power. In antenna systems, they are used for beamforming, signal distribution, and redundancy configurations. Power combiners also play a critical role in test and measurement environments, enabling the combination of signals from multiple sources for system characterization or multitone testing.

Key operational parameters of power combiners include bandwidth, port-to-port isolation, amplitude imbalance, and phase imbalance.

For applications requiring more than two outputs, N-way power combiners can be implemented using appropriate network topologies. In addition, power combiners may be designed to achieve unequal power division to accommodate asymmetric signal levels or specific system requirements [3].

6.7.6 Directional Coupler

A directional coupler, depicted in Figure 6.26(a), is a four-port RF component that enables the transfer of input power from port 1 (the input port) to port 2 (the through port) with minimal insertion loss [2]. Simultaneously, a fraction of the input power, determined by the coupling factor, is coupled to port 3 (the coupled port), while ideally no power is delivered to port 4 (the isolated port). Conversely, when power is applied to port 2, it is coupled to ports 1 and 4 but not to port 3, resulting in isolation between ports 1 and 4 as well as between ports 2 and 3.

Directional couplers are widely used to sample a portion of a signal propagating through an RF signal path and route it to an auxiliary circuit, such as a power detector. This enables real-time monitoring of amplifier output power, as illustrated in Figure 6.26(b). Directional couplers are also fundamental components in RF test and measurement systems, including network analyzers, where they are used to separate incident and reflected waves for accurate measurement of reflected power from a DUT.

The insertion loss of the forward (through) path, L, is defined as

$$L = 10\log\frac{P_1}{P_2} = -20\log|S_{12}|\,\text{dB} \tag{6.23}$$

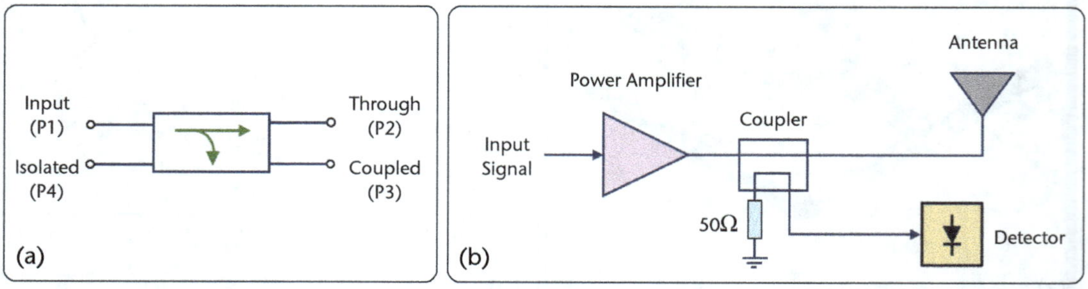

Figure 6.26 (a) Schematic of a directional coupler. (b) Application of a directional coupler for sampling the output power of a PA.

The fraction of power coupled from port 1 to port 3 is described by the coupling coefficient C, given by

$$C = 10\log\frac{P_1}{P_3} = -20\log\beta\,\mathrm{dB} \tag{6.24}$$

The leakage of power from port 1 to port 4 is characterized by the isolation

$$I = 10\log\frac{P_1}{P_4} = -20\log|S_{14}|\,\mathrm{dB} \tag{6.25}$$

The ability of the directional coupler to distinguish between forward and reverse traveling waves, equivalently, to isolate the coupled and isolated ports, is quantified by the directivity

$$D = 10\log\frac{P_3}{P_4} = 20\log\frac{\beta}{|S_{14}|}\,\mathrm{dB} \tag{6.26}$$

These quantities are related through the expression

$$I = D + C \tag{6.27}$$

Key operational parameters of a directional coupler include bandwidth, coupling factor, mainline insertion loss, directivity, coupling flatness, and maximum input power.

6.7.7 Circulator and Isolator

Figure 6.27(a, b) illustrates that an RF network is considered reciprocal if the response observed at one port remains unchanged when the ports of excitation and observation are interchanged. For a reciprocal component, the scattering matrix is

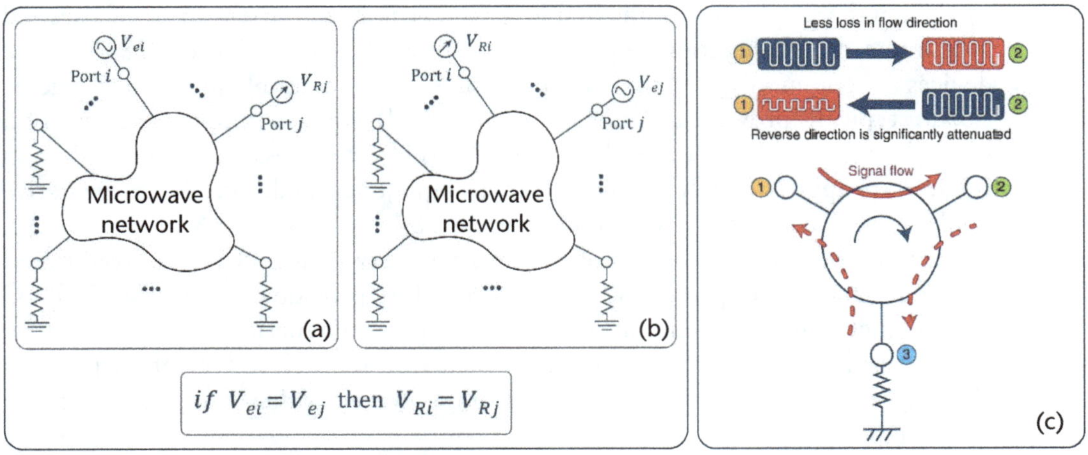

Figure 6.27 Reciprocity theorem. (a) Exciting at port i by V_{ei} and measure at port j (V_{Rj}). (b) Swapping the excitation and response if $V_{ei} = V_{ej}$, then $V_{Ri} = V_{Rj}$. (c) The operation of a circulator/isolator.

symmetric, such that $S_{ij} = S_{ji}$, indicating that the component's behavior is independent of the direction of signal propagation. Most passive components, including filters and attenuators, are reciprocal, whereas active devices such as amplifiers do not exhibit reciprocity.

Nonreciprocal components, including circulators, isolators, and gyrators, form an important class of RF devices. These directional components are typically realized using ferrimagnetic materials, whose electromagnetic properties can be controlled through the application of an external bias magnetic field. By adjusting the strength and orientation of this bias field, the interaction between the RF signal and the material can be tailored, enabling nonreciprocal behavior. This principle is also exploited in other RF devices, such as phase shifters, switches, and tunable resonators and filters.

Figure 6.27(c) shows the schematic of a circulator, which functions as a unidirectional RF valve. A circulator is a three-port nonreciprocal device designed such that a signal exits through the port immediately following the port at which it is applied. When a signal is injected at port 1, it is transmitted to port 2 with minimal attenuation or insertion loss. Conversely, when a signal is applied at port 2, it is strongly attenuated at port 1, while propagating with low loss from port 2 to port 3. Similarly, signals propagate with low loss from port 3 to port 1. This directional signal flow gives the device its name, a *circulator*. As discussed in Chapter 5, circulators are commonly used to allow a single antenna to be shared between transmit and receive paths.

An isolator can be realized by terminating one port of a circulator with a matched 50-Ω load, as illustrated in the same figure. In this configuration, signals traveling in the forward direction experience low insertion loss, while signals propagating in the reverse direction are absorbed by the termination rather than reflected back toward the source.

A single-junction isolator typically exhibits an insertion loss of less than 0.5 dB and an isolation of approximately 20 dB, meaning that signals traveling in the reverse direction are attenuated by 20 dB. Increasing the number of junctions improves isolation, leading to greater attenuation of reverse-propagating signals and enhanced noise-blocking performance.

When the output port of the first junction is cascaded into the input of a second junction, a double-junction isolator is formed, as shown in Figure 6.28(a). These devices are commonly used in cryogenic setups, where a double-junction cryogenic isolator typically provides isolation in excess of 40 dB while maintaining an insertion loss below 0.5 dB. In practice, the isolation and insertion loss of a double-junction isolator are approximately twice those of a single-junction device. A representative double-junction isolator is shown in Figure 6.28(b).

Circulators can be implemented using a variety of technologies. Traditional ferrite circulators are passive devices and therefore provide no gain. They rely on ferrimagnetic materials and an externally applied magnetic bias field to break time-reversal symmetry and achieve nonreciprocal signal flow. One of the primary challenges associated with ferrite circulators, however, is their relatively large size and weight.

Figure 6.28(c) shows a microstrip circulator implementation in which a ferrite slab is sandwiched between a microstrip Y-junction and a ground plane, with a

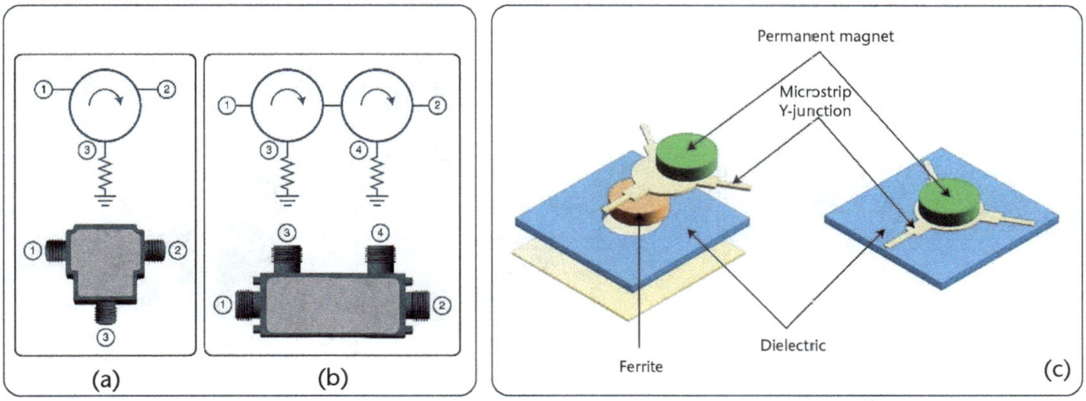

Figure 6.28 (a) Single junction circulator, (b) double junction circulator, and (c) ferrite circulator.

permanent magnet providing the required bias field. In qubit applications, magnetic shielding is critical to ensure proper circulator operation while preventing unwanted magnetic coupling to the qubit. Common shielding approaches include standard steel shields, double shielding configurations, and high-permeability mu-metal shields.

As an alternative, nonferrite circulators based on transistor technologies have been developed. These active circulators can provide gain but require external power to bias the active devices. However, compared to ferrite-based circulators, transistor-based implementations typically suffer from inferior noise performance and lower power-handling capability, which limits their use in high-performance or cryogenic RF systems.

6.7.8 RF Switch

Switches provide flexibility in RF circuits by enabling electronic signal routing without the need to manually connect or disconnect cables, allowing signal paths to be dynamically reconfigured during operation.

RF switches are commonly classified as reflective or absorptive based on their behavior at unused ports. As shown in Figure 6.29(a), reflective switches present an open or short circuit at the unused port, causing incident signals to be reflected back toward the source, which can result in high VSWR and poor return loss. In contrast, absorptive switches, shown in Figure 6.29(b), internally terminate the unused port in a matched load, resulting in low VSWR and improved return loss.

Absorptive switches are preferred in applications where reflections back to sensitive RF sources must be minimized. Care must be taken in high-power applications, however, since absorptive switches dissipate reflected power internally and therefore typically have lower power-handling capability, often requiring thermal management measures such as heat sinks. Reflective switches often achieve slightly lower insertion loss and higher isolation between ports because they do not include internal terminations.

In many low-to-moderate power applications, an absorptive switch can replace a reflective switch, whereas the reverse is not always possible due to the impact of reflections on system performance.

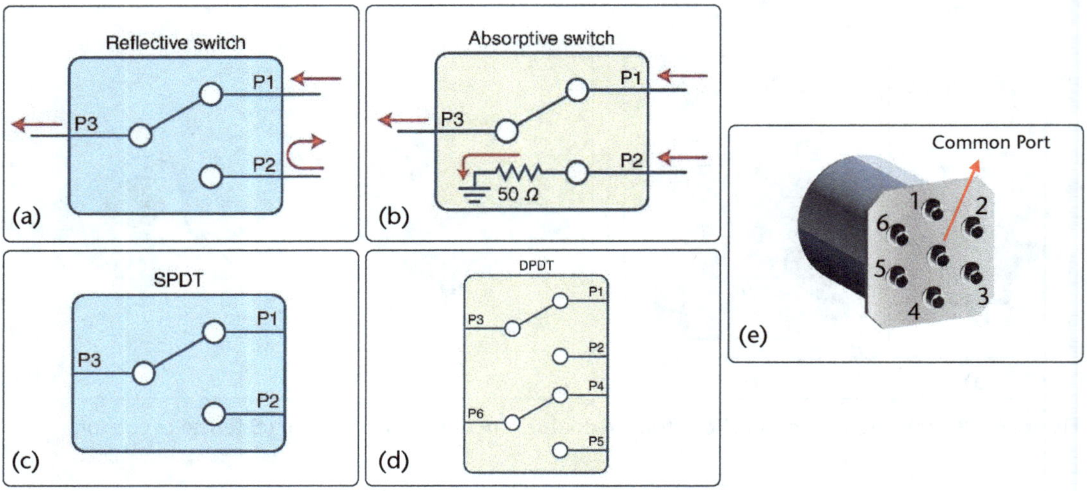

Figure 6.29 (a) Reflective and (b) absorptive switches, (c) SPDT and (d) DPDT switches, and (e) SP6T switch.

RF switches are available in a variety of configurations, including single-pole double-throw (SPDT), double-pole double-throw (DPDT), and multithrow multiplexers such as SP4T, SP6T, and SP8T, as illustrated in Figures 6.29(c–e). Multiplexer switches, such as the SP6T shown in Figure 6.29(e), are particularly valuable in test and measurement environments where switching between multiple DUTs is required, as they eliminate the need for repeatedly connecting and disconnecting RF cables.

RF switches are commonly controlled using digital control signals, often TTL-compatible logic. In this context, a logic "high" typically corresponds to a voltage of approximately 2V to 5V, while a logic "low" corresponds to a voltage below approximately 0.8V, depending on the specific device. Other switches are controlled using lower-voltage CMOS-compatible logic levels, such as 1.8V or 3.3V, and in some cases even sub-1-V levels, which are commonly used in modern low-power and cryogenic control electronics.

The unintended coupling of low-frequency control or bias signals into the RF signal path of a switch is known as video leakage. This results in baseband or low-frequency modulation artifacts on the RF output.

Switches can also be categorized as electromechanical or solid-state (e.g., PIN diode-based or FET-based). Electromechanical switches can handle higher power levels and typically exhibit lower insertion loss and higher isolation than solid-state switches. However, they suffer from slower switching speeds and limited mechanical lifetime. As a result, they are primarily used in high-power applications and precision measurement instruments. Solid-state switches, by contrast, offer faster switching speeds and longer operational lifetimes but typically have higher insertion loss and lower power-handling capability.

Key operational parameters of RF switches include isolation, insertion loss, video leakage, third-order intercept point (IP_3), 1-dB compression point (P_{1dB}), rise time, switching time, and repeatability.

Because solid-state switches are semiconductor devices, they can exhibit nonlinear behavior, making parameters such as P_{1dB} and IP_3 particularly important. When a switch is driven into compression, harmonic distortion increases and the effective insertion loss rises due to nonlinear conduction. Ensuring operation within the linear region is therefore critical, especially in high-dynamic-range systems.

The rise time of an RF switch characterizes how rapidly the RF signal envelope transitions during switching events. To accurately reproduce fast transitions, such as square-wave envelope modulation, the effective bandwidth associated with the rise time must be sufficient to pass at least the fifth to seventh harmonic of the modulation signal. The equivalent 3-dB bandwidth corresponding to a finite rise time can be approximated by

$$\mathrm{BW}_{3dB} \approx \frac{0.35}{\tau_R} \tag{6.28}$$

where τ_R is the rise time of the switch.

The switching time specifies how quickly the RF switch transitions between the on and off states. GaAs FET-based RF switches are among the fastest solid-state implementations, with switching times below 1 ns.

6.7.9 Balun

A balun (short for balanced-unbalanced transformer) is a passive component used to convert signals between differential (balanced) and single-ended (unbalanced) domains. The schematic of a balun is illustrated in Figure 6.30(a). Baluns are fundamental building blocks in both RF front-ends and mixed-signal subsystems, as many high-performance circuits operate differentially while others remain inherently single-ended.

As discussed in Chapter 5, modern mixed-signal systems rely heavily on differential ADC and DAC interfaces to achieve higher dynamic range, improved common-mode noise rejection, better linearity, and reduced susceptibility to digital substrate noise. In contrast, much of the surrounding RF signal chain, including mixers, LNAs, filters, PAs, and test interfaces, often employs a single-ended architecture. Baluns provide the necessary interface between these domains.

Specifically, baluns enable:

- Differential-to-single-ended (D2S) conversion for ADC inputs;
- Single-ended-to-differential (S2D) conversion for DAC outputs;
- Impedance transformation between RF stages and converter front ends;
- Common-mode rejection, improving SNR and SFDR;

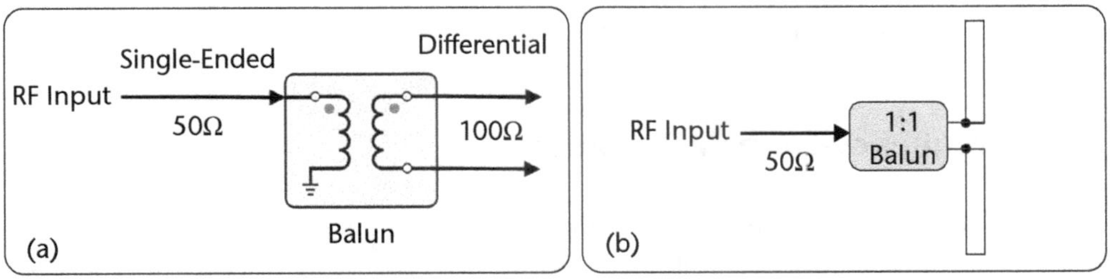

Figure 6.30 (a) Schematic of a balun. (b) Using a balun to feed a dipole antenna.

- Amplitude and phase symmetry, which is critical for low-distortion I/Q conversion.

Without an appropriately designed balun, converter performance can degrade significantly due to imbalance, incorrect impedance loading, or insufficient suppression of common-mode signals.

Baluns are also widely used in antenna systems, particularly when interfacing balanced radiators such as dipole antennas. The application of a balun with a dipole antenna is shown in Figure 6.30(b). A dipole naturally presents a balanced feed, whereas coaxial cables and most RF equipment are single-ended. A balun suppresses common-mode current on the outer surface of the coaxial shield, thereby preventing radiation pattern distortion, spurious radiation, and imbalance-induced impedance mismatch. Common examples include 1:1 baluns for dipoles, 4:1 baluns for certain loop antennas, and broadband transmission-line baluns for wideband antenna feeds.

6.7.10 Unun

An unun (unbalanced-to-unbalanced transformer) is a passive RF component used to transform impedance between two single-ended (unbalanced) ports. Unlike a balun, an unun does not perform balanced-to-unbalanced mode conversion; instead, it preserves the unbalanced signal reference while providing impedance transformation between interfaces.

Ununs are commonly used for impedance matching across a wide range of ratios, such as 50Ω to 75Ω, 50Ω to 300Ω, and 50Ω to 600Ω. By providing proper impedance transformation, ununs help to optimize power transfer, minimize reflections, and reduce VSWR. For example, in transmitters, this ensures that the PA operates into its intended load impedance, improving efficiency, linearity, and long-term reliability.

Common unun configurations include:

- *1:1 unun:* Used to preserve an unbalanced interface while providing DC continuity or minor mismatch tolerance without impedance scaling.
- *4:1 unun:* Widely used in high-frequency antenna systems for moderate impedance transformation.
- *9:1 unun:* Commonly employed with end-fed and random-wire high-frequency antennas to match high-impedance loads to standard coaxial feeds.

Across RF systems, ununs serve as practical and robust impedance-matching elements where both the source and load remain unbalanced, particularly in antenna feeds and transmitter output stages.

6.8 Frequency-Domain Techniques

The ability to modify the frequency content of a signal is crucial for reliable transmission and successful signal detection. *RF processing* refers to operations that alter or reshape the spectral content of an existing signal. Table 6.13 summarizes the

Table 6.13 Components Used for Frequency Manipulation

Operation	*Description*	*Relevant RF Components*
Filtering/multiplexing	Remove unwanted frequency components and define the desired passband within a system/ enable separation and combination of multiple frequency bands	Lowpass, highpass, band-pass, and bandstop (notch) filters/ diplexers, triplexers, multiplexers
Frequency translation	Shift signals between frequency bands for upconversion or downconversion without altering the modulation	Scalar mixers, IQ mixers, image-reject mixers, frequency-converter modules
Equalization	Compensate for frequency-dependent amplitude distortion caused by components, interconnects, or propagation effects	RF equalizers, tilt equalizers
Frequency division/ multiplication	Generate new frequencies that are integer multiples or fractions of a reference frequency for synthesis, clocking, or LO generation	Frequency dividers, frequency multipliers, prescalers, PLLs
Resonant components	Provide frequency selectivity or stability based on resonance phenomena; store energy at specific frequencies or define oscillation frequency	LC resonators, dielectric resonators, cavity resonators, SAW/BAW resonators

primary RF operations used for frequency processing, along with their functions and typical applications. In the following sections, we examine these techniques, filtering, frequency translation, equalization, frequency division and multiplication, and resonant frequency control, which together form the foundation of RF processing in modern RF systems.

6.8.1 Filtering and Multiplexing

A filter is a component that selectively attenuates specific frequency components of a signal, suppressing unwanted frequencies while preserving the desired ones. Filters are commonly used to suppress harmonics, spurious signals, and interference, limit noise bandwidth, and support mixed-signal applications such as anti-aliasing and reconstruction filtering.

Filter networks can also be arranged to provide frequency multiplexing functions, such as diplexers and triplexers, which are used to separate or combine signals occupying different frequency bands within a single RF path. The application of multiplexing for signal routing was discussed earlier in Chapter 5.

Filtering Mechanisms

Filtering can be realized through two primary mechanisms: reflective filtering and absorptive filtering. Reflective filtering rejects unwanted frequency components by reflecting them back toward the source, preventing them from reaching the filter output. Absorptive filtering, in contrast, dissipates unwanted frequency components as heat within the filter structure. Over its designed stopband, an absorptive filter presents an approximately matched load to the signal. Rather than encountering a strongly reactive impedance that would result in reflection, unwanted spectral components see a predominantly resistive path, where electromagnetic energy is converted into thermal energy through Joule heating.

As an example, a copper powder filter is an absorptive filter, whereas an LC filter typically operates as a reflective filter. Absorptive filters are preferred in applications where reflections back to sensitive sources must be minimized and impedance stability is critical. However, care must be taken in high-power applications, since absorptive filters dissipate rejected power internally and therefore require appropriate thermal management.

Consider an LC lowpass filter consisting of a series inductor L followed by a shunt capacitor C to ground. The shunt capacitor provides a low-impedance path at high frequencies $Z_C = 1/j\omega C$, so it diverts unwanted high-frequency components to ground. The series inductor presents low impedance at DC and low frequencies, but its impedance increases with frequency $Z_L = j\omega L$, increasingly blocking higher-frequency currents. As a result, low-frequency signals are transmitted while high-frequency components are attenuated.

Types of Filters

Filters can also be classified according to the shape of their frequency response. Figure 6.31 illustrates the four fundamental filter types: lowpass, highpass, bandpass, and bandstop. In a bandpass filter, frequency components within the passband, defined by the lower and upper cutoff frequencies, are transmitted with minimal attenuation, as shown in Figure 6.31(a), while frequencies outside the passband are strongly attenuated. The same principle applies to lowpass, highpass, and bandstop filters.

Figures 6.32(a, b) provide a visual illustration of the effect of a lowpass filter and a bandpass filter on the frequency spectrum of a signal.

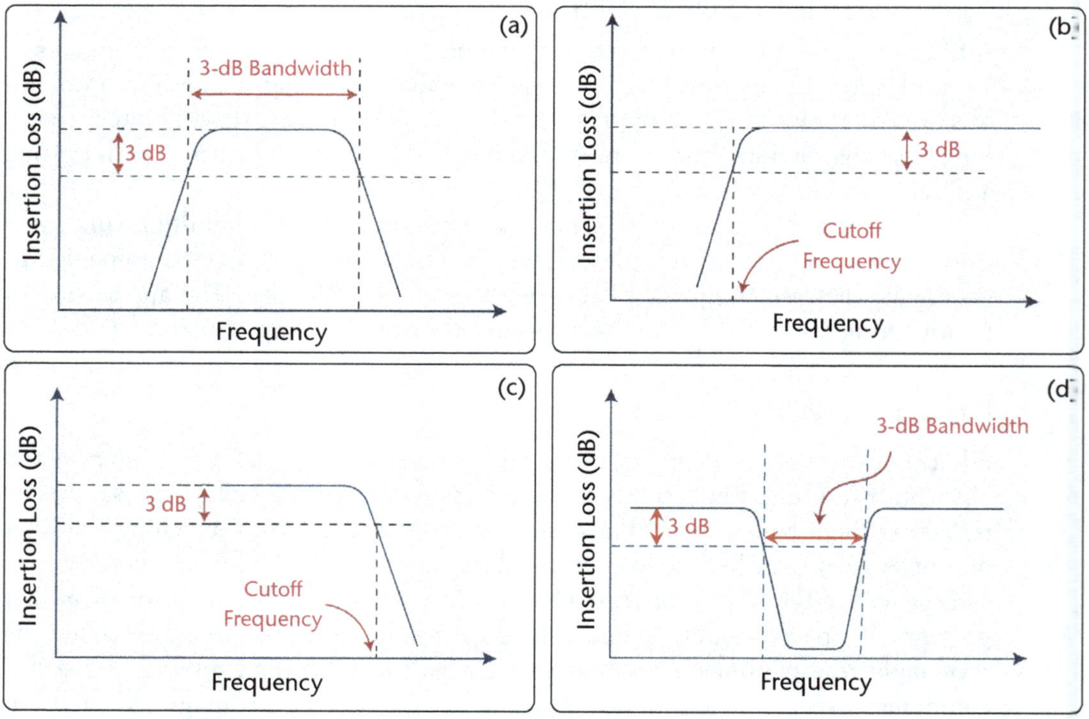

Figure 6.31 (a) Bandpass, (b) highpass, (c) lowpass filter, and (d) bandstop filter.

Filter Specifications

Filter specifications define the frequency-domain behavior of a filter and include parameters such as cutoff frequency, insertion loss, out-of-band rejection, roll-off, and group delay. Together, these metrics describe how the filter processes signals within the passband and attenuates unwanted frequency components.

The frequency response of a filter has a direct impact on signal integrity:

- Nonuniform amplitude response introduces amplitude distortion, altering the relative magnitudes of spectral components.
- Variations in group delay result in phase or timing distortion, causing different frequency components to arrive at different times.

Both effects can significantly degrade performance in broadband and modulated RF systems, particularly those employing wide instantaneous bandwidths or high-order modulation schemes such as QAM.

Out-of-band rejection is defined as the amount of attenuation applied to signals outside the passband. It is typically specified at one or more frequencies within the stopband relative to the passband reference level.

The roll-off of a filter describes the steepness of its frequency response in the transition region between the passband and the stopband. It is commonly expressed in decibels per decade (dB/decade). For example, a roll-off of 20 dB/decade indicates that the signal magnitude decreases by 20 dB for every tenfold increase in frequency beyond the cutoff frequency.

The achievable roll-off increases with the order of the filter, which is determined by the number of energy-storing elements (i.e., inductors and capacitors used in the filter implementation).

Multiplexers

We discussed multiplexers in Section 5.9.1. In RF systems, multiplexers are essentially combinations of filters, each covering a specific frequency band, used to route signals to different RF paths.

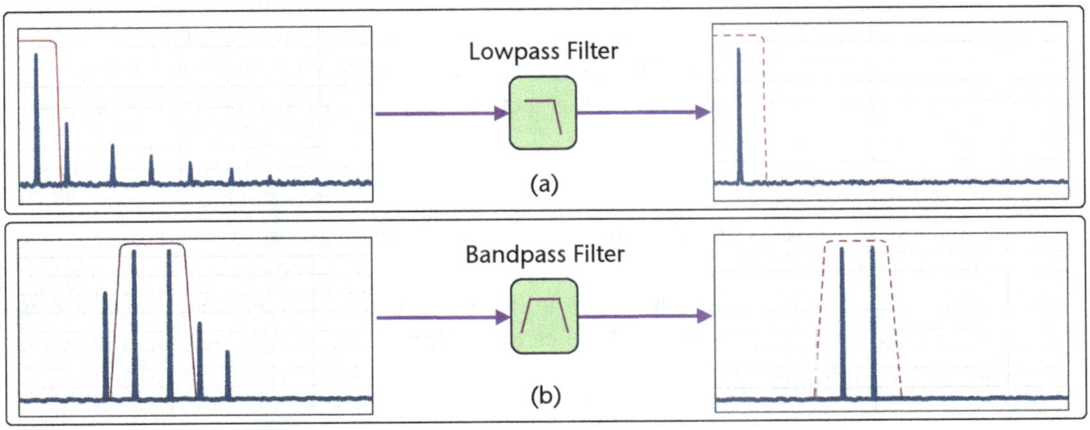

Figure 6.32 (a) Ideal lowpass filtering of a signal. (b) Ideal bandpass filtering of a signal.

Multiplexers are widely used in multiband systems, where multiple transmit and receive channels share a common antenna or RF front end. Examples include cellular base stations, satellite communication systems, and multiband radios, where frequency-selective routing is required to isolate signals operating in different bands.

Filter Technologies

RF filters can be implemented using a variety of technologies, each suited to specific frequency ranges, power levels, and integration constraints.

Common implementations include:

- Lumped-element filters, using discrete R, L, and C components, such as RC and LC filters, are typically employed at lower RF (up to a few gigahertz), where lumped-element parasitics do not significantly affect the frequency response. At these frequencies, distributed-element implementations based on transmission-line segments are impractical because their physical dimensions become excessively large relative to the wavelength.
- Distributed-element filters, based on microstrip, CPW, waveguide, or low-temperature co-fired ceramic (LTCC) structures, are prevalent at higher RF (microwave and millimeter-wave), where transmission-line effects dominate, and compact resonant structures can be realized. In this regime, the equivalent inductive and capacitive elements are implemented using transmission-line segments.

6.8.2 Frequency Translation

As discussed in Chapters 4 and 5, upconversion and downconversion are fundamental operations in modern RF systems, enabling translation of a signal's frequency content to support efficient transmission (upconversion) and reliable reception or detection (downconversion). These frequency translations allow signals to be placed in optimal spectral regions for propagation, amplification, filtering, and digitization.

Mixers remain the primary hardware components used for frequency conversion in RF systems. At the same time, digital frequency translation techniques, implemented using high-speed data converters, NCOs, and DSP, are becoming increasingly prevalent, particularly in SDRs, wideband communication systems, and modern radar architectures. In the following section, we will discuss the analog mixers.

Frequency Mixer

A frequency mixer is an electronic circuit with three ports: an LO port, an RF port, and an IF port, as shown in Figure 6.33. The mixer exploits device nonlinearity to translate signals in frequency. In an ideal mixer, the output spectrum contains components at the sum and difference of the input frequencies. For upconversion, this relationship can be expressed as

$$f_{RF} = f_{LO} \pm f_{IF} \tag{6.29}$$

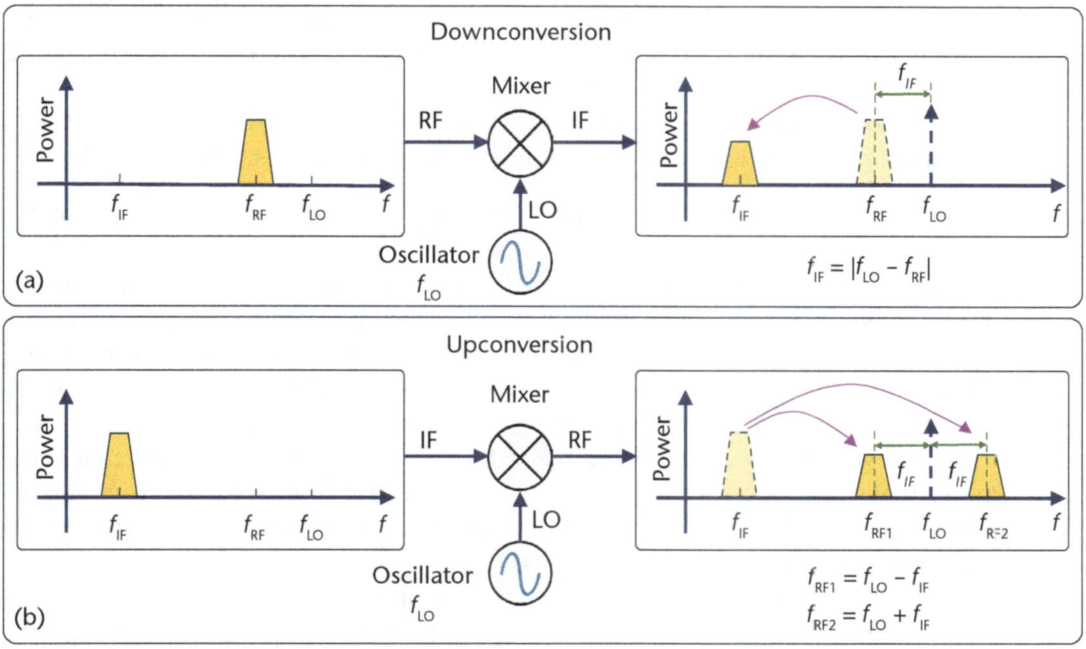

Figure 6.33 (a) Downconversion using a mixer, and (b) upconversion using a mixer.

Figure 6.33 illustrates the two fundamental modes of mixer operation. In downconversion, shown in Figure 6.33(a), an RF signal applied at the RF port is translated to a lower intermediate frequency at the IF port. In upconversion, shown in Figure 6.33(b), an IF signal applied at the IF port is translated to a higher RF at the RF port.

As shown in Figure 6.33(b), the upconversion process inherently produces two spectral components: an upper and a lower sideband, symmetric about the LO frequency. As discussed in Chapter 5 (see Figure 5.8), a similar ambiguity arises in downconversion: RF signals located symmetrically about the LO frequency are translated to the same IF. This phenomenon gives rise to the image frequency problem, in which undesired spectral components are also translated during frequency conversion. Consequently, image rejection becomes a critical consideration in both upconversion and downconversion architectures, as discussed in the following sections.

In principle, any nonlinear or time-varying device can be used to construct a mixer. A simple example is a diode-based mixer, in which the inherent device nonlinearity generates harmonics and intermodulation products, from which the desired upconverted or downconverted frequency components are selected through filtering. In practical RF systems, however, high-performance mixers are commonly implemented using GaAs FETs, Schottky diodes, or CMOS transistors. When properly designed, these technologies offer improved conversion efficiency, port isolation, and linearity compared to simple single-ended diode mixer implementations.

Mixer Parameters

Several key parameters influence mixer performance, including single-tone intermodulation distortion (IMD), multitone IMD, port isolation (LO-RF, LO-IF, and

RF-IF isolation), noise figure, image rejection ratio, and conversion loss. Collectively, these metrics determine the linearity, noise performance, spectral purity, and overall suitability of a mixer within an RF signal chain.

The conversion loss (CL), expressed in decibels, represents the difference between the input RF power level and the output power at the IF. It is defined as [7]

$$CL = P_{RF} - P_{IF} \tag{6.30}$$

where both power levels are expressed in dBm. The conversion loss of a typical mixer ranges from approximately 4 to 9 dB, depending on the mixer architecture and implementation. Conversion loss is closely correlated with other performance metrics, such as isolation and the P_{1dB}. As a result, a single measurement of conversion loss often provides a reliable first-order indication of overall mixer performance. If the conversion loss falls within a specified range, it generally implies that other performance parameters will also meet their respective specifications. However, the converse is not always true; for example, a mixer may exhibit good isolation while still suffering from poor conversion loss [8].

Isolation quantifies the amount of power that leaks from one port of the mixer to another. It is measured in decibels and defined as the difference between the applied input power and the leaked power observed at another port. For instance, if a signal is applied to the LO port and the resulting leakage is measured at the RF port at the LO frequency, the LO-to-RF isolation is given by

$$P_{ISO\,(L\text{-}R)} = P_{in}\,(\text{at LO}) - P_{out}\,(\text{at RF}) \tag{6.31}$$

Power leakage between mixer ports can lead to several practical issues. As discussed in Chapter 5, leakage of the LO signal to the RF port in a homodyne receiver results in a DC offset at baseband, which can degrade dynamic range and receiver sensitivity if not properly mitigated.

It is also important to note that LO drive levels for mixers can be relatively high, often requiring 10 to 20 dBm, particularly for passive diode-ring mixers. Proper LO drive is essential to achieve the specified linearity, isolation, and conversion loss, and insufficient LO power is a common cause of degraded mixer performance in practical implementations.

Image rejection will be discussed in the following sections. We now turn our attention to mixer spurious behavior, which plays a critical role in determining spectral purity and system-level filtering requirements as well as frequency planning.

Mixer Spur Table

Mixers are inherently nonlinear components and therefore generate spurious frequency components across the spectrum. A mixer spur table lists the frequency products created by the nonlinear mixing process, typically reporting the average spur level across the operating band.

Spur tables are a powerful tool for frequency planning, as they allow the designer to identify which spurious products fall inside or outside the band of interest and to assess whether they can be mitigated through filtering. Frequency planning was discussed in Chapter 5.

The frequencies of a mixer spurious products can be expressed as

$$f_{\text{spur}} = \left| m f_{\text{RF}} \pm n f_{\text{LO}} \right| \tag{6.32}$$

where $m,n \in \mathbb{Z}^+$ are positive integers. The case $m = 1$, $n = 1$ corresponds to the desired frequency conversion product, while higher-order values of m and n generate undesired spurious responses. Therefore, each entry in a mixer spur table corresponds to a specific nonlinear order of the mixing process.

Whether a spur matters depends on where it lands relative to the IF passband. Table 6.14 shows an example spur table measured for a downconverting mixer with RF = 6 GHz at –10 dBm and LO = 6.1 GHz at +20 dBm. With this frequency plan, the desired IF is $f_{\text{IF}} = |f_{\text{RF}} - f_{\text{LO}}| = |6.0 - 6.1|$ GHz = 0.1 GHz = 100 MHz. All spur levels in the table are reported in dBc relative to the IF power level.

IQ Mixer

An IQ mixer is a fundamental circuit that enables IQ modulation and demodulation. As shown in Figure 6.34(a), an IQ mixer consists of two mixers driven by LO signals that are 90° out of phase.

Depending on the signal direction, the same circuit topology can operate either as an IQ modulator, where IF1 and IF2 are the inputs and the RF port is the output (upconversion), or as an IQ demodulator, where the RF port is the input and IF1 and IF2 are the outputs (downconversion).

Figure 6.34(b) shows a COTS IQ mixer, highlighting the RF, LO, and IF ports.

In a digital IQ modulator, shown in Figure 6.7(a), much of the signal processing traditionally performed in analog hardware is moved into the digital domain. An interpolator increases the sample rate of the I and Q signals and applies digital lowpass filtering to suppress spectral images. The upsampled signals are then frequency-translated using digital mixers driven by a quadrature NCO, commonly implemented using DDS to generate phase-coherent sine and cosine waveforms. The digitally upconverted I and Q components are summed and applied to a high-speed DAC, which produces an analog IF signal or, if the DAC sampling rate is sufficiently high, directly generates the RF waveform through bandpass sampling.

Despite the increasing prevalence of digital IQ modulation, analog IQ mixers remain widely used, particularly in high-frequency and high-power applications,

Table 6.14 Spur Table for a Mixer (All Values Are in dBc Relative to the IF Power Level)

		nLO			
mRF	0	1	2	3	4
0	X	3	7	3	35
1	3	0 (desired conversion)	17	25	32
2	72	64	59	59	74
3	76	76	71	72	76
4	74	76	77	77	77

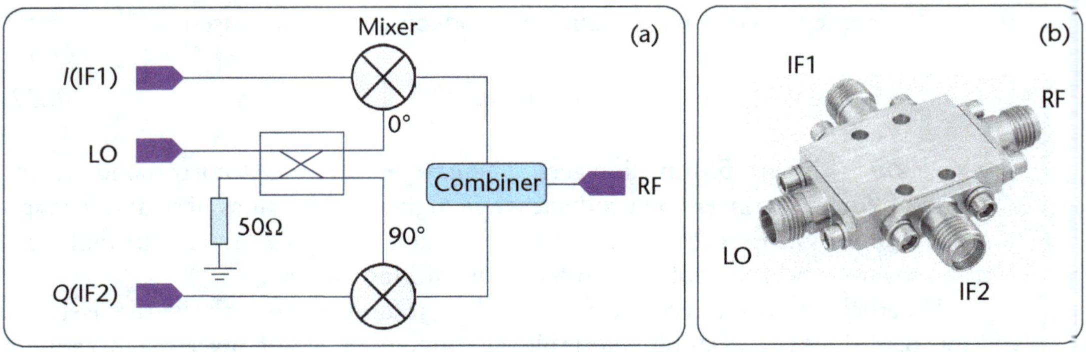

Figure 6.34 (a) Conceptual block diagram of an I/Q mixer architecture. (b) Photograph of a practical I/Q mixer implementation showing the corresponding LO, RF, IF1 (I), and IF2 (Q) ports.

where DAC speed, output bandwidth, spurious performance, or power consumption constraints make fully digital upconversion impractical.

Single-Sideband (SSB) and Image Reject (IR) Mixers

To address the issue of image frequencies in upconversion and downconversion, one effective approach is to generate two replicas of the input signal, manipulate their relative phases, and exploit constructive interference to preserve the desired signal while using destructive interference to suppress the unwanted image. The mixer architecture shown in Figure 6.35 plays a central role in this process.

This architecture consists of two scalar mixers driven by an LO that is split into two paths with a 90° phase difference, producing the in-phase (I) and quadrature-phase (Q) components of the signal. This configuration is commonly referred to as an IQ mixer.

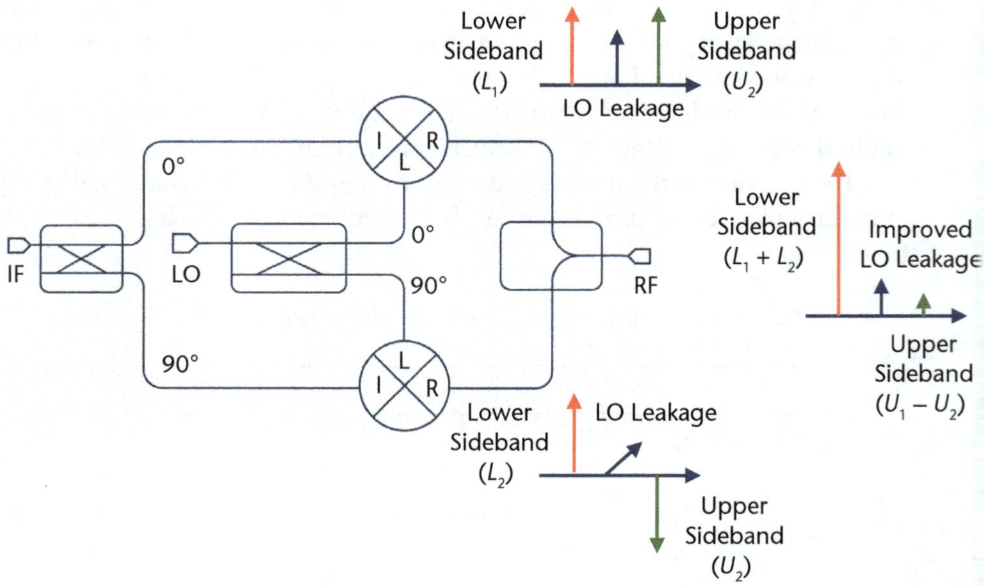

Figure 6.35 SSB mixer illustrating suppression of the upper sideband during upconversion through quadrature phase cancellation.

To transform an IQ mixer into an SSB mixer, the I and Q mixer inputs are driven with signals in quadrature phase using a 90° hybrid at the IF port. In combination with the quadrature LO signals applied to the mixer cores, the cumulative phase relationships cause one of the sidebands to experience a total phase difference of 180° between the I and Q paths [8]. As a result, that sideband undergoes destructive interference and is effectively canceled, while the desired sideband adds constructively, as illustrated in Figure 6.35. In practice, the cancellation is not complete, and there is always a residual image component, which is characterized by the image rejection ratio.

When an SSB mixer is used for downconversion, it is commonly referred to as an image-reject (IR) mixer, since the undesired image frequency is suppressed with the same principles of constructive and destructive interference.

Image Rejection Ratio (IRR)

To quantify the effectiveness of image suppression in an SSB or IR mixer, the IRR is defined as the ratio of the power in the desired sideband to the power in the image sideband, expressed in decibels:

$$\mathrm{IRR}\,(\mathrm{dB}) = 10\log_{10}\left(\frac{P_{\mathrm{desired}}}{P_{\mathrm{image}}}\right) \tag{6.33}$$

A higher IRR indicates stronger suppression of the image component and, consequently, improved spectral purity. In practical I/Q systems, however, the achievable IRR is limited by amplitude and phase imbalances between the I and Q signal paths. These imbalances arise from nonideal components such as hybrids, power combiners, and mixers.

In theory, perfect amplitude and phase matching between the I and Q channels would result in infinite image rejection. In practice, complete cancellation of the image sideband is not achievable, and a residual image remains, as illustrated in Figure 6.36. The presence of the unwanted image degrades modulation accuracy, typically quantified using metrics such as EVM.

Figure 6.37 quantifies the impact of amplitude and phase mismatches on image rejection. For example, an amplitude imbalance of 1 dB combined with a phase mismatch of approximately 6.5° results in an IRR of only about 25 dB. This example highlights how even modest amplitude and phase errors can significantly limit the achievable suppression of the unwanted image component.

When implemented using discrete components, image rejection levels on the order of 20 dB are typical, whereas commercially integrated solutions can achieve 40 dB or more. As discussed in the following section, quadrature error correction techniques, combining both analog and digital calibration, are often employed to mitigate amplitude and phase mismatches.

Quadrature Error Correction

Modern RF systems employ quadrature error correction to mitigate amplitude, phase, and DC offset mismatches between the I and Q signal paths. These correction techniques may be implemented in hardware, software, or a combination of both.

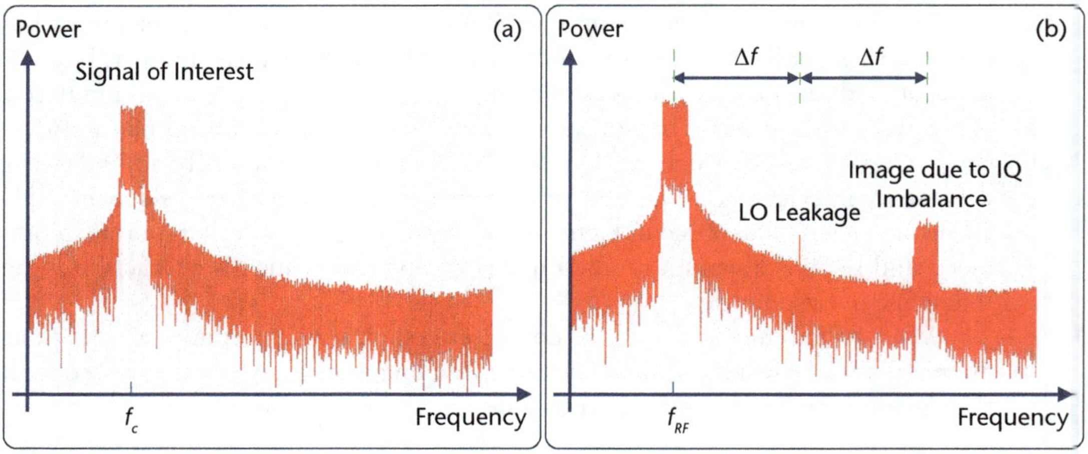

Figure 6.36 Spectral illustration of image rejection in an IQ upconversion system. (a) Desired signal centered at f_c. (b) Practical implementation showing residual image due to IQ amplitude and phase imbalance, as well as LO leakage.

Image rejection is particularly critical in the following scenarios:

- Wideband systems (e.g., OFDM, 5G), where even small I/Q imbalances translate into significant distortion across the signal bandwidth.
- Direct-conversion architectures, where the absence of an IF stage eliminates the possibility of filtering out image components.
- High-performance modulators and demodulators, where stringent EVM and spectral emission requirements must be met.

Quadrature error correction (QEC) methods can be broadly classified into analog and digital techniques. Analog correction relies on careful component matching

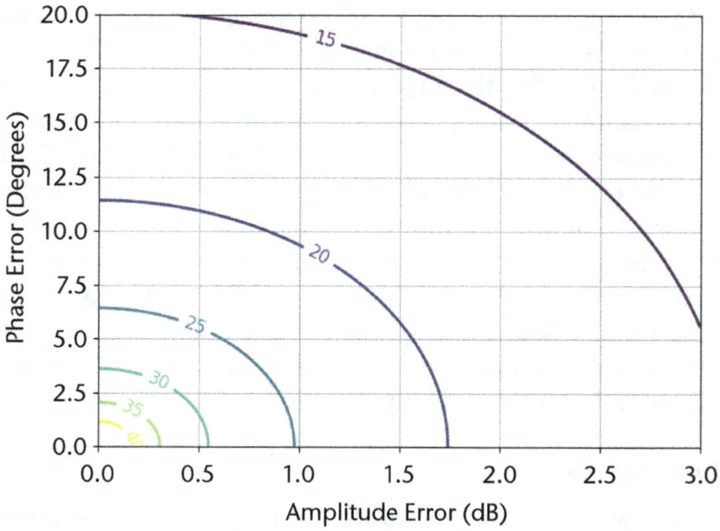

Figure 6.37 Contours for the IRR (in dB) versus amplitude and phase imbalance.

and manual or semiautomatic tuning of gain, phase, and DC offsets within the I/Q signal paths. While effective over narrow operating conditions, analog approaches increase system cost and complexity and are inherently susceptible to drift caused by temperature variations, aging, and bias instability.

In contrast, digital correction techniques compensate I/Q imbalance in the baseband using DSP. These methods estimate amplitude, phase, and offset errors, often through measurement-based calibration using reference tones, training sequences, or built-in self-test (BIST) mechanisms, and apply corrective coefficients digitally. Operating in the digital domain enables adaptive, repeatable, and highly stable image-rejection performance.

In traditional analog architectures, maintaining adequate image rejection often requires periodic recalibration to counteract component tolerances and long-term drift. Modern digital architectures largely eliminate this burden. By implementing digital upconversion (DUC) and digital downconversion (DDC) in FPGAs or RFSoCs, amplitude and phase alignment can be precisely controlled in the digital domain. This results in stable, repeatable image-rejection performance over time without manual adjustment, making digital architectures particularly attractive for wideband and precision-sensitive RF systems.

6.8.3 Equalizer

Equalization plays a vital role in modern RF systems, where wide instantaneous bandwidths are common in military, satellite, and wireless communication applications. As discussed in Chapter 4, frequency-dependent gain variations introduce amplitude distortion, resulting in unequal amplification of frequency components. This effect is particularly detrimental in high-order modulation schemes such as QAM, where amplitude inconsistencies directly increase BER.

Most transmit and receive chains naturally exhibit gain roll-off at higher frequencies due to component parasitics, dielectric losses, and frequency-dependent device behavior. To counteract this effect, equalizers are introduced to compensate for frequency-dependent loss and achieve a more uniform system response.

A typical equalizer, often referred to as a *lossy equalizer*, functions as a frequency-dependent attenuator [9]. It introduces greater attenuation at low frequencies and progressively less attenuation at higher frequencies, thereby producing a positive gain tilt, as illustrated in Figure 6.38(a). The magnitude of this correction is commonly quantified by the equalizer's slope, typically expressed in decibels across the operating bandwidth. Conversely, negative-slope equalizers are employed in certain waveguide-based systems, while parabolic equalizers are used in specialized applications such as traveling-wave tube (TWT) amplifiers. Regardless of the specific implementation, the fundamental objective remains the same: to flatten the overall system gain and ensure consistent performance across wide bandwidths.

Figure 6.38(b) illustrates the use of an equalizer in a two-stage amplifier chain, comparing the overall gain response with and without equalization. The inclusion of the equalizer significantly improves gain flatness, demonstrating its effectiveness in mitigating frequency-dependent loss and maintaining uniform amplification across the operating band.

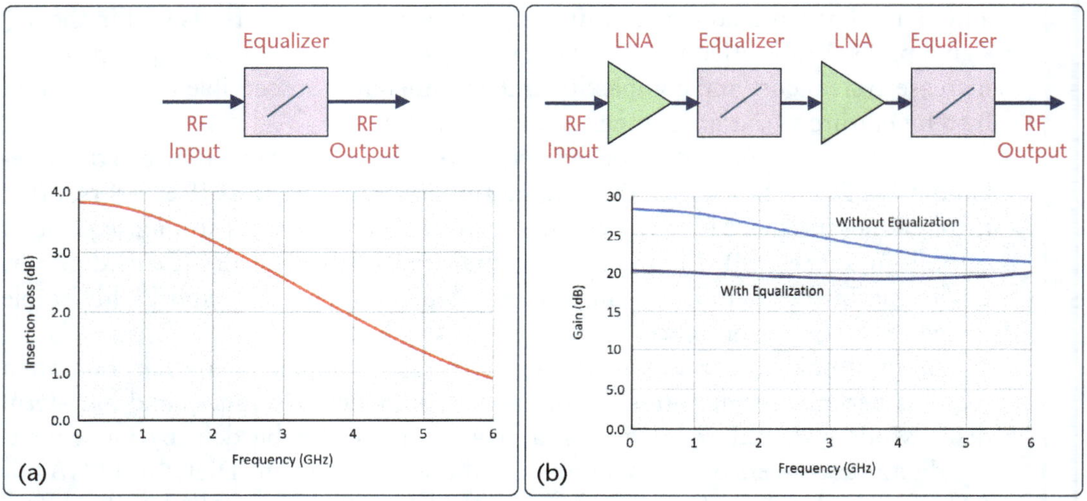

Figure 6.38 (a) Insertion loss of an equalizer as a function of frequency. (b) Application of an equalizer in a two-stage amplifier chain, showing the overall frequency response with and without equalization.

6.8.4 Frequency Division and Multiplication

Frequency dividers and multipliers were introduced in Section 6.4.4 in the context of nonlinear tone generation.

6.8.5 RF Resonator

RF resonators are fundamental building blocks in oscillators, filters, and tuned amplifiers, where they provide frequency selectivity, stability, and impedance control within the RF signal chain. They are also critical components in quantum computers, particularly in cavity quantum electrodynamics (cavity QED), where resonators enable qubit state readout while minimizing unwanted interactions with the environment.

Resonators appear across many domains of physics and engineering. A spring-mass system is a mechanical resonator, an LC circuit is an electrical resonator, and an atomic clock relies on an atomic resonator. What unifies all these examples is their ability to sustain oscillation through the periodic exchange of energy between two physical quantities. In a mechanical oscillator, energy alternates between potential and kinetic energy, whereas, in an LC resonator, it oscillates between electric and magnetic energy.

Figure 6.39(a) illustrates an ideal, lossless LC circuit. An initial charge on the capacitor produces an oscillating current $i(t)$ at the resonance angular frequency $\omega_r = 1/\sqrt{LC}$, as shown in Figure 6.39(b). At resonance, the time-averaged energy stored in the electric and magnetic fields is equal, $W_e = W_m$. The Fourier transform of this steady-state sinusoidal oscillation yields a spectral impulse, as illustrated in Figure 6.39(c).

In practical applications, all resonators exhibit losses. These losses are commonly modeled by introducing a resistor into the LC circuit, as shown in Figure 6.39(d). Energy dissipation in the resistor causes the oscillation amplitude to decay over

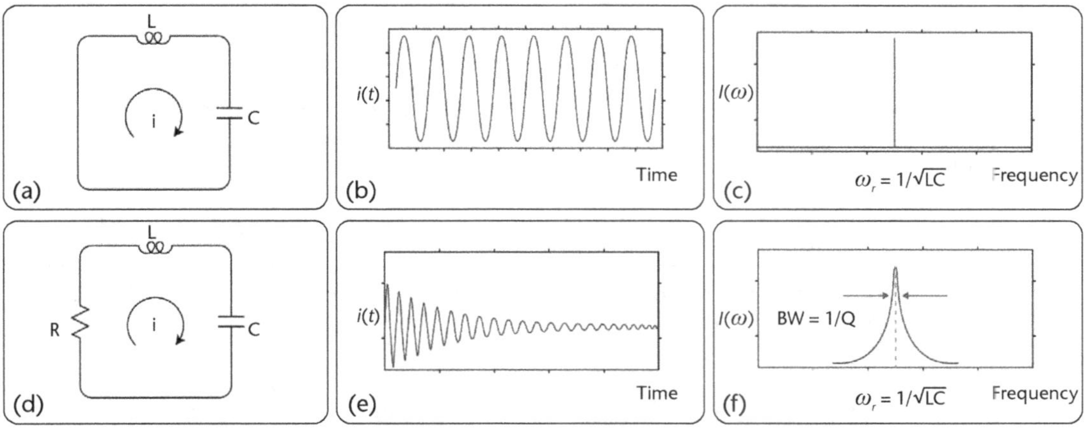

Figure 6.39 (a) LC circuit (lossless resonator). (b) Time-domain current. (c) Frequency-domain current. (d) RLC circuit (lossy resonator). (e) Time-domain current. (f) Frequency-domain current.

time, producing the exponentially damped response shown in Figure 6.39(e). The Fourier transform of this decaying oscillation is depicted in Figure 6.39(f), where the spectrum broadens relative to that of an ideal sinusoid. The spectral width is directly related to the decay rate: faster decay in the time domain corresponds to a wider frequency-domain response.

Figure 6.39(f) illustrates that the bandwidth (BW) of the resonator spectrum is inversely proportional to the quality factor (Q). The quality factor describes how effectively a resonator stores energy relative to how quickly that energy is dissipated due to losses. In practical terms, it quantifies the ability of an oscillator to sustain oscillations over time.

The quality factor is defined as the ratio of the energy stored in the resonator to the energy lost per cycle and can be expressed as [2]

$$Q = \omega \frac{\text{average energy stored}}{\text{energy loss per second}} = \omega \frac{W_m + W_e}{P_{\text{loss}}} = \frac{f_0}{B} \tag{6.34}$$

where W_m and W_e denote the average energy stored in the magnetic and electric fields, respectively, P_{loss} represents the power dissipated due to losses, f_0 is the resonant frequency, and B is the 3-dB bandwidth of the resonance.

A higher-quality factor corresponds to lower energy loss, resulting in a sharper frequency response and a narrower linewidth. Typical quality-factor values vary widely across physical systems. For example, a tuning fork exhibits a quality factor on the order of 10^3, an atomic clock can reach values as high as 10^{11}, microstrip resonators typically achieve $Q \sim 10^3$, and rectangular RF cavities can reach quality factors on the order of 10^5.

RF resonators exhibit behavior analogous to LC circuits when operated near their resonant frequency and can be realized using transmission-line structures. A common implementation is a half-wavelength ($\lambda/2$) transmission-line resonator with one or both ends either open or shorted, as illustrated in Figure 6.40.

The voltage distribution along the resonator follows directly from the boundary conditions. For an open-circuited end, the voltage reaches a maximum, whereas, for

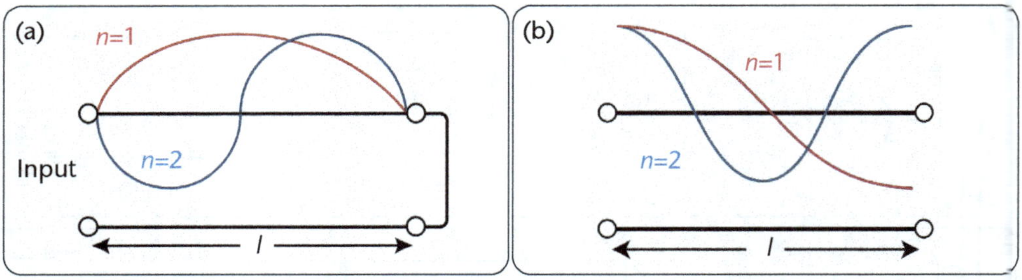

Figure 6.40 Voltage distribution for the first two resonant modes (n = 1, n = 2) of: (a) a short-circuit resonator and (b) an open-circuit resonator.

a short-circuited end, the voltage is 0. Along the length of the transmission line, the voltage varies sinusoidally, completing a half-wavelength variation over the physical length of the resonator. As a result, open-ended and short-ended resonators exhibit complementary voltage and current standing-wave patterns while sharing the same fundamental resonance condition.

Resonator Coupling

Resonators typically require coupling to the external environment to enable excitation. In Figure 6.41(a), a meander-line CPW resonator used in a superconducting quantum circuit is capacitively coupled to the input and output transmission lines. This weak capacitive coupling allows RF signals to probe the resonator without significantly degrading its quality factor. The state of the qubit is inferred by monitoring changes in the resonator's transmission response, most commonly through measurements of the S_{21} parameter.

Figure 6.41(b) illustrates the connection of a resonant circuit to an external load. Coupling the resonator to external circuitry introduces an additional energy-leakage path, which reduces the overall quality factor. To quantify the effect of this coupling, three quality factors are defined:

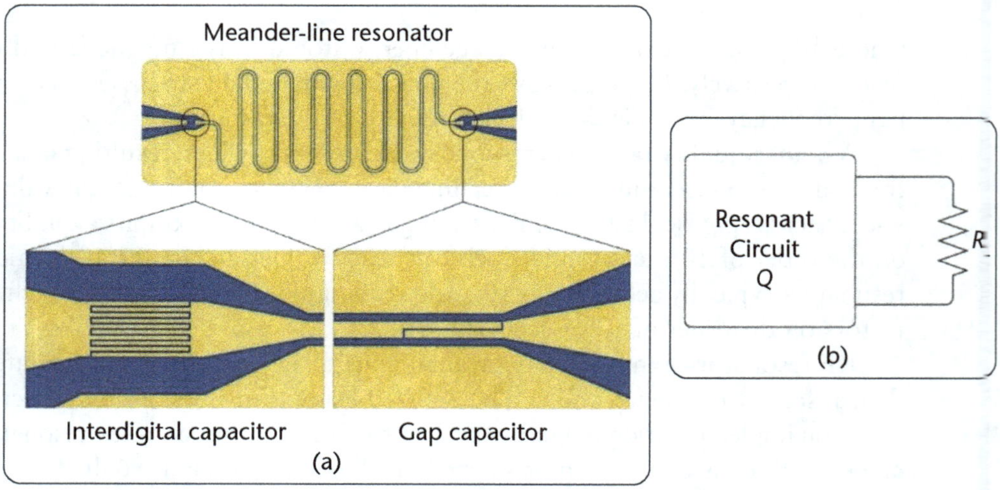

Figure 6.41 (a) Capacitive coupling to a meander line CPW resonator. (b) An external load, R, connected to a resonant circuit.

- The unloaded quality factor, Q_u, which characterizes the intrinsic losses of the resonator in the absence of external loading;
- The external quality factor, Q_e, which accounts for losses introduced through coupling to external circuits;
- The loaded quality factor, Q_L, which reflects the combined influence of intrinsic and external losses.

These quality factors are related by

$$\frac{1}{Q_L} = \frac{1}{Q_e} + \frac{1}{Q_u} \tag{6.35}$$

Table 6.15 summarizes the expressions for the quality factors of series and parallel resonant circuits. The external quality factor explicitly captures losses associated with coupling to the external load, represented by R_L. In contrast, the loaded quality factor incorporates both the resonator's internal dissipation and the additional losses introduced by the external coupling.

In the case of an open-ended half-wavelength ($\lambda/2$) transmission-line resonator, the unloaded quality factor is given by [2]

$$Q_u = \frac{\pi}{2\alpha l} = \frac{\beta}{2\alpha} \tag{6.36}$$

where l denotes the physical length of the transmission line and the attenuation constant $\alpha = \alpha_c + \alpha_d$ accounts for both conductor (metallic) losses α_c and dielectric losses α_d, assuming that radiation losses are negligible. The phase constant is $\beta = 2\pi f/v_p$ with v_p representing the phase velocity of the guided wave.

External circuits can be coupled to a resonator in three distinct regimes: undercoupled, critically coupled, and overcoupled. The coupling regime is determined by the coupling coefficient g, defined as

$$g = \frac{Q_u}{Q_e} \tag{6.37}$$

Table 6.15 Unloaded, External, and Loaded Quality Factors for Series and Parallel RLC Circuits

Parameter	*Series RLC resonator* $\left(\omega_0 = \frac{1}{\sqrt{LC}}\right)$	*Parallel RLC Resonator* $\left(\omega_0 = \frac{1}{\sqrt{LC}}\right)$
Unloaded Q	$Q_u = \frac{\omega_0 L}{R} = \frac{1}{\omega_0 RC}$	$Q_u = \omega_0 RC = \frac{R}{\omega_0 L}$
External Q	$Q_e = \frac{\omega_0 L}{R_L}$	$Q_e = \frac{R_L}{\omega_0 L}$
Loaded Q	$Q_e = \frac{\omega_0 L}{R + R_L}$	$Q_e = \frac{R_L \Vert R}{\omega_0 L} = \frac{R_L + R}{R R_L \omega_0 L}$

where Q_e is the external quality factor. The coupling regimes are classified as follows:

- $g < 1$: the resonator is undercoupled to the feedline.
- $g = 1$: the resonator is critically coupled to the feedline.
- $g > 1$: the resonator is overcoupled to the feedline.

When a resonator is connected to a transmission line with characteristic impedance Z_0, the coupling coefficient depends on whether the resonator behaves as a series or parallel resonant circuit. For a series resonator, the coupling coefficient is $g = Z_0/R$, whereas, for a parallel resonator, it is $g = R/Z_0$.

The coupling coefficient can be extracted experimentally from the measured transmission coefficient $|S_{21}|$ at the resonant angular frequency ω_0. For a two-port resonator that can be modeled as a series RLC circuit with symmetric input and output coupling, the external coupling coefficient is given by [3]

$$g = \frac{|S_{21}(\omega_0)|}{1 - |S_{21}(\omega_0)|} \tag{6.38}$$

where S_{21} is expressed in linear magnitude, not in decibels. For a resonator that exhibits parallel RLC behavior, the impedance response at resonance is inverted, and the corresponding expression is obtained by inverting (6.38).

Figure 6.42 illustrates the return loss of critically coupled and noncritically coupled (undercoupled and overcoupled) resonators.

Direct measurement of the unloaded quality factor Q_u of a resonator is generally not feasible because the measurement system itself loads the resonator and alters its intrinsic losses. However, Q_u can be extracted indirectly from measurements of the loaded resonator when it is connected to a transmission line.

The loaded quality factor, Q_L, can be obtained directly from the measured transmission response S_{21}. It is defined as $Q_L = f_0/\text{BW}$, where f_0 is the resonant frequency and BW is the 3-dB bandwidth, corresponding to the frequency separation

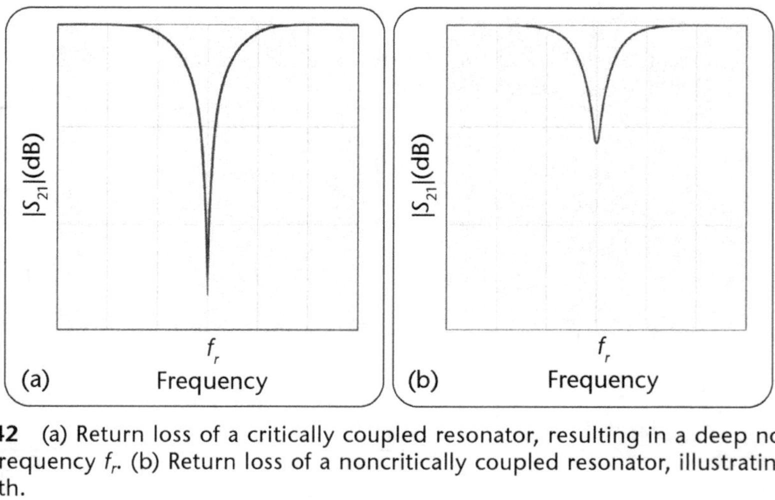

Figure 6.42 (a) Return loss of a critically coupled resonator, resulting in a deep notch at the resonant frequency f_r. (b) Return loss of a noncritically coupled resonator, illustrating reduced notch depth.

between the points at which the transmission magnitude drops by 3 dB relative to its peak value at resonance.

The unloaded quality factor is related to the loaded quality factor through the coupling coefficient g as $Q_u = (1 + g)Q_L$. This relationship enables the determination of Q_u by first extracting Q_L from the measured bandwidth and then calculating the coupling coefficient g from the measured S-parameters, as discussed earlier.

In addition to the two-port (transmission) measurement technique described above, a one-port (reflection) measurement approach can also be used to determine the quality factor of a resonator by analyzing the resonant behavior of the reflection coefficient S_{11} [10].

Resonator Technologies

Resonators can be implemented using a wide variety of structures, as summarized in Table 6.16. Each implementation offers distinct trade-offs in terms of quality factor, physical footprint, loss mechanisms, fabrication complexity, and ease of integration with active circuitry.

6.9 Phase Processing Techniques

Phase processing techniques play a central role in modern RF and microwave systems, enabling precise control of signal timing, beam steering, and waveform alignment. The primary components in this category are phase shifters and true time delay (TTD) elements, each offering distinct advantages.

Table 6.16 Resonator Technologies in RF Systems

Resonator Type	*Implementation*	*Typical Frequency Range*	*Key Advantages*	*Key Limitations/ Trade-Offs*
Lumped-element	Discrete inductors and capacitors (LC)	Low RF (up to a few gigahertz)	Compact, low cost, easy tuning	Limited Q, parasitics dominate at higher frequencies
Distributed-element	Transmission-line structures (microstrip, stripline, CPW, waveguide)	Microwave to millimeter-wave	Good predictability, higher Q than lumped elements	Larger size, limited tunability
Dielectric resonator	Low-loss dielectric material	Microwave	High Q, compact, good frequency stability	Limited tuning range, higher cost
Cavity resonator	Metallic waveguide or cavity structures	Microwave to millimeter-wave	Very high Q, excellent power handling	Large size, poor integration
Surface acoustic wave (SAW)/bulk acoustic wave (BAW)	Acoustic wave devices	RF to low microwave	Extremely compact, high selectivity	Limited power handling, fixed frequency

In addition to these single-path components, multiport phase control networks, such as Butler matrices and Rotman lenses, are widely used to implement spatial phase distributions across antenna arrays. These networks realize beamforming by imposing predefined relative phase shifts or time delays across multiple signal paths, enabling the generation of fixed or quasi-continuous beams without active control at each element.

The operating principles, implementation approaches, and practical trade-offs of these phase processing components and networks are discussed in the following sections.

6.9.1 Phase Shifter

A phase shifter is a two-port device that introduces a controlled phase offset to an RF signal while ideally preserving its amplitude. Phase shifters can be implemented using a variety of technologies, including RF diodes, field-effect transistors (FETs), and ferrite materials such as YIG or iron oxides, each offering different trade-offs in terms of bandwidth, insertion loss, power handling, and tuning speed.

Phase shifters are widely used in phased-array antennas, where the radiation beam is electronically steered by adjusting the relative phase of the signals feeding each antenna element. This principle is illustrated in Figure 6.43(a).

In addition to beam steering, phase shifters are employed in cancellation and feedback loops, such as interference or distortion cancellation circuits, where precise phase alignment is required to suppress unwanted signals or enhance amplifier linearity.

The phase of an analog tunable phase shifter is adjusted by applying a continuous control voltage, as illustrated in Figure 6.43(b). Compared to digitally controlled phase shifters, analog implementations typically exhibit lower insertion loss, making them attractive for low-loss and high-performance RF front ends.

However, a key consideration is that the resulting phase shift is generally a nonlinear function of the control voltage, as shown in Figure 6.42(b). This nonlinearity complicates precise phase control and often necessitates calibration, lookup tables, or closed-loop control in applications requiring accurate phase setting.

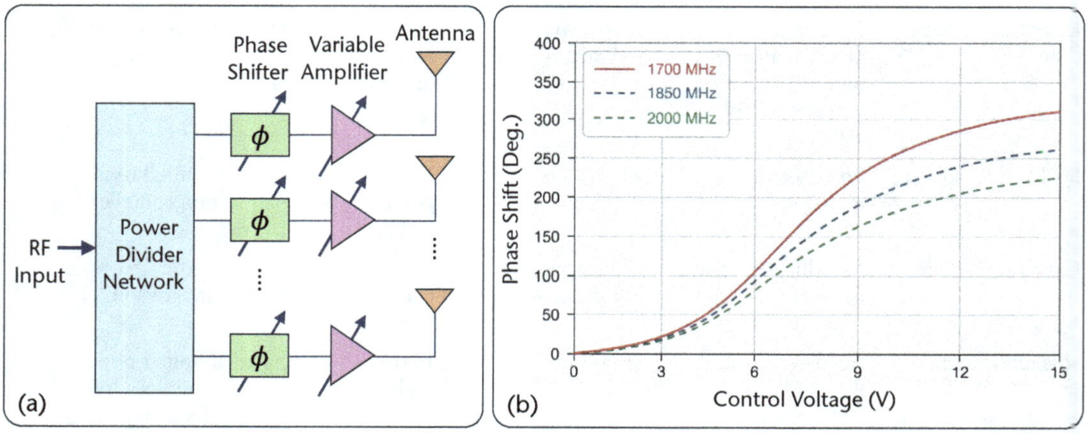

Figure 6.43 (a) Application of phase shifters in a phased array. (b) Phase shift versus control voltage.

Digital phase shifters offer finite phase resolution determined by their bit depth. For example, an 8-bit phase shifter provides a phase resolution of approximately 1.4° (360°/2^8).

An important performance consideration is insertion loss variation with phase setting, which quantifies how much the signal amplitude changes as the phase is adjusted. Ideally, a phase shifter would exhibit constant insertion loss across all phase states. This characteristic is critical in both phased-array antennas and cancellation circuits, as it avoids the need for gain compensation when the phase is changed. In practice, however, some amplitude variation is unavoidable.

Using a phase shifter can be problematic in phased-array systems because it applies a constant phase offset that is independent of frequency. Phase and time delay are related by $\tau = \phi/2\pi f$. When the phase ϕ is held constant, the effective time delay τ becomes frequency dependent. However, beam steering in phased arrays requires fixed relative time delays between antenna elements. As frequency changes, the varying delay causes the beam to point in different directions across the signal bandwidth. This frequency-dependent beam pointing is known as beam squint (see Figure 6.43(a)).

For wide instantaneous-bandwidth waveforms and narrow beamwidths, this frequency-dependent steering can deflect the beam off target, leading to degraded signal quality and reduced accuracy and resolution. For this reason, delay-line–based or true time delay beamforming architectures are preferred when the fractional bandwidth of the transmitted signal is large.

As a rule of thumb, phase-shifter–based beamforming is suitable for fractional bandwidths of approximately 5%–10%. Beyond this range, beam squint becomes significant and can noticeably degrade system performance. Accordingly, phase shifters are well suited for narrowband radar systems (typically <3%) and many satellite communication systems operating over 3%–7% fractional bandwidth. In contrast, wideband radar systems, EW platforms, and ultrawideband sensors, which often span 20%–50% or greater fractional bandwidth, generally require true time delay or delay-line–based beamforming architectures.

6.9.2 Delay Lines

A delay line introduces a constant time delay rather than a fixed phase shift, making it the preferred phase-control mechanism for wideband RF systems and true time-delay (TTD) beamforming architectures. By maintaining constant relative delays across frequency, delay-line-based beamforming preserves the intended beam direction over wide bandwidths and eliminates frequency-dependent beam-steering errors, commonly referred to as beam squint. Figure 6.44 compares beamforming using phase shifters and delay lines.

Time delay can be implemented using several approaches, depending on bandwidth, resolution, size, and system architecture:

- *Transmission-line delay:* Implemented using physical transmission media such as microstrip, stripline, coaxial cable, or optical fiber. The delay is set by the electrical length of the line, making this approach simple and broadband but often bulky at long delays.

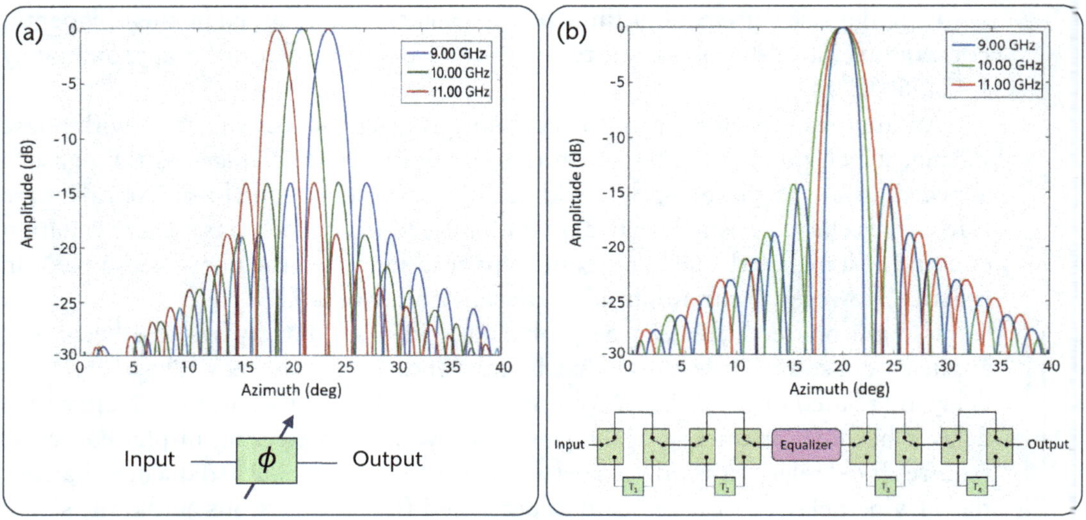

Figure 6.44 (a) Beam pattern steered using phase shifters as a function of frequency, illustrating beam squint due to the frequency-dependent phase-to-delay relationship. (b) Beam pattern steered using true time delay as a function of frequency, demonstrating the elimination of beam squint.

- *Switched delay lines:* Electronic switched-delay-line architectures, commonly implemented in MMIC technology, are widely used due to their compact size and favorable cost-performance trade-offs. Multibit time-delay units typically consist of RF switches, discrete delay elements, and equalizers, as shown in Figure 6.44(b). The overall delay range and resolution are achieved by selecting different signal paths. Delay elements are usually realized using transmission lines of varying lengths; as line length increases, insertion loss and frequency-dependent effects also increase, requiring equalization to maintain flat delay and amplitude response over frequency.
- *Digital delay (DSP/FPGA):* In digitally beamformed systems, time delay can be implemented in the digital domain using signal processing techniques within DSPs or FPGAs, offering high precision and flexibility. This approach requires high-speed data converters and is typically used in systems with digital beamforming architectures.

6.9.3 Multiport Phase Control and Beamforming Networks

Multiport phase control networks are passive structures that implement spatial phase distributions across multiple signal paths to realize beamforming without per-element active control. Unlike single-path phase shifters, these networks impose predefined relative phase relationships across an antenna array.

A Butler matrix is a well-known multiport phase control network used to generate multiple fixed, orthogonal beams in an antenna array. Each input port excites a distinct beam direction by applying precise relative phase shifts across the array elements. A Butler matrix is typically realized using a combination of hybrid couplers, fixed phase shifters, and transmission-line interconnects.

Another widely used multiport beamforming structure is the Rotman lens, which implements geometric time delay (often referred to as quasi-true time delay).

By exploiting path-length differences within the lens geometry, the Rotman lens generates multiple beams with reduced beam squint compared to phase-shifter-based networks, making it well-suited for wideband applications.

6.10 Signal Detection

In a broad sense, detection refers to the process of extracting information from a received signal. The output of a detection block may take several forms, including a continuous waveform (for example, audio), a scalar quantity such as amplitude, phase, or frequency, or a discrete representation such as symbols or bits. Demodulation is a form of detection in which the receiver extracts the transmitted information from a modulated signal.

From signal generation through transmission and subsequent RF signal-processing stages, such as filtering, amplification, and frequency conversion, each step ultimately serves a single overarching objective, accurate signal detection, namely the faithful recovery of transmitted information from the received waveform.

The definition of "successful" detection depends on system-level performance requirements and the chosen receiver architecture. In Chapters 3 and 4, key performance metrics such as minimum detectable signal (MDS) and BER were introduced to quantify receiver sensitivity and detection quality. As discussed in Chapter 5, receiver architectures, including heterodyne, direct-conversion, and low IF, offer different trade-offs in sensitivity, selectivity, linearity, and implementation complexity. Rather than revisiting receiver architectures, this chapter focuses on the components used for RF signal detection.

As illustrated in Figure 6.45, detection techniques can be broadly classified into scalar detection and vector detection. Scalar detection extracts a single signal parameter, amplitude, frequency, or phase and is typically implemented using power detectors, frequency counters, or phase detectors. Vector detection preserves both

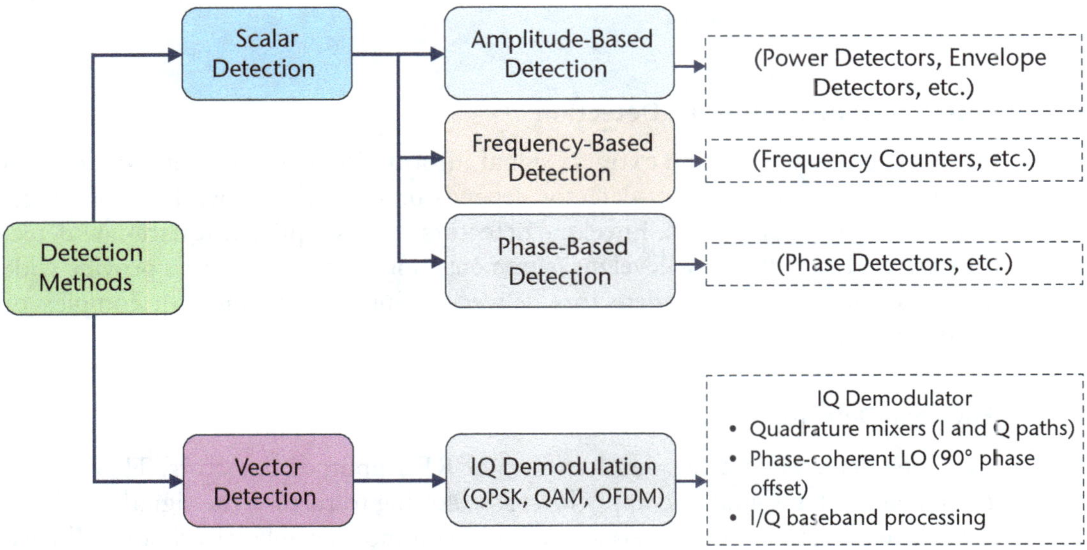

Figure 6.45 Detection method classification in RF systems.

amplitude and phase information through coherent I/Q demodulation and is required for complex modulation schemes such as QPSK, QAM, and OFDM.

6.10.1 General Detector Performance Metrics

Several performance metrics are common to all RF detectors, regardless of their internal architecture or operating principle. These parameters collectively define the detector's suitability for a given measurement or control application:

- *Sensitivity:* The minimum signal level that can be reliably detected above the noise floor.
- *Frequency response/bandwidth:* The range of input frequencies over which the detector maintains specified performance. Some detectors exhibit frequency-dependent sensitivity or calibration error.
- *Dynamic range:* The ratio between the maximum and minimum input power levels over which the detector operates within its specified performance limits.
- *Response time:* The speed at which the detector output responds to changes in the input signal, typically characterized by rise time or settling time.
- *Accuracy:* The degree to which the detector's output corresponds to the true value of the measured signal.
- *Precision (repeatability):* The ability of the detector to produce consistent results when measuring the same signal under identical conditions.
- *Linearity:* The extent to which the detector's output is proportional to the input over its operating range.
- *Environmental stability:* Long-term stability of the detector's performance under aging and environmental stresses such as temperature, humidity, supply voltage variations, and mechanical stress.

In the following sections, we review the different types of RF detectors. Vector detection, including IQ demodulators, was discussed in detail in Chapter 5 and is therefore not repeated here.

6.10.2 Amplitude-Based Detection

Amplitude-based detection extracts signal information from the magnitude of an RF waveform. The choice of detector depends on the signal type, dynamic range, and accuracy requirements. Envelope detectors offer simplicity, square-law detectors are well suited for low-level measurements, logarithmic detectors provide wide dynamic range, and RMS detectors deliver the highest accuracy for complex or modulated signals.

Envelope Detectors

Envelope detectors are the simplest form of RF amplitude detectors. They rectify the incoming RF signal and apply lowpass filtering to recover the signal envelope. A typical implementation consists of a diode, commonly a Schottky diode, followed by an RC filter.

Envelope detectors are widely used for AM demodulation, RF power monitoring, signal presence detection, and diode-based detector probes in test equipment, where simplicity and low cost are priorities.

Square-Law and Logarithmic Detectors

Square-law detectors operate in the region where the output voltage is proportional to the square of the input voltage (i.e., input power). This behavior makes them particularly effective for low-level signal measurements and forms the basis of many RF power sensors.

Logarithmic detectors, by contrast, produce an output proportional to the logarithm of the input power, enabling accurate measurements over a wide dynamic range. They are commonly used in AGC loops, RSSI circuits, and spectrum analyzers, where large variations in signal amplitude must be accommodated.

RMS Power Detectors

RMS detectors measure the true root-mean-square power of a signal, independent of waveform shape or crest factor. As a result, they provide accurate readings for modulated and complex waveforms where envelope or logarithmic detectors may introduce errors.

RMS detectors are widely used in test instrumentation, linear power control loops, and power monitoring applications in modern communication systems.

6.10.3 Frequency-Based Detection

Frequency-based detection extracts information by measuring the instantaneous or average frequency of an RF signal. The primary component used for this purpose is the frequency counter, which determines frequency by counting zero crossings or measuring the period of the input waveform over a defined time interval. Frequency counters are widely used in applications such as carrier identification, frequency stability measurements, frequency shift keying (FSK) demodulation, and calibration of RF sources.

In addition to standalone frequency counters, spectral analysis instruments can also be used for frequency-based detection. These include spectrum analyzers and SDRs, which estimate frequency by observing the signal spectrum or tracking spectral peaks in the frequency domain. SDR-based approaches are particularly attractive in modern systems, as they allow frequency detection, tracking, and demodulation to be implemented digitally using FFT-based processing.

6.10.4 Phase-Based Detection

A phase detector (PD) measures the phase difference between two signals and converts this difference into a DC voltage or digital value proportional to the phase error. In RF and microwave systems, this function is most commonly implemented using a double-balanced mixer, while XOR-based detectors or phase-frequency detectors (PFDs) are widely used in digital and frequency-synthesizer applications.

Phase detectors are fundamental building blocks in PLLs, frequency synthesis, synchronization, and coherent receivers. A typical RF phase detector consists of two RF inputs, one serving as a reference (often derived from a local oscillator) and one as the signal under test, and a DC output that represents the phase difference between them.

At each operating frequency, the DC output of a phase detector exhibits a statistical distribution characterized by a mean value and a variance. For example, at 50 MHz, the output may have a mean value of −850 mV with a variance of 17 mV. These variations arise from the combined effects of circuit nonidealities, port-to-port leakage, noise, and internal offsets.

The DC offset corresponds to the residual output voltage when the nominal phase difference is zero and represents a primary error source in precision phase measurements. Minimizing both the DC offset and output variance is essential for maintaining loop stability, measurement accuracy, and detection sensitivity, particularly in high-performance PLLs, phase-noise measurement systems, and coherent demodulation architectures.

6.11 Antennas

The antenna subsystem is sometimes underestimated during RF product development. Significant effort is typically devoted to RF circuitry and digital processing, while the antenna is sometimes treated as a secondary component. In practice, however, the antenna can make or break a product: it determines whether the intended link budget can be achieved, the desired coverage region is realized, and the target SWaP-C can be met. It can dramatically simplify or complicate the RF chain and strongly influence overall system performance.

A simple real-world example highlights this impact. As discussed in Chapter 4, the effective isotropic radiated power (EIRP) is the sum of the transmitter output power and antenna gain. Therefore, increasing antenna gain allows a reduction in required transmit power for a fixed EIRP. For example, a 3-dB increase in antenna gain enables a 3-dB reduction in PA output power, often resulting in a simpler amplifier design and thermal management, lower power consumption, improved linearity margins, and better SWaP-C.

However, higher antenna gain introduces important trade-offs:

- Narrower beamwidth, which reduces coverage and may require pointing, tracking, or beam steering at very high gain levels;
- Larger physical size, depending on operating frequency and antenna type.

This example illustrates why RF and antenna considerations must be addressed early in the product design process. In addition, antennas can be surprisingly expensive, particularly as operating frequency and power increase, or when high gain, wide bandwidth, or tight manufacturing tolerances are required. At millimeter-wave frequencies and above, even modest performance targets can drive complex fabrication methods, precision machining, and specialized substrates, all of which directly impact cost and feasibility during the product blueprint stage.

Although this book primarily focuses on RF circuits and systems, the antenna remains an integral part of the overall product. While a detailed treatment of antenna design is beyond the scope of this volume, we highlight the key antenna parameters most relevant during the RF product blueprint phase. Practical antenna concepts, advanced topics, and measurement techniques are covered in greater depth in Volume 2, which focuses on system-level integration and validation.

6.11.1 Key Antenna Parameters

The key antenna parameters and their fundamental relationships are summarized in Figure 6.46.

Figure 6.46(a) illustrates the radiation pattern of an antenna. The main lobe represents the direction of maximum radiation, while the sidelobes correspond to undesired radiation in other directions. The sidelobe level indicates how far the sidelobe amplitude is below the main lobe. The half-power beamwidth (HPBW) is defined as the angular separation between the points where the radiated power drops by 3 dB relative to the main lobe peak. A narrower HPBW corresponds to higher directivity and, consequently, higher antenna gain, as illustrated by the relationships shown in Figure 6.46(c).

Figure 6.46(b) shows the antenna return loss as a function of frequency. In practice, a return loss of at least 10 dB across the operating frequency range is

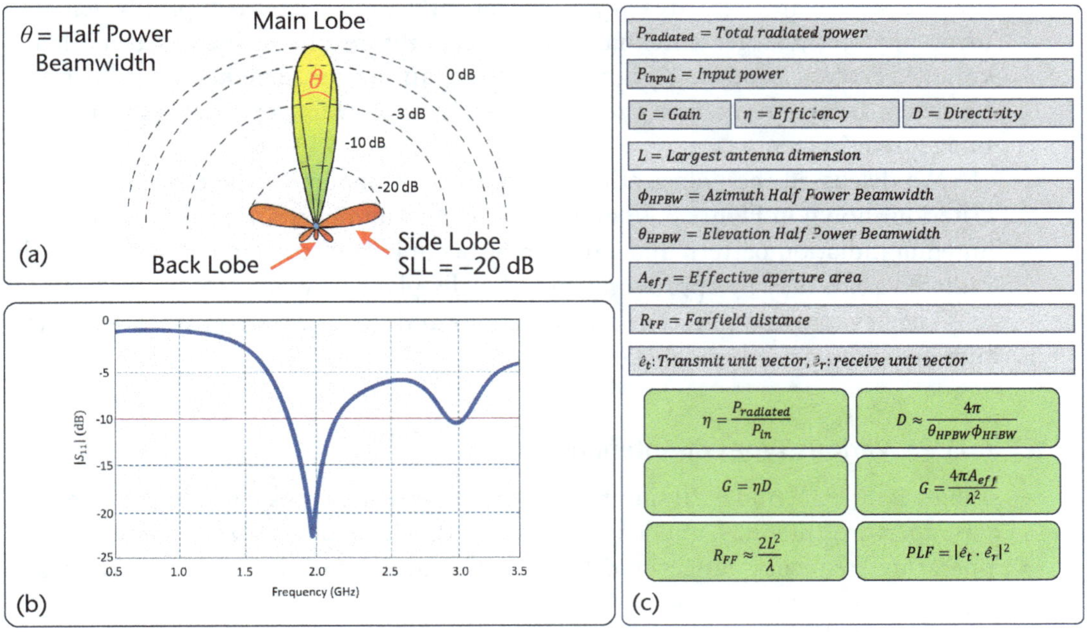

Figure 6.46 Antenna radiation pattern fundamentals and key performance metrics. (a) Normalized antenna radiation pattern illustrating the main lobe, side lobes, and back lobe, with the half-power beamwidth (HPBW) defined by the −3-dB points in elevation (θ) and azimuth (φ). The sidelobe level (SLL) is indicated relative to the main lobe peak. (b) Return loss of an antenna. A return loss better than 10 dB is typically considered acceptable. (c) Summary of common antenna definitions and relationships, including radiated and input power, radiation efficiency, directivity, gain, effective aperture, polarization loss factor, and the far-field distance criterion.

commonly accepted as a minimum requirement for good impedance matching and efficient power transfer. A 10-dB return loss indicates that approximately 90% of the input power is delivered to the antenna, with less than 10% reflected back toward the source.

However, good return loss alone does not guarantee good radiation performance. Power delivered to the antenna may still be dissipated through nonradiative mechanisms such as dielectric, conductive, or resistive losses, resulting in reduced radiation efficiency.

Figure 6.46(c) also illustrates another important antenna parameter, efficiency, which quantifies the fraction of input RF power that is converted into radiated electromagnetic energy. Practical antennas always exhibit losses, including dielectric losses in the substrate, conductive losses in the metal, and mismatch losses at the feed. Reduced efficiency directly degrades the effective radiated power and must therefore be accounted for when allocating transmit power and managing thermal dissipation.

Gain is a measure of how effectively an antenna concentrates radiated power in a given direction compared to an isotropic radiator [11]. Higher antenna gain directly improves the link budget; however, it comes at the cost of reduced beamwidth and, consequently, reduced spatial coverage. This fundamental trade-off must be carefully evaluated at the system level. Figure 6.46(c) summarizes the relationship between antenna gain and its defining parameters.

When selecting antennas for a product, physical size and weight are often critical constraints. As illustrated in Figure 6.46(c), antenna gain G is proportional to the effective aperture area A_{eff}. Consequently, higher gain is typically achieved using physically larger antennas, which generally results in increased mass and volume. From a product design perspective, an optimal balance must therefore be struck between antenna size, weight, achievable gain, and the resulting link-budget requirements.

We discussed the polarization loss factor in Chapter 4. The far-field distance criterion shown in Figure 6.46(c) defines the minimum distance beyond which the antenna radiation pattern and gain can be accurately characterized and where the radiated fields may be approximated as locally plane waves.

The key antenna selection considerations and their impact on system architecture are summarized in Table 6.17.

6.11.2 Various Types of Antennas

Tables 6.18, 6.19, and 6.20 summarize various antenna types, including wire antennas, aperture antennas, and lens antennas [11]. These tables highlight their typical operating frequency ranges and bandwidths, along with key characteristics and representative applications.

Figure 6.47 shows examples of wire, aperture, and lens antennas.

6.11.3 Radome and Packaging

In almost every RF product, antennas require radomes, RF-transparent structures that cover and protect the radiating elements. Radomes serve two primary purposes:

Table 6.17 Key Considerations When Selecting an Antenna for an RF Product

Category	*Key Considerations*	*Why It Matters for the Product Blueprint*
Frequency and bandwidth	Operating frequency, allocated bands, instantaneous bandwidth, fractional bandwidth	Determines antenna type, size, matching complexity, and regulatory alignment
Gain and directivity	Required gain, beamwidth, front-to-back ratio	Impacts PA requirements, coverage area, tracking needs, and EIRP
Polarization	Linear, circular, dual-polarized, polarization purity	Affects link reliability, fading performance, cross-polar isolation, and interoperability
Radiation pattern	Omni versus directional, elevation/azimuth patterns, nulls, sidelobe levels	Influences coverage, interference susceptibility, and system-level performance
Efficiency	Total radiation efficiency, mismatch loss	Determines actual radiated power and influences battery life and thermal design
Size and mechanical constraints	Physical size, form factor, housing integration, radome materials	Drives enclosure design, SWaP-C, and environmental robustness
Tuning and reconfigurability	Fixed versus tunable matching, beam steering, electronically controlled arrays	Supports multiband products, dynamic environments, tracking, or reconfigurable links
Co-location and isolation	Mutual coupling, spacing requirements, multiple antennas (GPS + LTE + Wi-Fi + radar)	Critical for MIMO, multiband systems, and minimizing self-interference
Integration with RF front end	Feed method, impedance, matching network, filter placement	Impacts PA/LNA performance, linearity, and overall link budget
Environmental requirements	Temperature range, humidity, shock/vibration, wind loading, outdoor exposure	Ensures reliability and durability under operational conditions
Regulatory and certification	FCC/CE compliance, SAR requirements, antenna gain limits	Prevents costly redesigns and ensures the product can be marketed legally
Cost and manufacturability	Unit cost, material selection, assembly tolerance, repeatability	Key for scaling from EVT → DVT → PVT and achieving target SWaP-C

Table 6.18 Wire Antennas

Antenna Type	*Typical Applications*	*Typical Frequency Range*	*Typical Bandwidth*
Dipole	IoT, handheld radios, Wi-Fi, Bluetooth, simple RF links	30 MHz–6 GHz	Narrow-moderate (5%–20%)
Monopole/whip	Mobile devices, automotive, base stations, SDR platforms	100 MHz–6 GHz	Moderate (5%–20%)
Loop antenna	RFID, NFC, LF sensing, magnetic near-field systems	10 kHz–30 MHz	Narrow (<5%)
Helical antenna	SATCOM, GNSS, space systems	100 MHz–8 GHz	Wide (10%–30%)
Yagi-Uda antenna	TV broadcast, point-to-point links, amateur radio	30 MHz–3 GHz	Moderate (~10%–20%)
Log-periodic antenna	Spectrum monitoring, EW, broadband comms, test labs	30 MHz–18 GHz	Very wide (up to decade bandwidth)

Table 6.19 Aperture Antennas

Antenna Type	*Typical Applications*	*Typical Frequency Range*	*Typical Bandwidth*
Patch/microstrip antenna	GNSS, Wi-Fi, SATCOM terminals, IoT, millimeter-wave	1–40 GHz	Narrow to moderate (1%–10%), wider if stacked
Horn antenna	Test and measurement, radar, millimeter-wave links	1–110 GHz	Very wide (up to 50% or more)
Slot antenna	Radar sensors, 5G, phased arrays, industrial sensing	1–40 GHz	Moderate (5%–20%)
Vivaldi (tapered slot) antenna	UWB radar, EW, broadband sensing, imaging, wideband arrays	0.5–110 GHz	Ultrawide (50–100%+)
Dielectric resonator antenna (DRA)	5G, millimeter-wave, compact radars	2–100 GHz	Wide (10%–30%)
Metasurface/RIS antenna	Beam steering, millimeter-wave, 6G, tunable apertures	10–100 GHz	Moderate (5%–20%), highly configurable
Phased array (active/passive)	Radar, 5G, EW, SATCOM-on-the-move	1–100 GHz	Wide-very wide (10%–40%)

Table 6.20 Lens Antennas

Antenna Type	*Typical Applications*	*Typical Frequency Range*	*Typical Bandwidth*
Parabolic dish	SATCOM, microwave backhaul, radar, radio astronomy	1–100 GHz	Wide (10%–30%)
Luneburg lens antenna	EW, wide-angle scanning, multibeam systems	5–100 GHz	Wide (10%–30%)
Rotman lens antenna	Beamforming networks, phased arrays, wide-angle scanning for radar, EW, and millimeter-wave communications	1–100 GHz (commonly 5–40 GHz)	Moderate-wide (10%–25%)
Dielectric/gradient-index (GRIN) lens antenna	Millimeter-wave, 5G, high-gain integrated terminals, compact SATCOM	20–100 GHz	Moderate-wide (5%–30%)

1. Environmental protection, shielding the antenna from rain, dust, UV exposure, ice buildup, and mechanical damage;
2. Aesthetics and integration, allowing the antenna to blend seamlessly into the product enclosure, vehicle, or platform without exposing fragile RF structures.

Radomes are commonly fabricated from materials such as fiberglass, PTFE-based composites, or specialized low-loss dielectric foams (e.g., ROHACELL). These materials are selected to provide mechanical robustness while introducing minimal attenuation, reflection, or distortion of the RF signal.

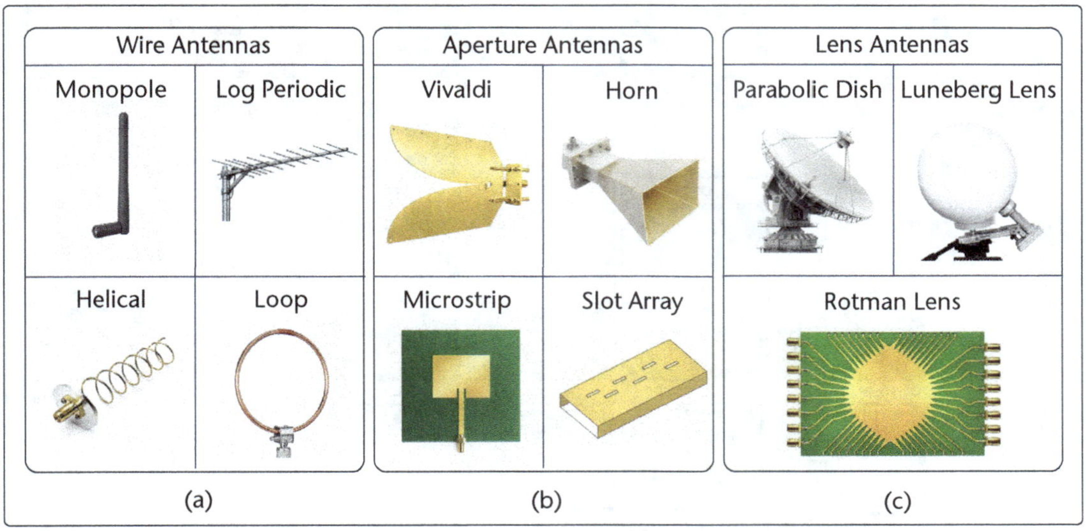

Figure 6.47 Examples of: (a) wire antennas (b) aperture antennas, and (c) lens antennas.

Codesign with packaging refers to the simultaneous design of the antenna and its surrounding mechanical and electronic enclosure to ensure optimal electromagnetic performance. This is particularly critical in compact systems and at higher frequencies, where nearby structures can significantly detune the antenna, degrade efficiency, or distort radiation patterns.

In some applications, such as aircraft, vehicles, or aerospace platforms, conformal antennas are required. These antennas follow the shape of the mounting surface rather than using a traditional planar or protruding geometry. While conformal antennas offer aerodynamic and integration advantages, they introduce additional design complexity and typically require extensive electromagnetic simulation and optimization. As a result, it is essential early in the product blueprint phase to determine whether a conformal antenna is truly necessary, as this decision has significant implications for cost, risk, and development effort.

6.12 A Function-Centric Periodic Table of RF Components

To provide a system-level view of RF building blocks, Figure 6.48 organizes common RF components according to their primary functional role. This function-centric perspective complements traditional signal chain descriptions and helps to guide early architectural decisions. Table 6.21 summarizes the acronyms used for the key parameters associated with each component.

6.13 What's Next?

In this chapter, we examined the critical RF components and the key operating parameters that govern their performance. In Chapter 7, we move beyond the RF domain to address non-RF system-level considerations, including mechanical design, DSP, software, and firmware, and explore how these elements come together to form an advanced RF product.

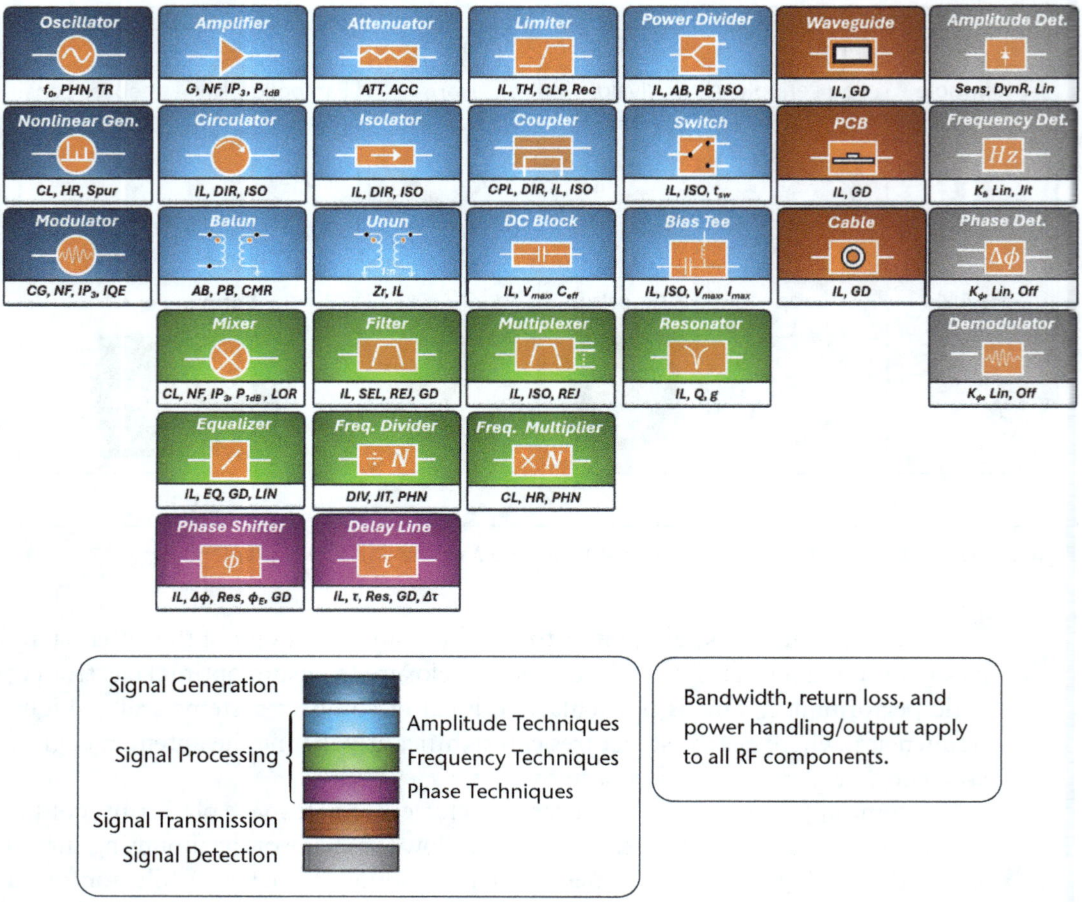

Figure 6.48 System-level classification of fundamental RF building blocks by primary functional role. Key performance parameters are indicated for each component.

Problems

6.1 You are designing the RF input chain of a 2–6-GHz wideband receiver intended for field deployment, where the signal environment is unknown and may include strong in-band and out-of-band interferers.

- Propose a complete RF input chain, starting at the antenna. Clearly indicate the ordering of each block.
- Select appropriate RF components for the input chain, including DC blocks, attenuators, filters (or preselectors), and limiters. For each component, specify realistic performance targets such as frequency range and insertion loss.
- Justify the placement and selection of each component in terms of its impact on:
  - Receiver noise figure;
  - Linearity and blocking performance;
  - Power handling and survivability;
  - Protection against accidental or hostile RF conditions.
- Discuss product-development considerations by identifying which components or architectural choices are most likely to change between

Table 6.21 RF Component Parameter Acronyms

Acronym	*Meaning*	*Acronym*	*Meaning*	*Acronym*	*Meaning*
AB	Amplitude balance	ACC	Accuracy	ATT	Attenuation
BW	Bandwidth	C_{Eff}	Effective capacitance	CG	Conversion gain
CL	Conversion loss	CLP	Clipping level	CMR	Common-mode rejection
CPL	Coupling factor	DIR	Directivity	DIV	Division ratio
DynR	Dynamic range	EQ	Equalization range	f_0	Center/oscillation frequency
G	Gain	GD	Group Delay	HR	Harmonic rejection
IL	Insertion loss	IP_3	Third-order intercept	ISO	Isolation
IQE	I/Q Error	I_{max}	Maximum DC	JIT	Jitter
K_f	Frequency detector gain	K_φ	Phase detector gain	LIN	Linearity
LOR	LO leakage	NF	Noise figure	Off	Output offset
PB	Phase balance	PHN	Phase noise	P_{1dB}	1-dB compression point
P_{out}	Output power	Q	Quality factor	Rec	Recovery time
REJ	Rejection	Res	Resolution	SEL	Selectivity
Sens	Sensitivity	Spur	Spurious content	t_{sw}	Switching time
TH	Threshold level	TR	Tuning range	τ	Time delay
$\Delta\varphi$	Phase shift range	φ_e	Phase error	$\Delta\tau$	Delay error
V_{max}	Maximum DC voltage	Zr	Impedance ratio		

Engineering Validation Testing (EVT) and Design Validation Testing (DVT) and explain why.

6.2 A fixed attenuator is added to improve matching and stability in an RF chain.
- Explain how this attenuator affects return loss, noise figure, and dynamic range.
- Determine where an attenuator is beneficial versus harmful in a receiver chain.
- Describe a practical scenario where an attenuator improves overall system performance.

6.3 You want to measure the reflected power at the output of a PA. The reflected wave is sampled using a directional coupler and routed to an ADC that is used for monitoring. The PA operates at 2.5 GHz with the following specifications:
- Maximum output power: +43 dBm (20W);
- Maximum allowable reflected power at PA output: −10 dBm;
- Protection ADC input range: −20 dBm to +5 dBm;
- Desired reflected-power measurement accuracy: ±1 dB.

You plan to use a directional coupler at the PA output.

a. Coupling factor selection:

   i. Determine the minimum coupling factor (in decibels) required so that the maximum forward power does not exceed the ADC input limit.

ii. Verify that the reflected signal remains above the ADC noise floor when the reflected power is −10 dBm.

b. Directivity requirement:

i. Calculate the minimum required directivity such that leakage from the forward wave into the reflected port contributes less than ±1-dB error in the reflected-power measurement.

ii. Comment on whether a 20-dB, 25-dB, or 30-dB directivity coupler is acceptable.

c. Insertion loss impact:

i. If the coupler insertion loss is 0.4 dB, calculate the effective reduction in PA-delivered power.

ii. Discuss whether this loss is acceptable for a production system and how it impacts efficiency.

6.4 A product supports four frequency bands using a switched RF signal path with the following requirements:

1. Operating frequency range: 0.8–3 GHz;
2. Maximum RF power: +30 dBm;
3. Allowable insertion loss per switch: ≤1 dB;
4. Minimum isolation between bands: ≥30 dB;
5. Switching time requirement: <1 μs;
6. Expected duty cycle: Continuous operation, $>10^9$ switching events over lifetime;
7. Control interface: Digital control preferred (GPIO).

Tasks:

a. Compare PIN-diode, FET, mechanical, and MEMS RF switch technologies.
b. Select the most appropriate technology and justify the choice based on insertion loss, isolation, switching speed, lifetime, and control complexity.
c. Identify key reliability risks for high-duty-cycle operation.

6.5 A receiver front end operates over 3.4–3.8 GHz with the following requirements:

- Minimum out-of-band rejection at ±500 MHz: 40 dB;
- Maximum allowable insertion loss: 1.2 dB;
- Production volume: 10,000 units/year;
- Operating temperature range: −20°C to +70°C.

You are evaluating three filter options, shown in Table 6.22.

Tasks:

a. Specification compliance:

i. Identify which filters meet all RF performance requirements.

Table 6.22 Filter Options

Filter Type	*Insertion Loss*	*Rejection*	*Unit Cost (10,000 Quantity)*	*Tuning Required*
Lumped LC	0.8 dB	30 dB	$0.80	Yes
Microstrip	1.0 dB	45 dB	$0.40	No
COTS SAW	1.4 dB	55 dB	$2.50	No

ii. Quantify how much each noncompliant option misses the specification.

b. Noise figure impact:

i. Compare the noise figure degradation introduced by each filter.

ii. Determine whether the SAW filter's rejection advantage offsets its higher loss.

c. Production risk analysis:

i. Estimate the yield risk associated with the lumped LC filter if manual tuning takes 3 minutes per unit.

ii. Calculate the annual labor cost impact, assuming \$60/hour technician cost.

d. Final selection: Choose the most appropriate filter for DVT/PVT and justify your decision quantitatively.

6.6 A transmitter employs an OFDM waveform with the following parameters:

- PA output average power: +27 dBm;
- Peak-to-average power ratio (PAPR): 9 dB;
- Operating frequency: 3.6 GHz.

The RF output chain is configured as follows: PA → 3 dB fixed attenuator → RF switch → bandpass filter. All components are assumed to be 50Ω and are each rated for +30-dBm average RF power.

For thermal calculations, assume the attenuator has a thermal resistance of 40°C/W (steady-state, junction-to-ambient). Treat the PAPR-defined peak power as a worst-case instantaneous value for conservative analysis.

Tasks

a. Peak power calculation:

i. Calculate the peak RF power at the output of the PA.

ii. Compare this peak power to the average power ratings of the attenuator, RF switch, and bandpass filter.

b. Thermal margin analysis:

i. Assuming that the attenuator provides 3 dB of attenuation, calculate the RF power dissipated as heat in the attenuator under average and peak conditions.

ii. Estimate the corresponding temperature rise of the attenuator under both conditions.

c. Linearity and reliability risk:

i. Explain why a passive RF component may exhibit compression or degradation even when its average power rating is not exceeded.

ii. Identify which of the three components is most likely to fail or degrade first and justify your choice.

d. Design mitigation: Propose two quantitative design changes (e.g., output power backoff, component power-rating upgrade, or waveform PAPR reduction) and calculate their impact on the peak RF power seen by the RF chain.

6.7 A matching network improves return loss at the center frequency but narrows bandwidth.
   a. Explain this trade-off using practical design considerations.
   b. Describe when a "good enough" match is preferable to an optimal match.
   c. Identify how over-matching can cause problems later in system integration.

6.8 Several RF components are mounted in close proximity inside a compact enclosure, including an LNA, PA, RF switch, bandpass filter, and PLL/VCO.
   - Identify which of these components are most sensitive to temperature drift.
   - Briefly explain how temperature variations affect gain, phase, and impedance matching.
   - Propose one simple thermal-mitigation strategy appropriate for PVT.

6.9 A datasheet claims excellent insertion loss and isolation for an RF component.
   a. List at least five parameters that are often optimistic or incomplete in datasheets.
   b. Describe how PCB layout, grounding, and connectors alter real-world performance.
   c. Explain why early lab measurements often differ from final system behavior.

6.10 You are integrating multiple RF components on a single PCB.
   a. Identify which components should have test access points and why.
   b. Explain how poor testability increases debug time and project risk.
   c. Propose a minimal set of test structures that add little cost but high development value.

References

[1] Salari, A., *Microwave Techniques in Superconducting Quantum Computers*, Norwood, MA: Artech House, 2023.

[2] Pozar, D. M., *Microwave Engineering*, Fourth Edition, New York: Wiley, 2012.

[3] Rubiola, E., *Phase Noise and Frequency Stability in Oscillators*, First Edition, Cambridge, UK: Cambridge University Press, 2008.

[4] Poisel, R. A., *RF Electronics: Design and Simulation*, Norwood, MA: Artech House, 1998.

[5] Lee, T. H., *Planar Microwave Engineering: A Practical Guide to Theory, Measurement, and Circuits*, Cambridge, UK: Cambridge University Press, 2004.

[6] Barrett, C., "Fractional/Integer-N PLL Basics," Technical Brief SWRA029, Texas Instruments, 1999.

[7] Razavi, B., *RF Microelectronics*, Second Edition, Boston, MA: Pearson, 2012.

[8] Jorgesen, D., "IQ/IR/SSB Mixer Primer," Marki Microwave, Technical Note, May 2020.

[9] Mini-Circuits, "RF Equalizers: An Essential Ingredient for the Modern System Designer." https://blog.minicircuits.com/rf-RF-equalizers-an-essential-ingredient-for-the-modern-system-designer/.

[10] Kamigaito, O., "Circuit-Model Formulas for External-Q Factor of Resonant Cavities with Capacitive and Inductive Coupling," arXiv:2005.05843 [physics.acc-ph], May 2020.

[11] Balanis, C. A., *Advanced Engineering Electromagnetics*, Second Edition, New York: Wiley, 2012.

CHAPTER 7

Beyond RF: System-Level Considerations

> A good engineer is someone who makes a design that works with as few original ideas as possible.
>
> —*Freeman Dyson*

In this chapter, we examine the non-RF building blocks that make a complete RF product possible. These include DC power management, software and firmware, processing platforms such as microcontrollers and central processing units (CPUs), digital signal processing (DSP) pipelines, and the mechanical and thermal infrastructure. Rather than treating these domains as isolated silos, we focus on the practical considerations, trade-offs, and integration challenges that arise at their interfaces with the RF chain. The objective is to show how power, code, computation, and mechanics collectively shape a product's performance, robustness, and readiness for both prototyping and production.

7.1 Introduction

While designing a high-performance RF chain is an essential part of any RF product, it represents only one piece of a much larger system, especially in modern designs where computation, reconfigurability, and multidimensional control play a central role. We introduced the high-level architecture of an RF product in Chapter 1 (see Figure 1.2), consisting of RF and non-RF building blocks. Later we examined the RF and the mixed-signal building blocks.

In this chapter, we shift our attention to the non-RF blocks that make an RF product functional, intelligent, and reliable. These include the mechanical subsystem, power management, integrated sensors, and the digital processing and control subsystem, which encompasses firmware, software, and networking. Together, these layers provide the coordination, monitoring, computation, and user interaction that transform an RF signal chain into a complete, deployable product. The following sections describe how each of these pieces operates, how they interconnect, and how they contribute to a cohesive and well-engineered product blueprint.

7.2 DC Power Management

Every RF product begins with the same fundamental concern: establishing how its power will be supplied, regulated, and distributed throughout the system.

Determining the required voltage rails, estimating total power consumption, and ensuring that the power solution fits within the system's SWaP-C constraints are foundational steps in developing a viable product.

DC power delivery is one of the most influential non-RF factors affecting RF performance, as explored in detail in Volume 2. Noise coupling from DC rails into sensitive RF paths can manifest as spurious tones, degraded phase noise, AM/PM distortion, and, most critically, electromagnetic interference (EMI) and electromagnetic compatibility (EMC) failures. These issues often present themselves as RF problems, yet their root cause can almost always be traced back to deficiencies in power-supply architecture, layout, or filtering.

To build a robust product blueprint, it is essential to understand the DC power architecture and how it must supply a diverse set of subsystems: RF front ends, data converters (ADCs and DACs), CPUs, field-programmable gate arrays (FPGAs), microcontrollers, sensors, cooling fans, and auxiliary electronics. Each rail must be designed with the appropriate balance of efficiency, noise performance, and load-transient behavior, typically requiring a hybrid approach using both switching regulators and low-noise linear regulators [1].

This is where the concept of a power tree becomes indispensable. A well-structured power tree clarifies how each subsystem is powered, shows the sequencing and dependencies between rails, and highlights where noise-sensitive or high-efficiency paths must be prioritized. Volume 2 will cover practical implementation details, component selection, sequencing, and noise-mitigation techniques; here, we focus on the architectural considerations needed to support a reliable product blueprint.

7.2.1 Common Power Sources for RF Systems

As illustrated in Figure 7.1, the power network begins with a primary AC/DC source that supplies energy to the entire product. Various types of voltage regulators are then used to step the voltage up or down, each selected to balance conversion efficiency with the amount of noise introduced into sensitive circuitry such as data converters (ADCs and DACs) or low-noise RF paths. In Figure 7.1, a high-efficiency switching regulator is followed by a low-noise linear regulator, a common combination that leverages the strengths of both devices.

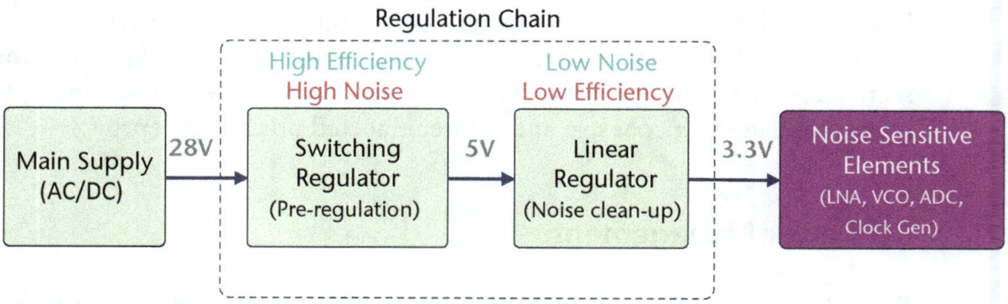

Figure 7.1 Typical power chain for sensitive RF and mixed-signal components: a switching regulator provides efficient bulk conversion, followed by a linear regulator that cleans up noise before feeding sensitive RF and data-converter circuits.

Because a linear regulator dissipates power proportional to the voltage drop across it, using one to directly convert from 28V down to 3.3V would be impractical, resulting in excessive thermal dissipation and poor efficiency. For large voltage reductions, a switching regulator is the preferred choice due to its much higher efficiency. Linear regulators are typically placed downstream, close to the load, where the input-to-output voltage difference is small, for example, using a switching regulator to step 28V down to 5V, followed by a linear regulator to generate a clean 3.3-V rail.

In RF product design, the choice of the primary power source depends heavily on system requirements and the intended application, whether the product is battery-powered and portable, infrastructure-based (such as radar or cellular equipment), or designed for automotive, aerospace, military, or consumer environments. This decision directly influences operational duration (especially for battery-driven systems) and has a major impact on SWaP-C constraints.

Powering RF circuits also introduces unique and demanding challenges. Even small amounts of voltage ripple, noise, or drift can degrade signal integrity, creating spurious tones, worsening noise figure and phase noise, or, in extreme cases, causing system instability. In high-performance designs, weak power management can compromise an otherwise solid RF signal chain, effectively turning a robust architecture into a fragile one.

In this section, we will review the major categories of power sources commonly used in RF systems.

Batteries

Batteries are essential in portable and remote RF systems, enabling operation without fixed power infrastructure. Selecting the appropriate battery chemistry involves trade-offs in energy density, voltage stability, lifetime, safety, and environmental tolerance. Table 7.1 shows the common battery types used in RF systems.

Table 7.1 Common Battery Types in RF Systems

Battery Type	*Key Characteristics*	*Typical RF Applications*	*Notes/Limitations*
Lithium-ion (Li-ion)	High-energy density; long cycle life	Handheld radios, portable transmitters, portable test equipment	Requires protection circuitry (over-voltage, under-voltage, thermal); sensitive to overheating; rechargeable
Lithium-polymer (Li-Po)	Similar performance to Li-ion; flexible form factors	Compact/UAV-mounted RF payloads, small portable devices	More mechanically fragile; requires strict thermal and short-circuit protection; rechargeable
Primary lithium (e.g., $Li\text{-}SOCl_2$)	Very long shelf life; extremely low self-discharge	Remote RF sensor nodes, emergency beacons, long-duration field deployments	Nonrechargeable; intended for long-duration or low-maintenance deployments
Alkaline	Low cost; widely available	Low-duty-cycle RF devices, simple transmitters, basic sensors	Lower energy density, nonrechargeable

Battery performance is highly sensitive to environmental conditions such as temperature, humidity, and vibration. Cold temperatures reduce capacity, heat accelerates aging, and mechanical stress can cause physical damage. RF product designers must account for:

- Voltage droop during discharge, which can shift bias points or detune sensitive RF circuits;
- Thermal stress from high transmit currents, especially in systems with bursty or high-power duty cycles such as handheld jammers;
- Battery aging, which increases internal resistance and reduces usable runtime;
- Environmental derating, ensuring reliable operation under extreme or fluctuating conditions.

AC-DC Converters

Stationary RF systems typically rely on AC-DC converters, such as wall adapters and external power bricks, to transform standard 120-V or 230-V mains power into stable DC rails for the RF circuitry. The two primary converter types, linear and switching, present different trade-offs in noise performance, efficiency, thermal behavior, and physical size.

- Linear AC-DC converters produce very low noise but are bulky, heavy, and inefficient. They are best suited for precision RF and test equipment where low ripple and clean supply rails outweigh concerns about size and heat dissipation.
- Switching AC-DC converters dominate modern RF products because of their high efficiency, compact form factor, and wide input-voltage compatibility. However, they generate high-frequency switching noise, making proper filtering, PCB layout, and shielding critical to prevent interference with sensitive RF paths.

Most RF products operate from single-phase AC sources. Three-phase AC power is used in some high-power RF systems, such as radars, high-power transmitters, and broadcast stations, where large amounts of power must be delivered. In these systems, the three-phase AC supply is typically rectified into a high-voltage DC bus before being stepped down for the RF electronics.

DC-DC Converters

In many RF systems, especially those powered by batteries, vehicle power buses, or DC infrastructure, DC-DC converters serve as the primary means of translating one DC voltage level to another. These switching-based regulators are especially advantageous when the voltage difference between input and output is large, providing far higher efficiency than linear regulators and greatly reducing thermal design constraints.

In power distribution, DC-DC converters form the backbone of efficiency, allowing high-voltage sources (such as 12-V, 24-V, or 48-V buses) to feed low-voltage rails (5V, 3.3V, 1.8V) without the excessive heat dissipation characteristic of linear regulation. However, this efficiency comes at the cost of increased switching ripple and EMI, which must be carefully managed to avoid degrading sensitive RF paths such as LNAs, synthesizers, mixers, and high-resolution data converters.

Common DC-DC converter types are:

- *Buck converters (step-down):* Reduce the input voltage to a lower output voltage with high efficiency and minimal heat. These are the most widely used regulators in RF systems.
- *Boost converters (step-up):* Increase the input voltage to a higher output voltage, common when a low-voltage battery must power circuits requiring higher supply rails.
- *Buck-boost converters:* Provide either step-up or step-down capability and may invert polarity depending on topology. They are ideal for battery-powered systems where the input voltage can move above or below the target output.

More broadly, DC-DC converters can be classified as nonisolated and isolated. Isolated converters are typically used in avionics, satellite payloads, sensitive receivers, and other systems that require safety isolation, ground separation, and fault containment between domains. Regardless of topology, all DC-DC converters introduce ripple and EMI, making proper filtering, layout, and cascaded regulation essential.

Energy-Harvesting Sources

Energy-harvesting systems extract power from environmental sources such as solar, thermal gradients, or vibration to sustain low-power RF operation over extended periods. In RF applications where long-term autonomy is essential and replacing or recharging batteries is impractical, such as hard-to-access or remote sensor deployments, energy harvesting becomes an attractive solution.

Despite these advantages, energy harvesting introduces significant design challenges. Harvested power levels are typically low and highly variable, requiring tight optimization of the RF system's overall power consumption. Energy storage must be carefully managed to buffer intermittent inputs, and RF transmission strategies must be adapted to align with the available energy budget and harvesting profile.

Vehicle Power Systems

Mobile RF systems, whether mounted on vehicles, UAVs, ships, or aircraft, typically rely on onboard DC power buses, most commonly 12V or 28V, depending on the platform class and power distribution standards.

- 12-V systems, standard in automobiles and many tactical ground vehicles, power RF equipment such as radios, fleet trackers, telemetry units, and sensor hubs. These buses exhibit significant electrical noise, including alternator ripple, ignition spikes, cranking brownouts, and load-dump events. Robust input protection, filtering, and well-designed regulation stages are essential to ensure clean, stable rails for RF subsystems.

- 28-V systems, used widely in aircraft, UAVs, and military platforms, support higher-power RF payloads such as radar modules, satellite communication (SATCOM) terminals, EW systems, and PAs. Sensitive RF circuits typically require staged step-down regulation, galvanic or DC-DC isolation, and careful filtering to maintain noise performance.

Designing for vehicle power requires addressing several key electrical and environmental challenges:

- Protection against large voltage surges, including load dumps and jump-start transients.
- Comprehensive EMI/EMC filtering, mitigating conducted and radiated noise from alternators, motors, actuators, avionics, or weapon systems.
- Environmental and electrical hardening to meet standards such as MIL-STD-1275 (28-V DC military vehicle power), MIL-STD-704 (aircraft electrical power systems), and ISO 7637/ISO 16750 (automotive electrical transients and environmental stress conditions).

These constraints directly influence the entire RF power architecture, from regulator topology and isolation strategy to filtering, sequencing, grounding, and placement, ultimately determining whether the RF subsystem can operate reliably in harsh and dynamic vehicle environments.

7.2.2 Power Tree Design

In RF products, managing multiple supply voltages across diverse functional blocks quickly becomes a complex task. Each block has its own noise sensitivity, load profile, and sequencing requirements, making ad hoc power distribution risky and error-prone. A power tree provides a structured way to organize, visualize, and plan how power is distributed throughout the system. It maps the flow of energy from the primary source to each subsystem, clarifying the relationships between regulators, protection stages, filtering elements, and the loads they support.

Within this framework, designers must balance noise performance, efficiency, size, and cost. Switching converters are typically well-suited for less noise-sensitive subsystems such as digital processors or PAs, where efficiency and power density dominate. In contrast, noise-critical sections such as receive RF front ends, frequency synthesizers, and oscillators often require additional post-regulation using linear low-dropout regulators (LDOs) to suppress residual ripple and broadband noise.

EMC further constrains power-architecture choices. Poorly filtered switching regulators can generate conducted and radiated emissions that violate EMI/EMC limits, particularly within RF bands of interest. As a result, selecting RF-friendly power supplies and applying proper filtering, grounding, shielding, and layout practices is essential to achieve regulatory compliance while preserving RF performance.

Structure of a Power Tree

As shown in Figure 7.2, at its core, a power tree represents the hierarchy of power conversion from the main input source down to each subsystem's specific voltage and current needs.

The structure of a power tree can be summarized as follows:

- *Primary input:* The primary input is the main power source for the system, such as a battery, an AC-DC converter, or a DC power supply. It typically provides the highest voltage in the power tree and must be capable of delivering enough current to meet the demands of all downstream components. The maximum current requirement for the primary input is determined through power budget analysis, which accounts for the power consumption of each subsystem and component within the design. The power budget analysis is discussed in the following section.
- *Regulators:* Switching regulators and LDOs are used to generate lower voltage levels from the primary source:
  - LDOs offer low noise but have lower efficiency, making them ideal for noise-sensitive applications.
  - Switching regulators provide high efficiency but typically generate more noise, making them suitable for efficiently creating intermediate voltage rails or powering components where higher noise levels are acceptable, such as digital circuits. These are particularly advantageous when there is a significant voltage difference between input and output, offering much higher efficiency than linear regulators and reducing thermal design constraints.

  A common strategy is to use a switching regulator to efficiently generate an intermediate voltage, followed by LDOs to create clean, low-noise supplies for sensitive circuits. Additionally, ultralow-noise switching regulators, such as the Silent Switcher series from Analog Devices, are available for cases where both efficiency and low noise are critical. Depending on system requirements, step-up (boost) or buck-boost converters may also be needed.
- *Load blocks:* The power tree also illustrates which voltage powers which subsystem or block, along with the associated load characteristics such as current consumption.

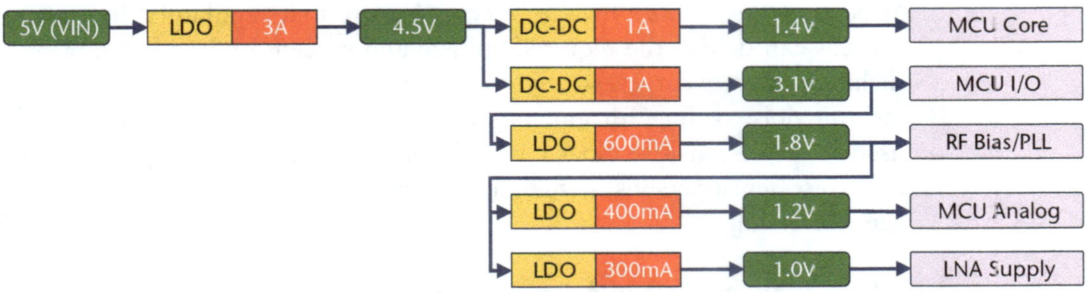

Figure 7.2 An example of a power tree. This power tree illustrates a typical multirail embedded RF system. A single 5-V input is converted into several low-noise and high-efficiency supply rails tailored for the microcontroller core, input/output (I/O) interfaces, analog blocks, RF bias networks, and LNAs.

Free applications are available for power-tree design and visualization (e.g., PTree). As we will see in the next section, a power-budget analysis is the essential first step before designing a power tree. Once the power tree is defined, a major architectural milestone is reached. The next phase involves selecting specific regulator part numbers, taking into account key parameters such as load and line regulation, quiescent current, dropout voltage, and other relevant performance metrics. These considerations are discussed in detail in Volume 2.

DC Power Budget Analysis

The first step in power planning is to accurately estimate the current consumption of each component and subsystem. This enables regulators and power supplies to be properly sized with sufficient margin for peak loads, startup events, and transient conditions. A well-defined power budget accounts for both steady-state and transient current draw, ensuring that each supply rail can meet demand while remaining within thermal, efficiency, and reliability limits. Without a disciplined power budget, RF systems become vulnerable to voltage droop, overheating, degraded RF performance, and unexpected system-level failures.

Here is a step-by-step approach for constructing a reliable power budget.

1. *List all power domains and loads:* Identify every subsystem that requires regulated power, RF front-end stages, baseband processing blocks, digital logic, converters, clocks, actuators, fans, sensors, and any auxiliary circuits. For each block (see Table 7.2), record:
   - Supply voltage (V).
   - Maximum or worst-case current (I_{max}), including peak loads and startup surges. For example, FPGA inrush current can be significantly higher than steady-state operation.
2. *Calculate power for each load:* Compute the power requirement using the formula: Power (W) = Voltage (V) × Current (A). Always size the power tree according to peak or worst-case values, not typical operating conditions.
3. *Calculate total power consumption:* Sum the maximum power values across all loads to determine the system's worst-case power consumption.
4. *Apply a safety margin:* Add a safety margin, typically 20%–30%, to account for manufacturing variations, component aging, temperature effects, and transient load conditions.
5. *Select the appropriate regulator type:* Ensure that regulators and power sources can deliver the required voltage and current under all conditions, including startup surges, steady-state operation, and transient events, while meeting noise, voltage-droop, and thermal constraints. LDOs are typically used for noise-sensitive blocks, while switching regulators are preferred for less noise-sensitive subsystems and for intermediate, high-efficiency voltage conversion.

Let's look at an example of power budget calculation for a small RF system. Suppose you are designing a small RF transceiver system with the blocks shown in Table 7.2. Each component's supply voltage and maximum current requirement are

Table 7.2 Power Budget Analysis for a Transceiver

Block	*Voltage (V)*	*Maximum Current (mA)*	*Power (W)*	*Noise Sensitivity*	*Recommended Regulator*
LNA	5.0	60	0.30	High	Switching + LDO
Phase-locked loop (PLL) synthesizer	5.0	300	1.50	High	Switching + LDO
PA driver	5.0	70	0.35	Medium	Switching only
PA	10.0	400	4.00	Low	Switching only
ADC	1.8	70	0.13	High	Switching + LDO
DAC	1.0	500	0.50	High	Switching + LDO
FPGA core	3.3	300	0.99	Low-medium	Switching only
FPGA I/O	3.3	50	0.17	Low	Switching only
Total		1,750	7.94		
With 25% margin		2,188	9.93		

shown. Accounting for a 25% design margin, the total maximum current reaches approximately 2.2A, with a total power consumption of around 10W. The highest required voltage is 10V, used by the PA. Therefore, a practical choice is to use a 12-V power source rated for 2.5A, ensuring an adequate margin for regulation losses and transient load conditions.

7.3 Sensors

In RF products, sensors play a critical role in monitoring, control, and protection of key subsystems. Although RF design is often focused on signal chains and electromagnetic performance, the reliability, safety, and efficiency of modern products depend heavily on a network of supporting sensors.

As discussed in Chapter 5, sensors act as the system's eyes and ears, continuously collecting real-time data that drives critical control decisions. They protect PAs from thermal stress, enable dynamic gain and bias adjustment, monitor supply voltages and currents, and track environmental conditions such as temperature, vibration, or humidity. Incorporating these feedback loops early in the design process enhances system robustness, reduces field failures, and improves performance across a wide range of operating conditions, ultimately extending the product's lifetime.

In this section, we examine the most commonly used sensors in RF systems and the communication interfaces that link them to microcontrollers, FPGAs, or CPUs for processing and decision-making. Volume 2 will expand on calibration methods, filtering techniques, and best practices for sensor placement to ensure accurate and reliable performance across the full operating envelope.

7.3.1 Types of Sensors in RF Systems

There are two main categories of sensors used in RF systems:

- *Direct RF sensors:* Measure quantities directly related to RF parameters, such as RF power, frequency, and phase.
- *Supporting sensors:* Monitor environmental and operational conditions that can impact RF performance, such as current, temperature, humidity, and pressure.

Table 7.3 summarizes the different types of sensors commonly used in RF systems, their use cases, and where they are typically employed.

7.3.2 Digital and Analog Sensors in RF Systems

Sensors can be broadly categorized by how they communicate their measurements: analog sensors and digital sensors. Understanding the differences between these two types is essential for selecting the right option for a given RF system and for ensuring proper interfacing with the rest of the architecture, such as the microcontroller.

Analog sensors output a continuous electrical signal, typically a voltage or current, proportional to the physical quantity being measured. Because of their fast response, they are often paired with analog circuits in applications where reaction time is critical. For example, excessive reflected power at the output of a PA can cause device failure within microseconds. In such cases, routing the sensor output through a microcontroller for processing before triggering a switch would be too slow. Instead, an analog comparator, typically implemented with an operational

Table 7.3 Different Types of Sensors Commonly Used in RF Systems

Sensor Type	*Purpose*	*Notes/Applications*
RF power sensors	Measure RF power (average, peak, pulse)	Essential for transmitter monitoring, PA protection, power calibration, and production testing
RF frequency counters	Measure the frequency of an RF signal accurately	Used for oscillator characterization, frequency stability monitoring, and test/validation setups
Phase detectors/ discriminators	Measure phase differences between two RF signals	Critical for PLLs, coherent receivers, and beamforming networks
VSWR/reflection coefficient sensors	Measure reflected power and impedance mismatch	Protects PAs from high VSWR; used in antenna tuners, protection circuits, and monitoring systems
Field probes (E-field/H-field)	Measure electric or magnetic fields in circuits or free space	Used for EMI/EMC diagnostics, near-field antenna measurements, and detecting RF leakage
Current and voltage sensors	Monitor power-supply rails feeding the system	Enable early fault detection; ensure stable biasing for active RF devices (PAs, LNAs, VCOs) and mixed signal components (DACs, ADCs)
Environmental sensors	Monitor temperature, humidity, and pressure	Protects temperature-sensitive RF components such as PAs and oscillators; supports derating and shutdown mechanisms
Mechanical and displacement sensors	Detect mechanical stress, strain, movement, or misalignment	Used in phased arrays, high-vibration platforms (UAVs, vehicles, aircraft), connector integrity monitoring, and tunable RF hardware such as cavity filters or mechanical phase shifters

amplifier, is combined with an analog power sensor to provide an immediate hardware-level reaction.

In applications where fast response is not critical, or where digital processing of the sensor output is required, such as temperature sensing, analog sensors can be interfaced directly with a microcontroller. Because these sensors produce a continuous analog voltage or current, their signals must be digitized using an ADC, either through the microcontroller's integrated converter or an external ADC component. The quality of this conversion stage is critical: ADC parameters such as resolution, dynamic range, sampling rate, input-referred noise, linearity, and reference stability directly determine the accuracy and fidelity of the sensor measurements.

Digital sensors perform measurements and output the results in digital form, typically over serial communication interfaces such as I^2C, SPI, UART, or other specialized buses. Unlike analog sensors, they eliminate the need for external ADCs and often include built-in filtering, calibration, or compensation functions.

Table 7.4 summarizes the key differences between analog and digital sensors across several design and implementation considerations.

Choosing between analog and digital sensors depends on factors such as the system bandwidth, latency and precision requirements, noise environment, and overall architecture. Analog sensors are preferred in high-speed or wide-bandwidth applications, such as fast power tracking or time-critical protection loops, where low latency and fine analog control are essential. They are also ideal for precision RF measurement setups or designs that require custom analog front-end optimization. Digital sensors, by contrast, are favored in electrically noisy environments where EMI immunity is crucial, as well as in low-speed environmental monitoring tasks such as temperature or humidity sensing. They integrate naturally into systems built

Table 7.4 Comparison Between Analog and Digital Sensors Used in RF Systems

Aspect	*Analog Sensors*	*Digital Sensors*
Output	Continuous analog signal	Discrete digital word
Interface	Requires an ADC and analog front end to interface with microcontrollers or processors	Direct interface to microcontrollers or FPGAs via digital buses (I^2C, SPI, UART)
Precision	Strongly influenced by system noise, ADC resolution, reference stability, and analog front-end design	Determined by the sensor's internal ADC and digital calibration; typically stable and consistent
Bandwidth/ speed	Can achieve very high bandwidth; suitable for fast or dynamic sensing	Limited by internal conversion rate and communication protocol; generally slower than raw analog signals
Susceptibility to noise	High, susceptible to cable noise, EMI, and ground shifts	Low, digital signaling is much more immune to noise
Long cable runs	Performance degrades with distance; requires shielding or buffering	More robust over long distances, especially with differential protocols (e.g., RS-485, CAN, USB)
Pros	Simple, low-latency, flexible analog signal processing	Noise-immune, easy to integrate, often factory-calibrated, reduced analog design overhead
Cons	Requires careful analog design; sensitive to noise and grounding; ADC adds complexity and cost	Limited bandwidth; dependent on communication timing; may require specific drivers or protocol support

around microcontrollers, FPGAs, or DSP chips. Moreover, they are advantageous in high-volume products where ease of integration, built-in calibration, and consistent performance are priorities.

7.3.3 Sensor Characteristics

Selecting the right sensor for an RF system goes beyond choosing the type. Equally important are the sensor characteristics, which determine the quality, reliability, and timeliness of the measurements. These parameters are critical for systems where calibration, protection, and control depend on precise readouts.

Table 7.5 shows the key sensor characteristics that must be considered during design and integration.

Choosing a sensor with poor characteristics can compromise the entire system. For instance:

- A highly accurate sensor with poor response time might protect too late.
- A fast but noisy sensor may trigger false alarms.
- A narrow dynamic range might miss edge cases in field conditions.

Table 7.5 Key Sensor Characteristics and Their Relevance in RF Systems

Characteristic	*Definition*	*Importance in RF Systems*	*Design Considerations*
Accuracy	Closeness of the measurement to the true physical value	Ensures correct calibration, thermal compensation, and protection thresholds; inaccurate data can degrade system performance or mask emerging faults	Prioritize for power, temperature, and voltage sensing where correct absolute values directly impact system reliability
Precision	Repeatability and consistency of measurements under identical conditions	Critical for detecting small trends such as drift, aging, or subtle bias changes, even if the absolute value is slightly offset	Important for predictive maintenance and control loops; high precision reduces noise in monitoring and improves stability
Sensitivity	The smallest detectable change in the measured quantity	Enables early detection of thermal drift, bias instability, and current variations linked to component aging	Essential for predictive maintenance, bias monitoring, and fine-grained control loops
Dynamic range	The ratio between the largest and smallest measurable values with acceptable error	Required to capture both small and large signals without clipping or masking	Critical for power, voltage, and current sensors; match the range to the expected operating envelope
Response time	How quickly the sensor reacts to changes in the input parameter	Determines whether protection circuits can respond before damage occurs (e.g., overcurrent, overtemperature, high VSWR events)	Must align with system dynamics, microseconds for PA protection, milliseconds to seconds for thermal control
Linearity	Consistency of the input-output relationship across the measurement span	Improves calibration accuracy and simplifies control logic; nonlinearity increases firmware complexity	Prefer sensors with high linearity to minimize correction algorithms, and lookup tables

Therefore, designers must carefully balance these characteristics based on the application, safety margins, and system requirements.

7.3.4 Interfacing Sensors

When working with sensors in RF and embedded systems, the measured data often needs to be transferred to a controller, such as a microcontroller, FPGA, or CPU, via a communication protocol for processing, monitoring, and decision-making. This flow is illustrated in Figure 7.3.

Choosing the right protocol impacts speed, noise immunity, scalability, and software complexity. Table 7.6 compares major communication protocols for sensors.

Table 7.7 shows practical considerations for selecting the right protocol.

Together, these considerations ensure that the sensor subsystem delivers reliable, accurate, and actionable data, forming a trustworthy foundation for system-level performance and informed engineering decisions.

7.4 Processing Platforms

Modern RF systems rely on a diverse set of processing platforms to handle everything from low-level hardware control to high-speed data processing. Figure 7.4 illustrates the functional roles of processing platforms in a modern RF system.

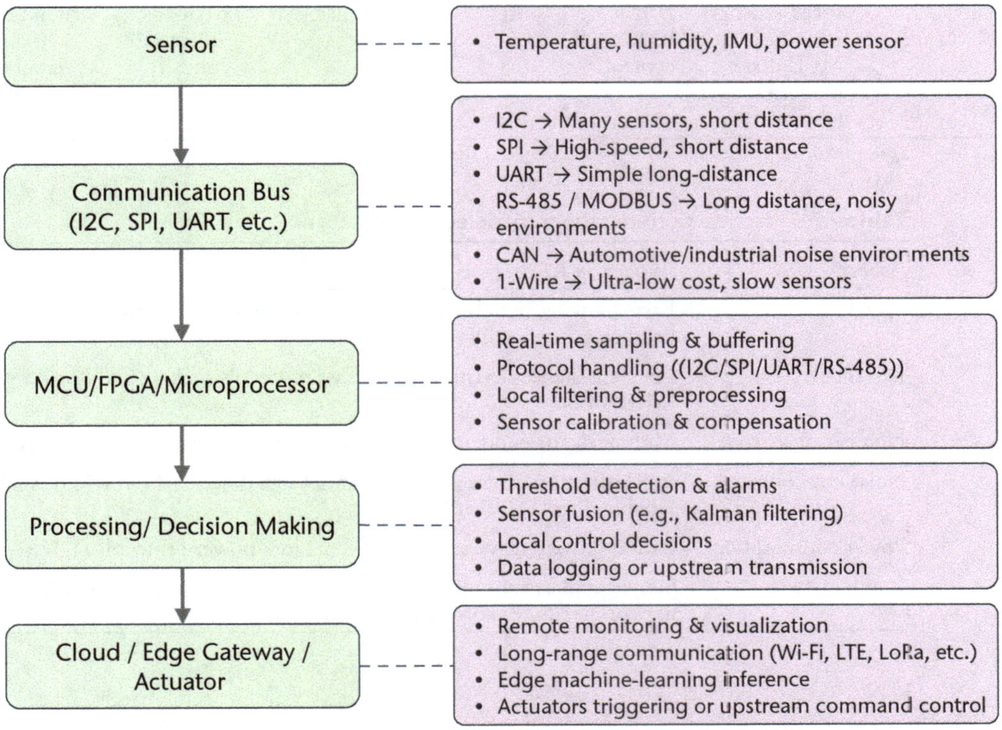

Figure 7.3 Sensor interface hierarchy and typical communication bus selection. The choice of interface depends on factors such as distance between the sensor and the controller, speed, noise environment, and system cost.

Table 7.6 Comparison of Communication Protocols for Sensor Integration

Protocol Attribute	*I^2C/I^3C*	*SPI*	*UART*	*1-Wire*	*CAN*	*MODBUS (RS-485)*
Key features	Two-wire (SCL, SDA), addressable, multidevice bus	Four-wire (MOSI, MISO, SCLK, CS), full-duplex	Asynchronous, two-wire (TX/RX), simple framing	Single data + ground, unique device IDs	Multimaster, differential signaling, EMI-robust	Master–slave, RS-485 or TCP/IP
Typical speed	Up to 1 Mbps/ up to 12.5 Mbps	10–50 Mbps	300 bps–few Mbps	~15 kbps	125 kbps–1 Mbps (5 Mbps CAN FD)	9.6 kbps–1 Mbps
Maximum devices	~100/many	Few (CS-limited)	1-to-1, multidrop with extra layers	Many	112+	32+ (247 typical)
Distance capability	Short (~0.5–1m)	Short (<0.5–1m)	Medium (15m native, longer with RS-232/ RS-485)	Long (10–100m)	Long (40m at 1 Mbps, up to 1 km lower rates)	Very long (up to 1,200m)
Noise immunity	Moderate	Low	Moderate	Moderate	Excellent EMI immunity	Good immunity
Power consumption	Low	Moderate	Low-moderate	Very low	Moderate	Moderate
Complexity	Low-moderate	Moderate	Low	Low	High	Low-moderate
Best for/ typical use cases	Multiple low to medium speed sensors near the MCU/dense, modern sensor networks	High-speed ADCs, DACs, RFIC control	GPS, Bluetooth modules, debug ports	Low-cost distributed sensing	Automotive, UAV payloads, industrial control	Industrial networks, building automation, remote RF monitoring

Table 7.7 Practical Considerations for Selecting a Protocol

Aspect	*Impact on RF Systems*
Speed	Fast-changing signals (e.g., RF envelope tracking) require high-speed protocols such as SPI
Distance	Longer distances (>1m) favor robust protocols such as CAN or RS-485 (MODBUS)
Number of sensors	Multidevice-friendly protocols (I^2C, CAN) are better for dense sensor arrays
Noise environment	In RF-noisy environments, differential signaling (CAN, RS-485) is more reliable than UART or I^2C
Power consumption	Battery-powered systems benefit from low-power protocols (1-Wire, I^2C)
Complexity	Simple applications may only need UART, while distributed or high-speed networks require more complex protocols

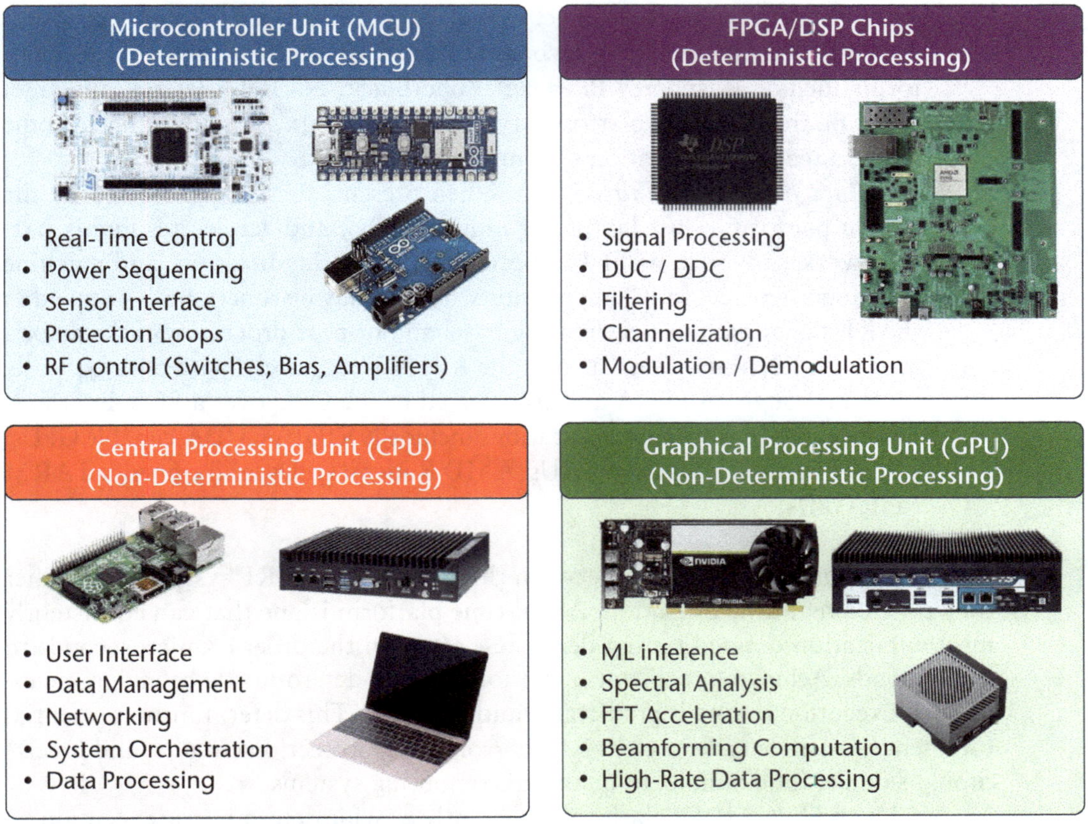

Figure 7.4 Functional roles of MCU, FPGA/DSP, CPU, and GPU in a modern RF system, highlighting deterministic and nondeterministic processing domains.

The main categories of processing platforms include:

- *Microcontroller unit (MCU):* Provides deterministic, real-time control of RF switches, bias networks, sensors, power sequencing, and protection loops, using platforms such as STM32, MSP430, Arduino (ATmega), PIC, or ESP32 when operated in the MCU mode.
- *DSP chips:* Specialized processors optimized for multiply–accumulate–intensive operations used in filtering, fast Fourier transforms (FFTs), modulation and demodulation, channel equalization, and baseband algorithms, with common families including TI C6000, Analog Devices SHARC, and Qualcomm Hexagon DSPs.
- *FPGA:* Enables deterministic, parallel DSP for high-speed data paths, including digital upconversion and downconversion (DUC/DDC), filtering, channelization, and wideband acquisition, using devices such as Xilinx Zynq, UltraScale/UltraScale+, Intel (Altera) Cyclone/Arria/Stratix, and RF system-on-chip (RFSoC) platforms with integrated ADCs and DACs.
- *CPU:* Typically hosted on a single-board computer (SBC), laptop, or industrial PC, the CPU executes user interfaces (UIs), protocol handling, general application logic, data management, and system coordination. An SBC integrates

the CPU with memory, storage, networking, and I/O on a single board but typically does not include a keyboard, display, or mouse. Representative platforms include Raspberry Pi, BeagleBone Black, NVIDIA Jetson, and Intel NUC Board. The host platforms typically run Linux, Windows, or, in some cases, a real-time operating system (RTOS).

- *Graphics processing unit (GPU):* When present, the GPU is hosted by the system platform (SBC, laptop, or industrial PC) and accelerates highly parallel workloads such as FFTs, spectral analysis, classification, and machine learning inference. In RF applications, a GPU may be functionally coupled to the RF signal path by accelerating baseband or post-processing tasks associated with RF transceivers such as the ADRV90xx family, with data typically routed through an FPGA and transferred to the GPU over a PCIe interface. Representative GPU implementations include NVIDIA CUDA-capable GPUs, NVIDIA Jetson integrated GPUs, AMD Radeon GPUs, and embedded ARM Mali GPUs.

One important distinction between these platforms in RF systems is whether they provide real-time operation. A real-time platform is one that can consistently meet application-defined timing deadlines, often on the order of microseconds to milliseconds. Achieving real-time operation requires deterministic behavior, meaning that execution timing is predictable and bounded. This determinism is essential for latency-sensitive tasks such as RF control loops, protection circuits, and precise timing signals, for example, in quantum computing systems.

MCUs and FPGA/DSP platforms inherently provide deterministic execution by operating with fixed, tightly controlled timing. In contrast, CPUs and GPUs running general-purpose operating systems such as Linux or Windows are typically nonreal-time platforms; although powerful, their execution timing is nondeterministic due to operating-system scheduling, interruptions, and context switching, making them unsuitable for timing-critical RF functions. In some systems, an RTOS is deployed on a CPU to provide bounded real-time behavior and prioritized task scheduling. While this improves timing predictability, it generally does not achieve the same level of determinism as MCU-based or FPGA-based implementations, which rely more heavily on hardware-level timing control.

7.4.1 Digital Dominance in the Modern RF Signal Chain

A clear industry trend is the ongoing shift of functionality from the analog domain to the digital domain. In many modern RF products, the analog portion of the RF chain has become remarkably minimal: on the receive side, it often consists only of low-noise amplification, occasional mixing, filtering, or signal routing; on the transmit side, signal routing, filtering, occasional mixing, and power amplification may be all that remain. Nearly everything else, channelization, frequency conversion (DDC/DUC), waveform generation, beamforming, modulation, demodulation, calibration, equalization, and system-level logic, now occurs in the digital domain.

This shift has been accelerated by highly integrated platforms such as mixed-signal transceiver ICs (e.g., the ADRV90xx series) and RFSoCs such as the RFSoC 4×2 and ZCU216 from Xilinx. These devices enable highly capable, compact, software-defined radio (SDR) architectures by tightly coupling high-speed converters (ADCs and DACs) with on-chip digital processing.

Functions that were once implemented using analog hardware now occur predominantly in the digital domain, with AI and machine learning increasingly integrated into signal processing, classification, and system optimization tasks. As a result, RF engineers developing modern systems and products must have at least a working understanding of software-defined architectures, firmware interfaces, embedded processing, and emerging AI-assisted workflows, topics that played only a minor role in the day-to-day work of RF engineers a decade ago.

7.4.2 Microcontrollers

MCUs play a central role in the control, monitoring, and coordination of RF systems. While they do not directly handle RF signals, they provide flexible, programmable control over key system functions in RF products, including:

- Power sequencing for PAs, LNAs, and bias circuits;
- Digital control of phase shifters, attenuators, or switches;
- Monitoring of temperature, voltage, current, and status signals;
- SPI/I^{2}C/UART interfaces to control RFICs or front-end modules;
- Interfacing with external systems, displays, or user inputs;
- Bootloading firmware into FPGAs, direct digital synthesis (DDS) chips, or RFSoC.

MCUs are especially useful when full-blown FPGAs or processors are unnecessary. You will find MCUs embedded in nearly every modern RF product, from compact modules to large systems. Here are some of the common applications of MCUs in RF systems:

- *Phased arrays:* Controls phase shifters, attenuators, and beamformer ICs, while monitoring bias currents, temperature, and output power.
- *RF front ends:* Sequences bias voltages, configures RF switches and attenuators, adjusts variable-gain amplifiers, and manages protection logic.
- *Measurement equipment:* Implements user-interface control, display management, SCPI/remote command handling, and coordination of internal calibration routines.

To integrate an MCU into an RF product, several infrastructure elements are required:

- *Power supply:* Stable, well-regulated rails (commonly 5V, 3.3V, and/or 1.8V) with proper sequencing if multiple domains are involved.
- *Clocking:* If the MCU does not rely solely on an internal oscillator, provide an external crystal/resonator or reference clock source to meet timing accuracy and communication requirements.
- *Programming and debug access:* A physical interface such as JTAG, SWD, or UART to support firmware loading, debugging, and production programming.

- *Isolation and decoupling:* Careful decoupling, grounding, and (when needed) isolation practices, especially when digital switching noise is near sensitive RF nodes.

Engineers integrating MCUs into RF systems should also be familiar with:

- *Embedded firmware development:* Proficiency in C/C++, along with an understanding of real-time constraints, timers, and interrupt handling, is essential for deterministic control and reliable system behavior.
- *Peripheral configuration:* Practical experience configuring and using general-purpose input/output (GPIO), SPI, I^2C, ADC, direct memory access (DMA), and pulse width modulation (PWM) peripherals to interface with RF components, sensors, and support circuitry.
- *Digital-to-RF interfacing:* Knowledge of how digital control interacts with RF hardware, including SPI timing for phase shifters or attenuators, synchronization of control pulses, and orchestration of calibration and alignment sequences.
- *Low-power design:* Use of sleep and deep-sleep states, wake-up sources, clock gating, and power-domain sequencing to minimize consumption in standby and battery-powered systems.
- *Bootloaders and firmware update mechanisms:* Implementation of bootloaders, in-system programming, and secure firmware update strategies to enable field upgrades, maintainability, and long-term product support.

While these are not traditionally considered RF skills, they are increasingly expected of RF engineers working on modern, integrated product development teams.

MCUs are typically small, inexpensive, and low-power, making them attractive in SWaP-constrained applications. However, poor planning, such as over-specifying the MCU, neglecting EMI considerations, or underestimating firmware complexity, can degrade system performance or introduce unnecessary design overhead. Table 7.8 summarizes the MCU's SWaP-C characteristics.

Table 7.9 summarizes widely used MCU families in RF systems, highlighting their typical applications, advantages, and limitations to guide engineers in selecting the right MCU for a given design.

MCUs typically lack native network connectivity, so external modules such as Ethernet shields (for example, Arduino Ethernet Shield) are often required. The

Table 7.8 SWaP-C Properties of an MCU

Metric	*Impact*
Size	Can be extremely compact (as small as 2 × 2 mm packages)
Weight	Negligible, often <1g
Power	Ranges from microwatts (in sleep) to hundreds of milliwatts active
Cost	Usually <\$5 in moderate volume, often <\$1 for the MCU itself; however, peripheral circuitry (regulators, clock sources, programming interface, protection components) must also be considered

Table 7.9 Common MCU Families and Their Role in RF Systems

MCU Family	*Typical RF Use Cases*	*Key Advantages*	*Limitations*
STM32 (STMicro)	SPI/I^{2}C control of beamformers, PLLs, RF switches; PA bias and protection	High performance, rich peripherals, widely supported	More power-hungry than ultralow-power MCUs
AVR/ATmega (Microchip)	Prototyping RF bias networks, temperature/current sensor readout, small control loops	Easy to program, rapid prototyping, low cost	Limited speed and memory, not suited for complex RF tasks
PIC (Microchip)	Filter banks, switch matrices, legacy RF control interfaces	Robust, mature ecosystem, low cost	Less flexible for modern high-speed interfaces
MSP430 (TI)	Wireless sensor nodes, low-power RF telemetry, duty-cycled monitoring	Ultralow power consumption, long battery life	Limited processing capability, small memory
ESP32/ESP8266 (Espressif)	Remote telemetry, wireless control of RF hardware, IoT-enabled RF devices	Integrated Wi-Fi/ Bluetooth, strong IoT ecosystem	Higher power draw than basic MCUs, limited real-time performance

NodeMCU platform, based on the ESP8266 or ESP32, provides a low-cost alternative by integrating Wi-Fi, and Bluetooth in the case of the ESP32, directly into the MCU. Compared to SBCs such as Raspberry Pi or BeagleBone, NodeMCU does not run an operating system (OS) such as Linux but excels in compact, low-power, network-enabled RF control and IoT applications.

7.4.3 General-Purpose Processors

CPUs and GPUs play an important role in RF systems that require more than basic control. They are typically chosen when:

- A user interface is needed to interact with the product, including integration with a graphical display, keyboard, mouse, or touchscreen.
- A full OS is required to run services, manage files, or support user applications.
- Complex software stacks are needed for data processing and interaction with the product (for example, Python, MATLAB runtime, or Qt-based GUIs).
- Network connectivity (Ethernet, Wi-Fi) and higher-level protocol stacks are required, for example, TCP/IP with HTTP/HTTPS for web interfaces, SSH for remote access, MQTT for cloud or IoT messaging, or FTP/SFTP for data transfer.
- The system must log data, host a web interface, or interact with databases or cloud services.

However, as mentioned earlier, these platforms are not real-time processors. They are therefore typically used for high-level control, user interaction, and data processing functions that operate outside the strict real-time control loop.

CPUs handle complex, multithreaded, data-heavy, or user-interface–intensive applications. CPUs excel at sequential, general-purpose processing and high-level system coordination.

GPUs are used when workloads demand massive parallel processing. In RF systems, this distinction becomes critical when handling large datasets, performing high-rate spectral or IQ analysis, or running machine-learning–based classification algorithms [2,3]. In constrained products, GPUs are rarely used unless essential (e.g., vision, machine learning, real-time classification).

Together, CPUs and GPUs provide high-throughput processing and broad software-ecosystem support for computationally intensive tasks that do not require tight real-time constraints, making them valuable for visualization, post-processing, protocol handling, data analytics, and edge-level AI functions.

Here are some applications of CPUs and GPUs in RF systems:

- *RF signal monitoring systems:* Real-time FFT computation, waterfall displays, and spectrum-mask analysis.
- *Cognitive radios:* Signal classification and dynamic spectrum access, including machine-learning inference for anomaly detection, interference classification, and localization [4].
- *EW systems:* Machine-learning–based emitter detection, classification, and identification.
- *Remote sensing and ground-penetrating radar (GPR) platforms:* Onboard data reduction, preprocessing, and feature extraction prior to storage or transmission.
- *Edge computing:* Local processing, such as spectrum analysis, performed instead of transmitting raw RF data to a remote server. This reduces latency, saves bandwidth, and enables real-time decision making in applications such as electronic warfare, remote sensing, and IoT-connected RF systems.

Remember that CPUs and GPUs are hosted on a platform such as a laptop, desktop, or SBC. These platforms require robust hardware support to operate reliably in RF products:

- *Stable power:* Typically 5-V, 12-V, or AC input with clean regulation and sufficient current margin, especially during boot and high-load conditions.
- *Mass storage:* SD card, eMMC, or SSD for the operating system, applications, and data logging.
- *Peripheral I/O:* GPIO, SPI, I^2C, UART, USB, and Ethernet to interface with RF hardware, sensors, and external systems.
- *Cooling:* Often passively cooled at low to moderate loads, but heat sinks or active cooling may be required under sustained processing or GPU acceleration.
- *Secure boot and file integrity:* Essential for products in regulated or security-sensitive environments to ensure trusted software execution.
- *EMI control:* Careful shielding, filtering, grounding, and physical separation, especially when placed near sensitive RF signal paths.

Table 7.10 shows some of the widely used SBCs in RF applications.

Designing and using CPU/GPU platforms effectively requires familiarity with the following areas:

Table 7.10 SBC Platforms and Their Applications in RF Systems

Platform/Family	*Typical RF Use Cases*	*Key Advantages*	*Limitations*
Raspberry Pi	User interfaces, RF data logging, lightweight software defined radio (SDR) control, educational/prototyping	Low cost, large community, easy to use	Not industrial-grade, limited real-time performance
BeagleBone Black	RF control interfaces, networked test setups, low-level I/O with programmable real time unit (PRUs)	Rich I/O, PRUs for real-time tasks, open-source	Lower performance than Pi/Jetson, smaller ecosystem
Nvidia Jetson (Nano/Xavier)	Edge machine learning for spectrum classification, real-time anomaly detection, SDR acceleration	Powerful GPU for AI/ machine learning, strong edge computing	Higher power draw, more expensive
Xilinx Zynq SoC (ZedBoard, UltraScale+)	SDR platforms, phased arrays, RFSoC-based systems, mixed FPGA–Linux tasks	Combines ARM cores with FPGA fabric, industrial-grade	Steeper learning curve, higher cost
Intel/AMD Embedded Boards (x86 SBCs)	High-end RF instrumentation, data acquisition, lab-grade control	PC-class performance, wide software compatibility	Power-hungry, less suited for mobile/ edge RF devices

- *Multithreaded software development:* Programming in C++, Python, or GPU frameworks such as CUDA or OpenCL.
- *Parallel computing models:* Concepts such as SIMD execution, thread scheduling, and workload partitioning.
- *DSP:* Practical use of NumPy, SciPy, or MATLAB for RF signal analysis and processing.
- *Machine-learning frameworks:* Experience with TensorFlow, PyTorch, or scikit-learn for inference and classification tasks.
- *OS administration:* System configuration, performance tuning, and software optimization.
- *Driver-level integration:* Developing or interfacing low-level drivers for custom hardware, sensors, or RF peripherals.

CPUs and GPUs are the most resource-hungry platforms, but also the most powerful in terms of computation. Table 7.11 shows the SWaP-C considerations.

Tables 7.12 and 7.13 show the recommended platforms for CPUs and GPUs in RF systems, respectively.

Table 7.11 SWaP-C Properties of CPUs and GPUs

Metric	*Impact*
Size	Varies from compact embedded modules (e.g., NVIDIA Jetson Nano) to full desktop-class or server-class PCIe cards
Weight	Moderate to high, depending on heat sink, cooling solution, and overall form factor
Power consumption	High, typically 5–100W for embedded GPUs and significantly higher for desktop-class devices; usually requires active cooling
Cost	Medium to high, approximately $50 to more than $2,000, depending on performance class and intended application

Table 7.12 Recommended Platforms for CPUs in RF Systems

Platform/Family	*Typical RF Use Cases*	*Key Advantages*	*Limitations*
x86 CPUs (Intel Core/i7/i9, AMD Ryzen/EPYC)	RF simulation tools, SDR host PCs, test instrumentation backends	High single-thread performance, large software ecosystem	High power draw, not suited for embedded edge
ARM Cortex-A (smartphones, SBCs, SoCs)	Embedded RF systems, protocol stacks, control, and user interface	Power-efficient, wide adoption, runs Linux/Android	Limited raw DSP compared to GPUs/FPGAs
Server-grade CPUs (Xeon, EPYC)	Large-scale RF data processing, cloud-based spectrum analysis	Multicore scalability, large memory bandwidth	Very power-hungry, high cost, not portable
RISC-V Emerging CPUs	Future RF SoC integration, low-cost IoT RF applications	Open-source instruction set architecture (ISA), customizable	Immature ecosystem, limited current adoption

Table 7.13 Recommended Platforms for GPUs in RF Systems

Platform/Family	*Typical RF Use Cases*	*Key Advantages*	*Limitations*
Nvidia CUDA GPUs	Massive parallel RF DSP, AI/machine learning for spectrum classification, MIMO channel simulation	Huge parallelism, strong machine learning/DSP libraries (CUDA, cuFFT)	High power draw, requires PCIe/host CPU
AMD GPUs (ROCm)	SDR acceleration, RF simulations, parallel DSP	Open ecosystem (ROCm), strong processing capability	Less mature machine learning/DSP ecosystem than Nvidia
Embedded GPUs (Nvidia Jetson, Qualcomm Adreno, ARM Mali)	Edge machine learning for RF anomaly detection, portable SDRs	Good balance of processing and power efficiency	Limited performance versus desktop GPUs
Intel GPUs (Arc/integrated)	Supplementary RF DSP, lightweight machine learning on laptops or SBCs	Broad availability, integrated in CPUs	Limited software ecosystem, weaker performance

7.4.4 DSP Chips

A DSP chip is a processor designed specifically for real-time signal processing tasks such as filtering, FFTs, modulation/demodulation, and control loops. Like a microcontroller or CPU, a DSP runs compiled software, but its architecture and instruction set are optimized for fast, deterministic math operations on data streams.

You can think of a DSP as a purpose-built signal-processing engine. It is less flexible than an FPGA but typically offers better SWaP-C efficiency for many RF signal-processing tasks, especially when the algorithms are well defined and stable.

Internally, DSPs include hardware features tailored for signal processing, such as:

- Multiply-accumulate (MAC) units for fast filtering and FFTs;
- Specialized arithmetic logic units (ALUs) and barrel shifters for bit shifting in FFT stages, automatic gain control (AGC) loops, and fixed-point arithmetic;
- Hardware loop support for efficient real-time execution;

- Circular and bit-reversed addressing for FIR filters and FFTs;
- Tightly coupled memory and DMA engines for efficient sample movement.

These features allow DSPs to process continuous data streams with low latency and predictable timing, which is critical in RF systems.

By contrast, as we will see shortly, an FPGA is a fully reconfigurable logic fabric made of lookup tables, flip-flops, routing resources, and DSP slices. Instead of running software, you define the hardware itself, building custom pipelines, filters, and data paths that operate in parallel and at precise timing.

From a design perspective, a DSP is programmed like a processor (C/C++ or assembly) but an FPGA is designed using hardware description languages (VHDL or Verilog). In short, a DSP gives you a fixed, highly optimized signal-processing engine, while an FPGA lets you build your own signal-processing engine.

DSPs are often a strong choice in RF products when real-time performance is required, the algorithms are well understood, and SWaP-C efficiency matters more than ultimate flexibility.

Here are some typical applications of DSP chips:

- Digital filtering (FIR/IIR) for channel shaping, equalization, and noise reduction;
- FFTs for spectrum analysis and signal monitoring;
- Modulation and demodulation for communication waveforms (QPSK, QAM, OFDM);
- Channelization and multicarrier processing in SDRs and telecom systems [5];
- Beamforming and array signal processing for phased arrays;
- Baseband processing in modems, SATCOM terminals, and radar backends;
- Adaptive algorithms such as least mean square (LMS) filtering, echo cancellation, and interference suppression for time-varying RF environments.

DSP chips typically require the following system resources:

- *Moderate power budget:* Ranging from hundreds of milliwatts to a few watts, depending on clock rate and processing load.
- *External memory:* Static RAM (SRAM) or dynamic RAM (DRAM) for buffering I/Q data, filter coefficients, FFT frames, and intermediate results.
- *Stable clocking:* Low-jitter clock sources to ensure deterministic timing through the DSP processing pipeline.
- *High-speed I/O interfaces:* SPI, parallel buses, or dedicated data ports to support efficient sample movement to and from ADCs, DACs, or FPGAs.
- *Thermal management:* Adequate cooling for sustained high-performance operation.
- *Mature toolchains:* Reliable compilers, libraries, and debuggers to generate optimized DSP code.

Using DSPs efficiently requires both algorithmic and system-level expertise:

- *Strong DSP fundamentals:* FFTs, digital filtering, modulation schemes, channelization, and adaptive algorithms.

- *C/C++ proficiency:* Including fixed-point and floating-point optimization techniques.
- *DSP development environments:* Experience with tools such as TI Code Composer Studio, ADI CrossCore, or Qualcomm Hexagon SDK.
- *Performance optimization:* Tuning MAC-heavy workloads using loop unrolling, vectorization, and memory-efficient coding practices.
- *Real-time programming:* Deterministic scheduling, interrupt handling, and low-latency pipeline design.
- *Memory architecture awareness:* Effective use of scratchpad memories, DMA engines, and double-buffering strategies.
- *Debugging and profiling:* Using simulators, performance analyzers, and hardware trace tools to identify bottlenecks.

Table 7.14 shows the SWAP-C characteristics of DSPs.

Table 7.15 shows some of the widely used DSP platforms for RF systems.

Table 7.14 SWaP-C Characteristics of DSP Chips

Metric	*Impact*
Size	Typically small IC packages (QFP/BGA), suitable for compact embedded boards
Weight	Very low, comparable to MCUs; negligible in system-level weight budgets
Power consumption	Low to moderate (hundreds of mW to a few watts), depending on processing throughput and clock speed
Cost	Low to medium (approximately \$5–\$50), varying by performance class and feature set

Table 7.15 DSP Platforms for RF Applications

Platform/Family	*Typical RF Use Cases*	*Key Advantages*	*Limitations*
TI C6000/C66x Series	Filtering, FFTs, modulation/demodulation, channelization in SDRs and RF instruments	Mature ecosystem, high DSP throughput, good development tools	Less flexible than FPGAs; aging architecture
Analog Devices SHARC (ADSP-214xx)	Audio/RF baseband processing, beamforming, multichannel filtering	Excellent floating-point performance, low latency	Limited integration with modern RF SoCs
Qualcomm Hexagon DSP (integrated in SoCs)	Mobile RF baseband, communication modems, AI-assisted signal classification	Low power, tightly integrated into mobile SoCs, efficient vector processing	Not available as standalone; limited programmability for custom RF applications
Cadence Tensilica HiFi DSP	Low-power RF baseband, IoT radio stacks, filtering/equalization	Configurable architecture, optimized for low-power DSP workloads	Often embedded inside SoCs, not standalone
NXP StarCore DSP	Legacy telecom, SATCOM modems, wideband filtering	Strong telecom pedigree, stable telecom toolchain	Older platform with limited modern ecosystem support

7.4.5 Programmable Logic: Deterministic High-Performance DSP

FPGAs bring reconfigurable, high-speed digital processing to RF systems, acting as glue logic, waveform processors, or complete digital front ends. While they can consume more power and resources than microcontrollers or DSP chips, they are essential when timing precision, parallelism, or data throughput exceed what a DSP chip or general-purpose processor can handle. FPGAs often sit between the RF chain and the CPU or directly connect to high-speed ADCs/DACs.

FPGAs are used when an RF product requires:

- *Real-time signal processing:* Digital upconversion or downconversion, channel filtering, FFTs, beamforming, and other latency-critical DSP functions.
- *Precise timing control:* Deterministic clocks, synchronous triggering, and pulse generation with subnanosecond alignment.
- *High-speed digital interfaces:* Support for standards such as JESD204B, LVDS, and high-speed SerDes links to ADCs, DACs, and RF transceivers.
- *Custom protocols and digital interfacing:* Implementing application-specific interfaces or fast alternatives to standard SPI or UART.
- *Multiple concurrent control paths:* Parallel state machines and control logic that are difficult or inefficient to implement in sequential firmware.
- *Partial or full reconfigurability:* The ability to modify data paths without changing hardware.

FPGAs are widely used in high-performance or highly flexible RF systems:

- *SDRs:* Implement modulation, filtering, digital upconversion or downconversion, and framing.
- *Radar and EW systems:* Perform real-time beamforming, matched filtering, and pulse compression.
- *5G/6G prototyping platforms:* Handle baseband physical layer processing and flexible RF front ends.
- *Quantum control systems:* Generate fast, coherent control pulses and precise timing triggers.
- *RF test equipment:* Enable configurable signal generation, signal capture, and deterministic triggering.

FPGAs typically require more system support than microcontrollers:

- *Stable multirail power supplies:* Separate rails for core logic, I/O banks, and high-speed transceivers, often with strict sequencing requirements.
- *Clocking infrastructure:* One or more external clock sources and internal PLLs for high-speed and low-jitter operation.
- *Configuration memory:* External SPI flash or configuration devices to load the FPGA bitstream at power-up.
- *Thermal management:* Heat sinks or airflow may be required in high-performance or high-density designs.

- *High-speed PCB design:* Controlled impedance routing, length matching, power integrity, grounding, and EMI mitigation.
- *Debug and bring-up tools:* JTAG access, external logic analyzers, or embedded logic analyzers for internal signal visibility.

Designing FPGA-based RF systems typically requires:

- *HDL development:* VHDL, Verilog, SystemVerilog, or high-level synthesis (HLS) for some platforms.
- *Digital design fundamentals:* Timing closure, pipelining, resource utilization, and metastability management.
- *Clock-domain crossing:* Safe transfer of data between asynchronous or independently clocked domains.
- *Toolchain proficiency:* Vendor tools for synthesis, place-and-route, timing analysis, and debugging.
- *RF-aware DSP implementation:* Digital filters, numerically controlled oscillators (NCOs), DDC/DUC, and fixed-point signal processing.
- *RF hardware integration:* SPI/I^2C for configuration, parallel or high-speed serial interfaces for data, and precise clock and trigger alignment.

FPGAs offer high performance at the cost of complexity and power. Table 7.16 shows the SWaP-C considerations.

FPGAs provide greater design flexibility and capability but should only be used when justified, particularly in cost-sensitive or power-sensitive products. Development and debugging are also significantly more time-consuming compared to MCUs.

In some cases, a lighter form of programmable logic is sufficient. A complex programmable logic device (CPLD) is a small, nonvolatile programmable logic device used for simple, deterministic digital control tasks. Unlike an FPGA, a CPLD powers up instantly with no configuration loading, offers predictable timing, and typically has lower cost and power consumption. For example, in phased array systems, a CPLD is often used to manage SPI communication with beamformer chips, enabling fast and reliable configuration.

Table 7.17 shows some of the FPGA platforms used in RF products.

7.4.6 Choosing the Right Processing Platform

We have discussed the SWaP-C trade-offs, capabilities, resource requirements, and typical applications of each processing platform. In practice, many RF products rely on multiple platforms working together rather than a single device. For example,

Table 7.16 SWaP-C Considerations for FPGAs

Metric	*Impact*
Size	Medium to large packages; BGA often required
Weight	Still low, but may drive the need for heavier cooling
Power	Can be high (0.5W to several watts); depends on logic use and clocks
Cost	Ranges widely from $10 (tiny FPGA) to >$1,000 (high-end FPGA with SERDES)

Table 7.17 FPGA Platforms in RF Systems

Platform/Family	*Typical RF Use Cases*	*Key Advantages*	*Limitations*
Xilinx Zynq UltraScale+/ RFSoC	SDRs, phased arrays, real-time beamforming, mixed-signal RFSoC platforms	Integrated high-speed ADC/DAC, FPGA + ARM cores, industrial-grade	Expensive, steep learning curve
Intel (Altera) Stratix/ Arria	High-speed transceivers, channelizers, radar front-ends, baseband processing	Strong DSP capability, high-speed SerDes	High cost, toolchain complexity
Microchip PolarFire/ SmartFusion	Low-power RF control, space RF systems, security-focused RF tasks	Radiation-tolerant options, low power, secure FPGA fabric	Less ecosystem support than Xilinx/Intel
Lattice FPGAs (ECP5, iCE40)	SPI/I²C expansion, RF front-end glue logic, portable RF devices	Low cost, small form factor, low power	Limited resources, not suitable for heavy DSP
QuickLogic FPGAs	Ultralow-power RF sensor interfaces, always-on control tasks	Very low power, small footprint	Niche use, limited performance, and ecosystem

in a drone-detection and mitigation system, an integrated transceiver may handle the RF front end, an FPGA may perform real-time DSP, and an NVIDIA Jetson or similar SBC may provide GPU-accelerated classification and decision-making. At the same time, a small MCU can monitor power amplifiers, fans, thermal sensors, bias circuits, and protection mechanisms.

In addition to raw technical capability, choosing the right processing platform also depends on factors such as:

- Team skills and development time;
- Life-cycle, supply chain, and production constraints;
- Security requirements, update strategy, and field support.

Ultimately, your system blueprint should reflect not just what is theoretically possible, but what is practical, maintainable, and sustainable throughout the product's entire life cycle.

7.5 DSP Blocks

DSP forms the computational backbone of modern RF systems. By transferring many traditionally analog functions into the digital domain, DSP enables flexibility, repeatability, and performance optimization that would be difficult or impractical to achieve with purely analog implementations. Key DSP functions in contemporary RF systems include [2–6]:

- *Digital downconversion and upconversion:* Translation of signals between RF/IF and baseband using numerically controlled oscillators, digital mixers, and digital filters.

- *Multirate processing:* Decimation (downsampling), interpolation (upsampling), and fractional resampling to adapt sample rates, reduce computational load, and align system clocks across heterogeneous subsystems.
- *Filtering (CIC, FIR, IIR, polyphase):* Channel shaping, anti-aliasing, noise suppression, and efficient channelization tailored to bandwidth, latency, and hardware constraints.
- *FFT/IFFT (inverse FFT) processing for multicarrier systems:* Conversion between time and frequency domains to enable orthogonal frequency-division multiplexing (OFDM), subcarrier-level equalization, and efficient spectral utilization in broadband standards.
- *Digital gain control and IQ correction:* Compensation for hardware impairments such as DC offsets, gain imbalance, and quadrature phase errors introduced by analog front-end components.
- *Equalization:* Mitigation of multipath propagation and frequency-selective fading using adaptive time-domain filters or per-subcarrier frequency-domain equalization.
- *Demodulation and symbol detection:* Recovery of transmitted symbols through carrier recovery, timing synchronization, phase tracking, and soft-decision metric generation.
- *Error-correction decoding:* Application of forward error-correction (FEC) schemes, such as convolutional, Turbo, LDPC, or polar codes, to improve link robustness and reliability under low SNR conditions.
- *Interference mitigation and diversity combining:* Enhancement of SNR through adaptive interference cancellation and diversity techniques such as maximal-ratio combining (MRC) or equal-gain combining (EGC).
- *Channel estimation:* Extraction of channel state information from pilot tones or training sequences to enable coherent detection, adaptive modulation, and accurate equalization.

Not every RF product requires heavy DSP, but when it does, the required processing blocks must be defined early in the blueprint phase because they directly determine the system's data rates, latency budgets, converter specifications, interface speeds, and ultimately the size and capability of the digital processing platform.

DSP blocks typically sit immediately after the ADC or just before the DAC, forming the bridge between the RF chain and the digital domain. DSP can be implemented on different platforms depending on performance and architectural needs:

- Dedicated DSP processors for deterministic, instruction-driven processing;
- FPGAs or RFSoCs for high-throughput, real-time parallel operations;
- CPUs or GPUs for higher-layer signal processing, post-processing, or machine learning-based classification.

The selection of DSP blocks influences whether processing is handled inside an integrated transceiver, implemented in an FPGA or RFSoC, or pushed to a CPU or GPU. Many modern RFICs such as ADRV90xx series integrate portions of the standard DSP functions, such as DDC/DUC, decimation filters, IQ correction, and calibration engines, reducing FPGA workload and improving overall system performance.

DSP choices also drive key architectural trade-offs, for example, deciding between DDC/DUC or direct RF sampling, selecting decimation and interpolation factors, sizing FFT or beamforming resources, and determining filter complexity relative to bandwidth and SWaP targets. These decisions affect hardware resource utilization, power consumption, and system scalability. They also guide team skill requirements and development timelines by clarifying which functions come pre-integrated in the transceiver, which must be engineered in the FPGA fabric, and which can be offloaded to higher-level processors.

Designing DSP architectures in RF systems typically requires:

- Knowledge of digital filtering and spectral transformation principles;
- Understanding of fixed-point and floating-point arithmetic effects, including quantization, overflow, and noise;
- Experience with VHDL/Verilog, C/C++, or MATLAB/Simulink for modeling, simulation, and verification;
- Awareness of pipeline depth, latency, and numerical scaling and their impact on system timing and performance;
- Familiarity with clock-domain crossing, synchronization, and buffer management for reliable real-time data flow.

DSP provides significant precision and flexibility, but it also introduces architectural complexity; defining the analog-digital partition early in the blueprint ensures that filtering, mixing, channelization, and analysis stages align cleanly with the RF hardware to deliver the required performance and efficiency.

7.6 How Software, Firmware, and Hardware Work Together

Before diving into the details of firmware, software, and networking, it is useful to first look at a simple end-to-end example of waveform generation with an RFSoC. This example illustrates how each layer, software, firmware, FPGA logic, and mixed-signal hardware, contributes to producing a signal. Seeing how these pieces interact provides a clearer understanding of the boundaries between layers and how design decisions in one layer affect the overall RF product.

Figure 7.5 illustrates the end-to-end process of generating a QPSK waveform on an RFSoC. The workflow begins at the user interface level, such as a GUI, web application, or Python notebook, where high-level parameters (for example, carrier frequency, output power, modulation type) are specified. These commands propagate through the application software and communication layer before reaching the embedded software running on the RFSoC's ARM cores.

In this architecture, firmware executes on the ARM processors (Cortex-A53 application processor or Cortex-R5 real-time processor) and provides low-level, deterministic control of clocks, data converters, mixers, and digital upconversion blocks. The FPGA fabric implements modulation-specific DSP functions, including

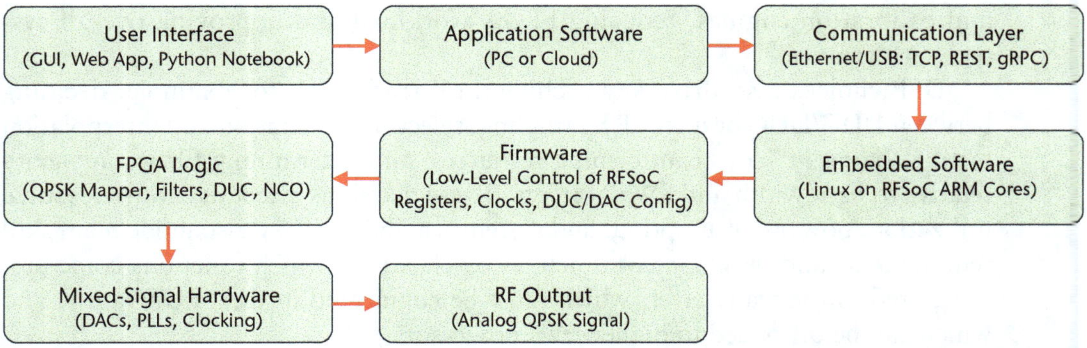

Figure 7.5 End-to-end interaction between software, firmware, FPGA logic, and mixed-signal hardware during waveform generation on an RFSoC.

QPSK symbol mapping, filtering, and frequency translation. Finally, the mixed-signal hardware converts the processed digital data into an analog RF output.

Table 7.18 compares firmware, software, and FPGA logic in RF systems. It is essential to recognize their differences, especially the fact that the FPGA portion represents hardware, not software. Although VHDL and Verilog use text-based

Table 7.18 Comparison of Firmware, Software, and FPGA Logic in RF Systems

Aspect	*Firmware*	*Software*	*FPGA Logic*
Primary role	Real-time hardware control and system bring-up	High-level control, user interface, networking, algorithms	Deterministic, parallel signal processing and timing-critical logic
Runs on	MCUs, embedded processors, SoCs	CPUs, OS-based platforms	Programmable logic fabric
Timing requirements	Hard/real-time; low latency	Soft real-time or nonreal-time	Ultra-deterministic; cycle-accurate
Typical tasks	Initialization, calibration, device configuration, sensor monitoring, protection circuits	System orchestration, data management, UI, networking, machine learning inference	DDC/DUC, filtering, beamforming, channelization, SERDES, timing alignment
Interaction with RF hardware	Direct access to registers, buses, and analog/digital front ends	Indirect via firmware application programming interface (APIs) or OS drivers	Direct control of high-speed digital interfaces and converters
Update frequency	Infrequent but critical (bootloader, OTA)	Frequent application updates	Rare; updates require recompilation and bitstream loading
Examples in RF products	RFIC configuration, PLL programming, PA protection logic	Remote control UI, telemetry dashboards, cloud integration	JESD204B/C handling, high-speed DSP chains, trigger logic
Failure sensitivity	Medium-high (can brick hardware if incorrect)	Low-medium (software crashes recoverable)	High (timing failures break data paths)
Design tools	C/C++, embedded toolchains (Keil, IAR, GCC)	Python, C++, JavaScript, Linux services	HDL (Verilog/VHDL), HLS, vendor tools (Vivado, Quartus)

syntax similar to programming languages, they are hardware description languages. They do not produce software or firmware; instead, they describe digital circuits that are synthesized directly into hardware gates, flip-flops, DSP blocks, and routing resources inside the FPGA fabric.

7.7 Firmware

Firmware is low-level code that directly controls, configures, and coordinates the hardware, directly manipulating registers, peripherals, timing, and I/O signals. For example, when you write an Arduino program that reads a temperature sensor through an ADC pin and toggles an output pin when a threshold is exceeded, the uploaded code is stored in the microcontroller's flash memory and becomes the firmware that directly drives and manages the hardware. This firmware remains on the device even when it is powered off, ensuring that the system behaves consistently every time it starts.

Running on embedded processors, microcontrollers, or internal control engines (for example, the controller embedded inside a beamformer IC), firmware provides the deterministic, real-time logic required for hardware to operate safely, predictably, and consistently. In simple terms, anything that interacts directly with physical hardware is handled by firmware, whereas functionality running on a full OS, such as Linux or Windows, is classified as software.

Table 7.19 summarizes firmware and software requirements across different processing platforms as well as sensor modules.

Table 7.19 Firmware Versus Software Requirements Across Processing Platforms and Sensor Modules

Platform	*Requires Firmware?*	*Requires Software?*	*Notes/Why*
MCU (Microcontroller)	Yes	No	Runs low-level control firmware (bare-metal or RTOS)
SBC (Jetson, Raspberry Pi)	Yes (via drivers)	Yes	Software environment (Linux); uses firmware for hardware interfaces and boot
CPU (Intel/AMD/ARM-A)	No	Yes	Runs operating systems, applications, drivers
GPU (NVIDIA/AMD)	No	Yes	Runs compute kernels through CPU-side software; no firmware-level control
DSP Chips	Yes	Sometimes	Uses firmware to load coefficients, manage modes; may run embedded RTOS
FPGA (Standalone)	Yes (control firmware)	No	FPGA fabric is hardware; firmware configures registers via AXI/SPI
RFSoC (FPGA + ARM)	Yes	Yes	ARM cores run firmware or Linux; FPGA fabric requires control firmware
Integrated Transceiver (ADRV9009, AD9361)	Yes (external)	No	Needs external MCU/FPGA/SoC firmware via SPI/I^{2}C for configuration; has no CPU
Power/sensor modules	Yes (drivers/ firmware)	No	Require firmware drivers to configure and read measurements

In RF products, firmware handles tasks such as power-up sequencing and initialization; configuring RF transceivers, PLLs, gain stages, data converters, and sensors; and managing communication buses such as SPI, I^2C, and UART. Firmware also executes calibration routines, including IQ imbalance correction, LO calibration, bias tuning, and temperature-dependent adjustments, while continuously monitoring temperature, power, current, and protection circuits. In addition, it provides deterministic timing, interrupt handling, and low-latency control, and exposes clean control and status interfaces to higher-level software or the OS.

From a product-blueprint perspective, firmware decisions dictate startup sequencing, timing, latency behavior, interrupt handling, and the level of determinism required for the system. They also define which components require custom drivers, how field updates will be delivered, and the depth of logging, monitoring, and protection that must be built into the product. Clearly allocating firmware responsibilities early in the design process ensures a clean handoff between hardware, DSP, and software teams, resulting in a robust, testable, and maintainable configuration and control architecture throughout the product's life cycle.

It is also important to recognize that each platform, whether an FPGA, a microcontroller, or a SBC, requires its own specialized expertise for firmware development. These skill sets are not interchangeable. Therefore, during the blueprint stage, ensure that your team either has the necessary in-house capability or the budget to bring in the required expertise. Overlooking this early can lead to delays, integration challenges, and costly redesigns later in the development cycle.

7.8 Software

Software defines how users interact with the product, how the system adapts to changing conditions, and how it integrates into larger workflows. Ranging from low-level drivers to high-level application programming interfaces (APIs) and user interfaces, software orchestrates system functionality and ultimately transforms the underlying hardware into a usable, configurable, and valuable product [7].

7.8.1 Core Software in RF Systems

Even if the product is mostly hardware-driven, software is necessary for:

- Configuring RF chains (for example, setting frequency, gain, and filters);
- Monitoring system performance (output power, center frequency) and the environment (temperature, humidity);
- Controlling sequences or measurements (for example, system shutdown in case of an emergency);
- Logging and visualizing results (received signal strength indicator (RSSI), targets in a radar system);
- Integrating with test benches, remote servers, or user workflows;
- Providing APIs and scripting hooks for automation.

Table 7.20 Key Software Layers in RF Products

Layer	*Examples*
Drivers	SPI and I^2C device control, ADC and DAC configuration, GPIO
Middleware	Protocol stacks, RTOS scheduler, task management
API layer	SCPI, REST APIs, Python APIs, MATLAB integration
User interface	Web-based graphical user interfaces (GUIs), command-line interfaces (CLIs), touchscreen applications
Automation hooks	Python scripts, LabVIEW virtual instruments (VIs), remote procedure calls

Key software layers are shown in Table 7.20.
Software impacts cost and power indirectly through:

- Processor overhead for UI/networking/processing;
- Memory/storage requirements;
- Licensing costs (especially for commercial stacks or OS).

Here are some of the essential APIs and protocol types:

- SCPI (Standard Commands for Programmable Instruments), commonly used in laboratory test equipment such as spectrum and network analyzers;
- REST/HTTP APIs, widely used in network-connected devices and enabling cloud integration;
- Python APIs, popular for scripting, automation, and test orchestration;
- MATLAB and Simulink interfaces, often used for research, modeling, and algorithm validation;
- WebSocket, used for real-time control and data streaming, for example, in network and spectrum analyzers;
- MQTT, used for lightweight, message-based telemetry in IoT systems.

The team may need developers or engineers with the following skills:

- Embedded firmware expertise (C, RTOS, HAL);
- Python/C++/MATLAB programming;
- Web technologies (HTML, JavaScript, REST, JSON);
- API design and documentation (OpenAPI, Postman);
- Software version control (GIT) and build systems (CMake, make);
- CI/CD practices if deploying firmware/software updates at scale.

Here are some best practices for software development.

- Use well-documented APIs to make your RF product programmable.
- Separate UI from core functionality using layered architecture.
- Ensure that firmware update mechanisms are robust and secure.
- Provide simulation modes or offline test environments for development.
- Maintain consistent command and status conventions across interfaces.

7.8.2 Machine Learning and Artificial Intelligence (AI) in RF Systems

Machine learning is already impacting RF engineering by enabling systems to extract insights from large volumes of spectral data and adapt in real time [3]. Frameworks such as TensorFlow, PyTorch, and scikit-learn provide ready-to-use libraries for model training and inference, allowing engineers to focus on RF applications rather than low-level algorithm implementation.

Typical use cases include:

- Spectrum classification, identifying signal types or emitters in crowded frequency bands;
- Anomaly detection, spotting interference, jamming, or equipment malfunctions;
- Adaptive RF systems, enabling cognitive radios, self-optimizing networks, and dynamic spectrum allocation.

By combining traditional DSP pipelines with machine learning inference at the edge or in the cloud, engineers can design systems that are smarter, faster, and more resilient in dynamic environments.

We will examine AI/machine learning in much greater depth in Volume 2, but, for now, it is important to recognize its role at the product blueprint stage. If a customer requires spectrum classification, anomaly detection, or adaptive RF features, then AI/machine learning capabilities must be considered early in architecture. Even if the algorithms themselves are developed later, the choice of processing platform, memory resources, and data interfaces should account for potential machine learning workloads.

7.9 Networking in RF Products

Modern RF systems increasingly rely on network connectivity for configuration, control, data transmission, and remote updates. Whether it is a sensor node, remote radio head, test instrument, or SDR platform, networking enables interoperability and scalability across platforms, users, and environments. Networking is no longer optional in most RF products, where it has become an expected feature, even in hardware-centric designs.

Networking allows RF products to:

- Transmit measurement or signal data to the cloud or another device.
- Receive remote control commands or configuration updates.
- Participate in distributed systems (e.g., phased arrays, radar clusters, defense systems).
- Enable remote debugging or diagnostics.
- Integrate with other digital ecosystems (e.g., SCADA, DAQ, or cloud AI platforms).

Here are some networking applications in RF products:

- *Remote RF labs and test instruments:* Enable access to instruments over LAN or the internet.
- *IoT sensor nodes with RF front ends:* Send data to the cloud or local gateways.
- *Distributed antenna systems (DAS):* Support synchronization, control, and data distribution.
- *SDRs:* Transmit IQ data and control information over Ethernet or USB.
- *Military communication systems:* Provide high-reliability, secure network backbones.
- *Quantum control systems:* Synchronize experiment timing and distribute control commands.

Here are some important design considerations for networking in RF products:

- Protocol stacks, including TCP/IP, UDP, MQTT, HTTP and REST, and WebSocket;
- Security, including TLS and SSL, authentication mechanisms, and firewalling;
- Time synchronization, using NTP or IEEE 1588 Precision Time Protocol (PTP), which is critical for phased arrays and quantum systems;
- EMI and RF isolation, particularly for Ethernet PHYs and SFP cages placed near sensitive RF circuits;
- Firmware updates, with robust support for secure and failsafe bootloaders.

Developing networking capabilities for RF systems requires a distinct set of skills, including:

- IP addressing and socket programming, along with protocol fundamentals;
- Embedded TCP/IP stack integration;
- Configuration interfaces, such as web-based user interfaces, command-line interfaces, and JSON over REST APIs;
- Network debugging, using tools such as Wireshark, ping, and traceroute;
- Secure networking, including SSH, VPNs, and TLS certificate management;
- Quality-of-service and bandwidth management for high-data-rate RF systems.

Table 7.21 shows some of the SWaP-C considerations for networking in RF systems.

Table 7.21 Networking Metrics and Impact

Metric	*Impact*
Size	Ethernet/Wi-Fi modules are compact; SFP connectors dominate PCB real estate
Weight	Minimal
Power	Ethernet PHY ~100 mW; Wi-Fi/5G >1W; SFP+ transceivers ~1–2W
Cost	Ethernet is inexpensive; Wi-Fi moderate; secure cellular modems \$10–\$50+; SFP modules \$20–\$200+

7.9.1 Communication Protocols

Another important consideration for networking is the choice of communication protocol, which must be defined early in the product blueprint phase. Networking in RF products begins with selecting a protocol whose characteristics directly influence latency, reliability, throughput, and overall system scalability. Depending on the product's complexity, data volume, and deployment environment, different technologies may be more suitable than others. Table 7.22 summarizes several common networking protocols and interfaces used in modern RF products.

Engineers should balance throughput against overhead when selecting protocols:

- For bulk data transfer (e.g., spectrum monitoring, IQ streaming), Ethernet, PCIe, or SFP is most suitable.
- For control and telemetry (e.g., switches, sensors), lightweight serial or message-based protocols are more efficient.

7.9.2 Remote Access and Control

Remote access is an enabler for distributed teams, remote labs, and field-deployed systems. Practical implementations range from SCPI over LAN for instrument control to web-based dashboards for visualization and configuration. The remote-control capability lets engineers adjust attenuators, tune filters, or monitor spectrum data without being physically present. For RF labs, it supports resource-sharing and faster collaboration. Key considerations include latency (especially for time-critical loops), synchronization across multiple devices, and fail-safe design so that loss of connection does not compromise hardware safety.

Table 7.22 Networking Protocols and Interfaces in RF Products

Interface/Protocol	*Typical Data Rate*	*Typical Use Case in RF Systems*	*Notes*
UART/I^{2}C/SPI	kbps–Mbps	Short-range control of RF switches, sensors	Lightweight, low power, minimal overhead
Ethernet (10/100/1,000 Mbps)	Up to 1 Gbps	General-purpose instrument control, data logging	Inexpensive, robust backbone for most systems
SFP (SFP/SFP+/SFP28)	1–25 Gbps	High-speed RF data streaming to servers/cloud	Scalable; copper or optical modules available
USB 2.0/3.0	480 Mbps/5 Gbps	SDR host control, lab instruments	Host-dependent, plug-and-play convenience
Wi-Fi (802.11ac/ax)	Hundreds of Mbps	Cable-free deployment, remote diagnostics	Flexible, but latency-sensitive and interference-sensitive
Cellular (4G/5G)	10–1,000 Mbps	Mobile or remote RF systems	Wide coverage, but higher power use and subscription costs
High-level (MQTT, REST, gRPC)	Protocol-dependent	Cloud integration, telemetry, remote APIs	Runs on top of Ethernet/Wi-Fi/5G; enables remote monitoring and automation

7.9.3 Edge Computing

Edge computing brings processing closer to the RF front end, reducing the need to stream high-rate raw data to remote servers. For example, instead of sending continuous I/Q samples to the cloud, an edge node can perform FFT analysis, anomaly detection, or signal classification locally, forwarding only essential results. This minimizes bandwidth use, lowers latency, and maintains functionality even when connectivity is limited.

In RF systems, edge computing enables faster closed-loop control, efficient network utilization, and scalable deployment of intelligent devices. Embedded Linux platforms and compact SoMs such as Jetson, Zynq, and Intel NUC are common hardware choices.

Edge machine learning extends this concept by running trained models directly on embedded devices, within a radio front end, phased array controller, or remote sensor. Instead of streaming all I/Q data, the system can:

- Classify signals in real time (e.g., friend versus foe in EW or interference detection).
- Detect anomalies or faults locally (e.g., PA overheating, antenna mismatch, spectrum occupancy shifts).
- Adapt immediately (e.g., beam steering, filter retuning, or power adjustment).

Platforms such as Raspberry Pi, BeagleBone, and Zynq SoCs support lightweight machine learning frameworks such as TensorFlow Lite, PyTorch Mobile, or ONNX Runtime, providing low latency, reduced bandwidth usage, and autonomy, crucial in RF systems where decisions must occur within microseconds to milliseconds.

7.10 Security Considerations

Security is a fundamental design consideration in modern RF products, particularly as systems become increasingly connected, programmable, and software-defined. Vulnerabilities can arise not only from network and software interfaces, but also from physical access to the hardware itself. As a result, effective security design requires a holistic approach that addresses cybersecurity and physical tampering together, ensuring that RF systems remain trustworthy, resilient, and safe throughout their entire operational life cycle.

7.10.1 Cybersecurity Considerations

As RF systems increasingly connect to networks for monitoring, control, and data exchange, cybersecurity has become a critical design factor. Networked RF products face the same threats as IT systems but often with higher stakes: unauthorized access can disrupt measurements, compromise sensitive data, or even damage hardware.

To mitigate these risks, engineers should implement secure communication protocols (TLS, VPN, SSH), apply strong authentication and role-based access control, and keep firmware and software regularly updated. For added protection, firmware should support secure boot and signed updates to ensure only trusted software runs on the device. Isolation of critical control paths (e.g., PA bias or high-voltage switching) from user-facing interfaces is essential to prevent malicious or accidental misuse.

For field-deployed RF products, additional measures such as intrusion detection, encrypted data storage, and redundant access paths help to strengthen resilience. Ultimately, cybersecurity must be treated as a design requirement on par with SWaP-C, not as an afterthought. Table 7.23 shows some of the common cybersecurity threats in networked RF systems.

Before deploying a networked RF product, verify that the design includes:

- Encrypted communication (TLS, VPN, SSH) for all data in transit;
- Strong authentication (certificates, keys, multifactor) and role-based access control;
- Secure boot and signed firmware updates to prevent tampering;
- Isolation of critical control paths (e.g., PA bias, high-voltage switching) from user interfaces;
- Regular software/firmware patching and update strategy for field devices;
- Intrusion detection and logging to monitor suspicious activity;
- Redundant or fallback paths for critical control in case of attack or outage.

While cybersecurity focuses on protecting RF systems from network and software-based threats, physical access to the hardware introduces a distinct class of risks that must be addressed separately. The next section discusses these physical security considerations in more detail.

7.10.2 Tamper Resistance in RF Products

Although tampering is largely a physical concern, it is strongly related to cybersecurity because physical access to the hardware can also lead to unauthorized access

Table 7.23 Common Cybersecurity Threats in Networked RF Systems and Mitigation Strategies

Threat	*Risk in RF Systems*	*Mitigation Strategy*
Eavesdropping/sniffing	Interception of control commands or measurement data	Use encryption (TLS, VPN, SSH) for data in transit
Spoofing/impersonation	Unauthorized device or user pretends to be trusted	Strong authentication (certificates, keys, multifactor)
Man-in-the-middle (MITM)	Attacker alters data or injects commands during transmission	Mutual authentication, integrity checks (hashing, digital signatures)
Denial of service (DoS)	Overloading the network link to disrupt the RF system availability	Rate limiting, firewalls, redundant links
Malware/firmware tampering	Insertion of malicious code into devices	Secure boot, signed firmware updates, regular patching
Unauthorized access	Compromise of user credentials or open ports	Access control lists, role-based permissions, port hardening

to firmware and software. In many sensitive RF products, such as EW systems, the hardware itself may not be subject to export control restrictions, such as ITAR, whereas the firmware and software are. This makes it essential to explicitly consider tampering as an integral part of the overall security strategy. Tampering can expose proprietary RF architectures, alter calibration constants, inject malicious firmware, or degrade system reliability in ways that are difficult to detect remotely.

Common RF applications where tamper resistance is essential include secure radios and SDRs, military and defense RF systems, and export-controlled or intellectual property-sensitive hardware.

Tamper resistance refers to the ability of an RF product to deter, delay, or reveal unauthorized physical access or modification. It does not imply that a system is impossible to access. Instead, tamper resistance ensures that tampering is difficult, detectable, or results in loss of sensitive data or system functionality. In RF systems, tamper resistance protects not only digital assets such as cryptographic keys and firmware, but also analog performance, calibration integrity, and overall system trust.

In practice, tamper resistance is achieved through a combination of mechanical design, hardware sensing, firmware logic, and software policy enforcement. While physical mechanisms provide the first line of defense, software and firmware define how tamper events are detected, logged, and handled at the system level.

Tamper resistance in RF products typically includes the following elements:

- Physical barriers, such as sealed enclosures, special fasteners, potting compounds, conformal coatings, or hardened casings that discourage access and probing;
- Tamper evidence, including seals, coatings, or mechanical features that provide a visible indication of opening or modification;
- Tamper detection, implemented using sensors that monitor enclosure access, voltage or clock anomalies, probing attempts, or abnormal temperature conditions;
- Tamper response, where firmware or software automatically takes protective action when tampering is detected, for example:
  - Erasing cryptographic keys or calibration data;
  - Disabling RF transmission or reception paths;
  - Logging security events for forensic analysis;
  - Permanently locking or bricking the device.

From a software perspective, tamper resistance influences system architecture decisions early in the product blueprint. Firmware must define how tamper events are detected, how quickly responses are triggered, and which actions are irreversible. Software layers determine how events are reported, logged, and integrated into monitoring or fleet-management systems.

Key considerations include:

- Secure handling and storage of cryptographic material;
- Integration of tamper events into logging and telemetry frameworks;
- Coordination between tamper response, secure boot, and firmware update mechanisms;

- Clear separation of responsibilities between hardware, firmware, and application software.

Tamper resistance is often driven by regulatory, contractual, or security requirements. Relevant standards and contexts include:

- FIPS 140-2 and FIPS 140-3 for cryptographic modules;
- Defense and secure communications programs;
- Anti-reverse-engineering and intellectual property protection strategies.

Designing tamper resistance into an RF product is a system-level decision that spans mechanical design, hardware sensing, firmware logic, and software policy, and it must be addressed early in the product blueprint to avoid costly redesigns and security gaps later in the development cycle.

7.11 Mechanical System

Mechanical systems play a multidimensional role in RF product design. They define physical enclosure, influencing both the aesthetics and SWaP-C. Also, mechanical considerations are critical to ensuring the product functions reliably in real-world environments, addressing factors such as thermal management [8–10], ingress protection, shock, vibration, and EMI/EMC.

Temperature is one of the most critical stress factors in RF systems. Active thermal management employs sensors, fans, heat sinks, and shutdown mechanisms to keep devices within safe operating limits. Beyond immediate protection, temperature has a direct impact on component lifetime. A widely cited reliability guideline, the 10° rule, states that for many semiconductor devices, every 10°C rise in junction temperature can reduce the expected lifetime by roughly half, while lowering the temperature by 10°C can double it. This relationship originates from the Arrhenius model of thermally activated failure mechanisms such as electromigration and dielectric breakdown. Consequently, thermal control loops are not only about preventing catastrophic failure but also about ensuring the long-term reliability of RF products.

Mechanical design also has a significant impact on EMI/EMC performance, particularly through the design of the enclosure, internal shielding, and the placement of components and PCBs. Even the choice of materials, gaskets, and fastening methods can influence signal integrity and compliance with regulatory standards.

We will explore these technical aspects in detail in Volume 2. For now, our focus is on how mechanical considerations influence the product blueprint, guiding system partitioning, thermal zoning, connector placement, and physical integration of subsystems. Therefore, in this section, we will focus only on the high-level architectural implications of mechanical design.

7.11.1 Mechanical Considerations at the Product Blueprint Stage

At the blueprint and block diagram stage, mechanical design is not about dimensions and materials yet, but about anticipating how each subsystem will physically coexist with others. Decisions made here about system partitioning, thermal zoning, connector locations, and shielding boundaries will shape the feasibility, performance, and robustness of the final product.

Figure 7.6 shows an example of system-level partitioning of a modern RF product, emphasizing the enclosure, external access points to internal subsystems, such as Ethernet, USB, SMA connectors, and power inputs, as well as thermal management using a fan. At the product blueprint level, a diagram of this type is sufficient to convey the overall structure and intended form factor of the product.

During PoC development and early prototyping, a design with fewer features is typically preferred. As discussed in Volume 2, speed of implementation is critical at this stage, making it essential to clearly understand which features must be retained and which can be deferred to later development phases.

The following are some of the critical considerations when designing the mechanical aspects of the product blueprint.

1. System partitioning and physical zoning:
   - Group subsystems based on thermal load, shielding requirements, or accessibility (e.g., keep RF front-end components close together to minimize trace length).
   - Determine if sensitive analog/RF and noisy digital sections need to be physically isolated.
   - Define where power supplies, processing units, RF subsystems, and antennas will reside relative to each other.
2. Thermal management strategy:
   - Identify high-power blocks (e.g., power amplifiers, FPGAs) early in the block diagram.
   - Leave space for heat sinks, thermal vias, or airflow paths.

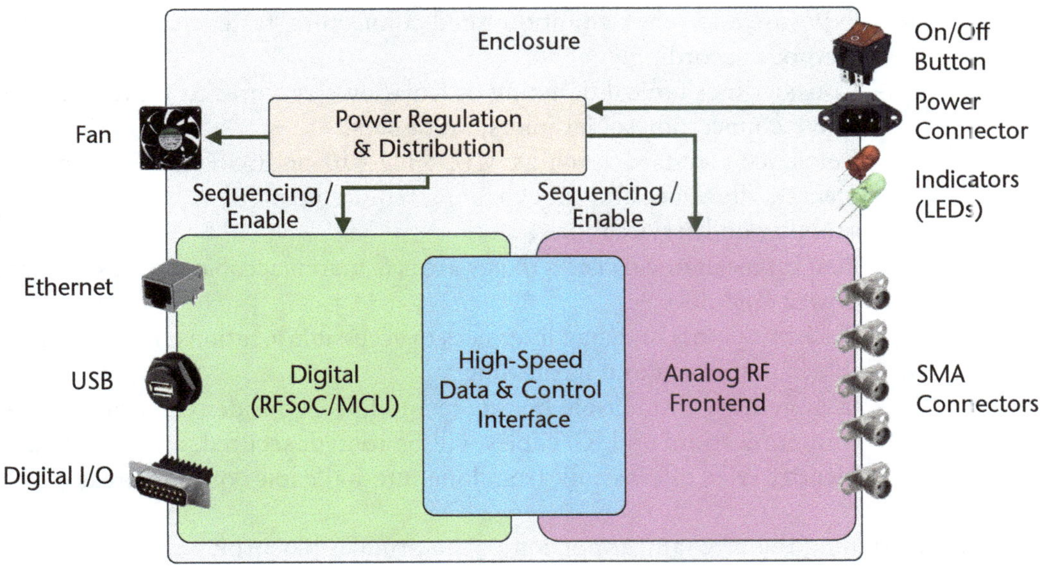

Figure 7.6 System-level partitioning of a modern RF product, emphasizing enclosure, thermal, and mechanical considerations.

- Consider thermal zones: Is there a need to physically separate hot and cold zones?

3. Connector placement and external interfaces:
   - Map where RF ports, power inputs, and communication interfaces (Ethernet, USB) will exit the enclosure.
   - Consider mechanical accessibility for test ports, calibration points, or reprogramming interfaces, such as SD card slots, UART headers, or JTAG access points.
4. Form-factor and mounting constraints:
   - Early decisions about whether the product is a rack-mounted unit, handheld device, or PCB card affect the blueprint.
   - The choice of enclosure material, such as plastic, metal, or additively manufactured (3D-printed) structures, affects mechanical strength, thermal performance, EMI shielding, weight, and cost.
   - Consider mounting holes, bracket areas, and mechanical attachment points in the layout logic.
5. EMI/EMC zones and shielding strategy:
   - Plan for shielded compartments or RF tight enclosures around sensitive circuits such as LNAs, LOs, and PAs.
   - Define clear zoning for digital, analog, and RF sections to minimize coupling and interference.
6. Ingress protection and environmental considerations:
   - For outdoor or mobile applications, blueprint-level planning should consider whether sealing is needed and how that affects connector types, venting, or heat dissipation.
   - Evaluate the required ingress protection (IP) rating early (e.g., IP65, IP67) to ensure adequate protection against dust and moisture without compromising thermal management.
   - For mobile, vehicular, or airborne systems, assess shock and vibration exposure and select vibration-rated connectors, fasteners, and mounting methods accordingly.
   - Consider mechanical damping or isolation structures to prevent damage to RF connectors, solder joints, and filters.
   - Reference standards such as MIL-STD-810 or IEC 60068 for environmental, vibration, and shock testing requirements.
7. Testability and serviceability:
   - Plan for mechanical access to key areas (e.g., replaceable filters, swappable power modules).
   - Place test points or debug interfaces in accessible locations from the outset.
8. Cable management and harnessing:
   - In systems with multiple boards or modules, consider early how power, communication, and RF cables will be routed, secured, and shielded.
   - Identify risks of crosstalk from long internal cables or loop areas.

Designing the mechanical portion of the product blueprint requires close collaboration between mechanical, RF, and digital engineers to ensure that all relevant aspects are properly considered.

7.12 What's Next?

We have now covered the key requirements for an RF product that extend beyond the RF chain itself, including power management, mechanical design, sensors, processing platforms, software, firmware, networking, and security. In Chapter 8, we will put these concepts into practice by developing the product blueprint for a phased array ground terminal, demonstrating how these multidisciplinary considerations come together in real-world designs.

Problems

7.1 A system requires 500-MHz instantaneous bandwidth, beamforming across 16 channels, and real-time signal classification. You are evaluating three processing architectures:
- Pure RF with minimal digital processing;
- RF with DSP-first processing with a dedicated DSP processor;
- Hybrid SoC or RFSoC platform.

Which architecture is most appropriate at the product blueprint stage? Justify your choice based on bandwidth, latency, scalability, power consumption, and development risk.

7.2 A receiver chain requires decimation, channel filtering, and FFT-based detection. For each function, decide whether it should be implemented in FPGA or RFSoC fabric, a dedicated DSP block, or on a CPU or GPU. Explain your choices in terms of latency, determinism, data movement, and flexibility.

7.3 You are designing a battery-powered handheld RF device that performs continuous signal monitoring and intermittent classification. Identify three architectural or algorithmic design decisions that would reduce DSP power consumption without sacrificing essential performance. Explain the trade-offs involved.

7.4 An RF system includes protection logic that must respond to over-power or over-temperature conditions within a bounded and repeatable time window. Which processing platform should implement this logic, and why? Discuss the implications for firmware, OSs, and system safety.

7.5 List three firmware responsibilities that must be clearly defined during the product blueprint phase to avoid integration gaps with the hardware and DSP teams. For each responsibility, briefly explain the risk of deferring it to a later development stage.

7.6 A product manager proposes moving from a traditional downconversion and DDC-based architecture to direct RF sampling. Discuss the implications of this decision on ADC performance requirements, FPGA or RFSoC resource usage, data throughput, and firmware complexity.

7.7 Identify two calibration routines commonly used in RF transceivers, such as IQ imbalance correction or LO calibration. For each routine, explain where it should be implemented, firmware, DSP blocks, or higher-level software, and justify your choice based on timing, repeatability, and system control.

7.8 An RF feedback loop must close within 5 μs under all operating conditions. Which processing platform is best suited to implement this loop: FPGA, CPU, GPU, or external DSP? Justify your answer based on latency, determinism, and system integration constraints.

7.9 An engineering team has strong experience in C and C++ and embedded systems, but limited FPGA expertise. How should this influence the choice of processing platform for a new RF product? Discuss the short-term execution risk, development timeline, and integration effort, as well as the long-term implications for scalability, performance, and maintainability. Consider whether building in-house FPGA capability or outsourcing FPGA development is more appropriate at different stages of the product life cycle.

7.10 A customer requires secure remote access and long-term field updates for a deployed RF system. How does this requirement influence decisions related to firmware architecture, processor selection, OS choice, and platform integration?

7.11 Rather than designing a fully tamper-proof system, identify three practical tamper-resistance measures suitable for a phased array RF product at the blueprint stage. For each measure, explain which layer it affects (mechanical, hardware, firmware, or software) and what risk it mitigates.

References

[1] Yu, Z., "A Comprehensive Guide to LDO Regulators: Navigating Noise, Compromises, Applications, and Trends," *Analog Devices*, July 2023.

[2] Lyons, R. G., *Understanding Digital Signal Processing*, Third Edition, Upper Saddle River, NJ: Prentice Hall, 2011.

[3] Kuzdeba, S., *Radio Frequency Machine Learning: A Practical Deep Learning Perspective*, Norwood, MA: Artech House, 2023.

[4] Haigh, K. Z., and J. Andrusenko, *Cognitive Electronic Warfare: An Artificial Intelligence Approach*, 2nd ed., Norwood, MA: Artech House, 2023.

[5] Reyland, J. M., *Software Defined Radio: Theory and Practice*, New York: Wiley, 2015.

[6] Crockett, L. H., D. Northcote, and R. W. Stewart, (eds.), *Software Defined Radio with Zynq UltraScale+ RFSoC*, Glasgow, UK: Strathclyde Academic Media, 2020.

[7] Martin, R. C., *Clean Architecture: A Craftsman's Guide to Software Structure and Design*, Boston, MA: Pearson, 2017.

[8] Kordyban, T., *Hot Air Rises and Heat Sinks: Everything You Know About Cooling Electronics Is Wrong*, Brookline, MA: ATC Press, 1998.

[9] Kordyban, T., *More Hot Air*, Brookline, MA: ATC Press, 2005.

[10] Steinberg, D. S., *Cooling Techniques for Electronic Equipment*, New York: Wiley-Interscience, 1980.

CHAPTER 8

Putting It All Together: End-to-End Product Blueprint

In theory, theory and practice are the same. In practice, they are not.

—*Benjamin Brewster*

This chapter brings together the concepts discussed throughout the book to form a complete RF product blueprint. We examine the signal chain layer, RF layer, and non-RF layer elements of the blueprint, including power management, firmware, software, processing platforms, and mechanical considerations, and provide a checklist to ensure all components are properly defined. We then introduce what a complete product-blueprint document looks like in practice. Finally, we walk through the product blueprint for a phased array ground terminal.

8.1 Introduction

So far, we have explored the fundamentals of building an RF product blueprint. We have examined the signal chain layer, RF systems and architectures, RF components, power management, processing platforms, software, firmware, and mechanical considerations.

Now it is time to put these principles into practice to create a complete product blueprint. This chapter represents the first substantial transition from theory to implementation, where conceptual understanding begins to take shape as a cohesive design. In many ways, this is where engineering becomes an art: integrating multidisciplinary knowledge to transform functional requirements into a tangible system that can be prototyped, tested, and ultimately manufactured at scale.

In this chapter, we walk through the process of turning a set of requirements into a detailed product blueprint. This chapter describes how to translate specifications into system-level block diagrams. By the end of this chapter, a clear framework is established for bridging the gap between concept and realization, applicable not only to RF systems but to any multidisciplinary hardware development effort.

8.2 Step-by-Step RF Product Blueprint Procedure

Throughout this process, your primary compass will be the system requirements, whether they originate from the customer or internal stakeholders or are defined by you when formal specifications are not yet fully established. We assume that al preparatory activities have already been completed, including the various forms o risk analysis (resource, schedule, technical, market, legal, and regulatory), surveys of available solutions, and feasibility studies discussed in Chapter 2.

8.2.1 Product Blueprint: Signal Chain, RF, and Non-RF Layers

As discussed earlier, the blueprint is organized into three layers: the signal chain layer, the RF layer, and the non-RF layer. Tables 8.1 through 8.3 list the key elements

Table 8.1 Signal-Chain Layer Checklist for a Communication Product

Signal Chain Layer	*Elements*
Physical-layer specification	Spectrum definition: operating frequency band, total allocated bandwidth, per-carrier/per-channel bandwidth
	Modulation: BPSK, QPSK, QAM, OFDM, chirp-based waveforms
	Waveform characteristics: pulse shaping, single-carrier versus multicarrier operation, peak-to-average power ratio (PAPR)
	Channel coding: LDPC, Turbo, Polar, BCH, CRC
	Multiple access/multiplexing: TDMA, FDMA, CDMA, OFDMA, SC-FDMA
	Duplexing scheme: frequency-division duplex (FDD), time-division duplex (TDD)
	Duplex capability: half-duplex, full-duplex
	Timing, latency, and synchronization:
	Latency requirements (end-to-end or control-loop)
	Timing alignment (uplink/downlink, slot and frame structure)
	Guard intervals and cyclic prefix
	Clocking and jitter budget
	Frequency and time synchronization (LO stability, GPS-disciplined oscillator (GPSDO), precision time protocol (PTP))
Propagation requirements and RX/TX performance	Propagation assumptions: line-of-sight (LOS) and non-line-of-sight (NLOS) conditions, indoor/outdoor environment, terrain and clutter, rain fade, atmospheric absorption
	Coverage and range targets: coverage area, cell or beam footprint, maximum and minimum communication range
	Link budget requirements: required SNR/SINR, link margin, fade margin, implementation loss
	Receiver performance parameters: sensitivity, noise figure, linearity (P1dB, IP3), dynamic range, gain budget
	Transmit performance parameters: output power, EIRP, spectral mask, adjacent-channel emissions, regulatory compliance
System architecture and frequency planning	Transceiver architecture: homodyne (direct conversion), superheterodyne, low IF, direct RF sampling, direct RF synthesis
	Frequency planning: LO plan, IF selection, image and spur avoidance, channel spacing, and guard bands
Control, monitoring, and calibration	Monitoring: sensors and detectors for RF power, temperature, supply voltage and current, VSWR, received signal strength indicator (RSSI)
	Control: digital predistortion (DPD), crest factor reduction (CFR), automatic gain control (AGC), gain and phase control
	Calibration: power calibration, phase calibration, I/Q imbalance correction, and LO leakage calibration

in each layer that should be addressed when developing the product blueprint. Note that these tables emphasize a communication product; the specific elements will need to be adapted for other types of products.

Note that although link-budget analysis and transceiver architecture are traditionally introduced in RF systems chapters, within the context of product blueprinting, they are classified under the signal chain layer. This is because they directly shape system-level signal flow, dynamic-range allocation, ADC/DAC requirements, filtering strategy, and gain architecture. In other words, these decisions influence the structure of the entire signal chain, RF and digital, and therefore must be considered at this earlier system-definition stage.

After defining the signal chain layer elements, the next step is to develop the block diagram for the RF layer. Depending on the selected architecture and performance requirements, the RF portion of the system can vary significantly. In direct-conversion architectures, the RF section is often relatively compact, typically consisting of amplification and filtering stages with no frequency-translation mixers. In contrast, traditional or high-performance multistage superheterodyne architectures require a more extensive RF chain, incorporating mixers, multiple LO stages, intermediate-frequency filtering, amplification, and additional signal-conditioning elements.

Table 8.2 summarizes the critical elements of the RF layer that must be defined at this stage.

The key elements of the non-RF layer, including power management, processing platforms, firmware, software, mechanical design, and environmental considerations, are summarized in Table 8.3.

A highly effective document for summarizing the overall system architecture is the Interface Control Document (ICD). The ICD defines how subsystems within

Table 8.2 RF Layer Checklist for a Communication Product

RF Layer	*Elements*
Signal generation	Tone generation: frequency synthesizers, fixed LOs Waveform generation: RFSoC-based generators, DSP-based RF signal generation LO distribution: buffering, fan-out, tuning range Frequency reference and clocking: TCXO, OCXO
Signal transmission	Wireless transmission: antennas Wired transmission: cables, connectors, waveguides
RF signal processing	Amplitude control: amplifiers, attenuators, signal routing elements (switches, circulators, isolators) Frequency control: filters (LPF, HPF, BPF, BSF), frequency converters (mixers, frequency multipliers) Phase control: phase shifters, delay lines
RF signal detection	Demodulation: analog and digital demodulators, baseband extraction Power detection: RMS detectors, envelope detectors, RSSI circuits
Compliance and coexistence	Electromagnetic compatibility: EMC/EMI compliance, signal integrity, spectral mask adherence Coexistence requirements: Interference constraints, out-of-band emissions, adjacent-channel performance

Table 8.3 Non-RF Layer Checklist for a Communication Product

Layer	*Elements*
Power management	Power-tree architecture: voltage rails, sequencing, inter-rail dependencies
	Regulation and conversion: LDOs, DC-DC converters, power-management ICs (PMICs)
	Current and power budgeting: all subsystems and operating modes
	Power sequencing: startup and shutdown behavior
	Efficiency and thermal impact
	Protection mechanisms: overcurrent protection (OCP), overvoltage protection (OVP), undervoltage lockout (UVLO), short-circuit protection (SCP), and overtemperature protection (OTP)
	Battery support and charging (if applicable)
	Power monitoring and telemetry: voltage, current, power
	EMI/EMC considerations: switching regulators and layout
	Redundancy or backup power (if applicable)
Processing platform	Processing selection: MCU, CPU, DSP, GPU, FPGA, RFSoC
	Latency and real-time requirements
	Data rate and throughput capability
	Memory architecture and storage
	High-speed interfaces: PCIe, JESD204, AXI, USB, Ethernet
	Clocking and synchronization requirements
	Power consumption and thermal constraints
	Toolchain and development environment
	Scalability and upgrade paths
Firmware	Scope and responsibilities: hardware control and orchestration
	Firmware architecture: bare-metal, RTOS, or embedded Linux
	Hardware interfaces: SPI, I^2C, GPIO, UART, triggers, control lines
	Host and software interfaces: command protocols, data formats, events
	Startup, sequencing, and calibration routines
	Real-time constraints: latency, update rates, synchronization
	Logging and diagnostics
	Firmware update and recovery strategy
	Test hooks and manufacturing diagnostics
Software	Application-level functionality and responsibilities
	Software architecture: modules, services, processes
	Operating system (OS) environment: Linux, Windows, embedded OS
	APIs and interfaces to firmware and hardware
	Data handling and processing requirements
	User interface and experience (UI/UX): local, remote, or web-based
	Configuration management and storage
	Error handling and exception management
	Logging, analytics, and visualization
	Update and deployment strategy
Networking	Supported interfaces: Ethernet, Wi-Fi, cellular, USB, CAN
	Network protocols: TCP/UDP, REST, gRPC, MQTT, WebSocket
	Addressing, discovery, and device identification
	Data transport requirements: bandwidth, latency, reliability

Table 8.3 *(Continued)*

Layer	*Elements*
	Time synchronization: network protocols (NTP, PTP), pulse-based references (PPS), and GNSS-derived timing sources Security mechanisms: TLS, authentication, encryption Firewall and port requirements Remote access and control Over-the-air (OTA) updates Network diagnostics and monitoring
Security	Cybersecurity: secure boot, authentication, encryption, key management Physical security and tamper resistance
Mechanical/thermal/ SWaP-C	Enclosure design: materials, form factor, assembly Mechanical layout: internal module placement Mounting interfaces: fasteners and attachment points Thermal design: heatsinks, airflow, conduction paths Connector placement and mechanical constraints Environmental requirements: shock, vibration, temperature, humidity Sealing and ruggedization: ingress protection (IP rating) Manufacturing tolerances and mechanical stack-up Mechanical reliability: fatigue, wear, stress CAD models and mechanical documentation

a product interact, ensuring that electrical, RF, digital, mechanical, and software interfaces are clearly specified and consistently implemented across teams. It captures signal definitions, connector and pin assignments, communication protocols, register maps, timing requirements, and control and status behaviors. While the detailed ICD is typically finalized during the implementation phase, a preliminary ICD is essential during the blueprint stage to enable parallel development and to prevent costly integration issues during EVT and DVT.

8.2.2 Contents of the Product Blueprint Document

The product blueprint document serves as the technical foundation for the product's implementation, including the prototyping and validation phases. It consolidates all key design decisions, trade-offs, and performance targets established during the blueprint stage. In practice, it functions both as a design reference and as a communication tool between the hardware, firmware, software, and test teams.

The contents of the blueprint document span system-level descriptions, engineering decisions, project planning elements, and supporting documentation. These elements ensure that all stakeholders share a unified understanding of the architecture, interfaces, constraints, and development path before entering the implementation stages. Table 8.4 summarizes the major elements that belong in a product blueprint document. The system block diagram with the selected architectures, signal paths, and control loops, derived from Tables 8.1 through 8.3, forms the core of the document and anchors the remaining elements listed in Table 8.4.

Table 8.4 Contents of the Product Blueprint Document

Category	*Elements Included in the Product Blueprint Document*
System-level content	Requirement summary (functional, performance, regulatory)
	Block diagram (architecture, signal paths, control loops)
	Product layout (physical placement of subsystems)
	Assumptions and design constraints (material or manufacturing constraints, SWAP-C, thermal, environmental limits)
Planning content	Milestones and Go/No-Go criteria
	Project timelines (PoC → MVP → EVT → DVT → PVT)
	Risk register (technical, resource, schedule, legal, and regulatory)
	Budgetary estimates (cost of hiring, components, manufacturing, testing, iteration)
Engineering content	Preliminary ICD (RF, electrical, digital, mechanical, control interfaces)
	Signal chain budgeting (NF, gain, linearity, power levels, dynamic range)
	RF layer (signal generation, transmission, RF signal processing, and detection)
	Frequency planning (LO plan, filtering, spurs, channelization)
	Power-tree concept (rails, sequencing, current estimates)
	Processing architecture (MCU/FPGA/DSP/RFSoC selection)
	Firmware/software/networking
	Thermal strategy (power density, airflow/conduction approach)
	Mechanical and packaging considerations (enclosure, connectors, antenna placement)
Documentation content	Initial bill of materials (BOM), including critical long-lead components
	Compliance and certification checklist (RF, EMC, safety, environmental)
	Verification strategy overview (test plan outline, key metrics, required equipment)
	Manufacturing considerations (Design for Manufacturability (DFM), Design for Testability (DFT), assembly notes, and any special manufacturing processes)
	Design rationale (key architectural and component decisions and their justification)

8.3 Blueprint of a Phased Array System

We discussed the fundamentals of signal chain layer and link-budget calculations and examined examples in Chapters 3 and 4. Building on that foundation, this section presents an example blueprint for a Ku-band phased array ground terminal.

In recent years, there has been rapid growth in low-Earth-orbit (LEO) satellite communication systems, particularly in the Ku-band, driven by companies such as OneWeb and SpaceX. These systems offer low-latency connectivity due to their proximity to Earth and enable emerging features such as direct-to-cell communication. Unlike geostationary (GEO) satellites, LEO satellites continuously move across the horizon, requiring ground terminals to track their motion. As a result, electronically steerable phased arrays, often implemented as active electronically scanned arrays (AESAs), have become essential for maintaining a stable link. In the following sections, we discuss how to build the product blueprint for such a system.

8.3.1 Requirements

In this section, we provide an overview of some of the key requirements for a LEO satellite terminal. While the complete list of requirements for such a system is extensive, our goal here is simply to illustrate how the main concepts come together without diving into unnecessary detail that would distract from the primary purpose of this example.

DC Power

Under worst-case operating conditions, including maximum transmit power, full beamformer activity, and all digital subsystems active, the terminal consumes approximately 150W, depending on overall efficiency, the specific beamformer ICs used, and architectural choices. Typical terminals are powered from an indoor AC supply in the 100-V to 240-V range, which is converted to a DC bus in the 24-V to 48-V range. This voltage level provides sufficient headroom for the power tree and downstream DC-DC conversion stages. At peak load, this corresponds to approximately 6A to 8A at 24V or 3A to 4A at 48V.

Frequencies and Bandwidth

Within the 10.7–12.7-GHz downlink band, a typical system provides eight channels, each with a 250-MHz bandwidth. On the uplink, covering the 14.0–14.5-GHz range, there are generally four channels, each occupying around 125 MHz of bandwidth [1]. These allocations are representative of commercial and government Ku-band satellite systems and are often shaped by regulatory and operator-specific frequency plans.

Ku-band terminals often employ an S-band IF, typically around 2 GHz on the receive path and 4 GHz on the transmit path. The conversion between RF and IF is performed within the frequency-conversion module, which interfaces with the modem and system controller to enable baseband processing (see Figure 8.1). Using an S-band IF satisfies several practical constraints in SATCOM terminals: it enables low-loss transmission over long coaxial cables (typically 5–30m), remains well below the cutoff frequency limits of inexpensive connectors and passive components (such as splitters), and provides sufficient bandwidth to support today's high-throughput modems.

Antenna and RF

OneWeb's LEO satellites operate at an altitude of approximately 1,200 km, whereas Starlink satellites orbit closer to Earth at roughly 550 km. The greater propagation distance to OneWeb's constellation increases free-space path loss by approximately 4 to 5 dB, which, in turn, requires user terminals to deliver higher EIRP, on the order of 35 dBW, and to employ larger antenna apertures to close the link. Starlink benefits from its lower orbital altitude, enabling smaller antenna panels and lower terminal EIRP for comparable data rates.

However, lower-altitude satellites traverse the visible sky more rapidly, which necessitates faster electronic beam steering and more frequent handovers. These satellites also experience greater atmospheric drag, leading to shorter orbital lifetimes and the need for more frequent replenishment. Higher-altitude systems such as OneWeb trade increased terminal EIRP and antenna size for slower handovers, reduced drag, and longer satellite lifetimes.

Ku-band AESA user terminals for LEO SATCOM typically support electronic scan angles of plus or minus 60° from boresight, enabling continuous tracking of fast-moving satellites across the visible sky. As the beam is steered off-axis, the effective projected aperture decreases approximately with the cosine of the scan angle, resulting in scan loss. This angle-dependent gain reduction becomes increasingly significant beyond scan angles of approximately 45° to 55°. Typical Ku-band flat-panel antennas exhibit a 3-dB beamwidth in the range of 3° to 5°, depending on aperture size, and maintain sidelobe levels between −20 dB and −13 dB. Tighter sidelobe control is often required to meet regulatory constraints in SATCOM bands. Last but not least, the G/T is around 11 dB/K [1].

Modulation and Coding

Modulation and coding employ adaptive strategies to maximize throughput while maintaining link reliability under varying atmospheric conditions and pointing errors. QPSK is commonly used under nominal or degraded link conditions due to its robustness to noise and phase distortion. When link quality permits, 8-PSK is employed to increase spectral efficiency and data rate without increasing bandwidth. In practice, adaptive modulation and coding schemes automatically switch between these modes based on measured signal-quality metrics such as SNR and EVM. Typical peak data rates are on the order of 75 Mbps on the uplink and approximately 15 Mbps on the downlink [1].

Maintaining signal integrity requires adherence to defined EVM and SNR targets. For QPSK, an EVM below 10% and an SNR in the range of 8 to 12 dB are typical design objectives. For 8-PSK, the system must achieve an EVM below 8% and an SNR between 12 and 15 dB to ensure reliable operation under clear-sky conditions. These requirements directly influence the RF chain noise figure, gain distribution, and linearity and serve as practical reference points during link validation.

Moreover, in LEO satellite communication systems, dual-carrier operation is commonly used on the uplink to increase throughput and refers to the simultaneous transmission or processing of two independent RF carriers. Instead of pushing a single carrier to very high symbol rates, which places stringent requirements on phase noise, PA linearity, and ADC and DAC performance, uplink traffic is distributed across two parallel carriers. As a result, each carrier operates within comfortable bandwidth and EVM limits, overall uplink throughput scales approximately with the sum of both carriers, and RF, DSP, and thermal design at the user terminal are simplified.

Multiplexing

The downlink, representing the satellite-to-user path, typically employs single-carrier time-division multiplexing (SC-TDM) on the downlink to allocate short transmission intervals among multiple users. This approach balances spectral efficiency and implementation simplicity, allowing multiple users to share a common transponder channel without significant mutual interference.

Multiple Access

Each channel is shared among multiple users using multiple-access techniques such as time-division multiple access (TDMA) or frequency-division multiple access (FDMA), depending on system architecture. This allows efficient use of the available spectrum while maintaining user isolation and predictable latency. The resulting design blueprint provides a holistic view of how frequency allocation, channelization, modulation, and performance targets interconnect, bridging the high-level system plan with the detailed RF, digital, and control design that follows in subsequent sections.

Software and Firmware

Responsibilities include GNSS acquisition and tracking, system initialization and calibration routines, data movement between subsystems, and coordination of control, monitoring, and communication functions.

Mechanical

The size, weight, and mounting fixtures of the terminal must allow it to be transported and securely installed in a variety of locations, including on the ground, walls, or rooftops. Because the unit is intended for outdoor deployment, it must meet appropriate ingress protection requirements and maintain reliable thermal performance over a typical operating temperature range of –40°C to +55°C. In some deployments, the terminal must also incorporate an integrated heater to prevent snow and ice accumulation.

8.3.2 Requirements for Commercial Versus Military Applications

LEO ground terminals for commercial and military applications differ significantly in design priorities, environmental requirements, security features, and operational capabilities. While both must acquire, track, and communicate with rapidly moving satellites, military systems impose far more stringent constraints driven by survivability, security, and mission-critical reliability. Table 8.5 summarizes these differences and considerations.

8.3.3 Overall System Architecture

The complete phased array ground terminal is composed of three primary subsystems, as illustrated in Figure 8.1:

1. *Modem and system controller:* Responsible for overall system management, handling modulation, demodulation, signal processing, beam control, configuration, and diagnostics. It sets TX/RX frequencies, processes IF signals, and interfaces with external networks or higher-level control systems.

Table 8.5 Considerations for Commercial Versus Military LEO Ground Terminals

Category	*Commercial LEO Terminal*	*Military LEO Terminal*
Performance requirements	High data rates for consumer/enterprise use Optimized cost per Mbps Moderate link margin for nonhostile environments Standard security (AES-based encryption, secure boot) Designed for typical interference levels Not optimized for jamming resistance	High link margin for operation under jamming/interference Antijam and antispoofing (frequency hopping, beam nulling, adaptive filtering) Use of secure, classified, or tactical waveforms with low probability of intercept (LPI) and low probability of detection (LPD) Operates in contested RF environments Robust synchronization under Doppler and interference
Antenna and tracking	Low cost Moderate beam-steering agility G/T and EIRP optimized for throughput Limited null-forming or anti-interference capability	High-performance phased arrays with fast beam steering Multibeam capability for redundancy and simultaneous links Null-forming to suppress jammers Wide-angle steering for low-elevation tracking Supports SATCOM-on-the-move (SOTM) and SATCOM-on-the-pause (SOTP)
Environmental, mechanical, shock, and vibration requirements	Weather-proof for rain, snow, and temperature swings Lightweight housings (plastic/aluminum) Designed for rooftops, RVs, and maritime yachts Basic shock and vibration tolerance (consumer grade)	Full MIL-STD-810 compliance High-shock loads (vehicle impact, drops, blast-induced shock) Severe vibration (tracked vehicles, helicopters, fast boats) Wide temperature (–40°C to +70°C) Dust, sand, humidity, salt, fog MIL-STD-461/464 EMI/EMP protection Reinforced rugged enclosures for battlefield durability
Power and thermal management	Power-efficient operation AC mains or 12–48-V DC Small passive/active cooling systems Optimized for low noise and low-energy usage	Wide DC input range for tactical power sources Ability to operate from vehicle/aircraft power buses Higher thermal budgets for high RF output power Advanced cooling solutions for extreme temperatures Silent cooling options if acoustic stealth is required
Cybersecurity, port locking, and information assurance	Standard encryption (AES-128/256, TLS) Firmware integrity checks Over-the-air (OTA) software updates Basic password protection and access control Standard physical port behavior (Ethernet, USB)	Hardware port locking: physical disablement/sealing of Ethernet/USB/debug ports Software port locking: access control, whitelisting, role-based authentication Anti-tamper mechanisms, including detection and shutdown Zeroize function to erase keys if compromised No-location-tracking mode: disables GPS broadcast, stops telemetry revealing position Compliance with military information assurance (IA) accreditation and cybersecurity policies Secure boot, signed firmware, and supply-chain validation

Category	*Commercial LEO Terminal*	*Military LEO Terminal*
Supply chain and manufacturing	Relies heavily on COTS components High-volume consumer manufacturing Cost-driven design and sourcing Standard quality assurance and testing processes	Secure, controlled, and vetted supply chain MIL-grade or ITAR-controlled components Low-volume, high-reliability manufacturing Extensive environmental and life-cycle testing High cost due to ruggedization and security requirements
Operational capabilities	Mainly static or low-mobility use Self-installation by end users Cloud-based monitoring and diagnostics Simple network management tools Operation in benign RF environments	Operation across land, sea, and air platforms High dynamic motion handling (airborne, maritime, ground vehicles) Integration with tactical radios and battle-management systems Robust handovers at high speed Resilient operation in contested and degraded environments Support for mission-critical availability and redundancy

2. *Frequency conversion module:* This block translates signals between IF and RF domains. It upconverts transmit IF signals to the desired RF and downconverts incoming RF signals to IF. It receives frequency commands from the modem and system controller.
3. *Phased array panel and RF front end:* This subsystem comprises the antenna array and integrated RF circuitry, including beamformer ICs. It is responsible for beamforming and electronic beam steering based on control commands received from the modem.

Figure 8.1 illustrates the interaction between the modem and system controller, frequency conversion module, and phased array panel and RF front end through power, signal, and digital control paths.

- *Transmit path (green):* The modem generates the transmit IF (TX IF) signal, which is sent to the frequency conversion module for upconversion to the RF band. The RF signal is then routed to the phased array panel for beamforming and transmission.
- *Receive path (red):* Incoming RF signals are captured by the phased array panel, downconverted by the frequency conversion module to receive IF (RX IF), and sent to the modem for baseband processing and demodulation.
- *Control and monitoring path (blue):* The modem sends digital commands to configure beamforming parameters, set transmit/receive frequencies and modes, adjust amplitude settings, and manage system states. In return, the phased array panel and frequency conversion module provide telemetry data, including sensor readings, monitoring status, and overall system health, for real-time diagnostics and fault detection.
- *Power distribution (orange):* Power is supplied from the modem and system controller to both the phased array panel and the frequency conversion module to drive all subsystems.

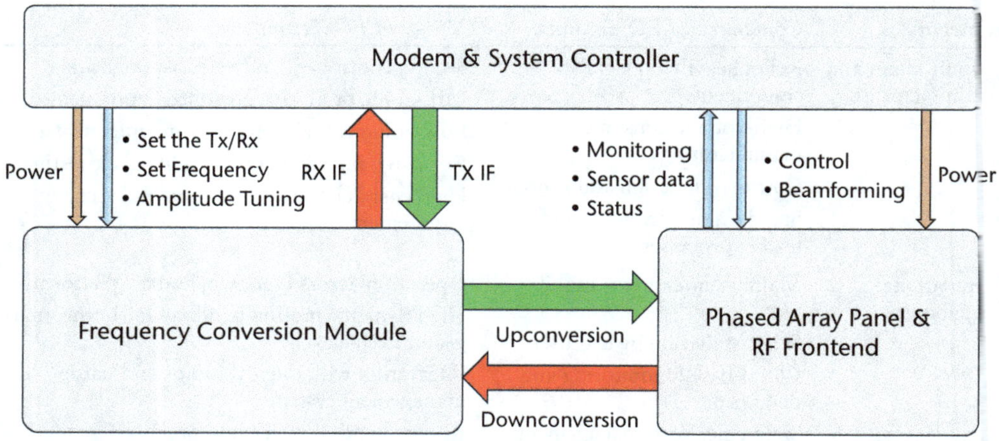

Figure 8.1 Phased array ground terminal architecture showing interaction between the modem and system controller, frequency conversion module, and the phased array panel with RF front end. Color-coded arrows represent transmit (green), receive (red), power (orange), and digital/control (blue) paths.

This modular architecture supports software-defined beamforming, with a clear separation between signal processing (modem), frequency translation (conversion module), and antenna-level control (phased array panel). It is designed for scalability, ease of integration, and flexibility across applications.

8.3.4 Product Layout

We began with the system requirements and then introduced the main building blocks in the previous section. A useful way to visualize how these subsystems come together is to include a simple sketch that highlights their physical arrangement within the final product. Figure 8.2(a) illustrates the product layout corresponding to the block diagram from the previous section, showing how the major hardware elements are positioned within the terminal.

Figure 8.2(a) presents a high-level overview of a typical flat-panel satellite terminal, including the phased array antenna panel, integrated radome and polarizer, heatsink, frequency-conversion module, and the modem and controller interface. In many modern terminals, the phased array panel, the frequency-conversion module, and the modem and system-controller electronics are integrated directly on the backside of the panel; together, these elements form the outdoor unit (ODU), which is mounted externally on a roof, vehicle, or wall.

The indoor unit (IDU), located inside the building, provides local user-network interfaces, including a Wi-Fi router and Gigabit Ethernet LAN ports, as well as MoCA support for Ethernet over coax. The IDU connects to the outdoor unit (ODU) through a single coaxial cable that carries DC power and MoCA-based data, enabling both power delivery and bidirectional communication over the same link. Alternatively, the IDU-to-modem connection can be implemented using a standard Ethernet interface, in which case the modem and controller are powered by a dedicated external power supply. Figure 8.2(b) provides a simplified external view

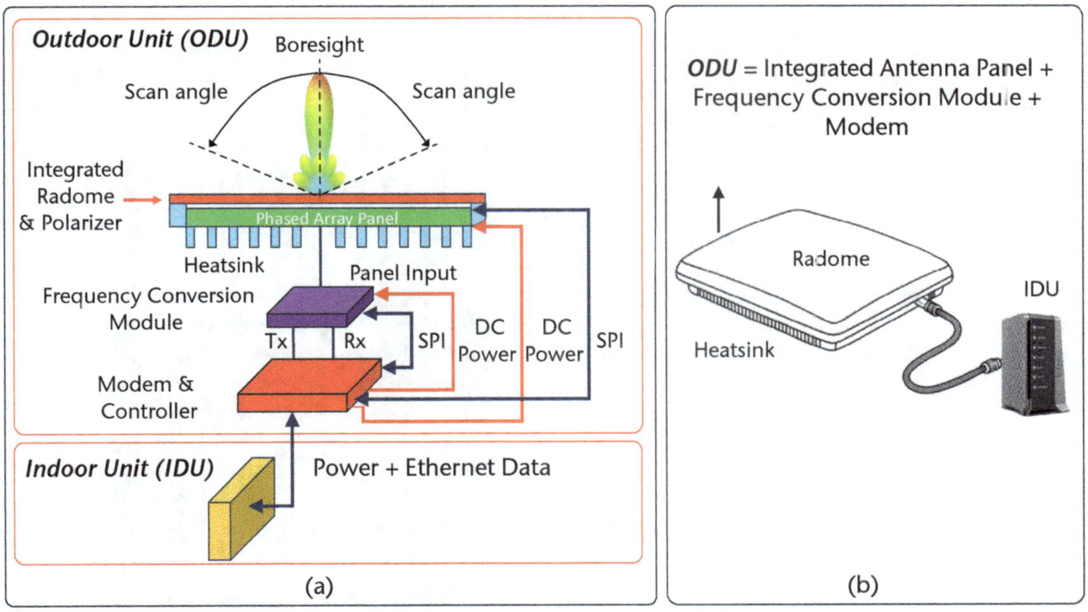

Figure 8.2 (a) System-level layout of a typical flat-panel satellite terminal. (b) Simplified external view of the terminal showing the ODU connected via a single coaxial link to the IDU.

of the terminal, showing the radome, integrated heatsink, and the indoor modem connected to the outdoor unit via the coaxial link.

8.3.5 Phased Array Panel

The 32 × 32 patch antenna array (1,024 elements) forms the primary radiating aperture of the system. Such arrays enable electronic beam steering in both azimuth and elevation, allowing the terminal to track satellites or ground nodes without any mechanical motion [2, 3].

The bottom side of the phased array PCB (Figure 8.3(a)) and the top side (Figure 8.3(b)) belong to a representative 16 × 16 phased array panel. Although smaller in scale, it contains all of the essential building blocks found in larger arrays.

The bottom side of the phased array PCB, shown in Figure 8.3(a), contains the electronics that power, control, and monitor the antenna array. These components include:

- *Connector for power and system controller:* The primary interface that delivers DC power from the modem and system controller module into the phased array panel and carries control and communication signals between the panel and the modem. This interface supports beamforming and scan-angle control, sensor data readout, and access to on-board memory.
- *Power distribution network (PDN):* Converts the input DC supply into the regulated voltage rails required by the beamformer ICs, digital control logic, memory devices, and sensors. The PDN also manages voltage sequencing and implements protection mechanisms.

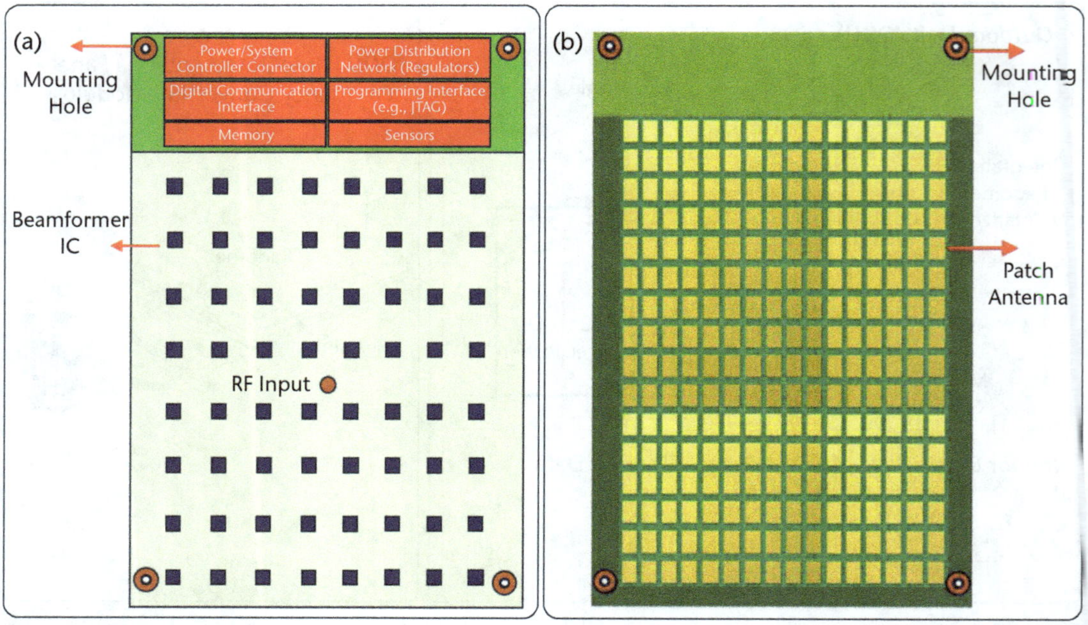

Figure 8.3 (a) Bottom side of a representative 16 × 16 phased array panel, showing the layout of the electronics layer. (b) Top side of the same panel, illustrating the 16 × 16 grid of patch antennas that form the radiating aperture. The antennas are fed by the underlying beamformer ICs, and mechanical mounting holes along the edges support integration with the enclosure.

- *Digital communication interface:* This block is commonly implemented using a CPLD or small FPGA. It provides SPI or similar control interfaces for configuring beamformer phase and gain settings, distributes timing signals, and reads sensor values from beamformer ICs.
- *Programming interface (JTAG):* Used to upload firmware to the CPLD.
- *Nonvolatile memory:* Stores the panel's serial number, calibration tables, bias settings, factory characterization data, and startup configuration required for autonomous initialization.
- *Sensors:* Temperature sensors and voltage and current monitors that enable system health monitoring and support closed-loop control functions.
- *Beamformer ICs:* The lower portion of the board contains active RF integrated circuits that provide per-element phase and amplitude control. Typical commercial devices support four transmit and four receive channels and include phase and gain control interfaces, integrated power detectors, and on-chip bias and temperature sensing.

The top side of the board, which faces the polarizer and radome shown in Figure 8.3(b), contains the patch antenna elements that form the radiating aperture. Sometimes, engineered metasurfaces are also used as the radiating element. Each patch is driven by a dedicated transmit/receive channel from a beamformer IC. Because each beamformer typically supports four channels, the total number of beamformer ICs is one-quarter of the antenna count. So a 16 × 16 array needs 8 × 8 beamformers.

The radiating elements are inherently linearly polarized. However, satellite links typically require circular polarization (CP), left-hand circular polarization (LHCP) for one direction (often downlink) and right-hand circular polarization (RHCP) for the other (uplink). Therefore, a polarizer is incorporated into the RF aperture to convert the linear polarization produced by the patches into the required circular polarization.

Polarizers in these systems are commonly realized using periodic printed structures, such as patterned copper grids or meandered lines on a dielectric substrate [4]. These structures introduce the required 90° phase shift between orthogonal field components, enabling efficient LHCP/RHCP conversion across the array's operating bandwidth.

Note that while circularly polarized patches are available, modern LEO phased array terminals rarely rely on them. These patches typically exhibit narrower axial-ratio bandwidth, making it difficult to maintain good circular polarization purity across the wide bandwidths required for Ku-band LEO systems. They also lose polarization quality at large scan angles, an issue for terminals that must track fast-moving LEO satellites, and they require dual feeds and tight phase balance, which increases cost and manufacturing complexity. In contrast, a simple linearly polarized patch combined with a broadband polarizer provides stable circular polarization over wide bandwidths and wide scan angles while keeping the antenna profile thin and the fabrication process more consistent. For these reasons, most Ku-band phased array terminals use a linear radiator with an added polarizer rather than a dedicated CP patch.

In addition to polarization conversion, a radome is required to protect the phased array panel from environmental exposure, including rain, snow, dust, moisture ingress, UV radiation, and aerodynamic loading. The radome must be engineered to provide this protection while maintaining low RF loss, stable dielectric properties over temperature, and minimal impact on the antenna's pattern, polarization purity, and scan performance.

The panel is typically half-duplex, meaning it switches between transmit and receive modes. To support full-duplex operation, transmitting and receiving simultaneously, there are two common design approaches. The first is to use separate panels, one dedicated to transmission and one dedicated to reception. While this simplifies the RF design, it increases the overall size, weight, and cost of the terminal by approximately a factor of 2.

The second approach, which is more compact but considerably more challenging, is to integrate both the transmit and receive antennas within the same aperture. Enabling full-duplex operation in a shared aperture requires careful electromagnetic isolation, frequency planning, and front-end linearity management to mitigate self-interference between the transmit and receive paths.

8.3.6 RF Front End

As shown in Figure 8.3(a), the array panel uses a single RF connector as the interface to the front end. Before examining the RF front end itself, it is important to understand how the signal propagates from this connector to each antenna element in the TX mode and, conversely, how the received signals from all elements are combined and routed back to the connector in the RX mode.

The RF input is distributed through a hierarchical network of wideband power dividers, such as Wilkinson dividers, as illustrated in Figure 8.4(a). This network feeds the signal to each beamformer chip. Within each chip, per-channel phase and amplitude control is applied before driving the corresponding antenna elements. These per-element adjustments collectively enable electronic beamforming and angular scanning across the array's field of view.

Each beamformer IC includes an internal one-to-four power divider, enabling support for four transmit and four receive antenna elements, as depicted in Figure 8.4(a).

Figure 8.4(b) illustrates a single beamformer channel. A transmit/receive switch routes the signal based on the operating mode, and each path includes an amplifier,

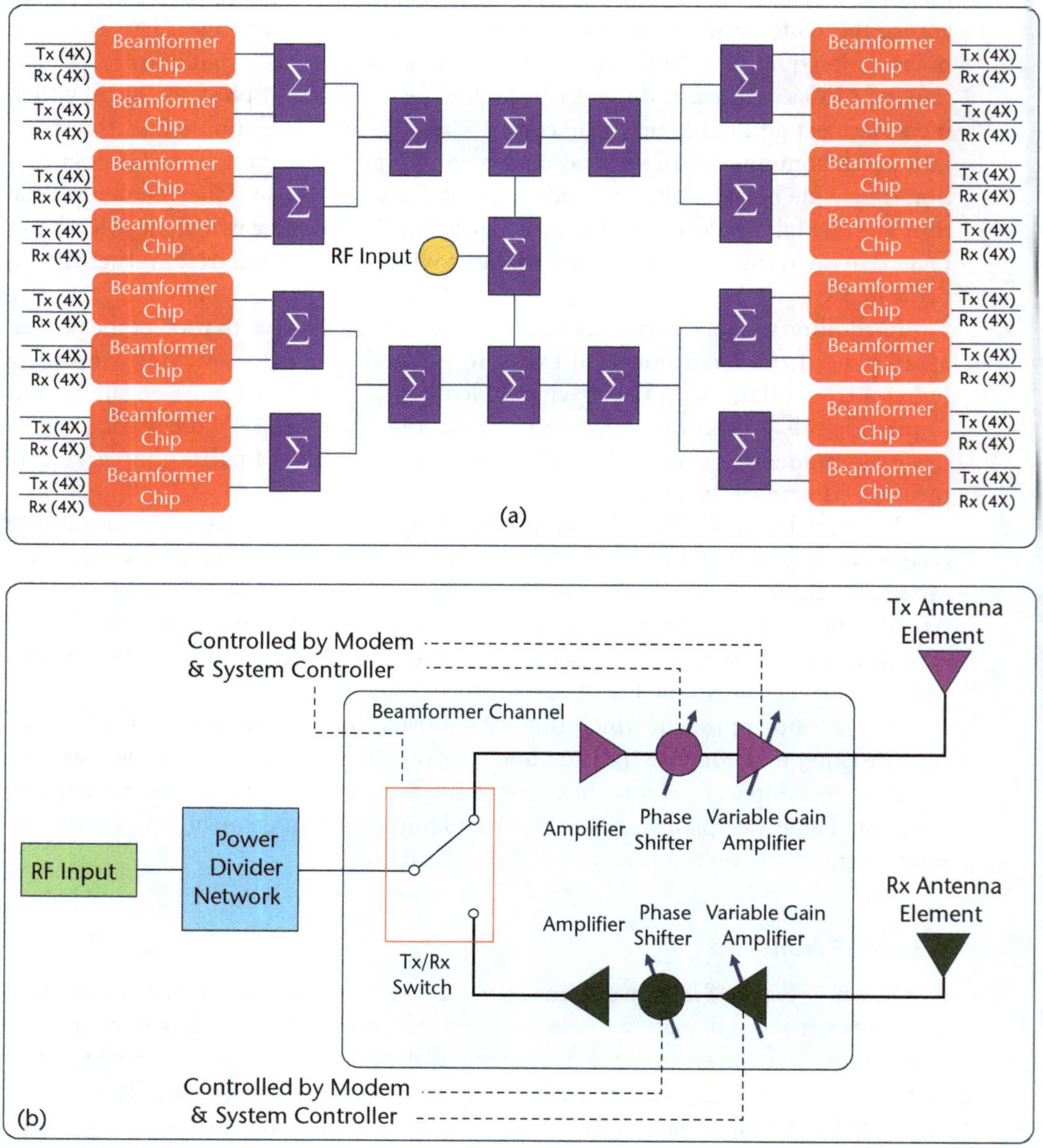

Figure 8.4 (a) Simplified representation of the RF power-distribution network for a phased array panel. (b) Block diagram of a phased array RF front-end chain.

a tunable phase shifter, and a variable-gain amplifier (VGA). These elements provide amplitude and phase control, enabling dynamic and directional beam steering in both transmit and receive modes.

Let's look at some numbers for the beamformer chip. The receive and transmit gain per channel typically falls in the 25–30-dB range. The RX noise figure is less than 2 dB, with practical implementations achieving values closer to 1.5 dB. The transmit P_{1dB} per channel is approximately 10 dBm, and the power consumption is on the order of 70–80 mW per channel. The system is designed to operate over a wide ambient temperature range of –40°C to +85°C, consistent with outdoor SATCOM terminal requirements.

The beamformer chips are configured and managed by the modem and system-controller module, which acts as the central interface for beamforming coordination, calibration, and overall system synchronization.

Several industry-proven beamformer ICs are available today for Ku-band phased array terminals. The most widely deployed devices are the Renesas F6521 (TX) and F6121/F6122 (RX) beamformers. The F6521 supports 14.0–14.5-GHz transmit operation with 4 independent TX channels, each offering 6-bit phase resolution and 6-bit gain control, with typical per-channel power consumption of ~70–80 mW. Its receive counterparts, the F6121/F6122, cover 10.7–12.7-GHz downlink operation and also provide four RX channels, 6-bit phase, 6-bit gain control, and an excellent ~1.5–2 dB noise figure. A widely used alternative, especially for prototyping and flexible designs, is the Analog Devices ADAR1000, a 4-channel TX/RX beamformer covering 2–18 GHz (fully spanning Ku-band) with 360° phase control, 31.5 dB of variable attenuation, and typical channel power consumption of ~250 mW in active mode. Together, these IC families represent the mainstream, field-tested solutions for building scalable Ku-band phased arrays.

The signal delivered to the RF input originates from the frequency-conversion module in TX mode or is routed back to the frequency-conversion module in RX mode. With the beamformer signal paths now defined, we next turn our attention to the frequency-conversion module, which serves as the interface between the modem/baseband and the phased array panel.

8.3.7 Frequency Conversion Module

As mentioned earlier, the modem operates at S-band intermediate frequencies (around 2 GHz for receive and 4 GHz for transmit). The frequency conversion module bridges this IF domain with the Ku-band RF domain of the phased array panel. It performs the necessary upconversion and downconversion so that signals from the modem are translated to Ku-band for transmission, and incoming Ku-band signals are translated back to S-band for demodulation.

Figure 8.5 illustrates a representative block diagram of such a module. It contains two distinct signal paths: an upconversion chain for the transmit (TX) path and a downconversion chain for the receive (RX) path. Each chain integrates the key building blocks commonly used in RF front-end design, including:

- LNAs in the RX path to keep the noise figure low and provide amplification;
- PAs in the TX path to increase signal strength prior to transmission;

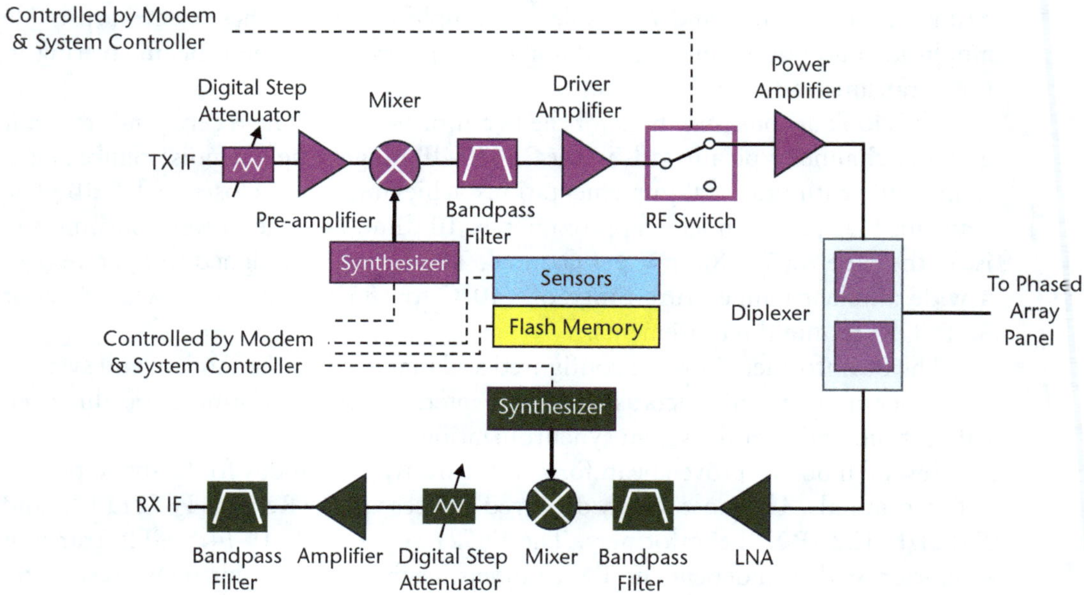

Figure 8.5 Architecture of a frequency conversion module.

- Mixers for frequency translation;
- Filters to suppress spurs, harmonics, and out-of-band interference;
- Digital step attenuators for amplitude tuning on both RX and TX sides;
- A synthesizer that generates the LO signals for the mixers, enabling channel selection (in this design, the LO subsystem must support eight receive channels and four transmit channels);
- A nonvolatile memory to store calibration data and the board information

DC power for the frequency conversion module is supplied by the modem and system controller. The digital communication between these two modules allows the modem and system controller to:

- Select transmit or receive mode via the T/R switch and associated biasing;
- Set attenuation values using on-board digital step attenuators for gain control and level management;
- Configure the synthesizer/LO to define the transmit and receive frequencies for channel selection;
- Set the PA bias level to control output power, linearity, and efficiency;
- Read sensor values (temperature, current, voltage, PA health) for monitoring and protection;
- Access flash memory to retrieve calibration tables, configuration parameters, and protection thresholds.

8.3.8 Modem and System Controller

Figure 8.6 illustrates a high-level block diagram of the modem and system controller. In addition to supplying DC power, the modem and system controller configure

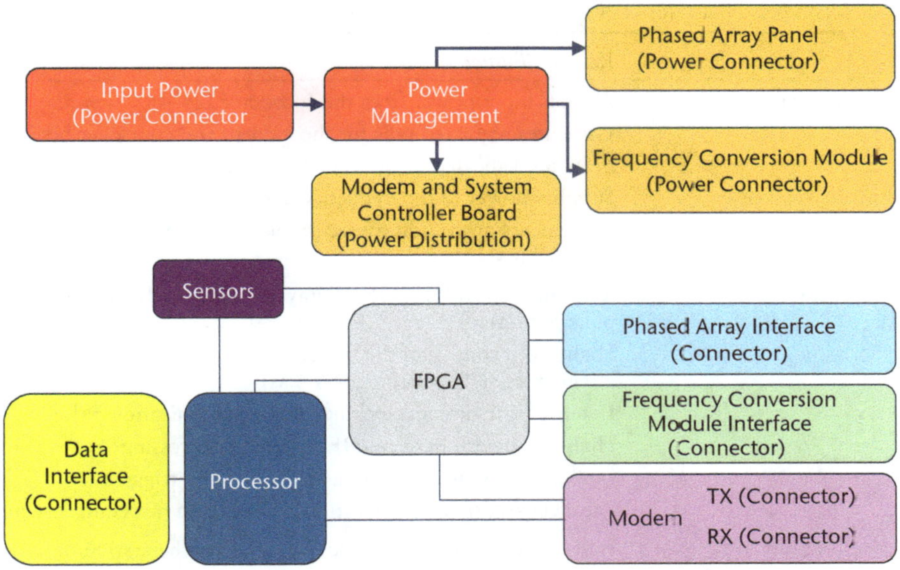

Figure 8.6 High-level system block diagram showing the integration of power management, processing, and RF subsystems. The FPGA acts as the central controller, interfacing with the phased array, frequency conversion module, modem, sensors, and processor to enable a fully functional RF communication system.

and control the frequency conversion module and the phased array panel, including beamforming parameters. The modem and system controller also receive sensor data from the phased array panel and the frequency conversion module, enabling real-time monitoring, status reporting, and fault detection for system health and diagnostics.

Beyond interaction with the phased array panel and the frequency conversion module, the modem and system controller also interface with the indoor unit and perform additional functions, including baseband processing, modulation and demodulation, and networking tasks. Table 8.6 summarizes the associated subsystems.

Thus far, the RF, power, and digital aspects of the product have been discussed. The following section focuses on the mechanical aspects.

8.3.9 Mechanical Subsystem

Figure 8.7 presents a 3D view of the final product. This figure is shown for illustration purposes only and should not be directly compared with the design described previously. As shown in Figure 8.7(a), the array panel typically incorporates aluminum heat spreaders and heatsinks to draw heat away from high-power devices such as beamformer chips. A thermally optimized heatsink is integrated beneath the panel to dissipate heat generated by the active components.

As illustrated in Figure 8.7(a), heat is conducted from the beamformer chips to the heatsink through extruded circular contact features on the internal surface of the enclosure. These features interface with the beamformer chips through layers of thermal interface material (TIM), ensuring efficient thermal conduction between the beamformer chips, internal heat spreaders, and the external heatsink. This thermal path helps maintain stable operating temperatures across the full ambient temperature range.

Table 8.6 Functional Responsibilities of the Modem and System Controller Subsystems

Subsystem	*Responsibilities*
Power management	Accepts input power from the connector (24–48V typical) Generates regulated rails for the modem and controller board Provides dedicated power rails for the phased array panel Supplies power to the frequency-conversion module Implements protection (overcurrent, overvoltage, thermal shutdown) and sequencing
FPGA	Real-time control of phased array beamformer ICs (phase, gain, amplitude, polarization) High-speed control of the frequency-conversion module (LO configuration, attenuators, T/R switching, PA bias) Deterministic timing, clocking, and synchronization pulses High-bandwidth baseband/IF data path processing Low-latency sensor sampling and protection triggers Parallel execution of calibration routines and fast control sequences
Processor	Runs system software, control logic, and orchestration of terminal operation Networking, routing, and modem protocol coordination Loads calibration tables, configuration files, and device parameters Aggregates sensor data and handles loop monitoring Communicates with external interfaces (Ethernet, management, cloud) System diagnostics, firmware updates, boot process, and fault reporting
Data interface (e.g., GigE)	Provides the external data connection to the modem/system controller Bridges user data traffic into the processor for further handling Supports management and diagnostic communication paths
Modem	Modulates the transmit signal according to the selected waveform Demodulates the received signal to recover baseband data Implements channel coding/decoding and synchronization functions Interfaces with the FPGA for real-time symbol and frame alignment
Sensors	Measure temperature, voltage, and current Provide telemetry to the FPGA for fast protection loops Provide averaged, long-term data to the processor for system monitoring Enable adaptive power control, thermal management, and fault detection
Phased array interface	High-speed communication with beamformer ICs (phase, gain, polarization settings) Executes beam commands and calibration sequences Synchronizes panel timing with the FPGA Reports panel status and errors
Frequency conversion module interface	Configures synthesizers/LOs for upconversion or downconversion Sets digital step attenuators, PA bias, and T/R mode Reads sensor values within the conversion module Accesses flash memory for calibration tables, parameters, and thresholds Manages timing and control handshakes with the FPGA

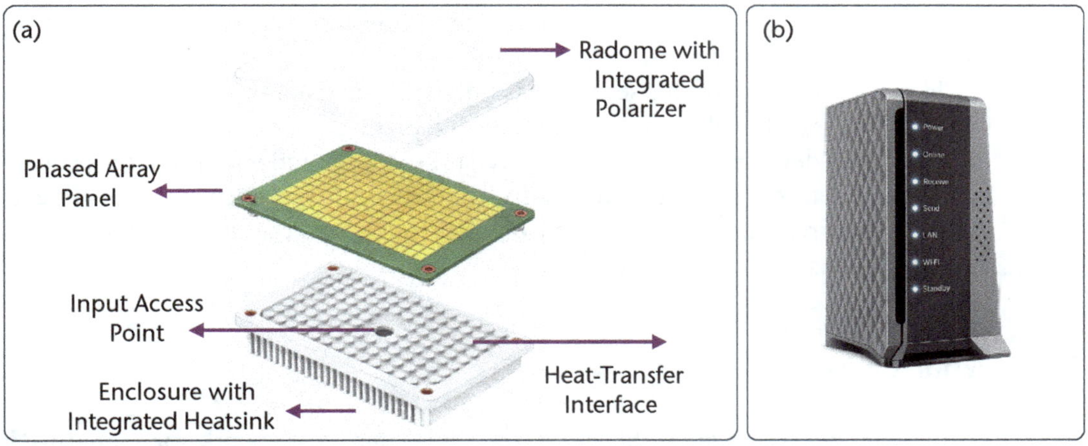

Figure 8.7 The 3D view of the final product: (a) ODU, and (b) IDU.

A lightweight plastic radome protects the phased array aperture from environmental exposure, including rain, snow, dust, ultraviolet radiation, and aerodynamic loading, while maintaining minimal RF degradation. In some designs, integrated heaters are also included to melt ice and snow on the surface of the panel.

The mechanical design must also provide a structurally robust interface between the phased array aperture, the frequency-conversion module, and the modem and system controller, enabling all three subsystems to be integrated within a single enclosure. In addition, the panel must support multiple mounting configurations, for example, ground mounts, wall mounts, or rooftop installations, without compromising structural integrity.

Finally, the size and weight of the terminal are critical system-level considerations. Excessive mass complicates shipping, handling, and installation, and, in many jurisdictions, occupational safety regulations limit the maximum load that a single person is permitted to carry. These constraints must be addressed early in the mechanical design process to ensure the final product remains practical, deployable, and compliant.

Figure 8.7(b) shows the IDU, which connects to the ODU using the interfaces described earlier. Unlike the ODU, the IDU is not subject to strict size, weight, or environmental constraints and generally follows standard commercial router and networking equipment requirements.

8.4 Conclusions

In this chapter, we examined the signal-chain layer as well as the RF and non-RF subsystems that make up a Ku-band phased array ground terminal. We reviewed the product layout, presented block diagrams illustrating the system blueprint, and described the key functions of each subsystem.

A more detailed blueprint is beyond the scope of this book, given the system's complexity and the breadth of topics involved. Elements such as complete power-tree design, digital signal-processing pipelines, firmware architecture, and software integration each require deep dives into subjects such as dynamic satellite positioning, timing, synchronization, and real-time control, topics that could fill entire chapters on their own.

However, the goal of this section was not to exhaustively cover every detail, but rather to illustrate the blueprint design flow and highlight the critical considerations that guide the development of a modern RF product. By understanding these interfaces, dependencies, and design decisions, engineers can approach complex systems with a structured methodology that carries through into the implementation and validation phases of product development.

8.5 What's Next?

In this volume, we have established the foundational framework for modern RF product development, moving from system-level thinking to a well-defined product blueprint. We began by identifying opportunities and risks, defining requirements, and understanding how constraints such as SWaP-C, regulatory compliance, and technology readiness shape design choices. We then examined how these considerations evolve into a complete system concept through architecture selection, link budgeting, frequency planning, and component trade-offs across the transmit, receive, and control paths.

The purpose of this volume has been to help you to think like a product designer, not just as a circuit or system engineer, but as someone who can balance performance, manufacturability, cost, and risk from the very beginning. By the end of this stage, you should be able to articulate a clear product vision supported by technical justifications and documented in a structured RF product blueprint.

In Volume 2, the focus shifts from the product blueprint to implementation and validation. The blueprint developed in this volume is carried forward into subsystem realization, system integration, and verification. Early design decisions are examined in the context of their impact on hardware performance, along with practical approaches for prototyping, testing individual functional blocks, and validating system-level requirements through structured iteration.

The blueprint marks the conclusion of the planning phase and the beginning of execution. Volume 2 continues this progression into building and validating the product.

Problems

8.1 A phased array panel uses beamformer chips that require 1.0-V and 2.3-V rails, while the remaining digital/control circuits require 2.5-V, 1.2-V, and 1.8-V rails. The terminal input supply is 24-V DC.

Design a power tree for the phased array panel that:

- Steps 24V down to intermediate rails (if needed) and then to each required voltage.
- Specifies where power-rail filtering is applied and what type of filters are used. (LC, π-filter, ferrite bead + decoupling).

- Indicates where protection elements (overcurrent protection (OCP), overvoltage protection (OVP), undervoltage lockout (UVLO), and thermal shutdown (TSD)) are implemented.

  Explain the rationale behind each major design decision.

8.2 Figure 8.6 and Table 8.6 describe the interfaces within the modem and system controller.

a. Create a 1-page mini-ICD covering the interfaces between:
   - The phased array panel and the modem and system controller;
   - The frequency-conversion module and the modem and system controller.

   Your ICD must specify:
   - Power interfaces;
   - Control and configuration signals;
   - RF/IF interfaces;
   - Monitoring and telemetry signals;
   - Timing and synchronization requirements;
   - Error, fault, or protection status indicators.
b. During EVT, the phased array panel intermittently fails to initialize. List five interface-related root causes that could explain this behavior.
c. Identify two specific ICD entries that would have prevented the failure and explain why.

8.3 Using the discussion in Section 8.3, evaluate the mechanical and thermal constraints for integrating: a 1,024-element phased array panel, the frequency-conversion module, and the modem and system controller. Answer the following:

- Why are aluminum heat spreaders and thermal interface materials essential?
- What materials can be used for the radome? Compare candidates in terms of RF performance, mechanical strength, environmental robustness, and cost.
- What practical weight limit would you impose on the final enclosure, and why must it comply with occupational-safety lifting limits?
- How would you estimate system weight during the blueprint phase?

8.4 Consider the three-block architecture shown in Figure 8.1:

Modem and System Controller → Frequency Conversion Module → Phased Array Panel

Propose one alternative architecture by merging or repartitioning subsystems. Then answer:

- What benefits does your architecture provide (e.g., reduced cable loss, improved timing, lower cost, simplified integration)?
- What drawbacks does it introduce (e.g., higher thermal density, reduced repairability, manufacturing complexity)?
- Would your architecture still support ±60° scan and S-band IF operation? Justify your answer.

8.5 Beamformer ICs with per-element phase and amplitude control dominate modern Ku-band phased arrays, but alternative beam-steering technologies exist.

a. List alternative technologies for implementing electronically steerable antennas without traditional beamformer ICs (e.g., metasurfaces, switched delay lines).
b. For each technology, describe principle of operation, key advantages, and key limitations.
c. Identify which alternative is most suitable for a low-cost flat-panel SATCOM terminal, and justify your choice in terms of bandwidth, scan range, efficiency, and robustness.

8.6 You are given the following receiver requirements for a product:
- Required demodulator SNR: 11 dB (QPSK);
- Required link margin: 3 dB;
- ADC full-scale input: 0 dBm;
- Target receiver sensitivity: −95 dBm.
a. Allocate a gain and noise-figure budget from antenna input to ADC.
b. Identify which blocks most strongly impact sensitivity and explain why.
c. How would the budget change if ADC dynamic range were reduced by 6 dB?

8.7 The frequency-conversion module uses:
- RX IF = 2 GHz;
- TX IF = 4 GHz;
- Ku-band RX = 10.7–12.7 GHz;
- Ku-band TX = 14.0–14.5 GHz.
a. Propose a single-LO frequency plan for both RX and TX paths.
b. Identify at least three potential spur mechanisms.
c. Recommend filtering or architectural changes to mitigate the most critical spur.

8.8 Each beamformer channel consumes 75 mW, and the panel contains 1,024 elements.
a. Estimate the total beamformer power dissipation.
b. Assuming a maximum allowable junction-to-ambient thermal resistance of 2°C/W, determine whether passive cooling alone is feasible.
c. Propose one system-level change that reduces thermal stress without reducing EIRP.

8.9 Late in the program, a customer imposes a 30% weight reduction requirement.
a. Identify which subsystems offer the largest weight-reduction potential.
b. Describe the performance, reliability, or cost trade-offs introduced.
c. Which subsystem should not be optimized for weight, and why?

8.10 You are asked to derive a military variant of the commercial terminal.
a. List the minimum changes required in RF architecture, power system, and firmware/software.
b. Identify which changes affect recurring cost versus nonrecurring engineering (NRE).
c. Explain why some military features cannot be retrofitted late in the design.

8.11 Rank the following risks from highest to lowest impact on program success:
- Beamformer IC supply-chain disruption;
- Thermal runaway at high scan angles;
- FPGA timing closure;

- Radome dielectric variation;
- Firmware update failure in the field.

Justify your ranking and propose one mitigation strategy for the top two risks.

References

[1] HL1100W Hughes LEO Terminal Datasheet.

[2] Gultepe, G., et al., "A 1024-Element K-Band Commercial SATCOM User Terminal with a G/T of 10.2 dB/K and Two Simultaneous Beams," *Proc. IEEE Int. Symp. Phased Array Syst. Technol. (PAST)*, Waltham, MA, October 2022, pp. 1–4.

[3] Low, K. K. W., "K/Ka-Band Receive/Transmit Silicon Planar Phased Arrays for SATCOM," PhD dissertation, Dept. Elect. Comput. Eng., University of California, San Diego, La Jolla, CA, 2021.

[4] Letizia, M., et al., "Oblique Incidence Design of Meander-Line Polarizers for Dielectric Lens Antennas," *IEEE Transactions on Antennas and Propagation*, Vol. 66, No. 7, July 2018, pp. 3565–3574.

About the Author

Alan Salari is a microwave engineer, quantum physicist, author, and entrepreneur with nearly two decades of experience in advanced RF and quantum technologies. He received degrees in electrical engineering (BS, MS, PhD) and physics (BSc, MSc) from the University of Bonn, RWTH Aachen University, and the University of Illinois at Chicago.

His professional work spans a broad range of application areas, including radar and satellite communication systems, microwave plasma generators, superconducting quantum processors, ultracold atomic gases, and particle detector systems for the Large Hadron Collider. He has held technical roles at leading industrial and research organizations, such as Microsoft, Axon, Hughes Network Systems, Quantum Circuits Inc., and the Fraunhofer Institute, where he contributed to the development of several first-of-their-kind products and platforms.

Dr. Salari is the author of the three-volume *Modern RF Product Development* series, which presents a structured, engineering-driven view of the complete product lifecycle, from concept and design through implementation, validation, and production. He is also the author of *Microwave Techniques in Superconducting Quantum Computers*, the first book dedicated to microwave hardware for superconducting quantum processors.

He is a Senior Member of the IEEE and a member of the Quantum Economic Development Consortium (QED-C), the Mid-Atlantic Quantum Alliance, and the Quantum Startup Foundry at the University of Maryland, College Park. His work has been recognized through awards and grants from organizations including the National Science Foundation, the Alexander von Humboldt Foundation, the German Ministry of Economic Affairs and Energy (EXIST program), the Discovery Partners Institute in Chicago, the Illinois Department of Transportation, and Siemens AG.

Index

G

H

I

Q

T

www.ingramcontent.com/pod-product-compliance
Lightning Source LLC
LaVergne TN
LVHW062214200926
848484LV00006B/1396